Topics in Applied Physics Volume 69

Topics in Applied Physics Founded by Helmut K. V. Lotsch

Volumes 1–56 are listed on the back inside cover

Unoccupied Electronic States

Fundamentals for XANES, EELS, IPS and BIS

Edited by J. C. Fuggle and J. E. Inglesfield

With Contributions by
P. T. Andrews I. R. Collins V. Dose J. Fink
J. C. Fuggle R. W. Godby J. E. Inglesfield
R. Schneider J. Somers D. D. Vvedensky
J. Zaanen R. Zeller

With 175 Figures

Springer-Verlag
Berlin Heidelberg New York
London Paris Tokyo
Hong Kong Barcelona Budapest

Professor John C. Fuggle, Ph. D.
Professor John E. Inglesfield, Ph. D.

Research Institute for Materials,
University of Nijmegen,
Toernooiveld,
6525 ED Nijmegen, The Netherlands

ISBN 3-540-54162-4 Springer-Verlag Berlin Heidelberg New York
ISBN 0-387-54162-4 Springer-Verlag New York Berlin Heidelberg

Printed in the United States of America

Typesetting: Thomson Press, New Delhi, India
54/3020 5 4 3 2 1 0 – Printed on acid-free paper

Preface

This book is devoted to the nature of electronic states above the Fermi level in solids, and the methods which are used to investigate them. We see the book as a companion to the two volumes of *Photoemission in Solids,* edited by *Cardona* and *Ley* and published in this series as volumes 26 and 27 more than ten years ago. At the time those books were published, photoemission had already been developed to a high level of sophistication; so much so that it tended to put techniques for studying unoccupied states in the shade and it certainly set new standards for experimental investigations of electronic structure.

Since 1980 the importance of photoemission has in no way diminished and the ideas about electronic structure have become more and more sophisticated. However, for the properties of materials, the unoccupied (mostly antibonding) electronic states of matter are just as important as the occupied (mostly bonding) states. We must also recognize that over the last decade studies of the unoccupied states have come of age and the concepts involved in their investigation have now reached a level on a par with photoemission. This had become clear by 1988, when a workshop was held at the International Centre for Theoretical Physics in Trieste on the topic of "Unoccupied States". Out of that meeting grew the idea that there was a need for a book with a philosophy similar to that of *Photoemission in Solids,* but directed at the unoccupied states. This volume is the result.

At the same time as we acknowledge our debt to *Photoemission in Solids,* we also note that our approach had to be somewhat different. The meeting in Trieste identified problems of electron localization and electron-electron interaction as central to the interpretation of spectra. However, these problems are common to spectroscopies of both occupied and unoccupied states and much of our insight has been achieved since 1980. This led us to avoid an artificial separation of the science of occupied and unoccupied states. Instead we attempt a more unified approach and devote three chapters (2, 3 and 4) to the theory of electronic states in solids and concepts like self-energies and the relationships between densities of states and spectral functions.

The second point that emerged in Trieste was that, for the studies of the unoccupied states, four techniques were of comparable importance (X-ray absorption, electron energy loss, inverse photoemission and the related technique of bremsstrahlung isochromat spectroscopy). This contrasts with the situation for the occupied states, where photoemission dominates the field. This difference led us to present the principles and possibilities of each of the techniques separately, rather than trying to separate the chapters in terms of, for instance, type of material under study. Thus Chaps. 5-10 consist of descriptions of the principles of the spectroscopies and illustrations of their contributions to the understanding of the electronic structure of condensed matter.

It remains for us to thank the people who have contributed to this work. Firstly we thank the authors of the various chapters for their cooperation. We tried to make the chapters blend together as a single book and also to avoid the standard "700-reference reviews". The aim was to achieve a more didactic approach illustrating the *principles* of the different areas. This approach demanded special flexibility from the authors, and sometimes a considerable sacrifice of their original concepts. We also wish to thank the authors for keeping to time and admit now that we actually told almost everyone that theirs was the only chapter missing as the deadline approached. Finally we express our thanks for many discussions over the years with colleagues too numerous to list here. These discussions have inevitably contributed to the philosophy we tried to develop here and are gratefully acknowledged for their scientific value as well as for the personal pleasure they brought.

Nijmegen, November 1991

J. C. Fuggle
J. E. Inglesfield

Contents

Contributors

Andrews, Peter T.

Department of Physics and Surface Science IRC, University of Liverpool, Oxford Street, P.O. Box 147, Liverpool L69 3BX, United Kingdom

Collins, Ian R.

Department of Physics and Surface Science IRC, University of Liverpool, Oxford Street, P.O. Box 147, Liverpool L69 3BX, United Kingdom

Dose, Volker

Max-Planck-Institut für Plasmaphysik, D-8046 Garching bei München, Fed. Rep. of Germany

Fink, Jörg

Institut für Nukleare Festkörperphysik, Kernforschungszentrum Karlsruhe, Postfach 3640, W-7500 Karlsruhe, Fed. Rep. of Germany

Fuggle, John C.

Research Institute for Materials, University of Nijmegen, Toernooiveld, 6525 ED Nijmegen, The Netherlands

Godby, Rex W.

Cavendish Laboratory, University of Cambridge, Madingley Road, Cambridge CB3 0HE, United Kingdom

Inglesfield, John E.

Research Institute for Materials, University of Nijmegen, Toernooiveld, 6525 ED Nijmegen, The Netherlands

Schneider, Ralf

Max-Planck-Institut für Plasmaphysik, W-8046 Garching bei München, Fed. Rep. of Germany

Somers, Joe

Fritz-Haber-Institut der Max-Planck-Gesellschaft, Faradayweg 4–6, W-1000 Berlin 33, Fed. Rep. of Germany

Vvedensky, Dimitri D.

The Blackett Laboratory, Imperial College, London SW7 2BZ, United Kingdom

Zaanen, Jan

Max-Planck-Institut für Festkörperforschung, Heisenbergstr. 1, W-7000 Stuttgart 80, Fed. Rep. of Germany

Zeller, Rudi

Institut für Festkörperforschung, Forschungszentrum Jülich, Postfach 1913, W-5170 Jülich, Fed. Rep. of Germany

1. Introduction

J.C. Fuggle and J.E. Inglesfield

With 12 Figures

1.1 What are Unoccupied States?

In our simplest, one-electron picture of the electronic states in a solid, energy levels are filled up to the Fermi energy in the ground state, with empty states above (see Fig. 1). There is then a complementarity between adding electrons to a system and removing them and hence both empty and filled states are equally important for understanding thermal properties and optical properties, which involve excitations from states below E_F to those above. Also, we know from semiconductor physics that electrons, occupying previously empty states, and holes below E_F can play equivalent roles in transport properties. The bonding of solids is intuitively thought of in terms of the energies of occupied states, but here too there is complementarity because unoccupied, antibonding states are as relevant to stability as occupied, bonding states.

Over the last 20 years or so our understanding of the occupied electronic states has undergone a revolution, as a result of techniques such as photoelectron spectroscopy coupled with advances in theoretical and computational methods. In many systems the calculated energy bands have been admirably confirmed by experiment, while in other cases the discrepancies have prompted work on the interaction between electrons. An equivalent understanding of unoccupied states has lagged behind the occupied states. However, recent advances in techniques such as X-ray absorption and inverse photoemission are changing the picture, and a more unified picture of unoccupied and occupied states is beginning to emerge, as will be discussed in this book.

The starting concept for understanding electronic properties is the band structure, which in a mean-field picture gives the energy versus Bloch wave vector, $\boldsymbol{k}$, of a single-particle wave function. This wave function describes an electron moving in the field of the ions and the charge density of the occupied electronic states. The bands of unoccupied states and their dependence on $\boldsymbol{k}$ (Fig. 1.1, right) can be probed by angle-resolved inverse photoemission. Techniques in which the electron wave vector is no longer well defined, such as XPS and BIS, can be related to the density of states, DOS, which is the energy distribution of the single-particle states (Fig. 1.1, left), and to some extent this is also true of angle-integrated photoemission and inverse photoemission.

Probes such as X-ray absorption excite an electron from an occupied state into an unoccupied state, and in the mean-field or independent-particle picture

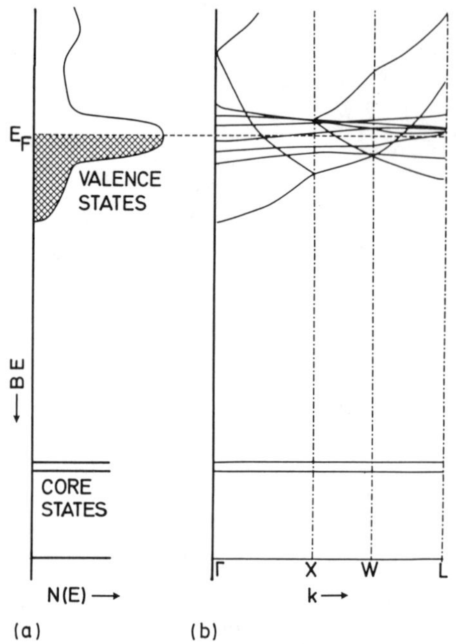

Fig. 1.1. (a) A schematic density-of-states curve showing core and valence levels. (**b**) A schematic representation of the dispersion of the bands as a function of the electron momentum $\boldsymbol{k}$ for various directions in reciprocal space

the excitations can be derived simply from Fig. 1.1b. The lowest energy excitations are from occupied to unoccupied valence states, and the shape of the excitation spectrum is the convolution of the respective single-particle densities of states; higher energy excitations are from a core state to the unoccupied states, giving an excitation spectrum which is the same as the partial density of unoccupied states.

The single-particle picture is very illuminating, and will be widely used throughout this book. Chapter 2 concentrates entirely on methods of calculating one-electron energy bands, and the associated excitation spectra. However, for many purposes it is inadequate and the correlated motion of the electrons must be taken into account, particularly in systems with fairly localized states. As we shall see in Chap. 3, the band structure still exists, but it must be reïnterpreted as a quasiparticle spectrum: the energy versus wave vector of an extra electron accompanied by its dynamic screening cloud (the exchange-correlation hole) for states above E_F, or of a hole with its screening cloud for states below E_F.

When the single-particle description is inadequate, it is sometimes sensible to use a total energy representation of the states of a solid. This is illustrated in Fig. 1.2 for the same solid illustrated in Fig. 1.1. The ground states in Fig. 1.2 are illustrated by single lines at the bottom of the diagram. By definition, the total energy is most negative for the ground state. In this solid there are two sets of excited neutral states which involve excitation of a single electron. Figure 1.2 illustrates this for a solid, such as a transition metal, with a band of states near E_F and a broad continuum at higher energies. For the first set of excited

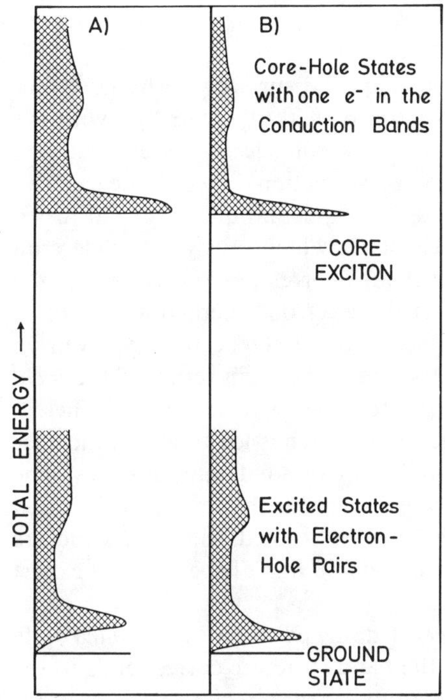

Fig. 1.2. (**a**) Total energy diagram of the charge-conserving excitations in a single-particle picture for the solid illustrated in Fig. 1.1. (**b**) Total energy diagram illustrating some possible distortions of the excited state distribution when electron–hole interactions are considered

states, in the independent-particle model, electrons are excited from the occupied to the unoccupied valence states. The width of this band of excited states is thus equal to the band width of this solid and its shape is given by the convolution of the shapes of the occupied with the unoccupied states. The second set of excited states consists of the excitation of a core electron to an unoccupied state. In the independent-particle model, the width of this continuum of unoccupied states in this case is given by the unoccupied band width only.

When we consider the interaction between the hole and the excited electron in Fig. 1.2, there are many possible variants in the behaviour and Fig. 1.2b gives only one illustration; more appear in Chaps. 4 and 5. Here we have assumed that the attraction between the electron and hole lowers the energy of many excited states, thus distorting the form of the continuum. In addition we have considered the possibility that a core hole may sufficiently distort the wave functions to pull a state right out of the band and form a core exciton. In most cases core excitons essentially have an extra electron tied to the site on which a core hole was created.

We ask the reader to note a difference between the situations represented in Figs. 1.2a and b. Figure 1.2a can be constructed by drawing vertical lines between all the occupied and all the unoccupied states in Fig. 1.1. This is not true for Fig. 1.2b. For instance, the core excitonic state in Fig. 1.2b cannot be constructed from Fig. 1.1 [1.1].

1.2 Theoretical Developments

Calculations of the electronic structure of solids and surfaces are by now very advanced. The major breakthrough occurred about thirty years ago with the development of a workable method to treat the electron–electron interaction – at least in the ground state. In this method, density-functional theory, the electron density of the ground state is treated as the fundamental quantity, and it can be calculated from occupied single-particle states which satisfy a one-electron Schrödinger equation. This adds an exchange-correlation potential to the Hartree potential to allow for the electron–electron interaction. What is remarkable is that this potential can be found quite accurately, even in inhomogeneous systems such as molecules and surfaces, in terms of the *local* electron density. Density-functional theory, together with the use of efficient and accurate basis sets in the Schrödinger equation, has led to a vast increase in our understanding of the bonding properties of solids and surfaces. The increase in computing power generally available means that more complicated structures, with many atoms per unit cell, can be studied; the band-structure methods themselves have reached the stage that you don't have to be an expert to apply them!

Even though density-functional theory is designed to give the *ground* state correctly, and there is no guarantee that the band structure corresponds to the excitations measured spectroscopically, it almost invariably forms the basis for comparison with experiment. The calculated bands can be compared directly with the occupied (hole quasiparticle) bands from photoemission experiments and the unoccupied (electron quasiparticle) bands from inverse photoemission, and joint densities of states are used to interpret optical and X-ray absorption experiments. However, until recently band-structure calculations were tailored to computation of occupied states, or low-lying unoccupied states relevant to near-ground-state properties. The techniques discussed in this book give new information on higher lying states (up to, say, 50 eV above the Fermi energy). The requirements needed to compute these states with accuracy leads to special problems not encountered with the occupied valence bands, which typically lie within 10 eV of E_F, and these will be discussed in Chap. 2. For comparison with core-level spectroscopies, conventional band-structure calculations can be modified to allow for the effect of the Coulomb interaction with the core hole on the unoccupied part of the density of states (Chaps. 4, 5). Efficient programs exist for including the matrix elements in photoemission and inverse photoemission and a full spectroscopic calculation is most useful for disentangling surface transitions from bulk, $\boldsymbol{k}$-conserving transitions. The same is true for X-ray absorption experiments such as EXAFS, XANES and SEXAFS, where matrix elements should be included for fitting an unknown atomic geometry to the experimental spectrum. In other cases direct comparison with features in the band structure or densities of states may be sufficient.

Taking into account the electron–electron interaction, the energy bands correspond to the motion of quasiparticles, i.e. of an extra electron or hole accompanied by its screening cloud. The effects of the electron–electron interaction are to shift the energy levels from the values given by the density-functional (or Hartree–Fock) theory, and to cause the state to decay in time, with consequent lifetime broadening of its energy. What is actually measured in an inverse photoemission or photoemission experiment is the spectral function $A(E,\boldsymbol{k})$, which is the probability of finding the system with an extra electron or hole in a state with energy E and wavevector $\boldsymbol{k}$. The spectral function has peaks corresponding to the quasiparticle energy bands with the form

$$A(E,\boldsymbol{k}) \propto \frac{\mathrm{Im}\{\Sigma\}}{(E - E_0(\boldsymbol{k}) - \mathrm{Re}\{\Sigma\})^2 + (\mathrm{Im}\{\Sigma\})^2}, \tag{1.1}$$

where $E_0(\boldsymbol{k})$ is the density functional or Hartree–Fock band structure and Σ is defined as the self-energy. The real part of the self-energy gives the energy shift of the bands and the imaginary part gives the lifetime broadening. Σ is energy dependent and there is a dispersion relationship between the real and imaginary parts, so that any real shift in energy bands has to be accompanied by lifetime effects. The spectral function generally contains additional structure, for example, satellites and quasiparticle peaks.

There has been real progress in recent years in understanding the quasiparticles and self-energies of *sp* metals and semiconductors, (Chap. 3) and in strongly correlated systems containing well localized electrons (Chap. 4). The quasiparticle wave functions satisfy the Schrödinger equation containing the self-energy, but rather than finding the individual wave functions, the corresponding Green's function is usually found. This describes the way that an extra electron or hole propagates in the system, and its imaginary part gives the spectral function. (At this level theoreticians prefer to talk about the single-particle Green's function, rather than individual wave functions.) At least for simple metals and semiconductors, the so-called GW approximation provides a manageable way of finding the self-energy; in this approximation the self-energy is built up from the Green's function for non-interacting particles (G), and the screened interaction between the electrons (W). A very successful research program based on this approach has provided a solution of the bandgap problem in semiconductors and the discrepancies between experimental and density-functional bandwidths in alkali metals (Chap. 3). Calculating the GW self-energy is a major computing task, however, and as yet there are no results for more complicated systems such as transition metals with more tightly bound electrons. We note here that a full program for calculating the quasiparticle self-energy involves calculating the frequency-dependent dielectric function, which screens the interaction between particles in W. This can be explored in optical absorption experiments, so that a really detailed (and symbiotic) collaboration between theory and experiment on the many-body problem is possible.

In strongly correlated systems such as rare earths, an atomic-like approach is successful in explaining X-ray absorption and BIS spectra, where characteristic atomic features show up. These are described in Chaps. 5 and 6, respectively. But there is still a major problem in trying to bridge the gap between the local atomic picture and the band picture, and this is particularly striking in experiments on Ce and the new superconductors described in Chaps. 5 and 8, respectively. Solutions of model Hamiltonians exist which are appropriate to some systems, (see Chap. 4) but this is very different from having an ab initio understanding of the electronic spectroscopies.

1.3 The Experimental Methods for Study of Unoccupied States

1.3.1 An Overview of the Spectroscopies

Before proceeding to a short discussion of the individual experiments, we give an account of the electronic transitions which provide the basis for experimental spectroscopic study of unoccupied states. These are illustrated schematically in Fig. 1.3. From left to right, the first involves optical transitions from occupied to unoccupied states. There has been an enormous amount of work in optical spectroscopy over the years, but its interpretation is complicated because the spectra must be related to the joint density of occupied and unoccupied states. This book does not attempt extensive discussion of optical spectra although some discussion is given in Chap. 8 and Sect. 10.8.

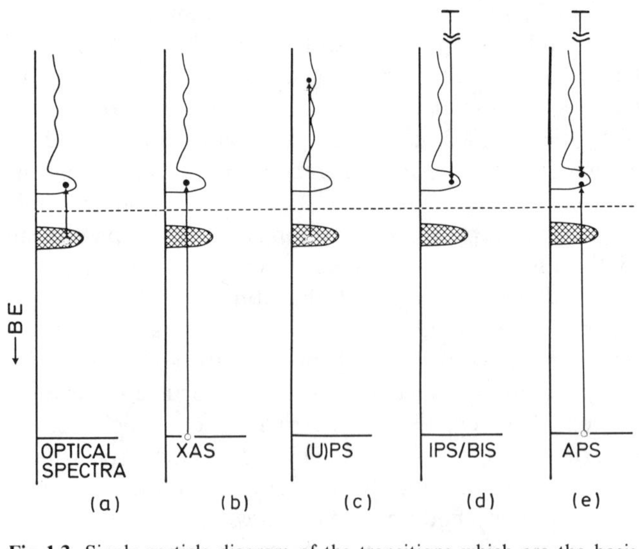

Fig. 1.3. Single-particle diagram of the transitions which are the basis of (**a**) optical transitions, (**b**) X-ray absorption and ELS, (**c**) photoemission, (**d**) inverse photoemission and BIS and (**e**) APS

The second transition in Fig. 1.3 is from a core to an unoccupied valence state and is the basis of X-ray absorption (XAS) experiments. In many cases the spectra can be related more or less directly to the unoccupied DOS. However the process is local and governed by dipole selection rules so that XAS gives data relating to the site- and symmetry-selected DOS and is thus complimentary to inverse photoemission, which samples all the unoccupied states. As explained in Sect. 1.1, the interaction between the excited electron and the core hole can severely distort the local unoccupied state density and even lead to core exciton states, as illustrated in Chap. 5, for instance. Furthermore, it must be noted that short core-hole lifetimes lead to broadening of XAS spectra.

As shown in Fig. 1.4, lifetime broadening can be significant, especially for deep core levels and core levels of low angular momentum quantum number, as illustrated in the figure. It is a natural consequence of Heisenberg's uncertainty principle and increases when the core holes decay rapidly. The decay channels are divided into fluorescent and Auger decay. In general, Auger decay dominates except for the 1s core level of very heavy elements. The Auger decay may be further sub-divided into normal and Coster–Kronig (CK) decay. In CK decay, at least one of the final state holes has the same principle quantum number as the initial state hole. CK processes can be much faster than normal Auger

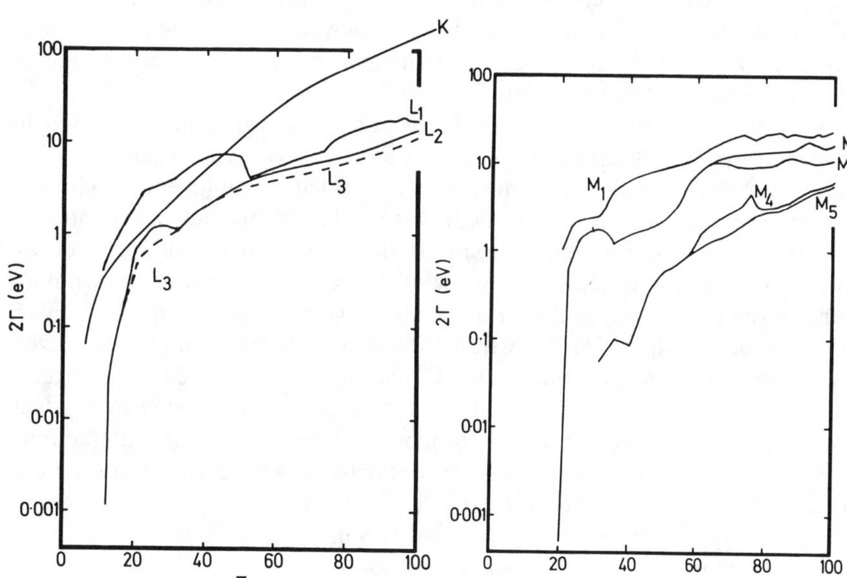

Fig. 1.4. Estimated lifetime broadening of the ionization energies of the K, L and M core levels. Reproduced with permission from [1.2]

processes, and can thus lead to very large lifetime broadening for sub-shells with low l quantum number. One should also note that the number of possible decay channels decreases within a principle shell as the l quantum number increases, so high-l levels are narrowest. A table summarizing our present knowledge of core level lifetimes is given in Appendix B.

The third and fourth transitions given in Fig. 1.3 are photoemission and inverse photoemission. In photoemission (PS), an electron is excited from a valence state to the continuum. $\boldsymbol{k}$-conservation rules apply to photoemission stimulated by UV photons so that angle-resolved photoemission can be used to probe the dispersion of unoccupied bands in $\boldsymbol{k}$-space. However, although this procedure has been successfully employed in some important work [1.20], the analysis is complicated and time consuming and this method will not be discussed extensively in this book. Of more importance to us is the inverse photoemission process in which electrons from a monochromatic beam fall into the unoccupied valence states with consequent emission of radiation. The energy distribution of this radiation gives information on the energy distribution of the unoccupied states. The inverse photoemission experiments can be done in the high-energy regime ($h\nu \geqq \sim 100\,\mathrm{eV}$, where the technique is normally known as bremsstrahlung isochromat spectroscopy or BIS), or at low energy ($h\nu \leqq \sim 50\,\mathrm{eV}$, where the technique is usually known as inverse photoemission or IPS). These two regimes bear the same relationship as XPS does to UPS. At low energies, in IPS, $\boldsymbol{k}$-conservation is important and from angle-resolved experiments one may obtain information relating to the dispersion of the bands, as described in Chaps. 7 and 9. At higher energies, the spectra are more closely related to the total density of states with appropriate allowance for matrix element effects, as described Chap. 6.

Finally, the transition on the right of Fig. 1.3 is typical of those used in appearance potential spectroscopies (APS). In this case, electrons of a monochromatic beam fall into unoccupied states with simultaneous excitation of a core electron to another unoccupied state. The transitions are monitored by the radiative or non-radiative decay of the excited states. Clearly the decay can only be monitored when the energy of the incoming beam is high enough to create the core hole, i.e. above a certain *appearance* potential which gives the technique its name. The APS transition matrix elements are short ranged so that the spectra in principle carry information on the local density of states at the core ionized atom. However the information is complicated by the fact that the spectra involve a self-convolution of the DOS, as well as the interactions between the two extra valence electrons and between valence electrons and the core hole (see e.g. [1.3]). Thus although APS may have its uses in special cases, it cannot compete seriously with XAS and IPS in investigations of unoccupied states. APS is not discussed further in this book.

We now proceed to a short summary of the experimental set-ups needed for the individual techniques.

1.3.2 X-ray Absorption Spectroscopy (XAS)

Most good modern X-ray absorption studies are made with the use of synchrotron radiation (SR), although information of similar quality is also obtained from electron energy loss (EELS). Special aspects of EELS specific to that technique are described in Chap. 7 so here we restrict ourselves to discussion of the experimental principles of the use of SR for XAS.

At photon energies above 800 eV the preferred method for high-resolution monochromatization is Bragg reflection from single crystals. The energy of the diffracted beam is set by the Bragg refraction condition

$$n\lambda = 2D \sin\theta, \tag{1.2}$$

where λ is the X-ray wavelength, D is the crystal lattice constant and θ is the angle of incidence with respect to the normal to the crystal planes.

The most common XAS monochromator is the double-crystal type, whose principle is illustrated in Fig. 1.5. As drawn, the beam is displaced by consecutive diffraction at two parallel crystals, but its direction of propagation remains constant. In all practical monochromators for spectroscopy, the distance between the crystals is varied as the crystals are turned in order to keep the beam output at a constant position. This avoids the need to move the experiment as the X-ray beam energy is changed, but it does require extreme precision in the movements. If beam focussing is required in a crystal monochromator it is normally achieved by a separate focussing mirror. In order to achieve high reflectivity the beam must be reflected at a glancing angle by this mirror (typically $\leqq 1.5°$).

A list of common crystals, with their lattice constants and the energy ranges they can serve is given in Table 1.1. At high energies many crystals are suitable, but problems occur at lower energies. Of the crystals listed, the inorganic crystal with the longest lattice constant is mica, but this is not very pure and gives poor resolution. Organic crystals such as KAP have even longer lattice constants but they are sensitive to radiation damage and cannot be used as the first crystal in synchrotron light. Thus the most useful crystal for long wavelength X-rays is beryl, which cuts off at about 800 eV. Various schemes have been used for double-crystal monochromators at longer wavelengths, including artificial multilayer crystals such as W/C, or multilayers matched to the KAP lattice

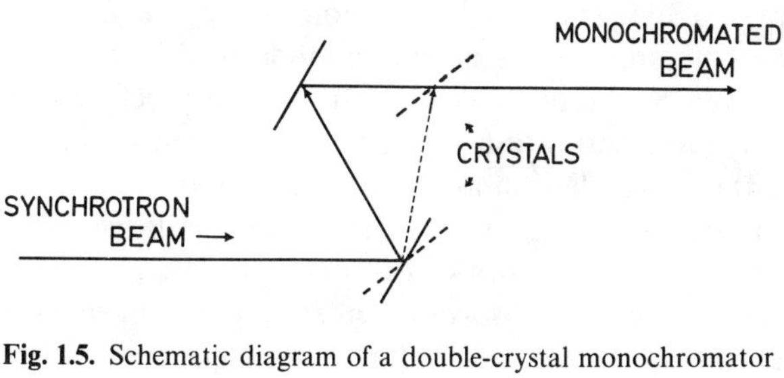

Fig. 1.5. Schematic diagram of a double-crystal monochromator

Table 1.1. Selected data for some crystals commonly used in X-ray monochromators

Crystal	Miller indices	2d (Å)	Useful wavelengths (Å)	Useful energies (eV)
InSb	(111)	7.4806	0.65–7.23	1,650–19,000
Silicon	(111)	6.271	0.55–5.98	2,100–22,200
Ge	(111)	6.532	0.57–6.23	2,000–21,700
α-Quartz	$(10\bar{1}1)$	6.687	0.58–6.38	1,950–21,300
Beryl	$(10\bar{1}0)$	15.954	1.4–15.22	800–8,800
Mica	(002)	19.84	1.73–18.93	650–7,100
KAP	(100)	26.632	2.32–25.41	490–5,300
Lead stearate	—	100.4	8.7–95.4	130–1,400

constant as a first crystal and KAP as a second crystal [1.4, 5]. However for spectroscopy these developments have been overtaken by the recent developments in grating monochromators.

For many years gratings have been the preferred monochromating elements in the energy region below 800 eV if high resolution is required. Only for some low-resolution, high-transmission experiments are large area artificial multilayers a serious competitor. The wavelength of light diffracted from a series of parallel lines perpendicular to the beam is given by [1.6].

$$\pm m\lambda = D(\sin\alpha + \sin\beta) \tag{1.3}$$

where α and β are the angle of incidence and exit of the beam with respect to the grating's surface normal. Focussing of the beam in a grating monochromator for the energy region of interest to us (50–1000 eV) is normally done by use of curved grating surfaces. Several reflections of the radiation are required in order to keep the exit slits of the monochromator constant when the energy of the radiation is changed by rotating the gratings. Figure 1.6 illustrates one of the more successful of such monochromators [1.7]. It consists of a horizontal (HFM) and vertical (VFM) focussing mirror and a single grating. The rms roughness of the mirrors has to be of the order of 6 Å and slope accuracy on the VFM is 0.5 arc sec. The radius of curvature of the three optical components is of the order of 50 meters, but because they are used at low angles of incidence, the beam from the synchrotron is focussed at a much shorter distance.

Once the monochromatic beam has been selected by the monochromator, it is necessary to measure the absorption of the photons by the sample. Traditional XAS spectra were measured in transmission mode, as illustrated in Fig. 1.7. The spectrum was given by the ratio of the beam intensity before and after the absorber. In order to avoid saturation effects it is desirable to use thin absorbers, so that the percentage of radiation absorbed in small, the spectrum is given as the ratio of two large numbers, and the shot noise is large. This mode is still commonly used for XAS studies of deep core levels. However, when the radiation energy is decreased, the absorption cross section is in general higher and the task of preparing suitable, free-standing, pin-hole-free absorber

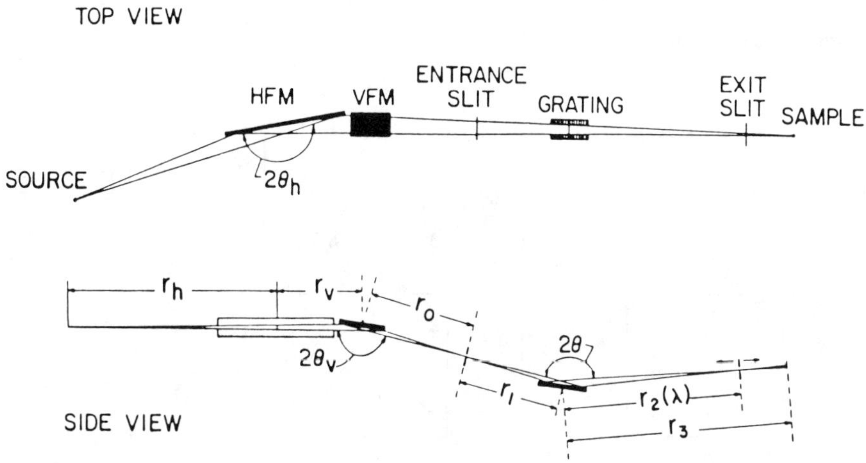

Fig. 1.6. The geometric layout of the "Dragon" cylindrical element monochromator. The figure is reproduced from [1.7]

foils is in most cases prohibitively difficult because of the extremely thin sample required (about 20–100 Å for energies below 1000 eV).

A second method of XAS measurement, known as photoyield, (Fig. 1.7) has largely replaced the traditional method for low energy X-rays. In the photoyield mode method, the radiation from the monochromator can be allowed to fall on a bulk sample, thus exciting a current of electrons and fluorescent radiation from the sample. The basis of the method is that if the mean absorption length is much longer than the mean escape depth of the electrons or fluorescent radiation, then the signal is proportional to the fraction of the incident radiation absorbed within the surface region [1.8]. This condition is generally met. The use of XAS in photoelectron yield mode is much more sensitive than photoelectron spectroscopy, because in XAS there is seldom any reason to energy select the electrons. This also results in a greater probing depth, because one also accepts electrons that have suffered many inelastic scattering events before escaping from the surface. The mean probing depth of total-electron-yield XAS in the soft X-ray region (200–2000 eV) is longer than the 5–20 Å typical of XPS. However evidence is accumulating that it is not always as long as several hundred angstroms, as is often thought. In some cases 20–40 Å seems to be a more reasonable estimate [1.9, 10]. For further ideas on mechanisms controlling the probing depth in total-electron-yield XAS we refer to the article by *Erbil* et al. [1.11].

It should also be mentioned that total and partial (i.e. energy selective) electron-yield modes are now extremely widely used in studies of absorbates and surfaces, as discussed in Chap. 6. The partial-yield mode allows suppression of the background from the substrate. An example of its use would be for studies of O or CO on, say, Ni single crystal substrates. Even when the X-ray energy

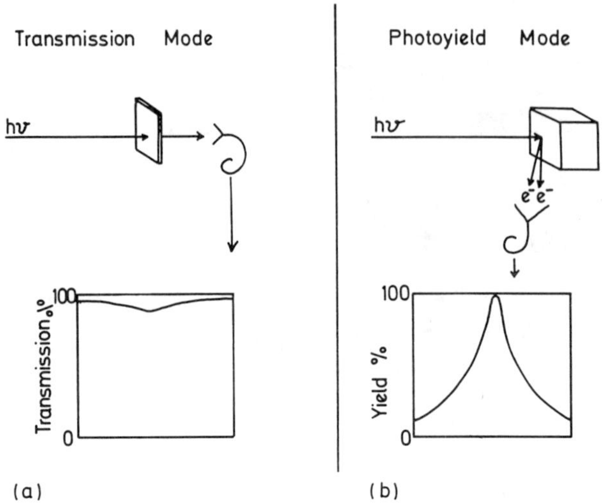

Fig. 1.7. (**a**) Schematic diagram of how XAS is measured in transmission, with a representation of the spectrum in which the transmitted signal should not be excessively attenuated by the absorption. (**b**) Schematic diagram of how XAS is measured in photoyield mode

is set on the O 1s absorption peak, most of the photoemitted electrons originate from the substrate. However, by setting a high throughput electron energy analyser to select only electrons at the O KLL Auger energy, one may suppress most of the electrons arising from the bulk and observe a signal dominated by the ionization at the oxygen site. Another variation of this theme is the use of fluorescence yield in conjunction with some energy filter to select the radiation emanating from a particular element and to make an analysis of the level and chemical state of trace elements.

1.3.3 Bremsstrahlung Isochromat Spectroscopy (BIS)

In inverse photoemission, electrons from a monochromatic beam relax into unoccupied valence states with simultaneous emission of X-rays or UV radiation, depending on the electron energy. There are two variants of this spectroscopy. One may keep the electron beam energy fixed and measure the spectrum of radiation emitted and this is often done by UV inverse photoemission. The second variant, known as bremsstrahlung isochromat spectroscopy (BIS) is normally used for high energy studies. As shown in Fig. 1.8, in the isochromat technique one varies the kinetic energy of the incident electrons, and by measuring the intensity of light at one frequency one obtains information on the number and position of states above the Fermi level.

The experimental set-up for BIS is nowadays based on the Rowland circle geometry, [1.12, 13] shown in Fig. 1.9. The reflector plate is a mosaic of single crystals cut parallel to a high symmetry surface and bent to twice the radius

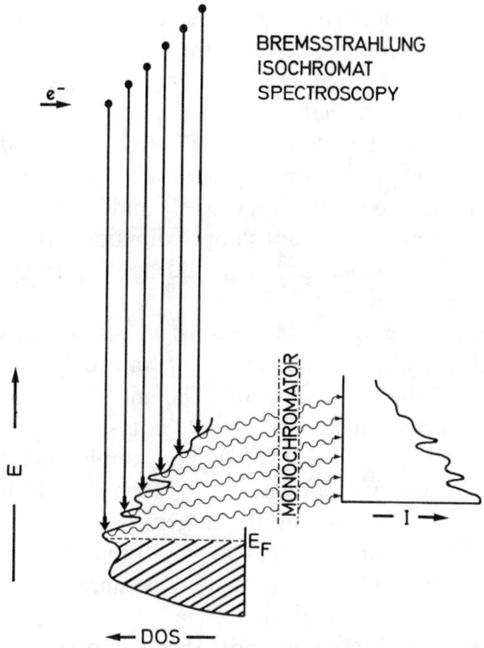

Fig. 1.8. Illustration of the principle of bremsstrahlung isochromat spectroscopy (BIS)

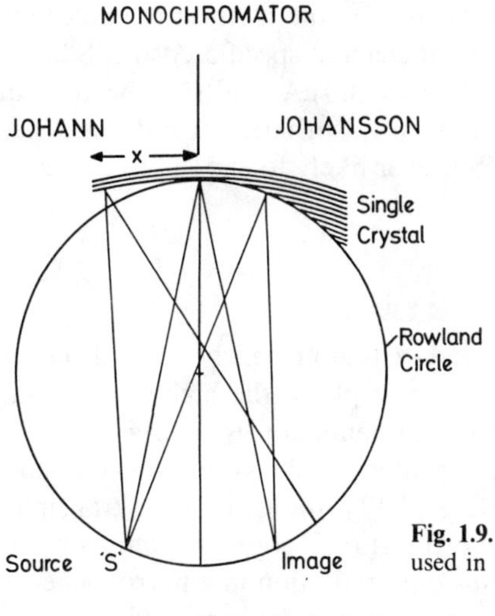

Fig. 1.9. Schematic diagram of the Rowland circle geometry used in BIS measurements. For further details, see text

of the Rowland circle. In this geometry radiation from any point on the circle will be specularly reflected and focussed at another point on the circle. For small reflecting surfaces the angle of incidence is the same on all points of the single crystal surface, so that Bragg reflection occurs for the same energy of radiation at all points. As shown on the left of Fig. 1.9, the set-up is similar to that of the Johann-type X-ray spectrometers [1.14]. This geometry has the disadvantage that the aberrations and monochromatization deteriorate as the size of the reflector plate is increased, essentially because the extremes of the reflector plate no longer lie on the Rowland circle. On the right of Fig. 1.9, the single crystal surfaces have been polished so that the crystal planes are still parallel to those of the Johann geometry but the surface still follows the Rowland circle. This is the Johansson geometry [1.15] which essentially eliminates aberrations in energy. The true Johansson geometry is too costly and is seldom, if ever, used for BIS. There are however many ingenious approximations to the Johansson geometry which can be used to increase the solid angle through which X-rays from the sample can be collected.

For experimentalists there are four other points in the construction of a BIS instrument. The electron gun should be fitted with a low-temperature emitter in order to limit the thermal spread of the electron beam. The gun should be of the Pierce type [1.16] in order to focus quite high currents in a small spot, as discussed in connection with Fig. 1.12 in Sect. 1.3.4 below. The sample should be set at an angle, as shown, to foreshorten the spot size seen by the monochromator, and the arc between sample and collector should be kept below $\sim 40°$ because the dispersion of the monochromator is higher for small Bragg angles. Of course, all the work must be done in ultra-high vacuum because of the surface sensitivity of the technique.

When assessing the applicability of BIS one must note that it is primarily a tool for fundamental research. It is excellent for studies of electronic structure and phenomena like self-energies, but would not be the method of choice for chemical analysis, for instance, because it is not element specific. Also BIS beam currents of ~ 1500 eV electrons are typically 100–300 μA, and such beams can be very destructive for insulating compounds, or adsorbates on surfaces. Thus one should not imagine that the use of BIS is ever likely to expand to the same level as the use of XAS.

1.3.4 Inverse Photoemission (IPS)

Inverse photoemission at low energy, in the UV region, can be carried out in three ways, two of which are very similar in principle to the high-energy BIS. The simplest method of measuring inverse photoemission is with a CaF_2 or SrF_2 window and a gas-filled proportional counter. As shown in Figs. 1.10 and 1.11, in the system first worked out by *Dose* [1.17], the CaF_2 only transmits radiation up to ~ 10.1 eV and the I_2 in the proportional counter is only ionized by radiation of energy above ~ 9.3 eV. Thus only radiation in a narrow-energy range produces a signal and the counter acts as a monochromator with a band pass of $10.1 - 9.3 = 0.8$ eV. Actually, the energy dependence of both the transmission function and the ionization probability are quite irregular, as shown in Fig. 1.10, so that the counter-response functions are also irregular. This is not such a disadvantage when one takes into account the smearing of the IPS response function by the energy spread of the electron beam. The advantages of the IPS system based on a gas cell include its simplicity, its low

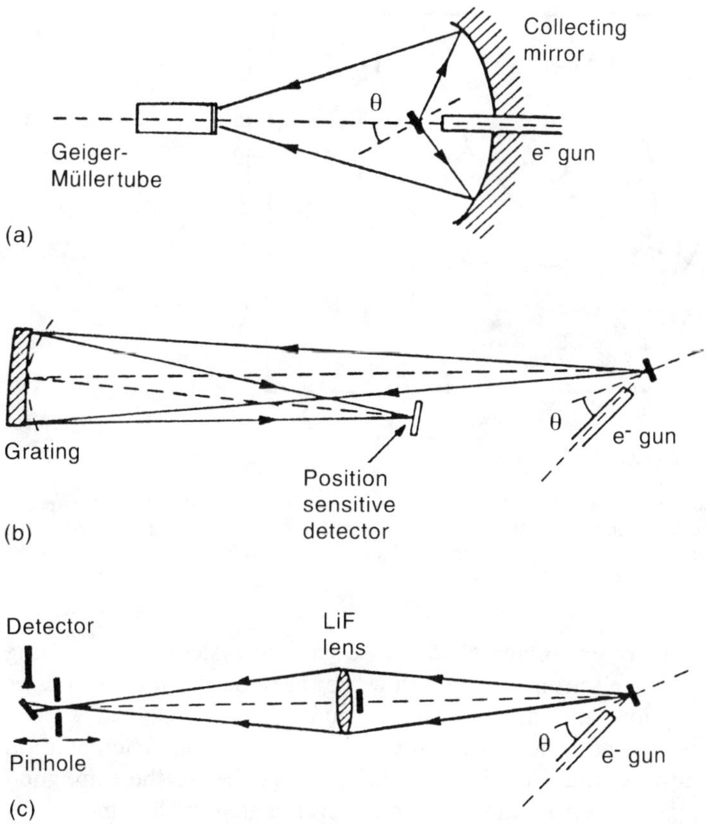

Fig. 1.10. Three basic methods for photon selection and detection in ***k***-resolved inverse photoemission (KRIPES): (**a**) collection mirror focussing into a band-pass Geiger–Müller tube [1.18], (**b**) grating spectrograph with parallel light detection throughout a wide energy range [1.19], (**c**) refracting monochromator [1.20]. The figure has been taken from [1.21]

cost and its compactness. This latter allows the cell to be set very close to the sample with the consequence that the radiation from the sample is collected over a large solid angle.

Dose's group has also thought out various improvements to this simple scheme [1.23]. The most important are replacement of the CaF_2 window by SrF_2 and heating the window. Both of these modifications have the effect of lowering and sharpening the cut-off of the radiation transmitted and hence narrowing the band pass. However both modifications cut down the intensities achieved with the systems. It is thus preferable to set a SrF_2 plate in front of the CaF_2 window in such a way that it can be heated, or removed entirely. In this way one avoids problems with gas leaks when the window is heated and one can use the higher sensitivity of the CaF_2 for setting up the experiment, or when the sample yields very low signal intensities.

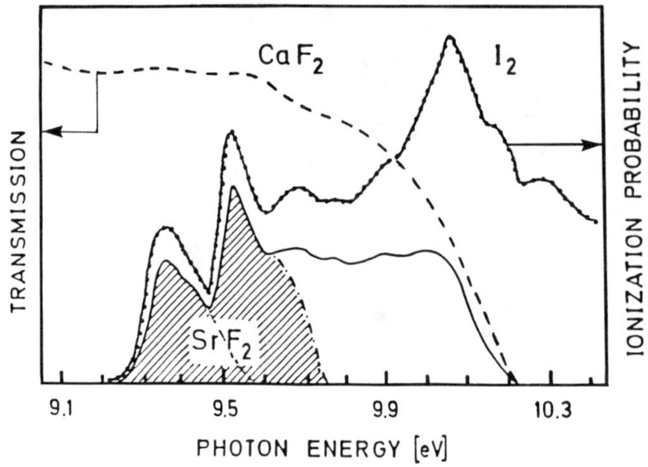

Fig. 1.11. Illustration of the principle of the proportional counter as used in IPS. The cell-response functions are related to the product of the ionization probability of I_2 and the transmission of the window. The response function is given for both CaF_2 and SrF_2 windows as a full line and the shaded area respectively. From [1.22]

There is one more advantage of the Dose gas cell system for measuring IPS: the energy of the electrons used is below the threshold for most processes leading to desorption and damage of compounds and adsorbed species [1.24–26]. Thus even the typical electron currents of about one microamp do not normally lead to serious sample degradation. Nevertheless, there are good reasons to use higher electron and radiation energies, as described more fully in Chaps. 6 and 7. This requires the use of gratings as monochromators, and experimental systems which are an order of magnitude more complex. The problems with gratings are that their reflectivity is low, it is hard to arrange for them to subtend a large solid angle, and they are expensive. Even with 10 μA of electron current and multichannel detection, count rates are likely to be below 10 counts per second. This means that the isochromat mode must be abandoned in favour of the spectrographic mode, i.e. instead of keeping $h\nu$ constant and obtaining the spectrum by sweeping the electron energy E, one must keep the incoming electron energy constant and measure the spectral intensity as a function of $h\nu$. In theory it is possible to achieve better resolution with the grating set-ups of the type shown in Fig. 1.10 than with the gas cell methods, but again the best resolution can only be gained at the cost of intensity.

Finally we note the ingenious use of a LiF lens [1.20] as the dispersive element for IPS in the isochromat mode (Fig. 1.10). Here, the energy dependence of the refractive index of LiF near its 11.7 eV cut-off, causes the focus of the lens to vary with photon energy. Thus by appropriate positioning of the exit slits one can obtain IPS spectra with reasonable resolution.

It is interesting to contrast the typical electron guns used for $\boldsymbol{k}$-resolved IPS and BIS. The IPS gun illustrated in the top half of Fig. 1.12 is the design of *Stoffel* and *Johnson* [1.27]. Just as for BIS, the gun must be equipped with

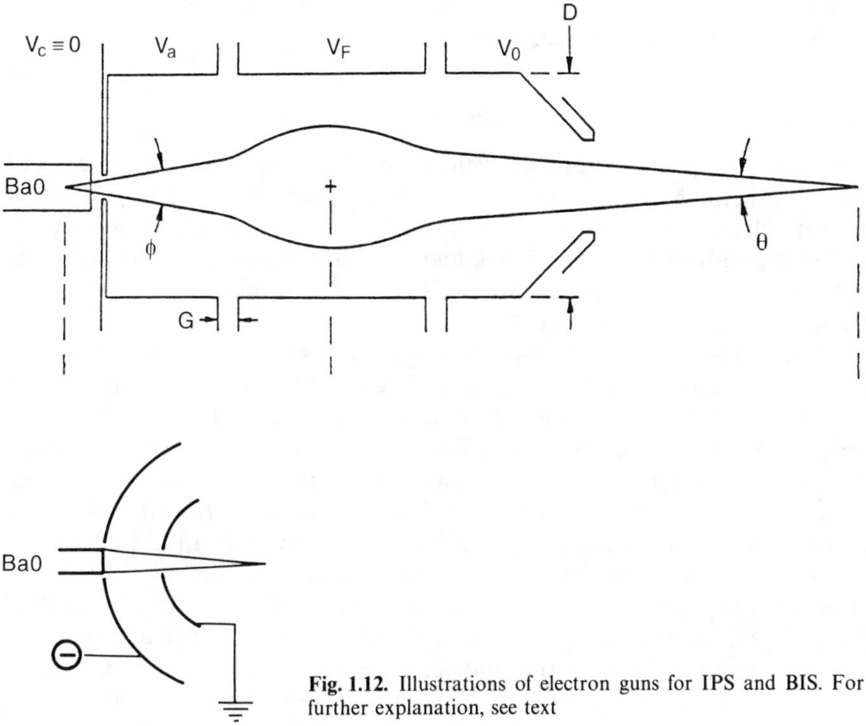

Fig. 1.12. Illustrations of electron guns for IPS and BIS. For further explanation, see text

a low-temperature emitter, such as BaO, LaB_6 or one of the special dispenser cathodes which work by diffusion of low work-function material to their surfaces. But the essential point of the low-energy guns is to focus a beam with small angular divergence on to the sample in order to define the momentum, or $\boldsymbol{k}$, as closely as possible. The philosophy of the Pierce gun for BIS is completely different. Pierce guns were originally designed to give maximum current density [1.16]. The electrons are first accelerated between two nearly spherical, or cylindrical electrodes. This gives them the required energy, and a converging path. They then pass through a small hole in the anode into a field-free space around the sample. For BIS the arrangement has the advantage that the electrons only focus once, when they have a maximum energy. This configuration thus minimizes the energy spread due to space charge effects and the interaction between the electrons in the beam (the Boersch effect [1.28]).

1.4 Historical Perspectives

In this book we consider four experimental techniques for study of unoccupied states, of which three have a long history (XAS since ~ 1913, BIS since ~ 1917 and EELS since ~ 1929) and one quite short (IPS since ~ 1970). In the following we give a short account of the progress in these areas and sufficient references for the interested reader to continue further studies.

1.4.1 The Development of X-ray Absorption Spectroscopy

X-rays have played a major role in the development of science in this century and have been the basis of numerous Nobel prizes. Good summaries of the early developments are accessible in the fascinating accounts of *Compton* [1.29] and of *Stumm von Bordwehr* [1.30], and these are highly recommended. (The latter has also the advantage of omitting Compton's bias against the "new" theories of quantum mechanics.)

The first X-ray absorption spectroscopic studies were made using thin films as filters but these measurements were difficult and it is a tribute to the skill of people such as Barkla that they were able to observe the *K*, *L*, *M* and *N* series of X-rays in such "spectra". The most significant experimental breakthrough in those early years was the development of X-ray diffraction by Bragg and Laue, so that in about 1912–1915, *de Broglie* and *Herweg* could start accurate photographic recording of X-ray absorption spectra [1.31, 32].

There were many experimental developments in the field of X-ray absorption in the decades of the 1920s and after, such as better X-ray tubes, focussing spectrometers, digital detection and counting devices, and development of high-energy electron transmission experiments in which core excitations give data equivalent to XAS [1.33]. Thousands of experimental papers on XAS were published using these crude techniques and many of today's theoretical concepts were developed to a surprising level of sophistication before 1970 [1.34, 35]. Indeed the basic ideas about EXAFS, the extended fine structure in X-ray absorption, were sufficiently clear to motivate a whole range of technical developments to make more sophisticated measurements possible. EXAFS is still a major research tool, although it is outside the scope of this book. However, the experimental developments since 1970 have revolutionized the practise of X-ray absorption spectroscopy and opened up new vistas for the near edge X-ray fine structure (NEXAFS, also known as XANES). The relevance of these developments, which include the advent of synchrotron radiation, photoyield techniques, routine availability of ultra-high vacuum equipment, and grating monochromators to cover the region up to $\sim 1000\,\mathrm{eV}$, will now be explained.

Synchrotron radiation (SR) has achieved almost total domination of XAS work in a comparatively short time. SR from electron or positron storage rings has many unique properties including: (a) high intensity over a wide spectral range with very smooth and predictable changes as a function of energy; (b) very small divergence of the beam and (c) strong and predictable polarization.

The high intensity and small divergence of synchrotron radiation have allowed an enormous improvement in productivity for XAS. This is illustrated by spectral studies over a wide energy range, such as those used for extended X-ray absorption fine structure (EXAFS) which used to take days in a laboratory with sophisticated rotating X-ray anodes, and even then the signal to noise and sensitivity were severely limited. Now, at a reasonable synchrotron, the X-ray flux density at the sample may be 10^5–10^7 higher and spectra may take only minutes. In addition, one can now study highly dilute systems and sub-monolayer studies of surfaces.

The small beam divergence of synchrotron radiation is especially advantageous for spectral studies because it allows new concepts in monochromator design, with a minimum of compromise on the optical aberrations and resolution in order to get sufficient intensity for measurements. This development has been most dramatic, and slowest, in the soft X-ray region below 800 eV, which is served by grating monochromators. The resolving power at 500 eV has increased from about 400 in 1980 to 1000 in 1985 and 10,000 in 1989. This is highly relevant because it is precisely in this spectral region that the sharpest core levels are found for many of the most important elements of the periodic table [1.36–39] (see also Appendices I and II). The improvement in resolution has yielded XAS spectra with superb structure, as discussed in Chaps. 5 and 6, and is creating a great deal of excitement at the time of writing [1.40, 41].

The final property of SR listed is its polarization. The radiation emitted parallel to the orbit from a bending magnet in a storage ring is 100% linearly polarized with the electric vector in the plane of the ring. Above and below the ring, high degrees of circular polarization are found, and in addition crossed polarized insertion devices are now being developed and incorporated in many of the world's synchrotrons to give even brighter sources of circularly polarized X-rays. The utilization of the polarization properties of SR has taken longer to develop, but the possibilities it opens for study of the electronic structure of low symmetry systems, such as high T_c superconductors [1.42] (Sect. 7.4.3) and surfaces (Chap. 6), are now rapidly being recognized. Also, the development of magnetic X-ray dichroism experiments in XAS has attracted much attention [1.43–47]. Indeed, many crossed polarized insertion devices are now being constructed at synchrotrons around the world to produce circularly polarized X-rays. Thus the use of the polarization properties of SR in XAS must be considered as one of the fastest growing areas of science in 1990.

The second major technical development listed above is the use of photoyield techniques. This has been described in Sect. 1.3.2, and is a deceptively simple trick to get good data in the soft X-ray region. However it has contributed enormously to the growth of soft XAS studies in the last decade, which is very important for studies of unoccupied states because only shallow core levels have narrow line widths. Hard X-ray studies have a much more limited potential for electronic structure studies because the large core level line widths smear out all relevant details.

The influence of the routine availability of ultra-high vacuum (UHV) equipment is simple to explain. Without this it would not be conceivable to make well-defined studies of surfaces, or to use surface-sensitive techniques, such as soft XAS. In addition, UHV is a factor permitting the construction of electron and positron storage rings; without UHV we would still be limited to true synchrotrons, with all their associated radiation hazard problems, and we would still have the problems of severe contamination of the surfaces of optical elements in the monochromators needed for XAS. The developments in monochromator technology were discussed in connection with the small divergence of SR.

A final factor in the experimental development of XAS should be mentioned, although it actually refers to the theory. There have been phenomenal developments in the theory of XAS and in computational facilities in the last two decades. Thus it is no longer considered impossible for experimentalists to perform model and ab initio calculations as a tool for the *interpretation* of their data. This development is also opening up entirely new possibilities for the application of XAS.

1.4.2 The Development of Electron Energy Loss Spectroscopy

The development of electron energy loss measurements (EELS) in reflection started in 1929 [1.48] and the first transmission measurements were in 1941 [1.49]. It was recognized relatively early that EELS could yield information on the momentum and energy transfer associated with excitations [1.33], in contrast to measurements with X-rays which only give an energy transfer. This made EELS the method of choice for studies of plasmons [1.33]. However, one does need very thin samples (500–1000 Å) for EELS studies and this has always limited the use of the technique. In this sense the absence of UHV techniques was also a disadvantage because evaporation techniques yielded impure films. It is only in the last few years that these problems have been overcome and it has been possible to make experiments as sophisticated as those described in Chap. 7.

A second major development in recent years has been the increasing use of electron microscopes to measure EELS spectra. These obviously have advantages for high spatial resolution studies but do not have the ultimate energy and momentum resolution of purpose-built spectrometers. Nevertheless they have been able to produce several landmark papers in the field of energy loss (see e.g. [1.50]) and are widely used at present. Also, the energy resolution attainable in such instruments is improving as their design becomes more flexible and as the potential of the EELS measurements of core excitations for quantitative and qualitative analysis becomes widely recognized.

1.4.3 The Development of Bremsstrahlung Isochromat Spectroscopy

The history of BIS can be traced back to the work of *Duane* and *Hunt* in 1915 [1.51]. In the early years the resolution was poor and the measurements were crude. Just as in photoemission experiments, [1.52] it was not realised how surface sensitive the measurements were and in the absence of UHV facilities, the spectra seldom showed much structure. Indeed it was probably 1942 before the first observation, by *Ohlin* [1.53], of structures in the short-wavelength limit of the bremsspectrum emitted when a monoenergetic beam of electrons bombarded a solid. Most of the interest in the short-wavelength bremsspectrum was related to the possibility of using the data to determine Planck's constant via the de Broglie relationship, so it was 1946 before Ohlin's structure was interpreted as DOS features by *Nijboer* [1.54].

In the 35 years following Nijboer's work, BIS was periodically rediscovered as a spectroscopic tool whose use was limited partly by vacuum capabilities in that era [1.55–58]. The regenesis of the technique dates from the start of ultra-high vacuum (UHV) measurements (see e.g. [1.59–65]) and especially the realization that an X-ray photoelectron spectrometer monochromator could also be used for BIS studies [1.64, 65]. It is probably fair to say that the majority of BIS studies now published involve measurements at 1486.7 eV, the energy of the Al $K\alpha$ radiation used in XPS monochromators. The resolution is typically 0.6 eV. In addition, advances in monochromator design in the last ten years have increased the solid angle of X-ray collection to $\sim$0.1 ster and decreased counting times considerably.

1.4.4 The Development of Low-Energy Inverse Photoemission (IPS)

The history of IPS is much shorter than that of the other techniques discussed here. It really started with the experimental work of *Dose* [1.17] and independent theoretical work of *Pendry* [1.66]. The experimental advances since then have been rapid. The first measurements were made with a proportional gas counter, as described in Sect. 1.3.4 and this scheme is still widely used. Also, since then several groups have built other apparatus based on LiF lenses for monochromation, and especially on grating monochromators. These latter are much more expensive to build, and collect radiation over a smaller solid angle, but allow the use of different light energies. Another development in recent years has been the use of electron beams with good angle definition to permit mapping of the unoccupied state bands in **k**-space. The range of IPS applications is thus expanding very rapidly at the present time and this is summarized in Chaps. 8 and 9.

References

1.1 Note also that neither Fig. 1.2a or b represents a true spectral function because both the single-particle matrix elements giving the *probability* of transitions, and the many-particle excitations have been neglected

1.2 J.C. Fuggle: In *Electron Spectroscopy*, Volume IV, ed. by C.R. Brundle, A.D. Baker (Academic, New York 1981) p. 85ff and references therein

1.3 H.J.W.M. Hoekstra, J.C. Fuggle, W. Speier, D.D. Sarma: J. Electron Spectr. Rel. Phen. **42**, 27 (1987)

1.4 G. van der Laan, B.T. Thole, G.A. Sawatzky, J.C. Fuggle, R. Karnatak, J.-M. Esteva, B. Lengeler: J. Phys. C **19**, 817 (1986)

1.5 G. van der Laan, J.B. Goedkoop, J.C. Fuggle, M.P. Bruijn, J. Verhoeven, M. van der Wiel, A.A. MacDowell, J.B. West, I.H. Munro: Nucl. Instrum. Methods **A 225**, 592 (1987)
1.6 J.A.R. Samson: *Techniques of Vacuum Ultraviolet Spectroscopy*, (Wiley, New York 1967)
1.7 C.T. Chen: Nucl. Instrum. Methods **256**, 595 (1987)
1.8 W. Gudat, C. Kunz: Phys. Rev. B **29**, 169 (1972)
1.9 M. Abbate, J.C. Fuggle et al.: In preparation
1.10 G. Kaindl, O. Strebel et al.: Private communication
1.11 E. Erbil, G.S. Cargil III, R. Frahm, F. Böhme R: Phys. Rev. B **37**, 2450 (1988)
1.12 H.R. Rowland: Phil. Mag. **13**, 349 (1882)
1.13 H.R. Rowland: Phil. Mag. **13**, 197, 210 (1883)
1.14 H.H. Johann: Z. Phys. **69**, 185 (1931)
1.15 T. Johansson: Naturwiss. **20**, 758 (1932)
1.16 J.R. Pierce: J. Appl. Phys. **11**, 548 (1940)
1.17 V. Dose: Appl. Phys. **14**, 117 (1977)
1.18 K. Desinger, V. Dose, M. Glöbl, H. Scheidt: Sol. St. Commun. **49**, 479 (1984)
1.19 S.L. Hulbert, P.D. Johnson, N.G. Stoffel, W.A. Royer, N.V. Smith: Phys. Rev. B **31**, 6815 (1985)
1.20 T.T. Childs, W.A. Royer, N.V. Smith: Rev. Sci. Instrum. **55**, 1613 (1984)
1.21 N.V. Smith: Rep. Prog. Phys. **51**, 1227 (1988)
1.22 A. Goldmann, V. Dose, G. Borstel: Phys. Rev. B **32**, 1971 (1985)
1.23 G. Denninger, V. Dose, H. Scheidt: Appl. Phys. **18**, 375 (1979)
1.24 D. Menzel: Desorption Methods in *Interactions on Metal Surfaces*, Topics Appl. Phys. Vol. 4, ed. by R. Gomer (Springer, Berlin, Heidelberg 1975) p. 101
1.25 See e.g. T.E. Madey, R. Stockbauer: In *Methods of Experimental Physics*, Vol. 22, ed. by R.L. Park, M.G. Lagally (Academic, New York 1985) and references therein
1.26 E. Umbach, J.C. Fuggle: In *Laboratory Methods of Surface Science*, ed. by C.E. Powell, T.E. Madey (Plenum, New York 1992) and references therein
1.27 N.G. Stoffel, P.D. Johnson: Nucl. Instrum. Methods **A234**, 230 (1984)
1.28 H. Boersch: Z. Phys. B **139**, 115 (1954)
1.29 A. Compton: *X-Rays in Theory and Experiment* (Van Nostrand, New York 1926). A second edition, by A. Compton and S.K. Allison, was published in 1935 and reprinted many times up to 1960
1.30 R. Stumm von Bordwehr: Ann. Phys. Fr. **14**, 377 (1989)
1.31 See e.g. M. de Broglie: Philos. Mag. **23**, 987 (1912); C.R. Acad. Sci. Paris **157**, 924, 1413 (1913)
1.32 J. Herweg: Verh. Dtsch. Phys. Ges. **15**, 555 (1913)
1.33 H. Raether: Springer Tracts in Modern Physics, Vol. 38 (Springer Berlin, Heidelberg, 1965); ibid. **88**, 1 (1980) and references therein
1.34 See e.g. L.G. Parratt: Rev. Mod Phys. **31**, 616 (1959) and references therein
1.35 See e.g. L.V. Azároff: *X-Ray Spectroscopy* (McGraw-Hill, New York 1974) and references therein
1.36 O. Keski-Rahkonen, M.O. Krause: At. Nucl. Data Tables **14**, 139 (1974)
1.37 M.O. Krause: J. Phys. Rev. Ref. Data **8**, 307 (1979)
1.38 M.O. Krause, J.H. Oliver: J. Phys. Rev. Ref. Data **8**, 328 (1979)
1.39 J.C. Fuggle, S.F. Alvarado: Phys. Rev. A **22**, 1615 (1980)
1.40 C.T. Chen, F. Sette: Phys. Scr. **40**, 119 (1990)
1.41 C.T. Chen, Y. Ma, F. Sette: Phys. Rev. A **40**, 6737 (1990)
1.42 See F. Al Shamma, J.C. Fuggle: Physica C, **169**, 325 (1990) and references therein
1.43 G. van der Laan, B.T. Thole, G.A. Sawatzky, J.B. Goedkoop, J.C. Fuggle, J.-M. Esteva, R. Karnatak, J.P. Remeika, H.A. Dabkowska: Phys. Rev. B **34**, 6529 (1986)
1.44 G. Schütz, W. Wagner, W. Wilhelm, P. Kienle, R. Zeller, R. Frahm, G. Materlik: Phys. Rev. Lett. **58**, 737 (1987)
1.45 J.B. Goedkoop, B.T. Thole, G. van der Laan, G.A. Sawatzky, F.M.F. de Groot, J.C. Fuggle: Phys. Rev. B **37**, 2086 (1988)
1.46 J.B. Goedkoop, J.C. Fuggle, B.T. Thole, G. van der Laan, G.A. Sawatzky: J. Appl. Phys. **64**, 5595 (1988)
1.47 G. Schütz, M. Knülle, R. Wienke, W. Wilhelm, W. Wagner, P. Kienle, R. Frahm: Z. Phys. B **73**, 67 (1988)
1.48 E. Rudberg: K. Svenska Vet. Akad. Handl. **7**, 1 (1929)
1.49 G. Ruthemann: Naturwissen **29**, 648 (1941)
1.50 L. Grunes, R.D. Leapman, C.N. Wilker, R. Hoffmann, A.B. Kunz: Phys. Rev. B **25**, 7157 (1982)
1.51 W. Duane, F.L. Hunt: Phys. Rev. **6**, 166 (1915)

1.52 J.G. Jenkin, R.C.G. Leckey, J. Liesegang: J. Electron Spectrosc. Relat. Phen. **12**, 1 (1977)
1.53 P. Ohlin: Ark. Mat. Astron. Fys **A 29**, 3 (1942)
1.54 B.R.A. Nijboer: Physica (Utrecht) **12**, 461 (1946)
1.55 K. Ulmer: Phys. Rev. Lett. **3**, 514 (1959)
1.56 P. Johansson: Ark. Phys. **18**, 329 (1960)
1.57 H. Merz, K. Ulmer: Z. Phys. **210**, 92 (1968)
1.58 J. Auleytner, B. Lidén: Ark. Fysik **23**, 41 (1962)
1.59 R. Turtle, R. Liefield: Phys. Rev. B **7**, 3411 (1973)
1.60 P.E. Best, C.C. Chu: Jpn. J. Appl. Phys. **17**, 317 (1978)
1.61 F. Riehle: Phys. Status Solidi B **98**, 245 (1978)
1.62 C.C. Chu, P.E. Best: Phys. Rev. B **12**, 4575 (1975)
1.63 C.C. Chu, P.E. Best: Phys. Rev. B **19**, 3414 (1979)
1.64 J.K. Lang, Y. Baer: Rev. Sci. Instrum. **50**, 221 (1979)
1.65 J.K. Lang, Y. Baer, P.A. Cox: J. Phys. F **11**, 121 (1981)
1.66 J.B. Pendry: J. Phys. C **14**, 1381 (1981)

2. Band-Structure Methods

R. Zeller

With 3 Figures

The theoretical determination of the unoccupied electronic states is an important tool for understanding the results of experimental techniques such as X-ray absorption, inverse photoemission and bremsstrahlung isochromat spectroscopy. The availability of reliable calculations always serves as a starting point to interpret the measured spectra in the one-electron picture. In periodic crystals the electronic states have historically been named band structures, and their calculation is the aim of band-structure methods. A large variety of methods have been invented and an enormous amount of literature exists on the calculation of occupied and low-lying unoccupied states. There is much less work on higher lying unoccupied states and particular problems arising from the higher energy range are seldom discussed. In this chapter I describe several modern band-structure methods and discuss their applicability to calculate the unoccupied electronic states within an energy range of about 10 Rydbergs above the occupied states. To simplify my presentation, I consider only lattices with one atom per unit cell and I omit the discussion of spin-polarisation and relativistic effects and of group theory to exploit symmetry.

The multiple-scattering method will be emphasized because it leads to the smallest matrices in the computations and it can be developed into the linear muffin-tin orbital (LMTO) method, the fastest band-structure technique. The multiple-scattering method is in principle exact and easily extended to non-periodic systems, such as crystals locally perturbed by impurities or excitations, clusters of atoms, surfaces and disordered alloys.

2.1 One-Particle Equations

The fundamental objective of band theory is to study the behaviour of the electronic system in the field of the nuclei. This is essentially a many-body problem and obviously a hopeless task without simplifying approximations. By the Born–Oppenheimer approximation the motion of the electrons is assumed to be independent of the motion of the nuclei, and the problem is reduced to solve a many-electron Schrödinger equation with the Hamiltonian

$$H = -\sum_{i}^{N}\nabla_i^2 + \sum_{i}^{N} V_{\mathrm{e}}(\boldsymbol{r}_i) + \frac{e^2}{2}\sum_{i\neq j}^{N}\frac{1}{|\boldsymbol{r}_j - \boldsymbol{r}_j|}, \tag{2.1}$$

which describes N interacting electrons moving at positions $\boldsymbol{r}_i$ in the external potential $V_e(\boldsymbol{r})$ produced by the nuclei. (Throughout this chapter I use Rydberg atomic units $\hbar = 1, 2m = 1, e^2 = 2$, but keep the factor e^2 in the expressions for clarity.) The second important approximation is to reduce (2.1) to a self-consistent field approach where each electron independently moves in the average field of all the other electrons and the nuclei. Most recent electronic-structure calculations use density-functional theory [2.1–11] for this purpose. Compared to the traditional Hartree and Hartree–Fock approximations, density-functional theory provides (in principle) an exact all-electron formalism for determining the density and the energetics of the electrons. The method relies on a theorem proven by *Hohenberg* and *Kohn* [2.1] which states:

For non-degenerate ground states the external potential $V_e(\boldsymbol{r})$ is a unique functional (to within an additive constant) of the ground-state density $n(\boldsymbol{r})$ and there exists a unique functional $E[n(\boldsymbol{r})]$, which under the constraint that the number of electrons is conserved, is minimized by the ground-state density. This minimum is the ground-state energy.

This functional $E[n(\boldsymbol{r})]$ for the total energy can be written as

$$E[n(\boldsymbol{r})] = E_0[n(\boldsymbol{r})] + \frac{e^2}{2}\iint \frac{n(\boldsymbol{r})n(\boldsymbol{r}')}{|\boldsymbol{r}-\boldsymbol{r}'|} d\boldsymbol{r} d\boldsymbol{r}' + E_{xc}[n(\boldsymbol{r})]. \tag{2.2}$$

Here $E_0[n(\boldsymbol{r})]$, the known functional for the total energy of a system of non-interacting electrons, and the classical electrostatic interaction energy are separated out. Only the remaining exchange-correction part $E_{xc}[n(\boldsymbol{r})]$ must be approximated. This is often done by the local-density approximation [2.3–14]

$$E_{xc}[n(\boldsymbol{r})] = \int n(\boldsymbol{r})\varepsilon_{xc}(n(\boldsymbol{r}))\, d\boldsymbol{r}, \tag{2.3}$$

in which ε_{xc} is the exchange-correlation energy for the homogeneous electron gas of density n. The function $\varepsilon_{xc}(n)$ can be obtained from many-body calculations [2.13–16].

To minimize (2.2) the density is written in terms of one-particle wave-functions of non-interacting electrons as

$$n(\boldsymbol{r}) = 2 \sum_{i \leqq E_F} |\Psi_i(\boldsymbol{r})|^2. \tag{2.4}$$

States i are filled up to the Fermi level E_F and the factor 2 comes from the electronic spin. The minimisation of (2.2) leads to one-particle Schrödinger equations for the Ψs:

$$(-\nabla_r^2 + V(\boldsymbol{r}) - E_i)\Psi_i(\boldsymbol{r}) = 0, \tag{2.5}$$

in which the potential $V(\boldsymbol{r})$ is the sum of the nuclear potential $V_e(\boldsymbol{r})$, the electrostatic (Hartree) potential

$$V_H(\boldsymbol{r}) = e^2 \int \frac{n(\boldsymbol{r})}{|\boldsymbol{r}-\boldsymbol{r}'|} d\boldsymbol{r}, \tag{2.6}$$

and the exchange-correlation potential $V_{xc}(\boldsymbol{r}) = \delta E_{xc}[n(\boldsymbol{r})]/\delta n(\boldsymbol{r})$.

One important fact of density-functional theory must be realized in comparisons of calculated results with experiments. The one-particle Hamiltonian in (2.5) described an "electron" in an effective mean-field potential. However, there is no formal justification to interpret the solutions of (2.5) as wavefunctions of actual electrons or the distribution

$$n(E) = \sum_i \delta(E - E_i) \tag{2.7}$$

of energy eigenvalues as a measurable density of states. Nevertheless, this interpretation is often in close correspondence to experimental results, and one of the themes of this book is how useful this picture can be in interpreting experimental spectra. Deviations from the density-functional picture (in practice with the local-density approximation) are due to self-energy effects. The next chapter by Godby contains a detailed discussion of the differences between density-functional eigenvalues and the true single-particle excitations.

2.2 Methods

Many methods for solving the Schrödinger equation (2.5) expand the wave-functions in some basis set and use the Rayleigh–Ritz variational principle. Examples are the pseudopotential method which uses the most simple and convenient set, i.e. plane waves, and the modern linear band-structure methods which use more efficient sets constructed from solutions of the Schrödinger equation for the atomic potential at a single crystal site. A disadvantage of basis-set methods is that for very accurate results the number of basis functions rapidly increases unless they are really well chosen for the physical system under consideration.

2.2.1 Multiple-Scattering (Korringa–Kohn–Rostocker) Method

An alternative to basis-set methods is the KKR method, which uses scattering theory [2.17–28] to provide an exact solution of the Schrödinger equation. This is particularly appropriate for considering the unoccupied states which can have arbitrarily large energies. To derive this, the Lippmann–Schwinger equation

$$\Psi(\boldsymbol{r}, E) - \int_{\infty} g(\boldsymbol{r} - \boldsymbol{r}', \kappa^2)[V(\boldsymbol{r}') - E + \kappa^2]\Psi(\boldsymbol{r}', E)d\boldsymbol{r}' = 0, \tag{2.8}$$

which represents the Schrödinger equation in integral equation form, can be used as a starting point. Here $g(\boldsymbol{r} - \boldsymbol{r}', \kappa^2)$ is the free-particle propagator, or Green function:

$$g(\boldsymbol{r} - \boldsymbol{r}', \kappa^2) = -\frac{\exp(i\kappa|\boldsymbol{r} - \boldsymbol{r}'|)}{4\pi|\boldsymbol{r} - \boldsymbol{r}'|}, \tag{2.9}$$

which satisfies the inhomogeneous free-particle Schrödinger equation

$$[\nabla_r^2 + \kappa^2] g(\boldsymbol{r} - \boldsymbol{r}', \kappa^2) = \delta(\boldsymbol{r} - \boldsymbol{r}'). \tag{2.10}$$

Solving either (2.5) or (2.8) is not straightforward, because the potential $V(\boldsymbol{r})$ is singular at each nucleus. However, the problem is simplified by the fact that $V(\boldsymbol{r})$ is almost spherically symmetric near the nuclei, and is rather flat in the interstitial region between the atoms. The solution of (2.8) becomes particularly simple if this form of the potential is approximated by the muffin-tin potential—one which is spherical inside muffin-tin spheres around each atomic site, and constant in the interstitial region. The interstitial potential can be taken as the zero of energy, so that $E = \kappa^2$ in (2.8). Then the integration becomes confined to the muffin-tin spheres. Within the sphere centered at $\boldsymbol{R}^n$, the wavefunction can be expanded in spherical harmonics $Y_L(\hat{\boldsymbol{r}})$ and radial functions $R_l^n(r, E)$ as

$$\Psi(\boldsymbol{r} + \boldsymbol{R}^n, E) = \sum_L Y_L(\hat{\boldsymbol{r}}) R_l^n(r, E) C_L^n(E), \tag{2.11}$$

where L stands for both angular-momentum quantum numbers l and m. The radial functions are solutions of the Schrödinger equation inside the muffin-tin spheres. They are well behaved at the nucleus and satisfy the Lippmann–Schwinger equation

$$R_l^n(r, E) = j_l(\kappa r) - \mathrm{i}\kappa \int_0^{S_n} j_l(\kappa r_\langle) h_l^{(1)}(\kappa r_\rangle) V^n(r') R_l^n(r', E) r'^2 \, dr', \tag{2.12}$$

where the arguments $r_\langle$ and $r_\rangle$ are the smaller and larger, respectively, of r, r', and S_n is the muffin-tin radius. The first term $j_l(\kappa r)$ provides the correct boundary conditions for $R_l^n(r, E)$ at $r \to 0$ and $r > S_n$.

The central result of KKR theory is the transformation of the complicated integral equation (2.8) into a system of linear algebraic equations

$$\sum_{n'} \sum_{L'} (\delta_{LL'}^{nn'} - g_{LL'}^{nn'}(\kappa^2) t_{l'}^{n'}(E)) C_{L'}^{n'}(E) = 0, \tag{2.13}$$

where the t matrices are defined as

$$t_l^n(E) = \int_0^{S_n} j_l(\kappa r) V^n(r) R_l^n(r, E) r^2 \, dr. \tag{2.14}$$

These single-site t matrices describe the scattering produced by a single muffin-tin potential $V^n(r)$. The interplay of all scattering processes is described by the potential-independent structure constants $g_{LL'}^{nn'}(\kappa^2)$ which, derived from the free-particle propagator (2.9), describe the connection of the different sites in the crystal. The structure constants depend only on the difference $\boldsymbol{R}^{nn'} = \boldsymbol{R}^n - \boldsymbol{R}^{n'}$, vanish for $\boldsymbol{R}^n = \boldsymbol{R}^{n'}$, and are given for $\boldsymbol{R}^n \neq \boldsymbol{R}^{n'}$ by

$$g_{LL'}^{nn'}(\kappa^2) = -4\mathrm{i}\pi\kappa \sum_{L''} \mathrm{i}^{l-l'+l''} h_{l''}^{(1)}(\kappa R^{nn'}) Y_{L''}(\hat{\boldsymbol{R}}^{nn'}) \int_{4\pi} Y_L(\hat{\boldsymbol{r}}) Y_{L'}(\hat{\boldsymbol{r}}) Y_{L''}(\hat{\boldsymbol{r}}) \, d\boldsymbol{r}. \tag{2.15}$$

The angular integral in (2.15) is normally written in terms of Gaunt coefficients as $C_{LL'L''}$.

In periodic crystals, the sums over all atoms can be reduced to sums over the atoms in a single unit cell. For simplicity, let us assume one atom per unit cell, so that all potentials are equal: $V^n(\boldsymbol{r}) = V(\boldsymbol{r})$. Lattice Fourier transforms

$$C_L(\boldsymbol{k}, E) = \sum_n \exp(\mathrm{i}\boldsymbol{k}\cdot\boldsymbol{R}^n) C_L^n(E), \tag{2.16}$$

$$g_{LL'}(\boldsymbol{k}, \kappa^2) = \sum_n \exp(\mathrm{i}\boldsymbol{k}\cdot\boldsymbol{R}^n - \mathrm{i}\boldsymbol{k}\cdot\boldsymbol{R}^{n'}) g_{LL'}^{nn'}(\kappa^2) \tag{2.17}$$

can be introduced and the secular equation (2.13) reduces to

$$\sum_{L'} [\delta_{LL'} - g_{LL'}(\boldsymbol{k}, \kappa^2) t_{l'}(E)] C_{L'}(\boldsymbol{k}, E) = 0. \tag{2.18}$$

For non-trivial solutions of the C_Ls, the determinant of the secular matrix in the brackets of (2.18) must vanish. At each wave vector $\boldsymbol{k}$ the condition

$$\det[\delta_{LL'} - g_{LL'}(\boldsymbol{k}, \kappa^2) t_{l'}(E)] = 0 \tag{2.19}$$

is satisfied for an infinite number v of energies $E = E_v(\boldsymbol{k})$, giving the energy bands. For actual calculations, the form (2.17) of the KKR structure constants is not useful because the sum converges only if $\kappa^2 < 0$, but (2.17) can also be evaluated by a sum in reciprocal space

$$g_{LL'}(\boldsymbol{k}, \kappa^2) = \frac{\mathrm{i}\kappa h_l^{(1)}(\kappa r')}{j_l(\kappa r)} \delta_{LL'} - \frac{(4\pi)^2 \mathrm{i}^{l-l'}}{\tau j_l(\kappa r) j_{l'}(\kappa r')} \sum_n \frac{j_l(k_n r) Y_L(\hat{\boldsymbol{k}}_n) j_{l'}(k_n r') Y_{L'}(\hat{\boldsymbol{k}}_n)}{k_n^2 - \kappa^2} \tag{2.20}$$

where r and r' are arbitrary radii [2.29] such that $r < r' < S$. Here τ is the volume of the real-lattice unit cell and $\boldsymbol{k}_n = \boldsymbol{k} + \boldsymbol{K}_n$ is the sum of wave vector $\boldsymbol{k}$ and the reciprocal lattice vector $\boldsymbol{K}_n$. For large arguments $j_l(k_n r)$ and $j_{l'}(k_n r')$ behave as $(k_n r)^{-1}$ and $(k_n r')^{-1}$, and the sum in (2.20) absolutely converges because the terms decrease as k_n^{-4}. But the convergence is too slow for practical purposes and a splitting factor must be introduced by an Ewald procedure [2.30] or by other means. The secular equation (2.18) and the secular determinant (2.19) rapidly converge with the number of angular momenta used (Sect. 2.4.1), but the matrix elements are singular for $\kappa^2 = k_n^2$ (Sect. 2.4.2). For actual calculations it is useful to remove the singularities by a folding technique as described by *Williams* et al. [2.27]. The dimension of the secular matrix is slightly increased by a few reciprocal-lattice-site indices to describe the free-electron-like bands. The resulting fast KKR method has been used [2.31] by these authors in KKR calculations for all the metallic elements with atomic number $Z \leqq 50$.

The last part of this section is used to show that (for the muffin-tin approximation with $E = \kappa^2$) the wavefunction (2.11) solves the Lippmann–Schwinger equation (2.8) if the condition (2.13) is satisfied. In (2.8) the potential is written as a sum of non-overlapping muffin-tin potentials $V(\boldsymbol{r}') = \sum_{n'} V^{n'}(|\boldsymbol{r} - \boldsymbol{R}^n|)$, the variables are changed $\boldsymbol{r} \to \boldsymbol{r} + \boldsymbol{R}^n$, $\boldsymbol{r}' \to \boldsymbol{r}' + \boldsymbol{R}^{n'}$, and the integral is split into a sum of integrals over the spheres. This leads to

$$\Psi(\boldsymbol{r} + \boldsymbol{R}^n, E) = \sum_{n'} \int_{n'} g(\boldsymbol{r} + \boldsymbol{R}^n - \boldsymbol{r}' - \boldsymbol{R}^{n'}, \kappa^2) V^{n'}(\boldsymbol{r}') \Psi(\boldsymbol{r}' + \boldsymbol{R}^{n'}, E) d\boldsymbol{r}'. \tag{2.21}$$

The wavefunction expansion (2.11) is inserted and the spherical expansion of the free-particle Green function (2.9), defined by

$$g(\boldsymbol{r} + \boldsymbol{R}^n - \boldsymbol{r}' - \boldsymbol{R}^{n'}, \kappa^2) = \sum_{LL'} j_l(\kappa r) Y_L(\hat{\boldsymbol{r}}) g^{nn'}_{LL'}(\kappa^2) j_{l'}(\kappa r') Y_{L'}(\hat{\boldsymbol{r}}') \tag{2.22}$$

for $n' \neq n$ and by

$$g(\boldsymbol{r} - \boldsymbol{r}', \kappa^2) = -\mathrm{i}\kappa \sum_L Y_L(\hat{\boldsymbol{r}}) j_l(\kappa r_<) h_l^{(1)}(\kappa r_>) Y_L(\hat{\boldsymbol{r}}') \tag{2.23}$$

for $n' = n$ is used. The orthogonality of the spherical harmonics allows us to integrate over the angular variables $d\hat{\boldsymbol{r}}'$, which leads to

$$\begin{aligned} &\sum_L R^n_l(r, E) Y_L(\hat{\boldsymbol{r}}) C^n_L(E) \\ &= -\mathrm{i}\kappa \sum_L Y_L(\hat{\boldsymbol{r}}) \int_0^{S_n} j_l(\kappa r_<) h_l^{(1)}(\kappa r_>) V^n(r') R^n_l(r', E) C^n_L(E) r'^2 dr \\ &\quad + \sum_L j_l(\kappa r) Y_L(\hat{\boldsymbol{r}}) \sum_{n' \neq n} \sum_{L'} g^{nn'}_{LL'}(\kappa^2) \int_0^{S_{n'}} j_{l'}(\kappa r') V^{n'}(r') R^{n'}_{l'}(r', E) C^{n'}_{L'}(E) r'^2 dr. \end{aligned} \tag{2.24}$$

This result is simplified by the definitions (2.12) for radial functions and (2.14) for the t matrices as

$$\sum_L j_l(\kappa r) Y_L(\hat{\boldsymbol{r}}) C^n_L(E) = \sum_L j_l(\kappa r) Y_L(\hat{\boldsymbol{r}}) \sum_{n' \neq n} \sum_{L'} g^{nn'}_{LL'}(\kappa^2) t^{n'}_{l'}(E) C^{n'}_{L'}(E). \tag{2.25}$$

The orthogonality of the spherical harmonics is used to remove the common factor $j_l(\kappa r) Y_L(\hat{\boldsymbol{r}})$ and the secular equation (2.13) is obtained

2.2.2 Pseudopotential Method

The basic physical idea of the pseudopotential method [2.26, 32–36] is to remove the chemically inert core electrons from the calculations. This possibility is based on the following observation. The only potential-dependent terms in the secular equation (2.19) are the single-site t matrices. They describe the scattering properties of the single-site potential and are uniquely determined by it. The opposite is certainly not true and for each energy E many different potentials exist which scatter the wavefunctions identically. This fact can be exploited to choose a pseudopotential which scatters the valence electrons in the same way as the actual potential but which does not have any core states. In general, pseudopotentials are non-local and energy dependent: $V_{\mathrm{ps}}(\boldsymbol{r}, \boldsymbol{r}', E)$, but for the calculation of the valence states it has been found that energy independent pseudopotentials exist for most elements which reproduce the eigenvalues of the valence states over a typical energy range of about 1 Rydberg.

Most older work is based on empirical pseudopotentials [2.37], but the methods for constructing first-principles pseudopotentials [2.38–41] have evolved considerably over the years. A typical construction starts from self-consistent spherical atomic potentials $V^n(r)$ and replaces them in a first step by $[1-f(r/r_l)]V^n(r)+c_l f(r/r_l)$, thus removing the Coulomb singularity at the nuclei. The function $f(x)$, usually $f(x)=\exp(-x^{3.5})$, is chosen so that the pseudopotential and the resulting radial pseudowavefunctions rapidly approach the real potential and radial wavefunctions outside the cut-off radii r_l (to within the normalization of the wavefunctions). The angular-momentum-dependent values for r_l are chosen so that the pseudowavefunctions have no nodes and the constants c_l are chosen so that the pseudopotential reproduces the atomic eigenvalues for the valence states of the real potential. Schematically this procedure results in potentials which are shown in Fig. 2.1. They are angular-momentum-dependent, hence non-local. For density-functional calculations it is essential that the integrated charge is correct also in the core region, which is achieved by renormalizing the pseudowavefunctions. As discussed in [2.40, 41], this conservation of the norm implies that the pseudopotential produces the same scattering properties as the full atomic potential to zeroth and first order in energy. For the valence levels this makes the pseudopotential energy independent and transferable. From the renormalized pseudowavefunctions the final *norm-conserving* pseudopotentials are constructed by inverting the Schrödinger equation and by unscreening, i.e. by removing exchange-correlation contributions. More details can be found in a recent review [2.36].

The fact that pseudopotentials and pseudowavefunctions are smoother near the nuclei than real potentials and wavefunctions is exploited by expanding

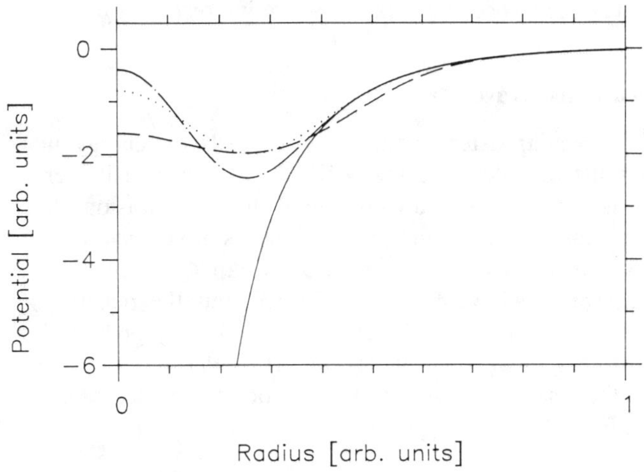

Fig. 2.1. Schematic representation of a l-dependent pseudopotential ($l=0$: dashed line, $l=1$: dotted line, $l=2$: dash-dotted line) and the real potential (solid line)

the pseudowavefunctions in the most simple and convenient basis set, in plane waves

$$\Psi_{ps}(\boldsymbol{r},E)=\sum_n C_n(\boldsymbol{k},E)\exp(-\mathrm{i}\boldsymbol{k}_n\cdot\boldsymbol{r}). \tag{2.26}$$

The secular equation in the pseudopotential method

$$\sum_{n'}[(k_n^2-E)\delta_{nn'}+V_{nn'}(\boldsymbol{k})]C_{n'}(\boldsymbol{k},E)=0 \tag{2.27}$$

with matrix elements

$$V_{nn'}(\boldsymbol{k})=\frac{1}{\tau}\int_\tau\int_\tau \exp(-\mathrm{i}\boldsymbol{k}_n\cdot\boldsymbol{r})V_{ps}(\boldsymbol{r},\boldsymbol{r}',E)\exp(\mathrm{i}\boldsymbol{k}_{n'}\cdot\boldsymbol{r}')d\boldsymbol{r}\,d\boldsymbol{r}' \tag{2.28}$$

is obtained from (2.5) by inserting (2.26), multiplying with $\exp(\mathrm{i}\boldsymbol{k}_{n'}\cdot\boldsymbol{r}')$, and integrating over $d\boldsymbol{r}$ and $d\boldsymbol{r}'$. Non-trivial solutions of (2.27) require

$$\det[(k_n^2-E)\delta_{nn'}+V_{nn'}(\boldsymbol{k})]=0. \tag{2.29}$$

For energy independent pseudopotentials, all the energy bands at each wave vector $\boldsymbol{k}$ are obtained by *one* matrix diagonalization. This considerably simplifies the calculations.

A disadvantage of the plane-wave pseudopotential method is that for nodeless valence states, e.g. for p states of B to F, for d states of Sc to Zn or for f states of the rare earths, the pseudowavefunctions are not much smoother than the real wavefunctions. Thus, too many plane waves may be required in the calculations. The application of local or mixed basis sets instead of plane waves is possible for these systems (see references given in [2.36]), but the natural simplicity of the plane-wave approach is lost. Moreover, the energy independent pseudopotential is only valid over a restricted energy range, not large enough to cover the unoccupied states we are interested in here.

2.2.3 Linear Augmented-Plane-Wave Method

Unlike (2.29), the KKR secular determinant (2.19) depends on energy in a complicated way and must be calculated many times to determine its zeros. The problem mainly arises because the functions used in the expansion (2.11) depend on energy. Many authors have tried to introduce a simple *linear* energy dependence in the KKR method [2.42–49 and 2.50, Chap. 4] or in *Slater's* [2.51] augmented-plane-wave (APW) method [2.52–55], but the real breakthrough for the linearization only came when *Andersen* [2.56] presented a combined view of the linear augmented-plane-wave (LAPW) and the linear muffin-tin orbital (LMTO) method. The LAPW method is well described in [2.56–62]; it uses basis functions

$$\phi(\boldsymbol{r},\boldsymbol{k}_n)=\begin{cases}\exp(\mathrm{i}\boldsymbol{k}_n\cdot\boldsymbol{r}) & \text{for } r>S\\ \sum_L Y_L(\boldsymbol{r})\mathrm{i}^l[R_l(r,E_l)a_L(\boldsymbol{k}_n)+\dot{R}_l(r,E_l)b_L(\boldsymbol{k}_n)] & \text{for } r<S\end{cases} \tag{2.30}$$

constructed from plane waves in the interstitial region between the muffin-tin spheres and from radial solutions $R_l(r, E)$ and their energy derivatives $\dot{R}_l(r, E)$ in the spheres. The amplitude and slope of (2.30) are made continuous at the surface of the muffin-tin spheres by proper choice of the coefficients $a_L(\boldsymbol{k}_n)$ and $b_L(\boldsymbol{k}_n)$, contrary to the usual augmented plane waves, e.g. given in [2.26, (11.1)], which are defined with $b_L(\boldsymbol{k}_n) \equiv 0$ and which have discontinuous slopes. The radial functions in (2.30) are computed at fixed l-dependent pivot energies and do not depend on energy whereas the usual augmented plane waves contain energy dependent radial functions $R_l(r, E)$. By the energy derivatives $\dot{R}_l(r, E_l)$ the basis functions (2.30) are essentially correct to linear order in energy. Thus the main approximation in (2.30) is the neglect of the non-linear energy dependence of the radial functions. The LAPW method leads to a secular equation

$$\sum_{n'} [H_{nn'}(\boldsymbol{k}) - EO_{nn'}(\boldsymbol{k})] C_{n'}(\boldsymbol{k}, E) = 0 \tag{2.31}$$

similar to the one of the pseudopotential method (2.27) with both matrices $H_{nn'}$ and $O_{nn'}$ independent of E. Contrary to the pseudopotential method, the overlap matrix $O_{nn'}$ is not given by unit matrix $\delta_{nn'}$, but this represents no problem in the calculations.

2.2.4 Linear Spherical-Wave Methods

The LMTO method is the linearized version of the KKR method. It is very fast and its formalism can be made simple and transparent. It was recently reviewed in [2.63] and an excellent account with computer programs is given in [2.64]. In the so-called atomic sphere approximation (ASA), the LMTO secular determinant can be derived from (2.8) if the choice $\kappa \to 0$ is made and if the integral in (2.8) is replaced by a sum of integrals over the atomic Wigner–Seitz spheres [2.65]. Divergences for $\kappa \to 0$ must be properly renormalized, e.g. the structure constants (2.20) must be multiplied with $\kappa^{l+l'}$. The resulting secular determinant

$$\det [S_{LL'}(\boldsymbol{k}) - P_{l'}(E)\delta_{LL'}] = 0 \tag{2.32}$$

contains energy independent structure constants $S_{LL'}(\boldsymbol{k})$. For the definition and physical significance of the potential functions $P_l(E)$ and for a discussion about the powerful concept of canonical band structures, I refer to the original literature [2.56, 60, 64–65]. The secular determinant (2.32) can be derived variationally by using basis functions similar to (2.30) except that the plane waves in the interstitial region ($r > S$) are replaced by spherical waves $h_l^{(1)}(\kappa r) Y_L(\hat{r})$. As shown in [2.65], the secular determinant (2.32) can be transformed into the form of (2.31).

The augmented-spherical-wave (ASW) method [2.66] and the linear Slater-type-orbital (LSTO) method [2.67, 68] are very similar to the LMTO method. The LSTO method applies Slater orbitals $r^{n-1} \exp(-\xi r)$ instead of

Hankel functions $h_l^{(1)}(\kappa r)$ and the ASW method applies two radial functions at different energies $E_l^{(1)}$ and $E_l^{(2)}$ in (2.30) instead of one radial function and its energy derivative. (A minor difference is the choice $\kappa^2 = 0$ in the LMTO method whereas a small negative value for κ^2 is chosen in the ASW method). Because of its economical basis set, the ASW method has been widely used in spectroscopic applications.

2.2.5 Non-Muffin-Tin Potentials

In basis-set methods with complete basis functions it is easy to include general potentials beyond the muffin-tin approximation. It is only necessary to calculate the matrix elements $V_{nn'}(\boldsymbol{k})$ with the full potential. This is obvious for the plane waves in the pseudopotential method because they are potential independent, but also true in the LAPW, LMTO, ASW and LSTO methods where the basis functions are constructed from the spherical part of the potential in the muffin-tin spheres only. This construction is sometimes viewed as an approximation in full-potential calculations, but in fact it is valid variational treatment. It is only a question of convergence whether the basis functions of the muffin-tin potential are also well suited for the full potential [2.69]. For calculations of total energies or densities of states no convergence problems have been found, but for calculations of other physical quantities such as electric-field gradients or hyperfine fields it may be better to include full-potential effects in the basis functions (2.30). That can be done by replacing the radial functions with solutions of the Schrödinger equation for the full potential in the muffin-tin spheres, but is seldom used because the resulting coupled differential equations [2.70] require considerable numerical effort. For references containing details of full-potential schemes for the LAPW, LMTO and LSTO methods I refer to [2.59, 61, 62, 68, 71–73].

Contrary to the basis-set methods it is difficult to include the full non-muffin-tin potentials in the KKR method. Its plane-wave form, originally derived by *Ziman* [2.26, 29], has been extended to non-muffin-tin potentials in [2.74, 75], but these formulations lose the advantages of the angular-momentum form (2.19), the potential independent structure constants (2.20) and the small matrices (2.18). *Williams and Morgan* [2.70] proposed that (2.19) can be used for full potentials with *unchanged* structure constants (2.20), but with more complicated non-diagonal t matrices to be obtained from coupled radial equations. Their work was questioned and the existence of "near-field" corrections suggested [2.28, 76]. In the literature this issue remains controversial [2.77].

2.3 Computational Techniques

Electronic-structure calculations for periodic crystals require essentially four computational steps. It is necessary to determine the energy bands $E_\nu(\boldsymbol{k})$ and expansion coefficients $\boldsymbol{C}(\boldsymbol{k}, E)$ from the secular equations and the determinantal conditions, to obtain the density (2.4) by Brillouin-zone integrations, to calculate the electronic potential by the Poisson equation (2.6), and to make density and potential self-consistent.

2.3.1 Secular Equation

In the secular equations (2.19, 29, 31), there is an important difference between the pseudopotential and linear methods on the one hand and the KKR method on the other. Secular matrices such as (2.29, 31), with positive definite overlap matrices, can be transformed into eigenvalue problems and *one* matrix diagonalization is enough to obtain *all* energy bands.

However, in the KKR method the calculation of the energy bands is not so easy. The matrix elements in (2.19) do not depend on energy in a simple manner and even have singularities, poles occur for $\kappa^2 = k_n^2, j_l(\kappa r) = 0$, or $j_{l'}(\kappa r') = 0$. The zeros of the determinant (2.19) must be found by evaluating it as function of energy. This is time consuming because the determinant varies rapidly with energy [2.78, 79] and must typically be calculated 10 to 20 times to find *one* energy band.

Another problem is the calculation of the expansion coefficients $\boldsymbol{C}(\boldsymbol{k}, E)$. In the pseudopotential and linear methods they are easily obtained from the eigenvectors of the secular matrices. The calculation of eigenvectors is a standard problem [2.80–82] and requires similar computational work as eigenvalue determinations. But in the KKR method, the calculated value $E_\nu(\boldsymbol{k})$ for an energy band must be inserted for E into the secular equation and the resulting homogeneous systems of linear equations such as (2.18) must be solved, e.g. by Gaussian elimination with pivoting [2.80, 83] as proposed in [2.22, 84].

All methods except the KKR method define the wavefunctions throughout the unit cell and the expansion coefficients are easily used to normalize the wavefunctions. In the KKR method the wavefunctions are only specified in the muffin-tin spheres and special techniques must be used to normalize them. Either the wavefunction in the interstitial region is represented in a plane wave form [2.21, 22] or the energy derivative of the KKR secular matrix is used [2.85].

2.3.2 Brillouin Zone Integration

The solution of the secular equations provides the energy bands $E_\nu(\boldsymbol{k})$ and the wavefunctions $\Psi(\boldsymbol{r}, E_\nu(\boldsymbol{k}))$ as function of wave vector $\boldsymbol{k}$ and band index ν. For a comparison with $\boldsymbol{k}$-conserving measurements such as angle-resolved photoemission or inverse photoemission, this information is enough. For a comparison with other experiments such as X-ray, electron energy loss or bremsstrahlung isochromat spectroscopy, and for the density (2.4) needed in density-functional calculations, an integration over $\boldsymbol{k}$ in the first Brillouin zone (BZ) and a summation over ν is necessary to obtain the distribution of electrons in space and energy. This is conveniently given in terms of the local density of states

$$n(\boldsymbol{r}, E) = \frac{1}{V_{\mathrm{BZ}}} \sum_\nu \int_{\mathrm{BZ}} \delta(E - E_\nu(\boldsymbol{k})) |\Psi(\boldsymbol{r}, E_\nu(\boldsymbol{k}))|^2 d\boldsymbol{k}. \tag{2.33}$$

Because $\Psi(\boldsymbol{r}, E_\nu(-\boldsymbol{k}))$ and $\Psi^*(\boldsymbol{r}, E_\nu(\boldsymbol{k}))$ are equal by time-reversal symmetry, the integration volume is always restricted to one half of the BZ. Space-group symmetry further reduces the integration volume, e.g. in cubic and zincblende type crystals only 1/48th of the BZ is necessary. Nevertheless, because of the δ-function the numerical integration in (2.33) is difficult. Because the determination of the energy bands $E_\nu(\boldsymbol{k})$ and the wavefunctions is expensive (Sect. 2.3.1), the number of different $\boldsymbol{k}$-points must be kept as small as possible. Many ideas have been suggested to achieve this goal and depending on the material and the physical property of interest, several methods are now in use to evaluate (2.33). They include sampling techniques [2.86], the methods of special points [2.87–90] or special directions [2.91–93], and the use of microvolumes such as cubes [2.94] or tetrahedrons [2.95–98]. In the microvolumes (2.33) is calculated analytically by approximating $E_\nu(\boldsymbol{k})$ linearly

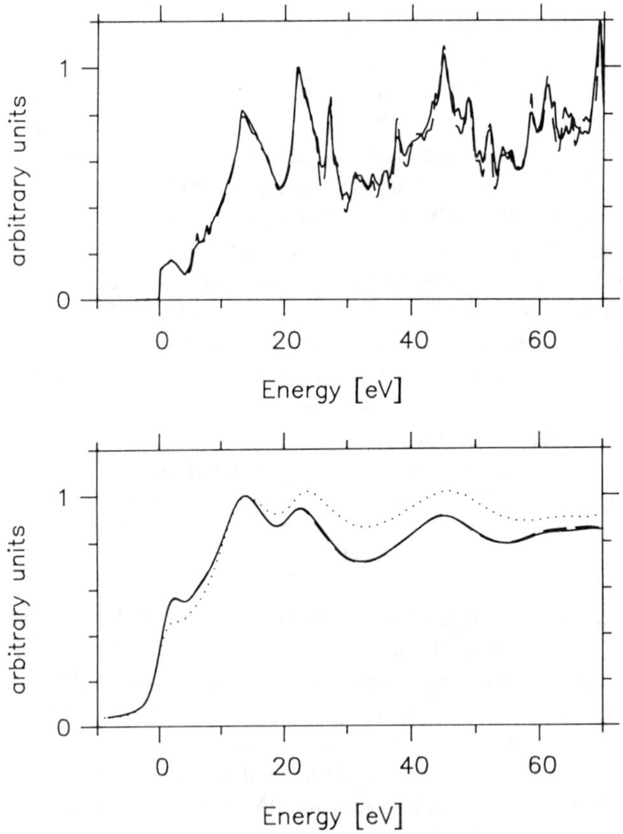

Fig. 2.2. Calculated K edge absorption spectrum for Cu for 768 (dashed lines) and 12288 (solid lines) tetrahedrons used in 1/48th of the Brillouin zone. The upper panel contains the unbroadened spectra, the lower panel the realistically broadened ones. The experimental result (by B. Lengeler, private communication) is also included (dotted line)

or quadratically [2.99] in $\boldsymbol{k}$. In practice, the choice of a good BZ-integration method requires to distinguish whether energy-integrated quantities such as the density $n(\boldsymbol{r})$ or energy-resolved quantities such as the density of states are to be calculated. Furthermore, the choice depends on whether insulators, semiconductors or metals are considered. At the expense of more $\boldsymbol{k}$-points, the microvolume techniques and the special-direction methods (with accurate numerical integration along the directions) are particularly suited to obtain the fine details in the density of states. These details are smeared out in the other methods which are more suited to obtain energy-integrated quantities, but have the advantage that fewer $\boldsymbol{k}$-points are necessary. Metals differ from semiconductors and insulators because details in the density of states near the Fermi level also considerably affect energy-integrated quantities such as the density $n(\boldsymbol{r})$. Therefore, in general, for metals more $\boldsymbol{k}$-points are necessary and micro-volume and special-direction techniques are more often applied.

When calculations are compared with experiments, several broadening mechanism, e.g. various life-time effects and the experimental resolution, must always be included. The broadening is essentially an energy integration (convolution) with Lorentzian weight functions to simulate life-time effects and with Gaussians to account for the experimental resolution. Because of this, the Brillouin-zone integrations can be obtained with fewer $\boldsymbol{k}$-points. As an example for the dependence on the number of $\boldsymbol{k}$-points, I consider the X-ray absorption spectra for the K edge of Cu, found by multiplying the p partial density of states obtained from a KKR band-structure calculation with the appropriate matrix elements as described in [2.100]. The BZ integration was done by the linear tetrahedron method. In Fig. 2.2 the spectrum calculated without any broadening is compared with one including an experimental resolution of 2.40 eV, a core-hole life time of 1.32 eV, and an energy-dependent life time of the excited state [2.100]. Whereas the unbroadened spectra show appreciable changes between 768 and 12288 tetrahedron used in the 1/48th of the fcc Brillouin zone, requiring 264 and 3364 $\boldsymbol{k}$-points, the broadened spectra are very similar.

2.3.3 Poisson Equation

In the muffin-tin approximation it is straightforward [2.101, 31] to calculate the electrostatic (Hartree) potential (2.6). For non-spherical densities $n(\boldsymbol{r})$ and potentials, an accurate and efficient calculation of $V_{\mathrm{H}}(\boldsymbol{r})$ is difficult in spite of the fact that (2.6) represents the explicit solution. In fact, numerically the three-dimensional integral (2.6) requires similar effort to the three-dimensional integral equation (2.8). In both equations the integrals arise from partial differential equations containing the same Laplace operator ∇_r^2. Whereas an efficient numerical treatment of the quantum-mechanical problem (2.8) has been studied in numerous publications, the classical problem (2.6) has received less attention, though notable exceptions are [2.73, 102–105]. One way of solving the problem is to use the muffin-tin concept of the APW method [2.51] and to represent the charge density as a sum of two contributions, a rapidly varying contribution inside the spheres and a slowly varying one between the spheres. Outside each sphere, the Coulomb potential produced by the nuclear charge and the electron density within that sphere depends only on the multipole moments of the density. A much smoother pseudodensity in the spheres can have identical multipole moments. This allows to Fourier-expand the pseudodensity together with the smooth density between the spheres. The resulting potential (2.6) is easily obtained in Fourier space and is correct in the interstitial region. This potential can be continued into the spheres by the standard methods of electrostatics. For self-consistency it is, in the author's experience, important to solve the Poisson equation (2.6) at least as accurately as the Schrödinger equation (2.5). For example, in our calculations [2.106] for dilute alloys where we treat the influence of the atomic displacements around the impurity atoms, we found that the overlapping-atomic-spheres approximation (ASA) with displaced spheres is reasonably valid for (2.5) but inadequate for (2.6), pointing out that approximations are more severe for the Poisson equation.

2.3.4 Self-Consistency

The density (2.4), the central quantity in density-functional electronic-structure calculations, is given in terms of the one-particle wavefunctions $\Psi_i(\boldsymbol{r})$. From (2.5) they depend non-linearly on the potential $V(\boldsymbol{r})$, and this potential again depends on the density by its electrostatic (2.6) and exchange-correlation parts. This results in a non-linear self-consistency problem which can only be solved by iteration techniques. Starting with a reasonable input density, e.g. as described in [Ref. 2.36, pp. 145, 146] by overlapping the densities of neutral atoms, one calculates the electrostatic and exchange-correlation part of the potential for the Hamiltonian in (2.5), the wavefunctions by (2.5) and a new density by (2.4) as input to the next iteration. Because of the long-range nature of the Coulomb interaction, a small change in the input density can lead to a relatively large change in the output density and, often, the straightforward iterations oscillate and diverge. Then input and output densities n_i^{in} and n_i^{out} of iteration i must be mixed to obtain a damped input density n_{i+1}^{in} for iteration $i+1$:

$$n_{i+1}^{\text{in}}(\boldsymbol{r}) = \alpha n_i^{\text{out}}(\boldsymbol{r}) + (1-\alpha) n_i^{\text{in}}(\boldsymbol{r}). \tag{2.34}$$

By suitably choosing the parameter α, convergence is usually achieved. It can be shown [2.107], by using the minimum principle for $E[n(\boldsymbol{r})]$, that parameters $\alpha > 0$ always exist [2.107] with which the iterations converge to a local minimum of $E[n(\boldsymbol{r})]$. The larger the value of α which is used, the fewer iterations are necessary. However, it is difficult to determine the possible maximum value of α, because it critically depends on the particular physical system, and we are normally satisfied to have convergence without optimisation of α. For simple and transition metals α can be as large as 0.1 to 0.2, but for rare-earth metals (with f electrons) α must be smaller, typically 0.01. Then hundreds of iterations are necessary. The situation is even worse in compounds where charge can rapidly oscillate from one atom to the others. Better iteration schemes which transcend the simple linear mixing (2.34) are highly desirable. A detailed study and discussion of some of them can be found in [2.107]. In particular, it is demonstrated there that the slow convergence which occurs in spin-polarized calculations, if magnetic and non-magnetic solutions are energetically close, can be overcome by mixing charge and magnetisation densities differently, with much larger mixing factors for the magnetisation. For a recent review of the self-consistency-iteration problem and newer developments for iteration techniques, e.g. Broyden's gradient method, I refer to [2.36, pp. 146–155, 190].

2.4 Computational Problems at Higher Energies

The standard band-structure techniques and computer programs have usually been developed with the aim of calculating the occupied states and the low-lying unoccupied states. When applying these techniques to the higher lying states, which are studied in X-ray absorption or bremsstrahlung isochromat experiments, it is found that the convergence with respect to the size of the secular matrix is slower, singularities in the KKR method occur more often; and basis sets are less adequate. For the higher states these problems are rarely considered in the literature and therefore discussed in detail in this section.

2.4.1 Convergence

The infinite-dimensional secular matrices are always truncated in actual calculations. It is very important to make these matrices as small as possible, without losing the required accuracy, because the computational work scales with the third power of the matrix size. This leads to the question of how many angular momenta are needed in the KKR and linear spherical-wave methods, and how many plane waves are needed in the pseudopotential and in the LAPW methods. In the linear spherical-wave methods the question of the number of basis functions also arises, because for the higher states it may be necessary to use more than one value of κ and of E_l to represent the wavefunctions outside and within the spheres, respectively.

For the convergence properties of the KKR and linear spherical-wave methods, it is important to distinguish between the l-convergence of the energy bands obtained by truncating (2.19), and of the wavefunctions obtained by truncating (2.11). The convergence of the energy bands with l_{max} is rapid because the centrifugal repulsion $l(l+1)/r^2$ dominates in the Schrödinger equation over the atomic potential $V^n(r)$ for higher l values. At lower energies the high l electrons cannot penetrate to the core region, the radial functions $R_l^n(r,E)$ approximate the free-space solutions $j_l(\kappa r)$, the t matrices (2.14) tend to zero and small values of l_{max} are possible. Usually, the occupied bands are adequately calculated with $l_{max}=2$ for transition metals and with $l_{max}=3$ for elements with

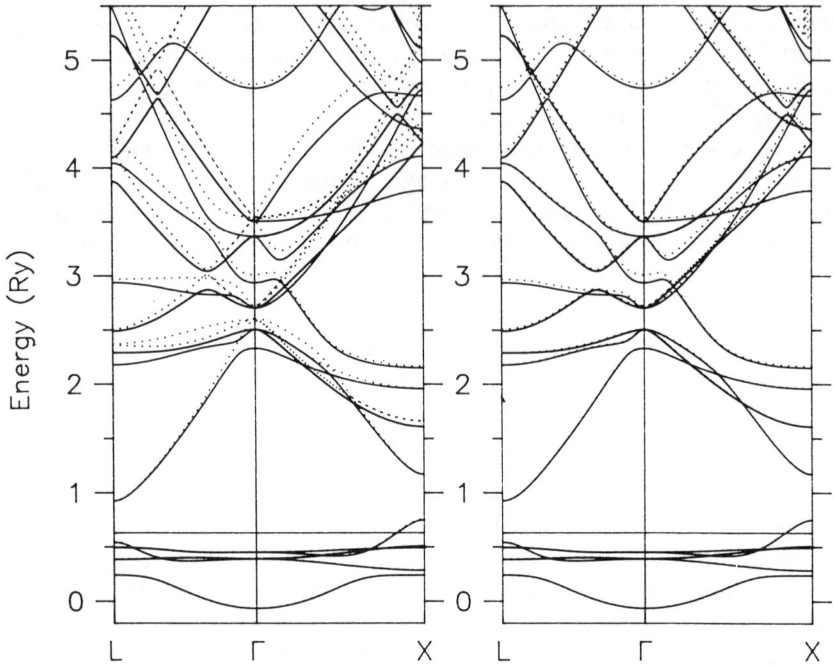

Fig. 2.3. Band structure of Cu in $L-\Gamma$ and $\Gamma-X$ directions in the Brillouin zone calculated for the muffin-tin potential given in [2.31]. Results are shown for $l_{max}=4$ (full lines), for $l_{max}=2$ (dotted lines) in the left panel and $l_{max}=3$ (dotted lines) in the right panel

f electrons; rarely is $l_{max}=4$ needed. At higher energies the electrons penetrate more to the core region and l_{max} should be higher. This point is illustrated in Fig. 2.3 which shows a band structure calculated with the Cu muffin-tin potential of [2.31] for $l_{max}=2$, 3, 4. The occupied bands below the Fermi level are accurately described already for $l_{max}=2$. For $l_{max}=3$, good bands are obtained for energies below 2 Ry and reasonable bands below 4 Ry. Above 4 Ry, an $l_{max}=4$ calculation should be used.

Compared to (2.19) the convergence of (2.11) is not so fast. It is similar to that of the plane wave expansion

$$\exp(\mathrm{i}\boldsymbol{k}_n\boldsymbol{r}) = 4\pi \sum_L Y_L(\boldsymbol{r})\mathrm{i}^l j_l(k_n r) Y_L(\hat{\boldsymbol{k}}_n). \tag{2.35}$$

The importance of higher angular momenta for the density (2.4) can be estimated by multiplying (2.35) with its complex conjugate and averaging over the angles $\hat{\boldsymbol{r}}$. The result on the left hand side is equal to one and on the right hand side this leads to

$$4\pi \sum_L j_l^2(k_n r) Y_L(\hat{\boldsymbol{k}}_n) Y_L(\hat{\boldsymbol{k}}_n) = \sum_l (2l+1) j_l^2(\kappa r), \tag{2.36}$$

where $k_n = \kappa$ and $\sum_m Y_{lm}(\hat{k}_n) Y_{lm}(\hat{k}_n) = 2l+1$ are used (remember $L = lm$). As an example for the convergence of the right hand side of (2.36), I choose r as the muffin-tin radius of Cu, $r = 2.39$, and truncate the sum in (2.36) at $l'_{max} = 4, 6, 8$. For $E = 1$ Ry the convergence of (2.36) is rapid and the error is 4×10^{-4}, 5×10^{-7}, 2×10^{-10} for $l'_{max} = 4, 6, 8$. For $E = 4$ Ry the convergence is slower and the error is 1×10^{-1}, 3×10^{-3}, 3×10^{-5}. Thus higher angular momenta are more important for higher energies. Fortunately, as described in [2.108], the coefficients for $l > l_{max}$ are easily calculated from those for $l \leqq l_{max}$.

In the spherical-wave methods the angular-momentum convergence is similar to the KKR method. It may be thought that their derivation from the Rayleigh–Ritz minimum principle ensures faster convergence in l for the wavefunctions, but the KKR method can also be derived from a variational principle [2.20], although not with the minimum property, and in this variational sense the truncated KKR wavefunctions are the best ones obtainable.

For the LAPW method a detailed convergence study is given by *Müller* and *Wilkins* [2.98]. They state that for the unoccupied states of Pd for an energy range of 10 Ry it is necessary to include about 160 plane waves for an accuracy of 0.015 Ry, whereas an accuracy of 0.001 Ry requires 320 plane waves, thus leading to secular matrices of size 320 × 320.

In the pseudopotential method the number of plane waves and the resulting size of the secular matrix are necessarily higher than in the LAPW method, because the plane waves are also used to expand the more rapidly varying wavefunctions inside the pseudo-atoms. Up to 500 plane waves have been used in calculations for Si [2.109], and for crystals containing first-row elements such as N or O, more may be required. For occupied states in the transition metals Mo and W, as many as 2000 plane waves have been used [2.110]. Other difficult elements are Cs and Ba because they have narrow f resonances at positive energies which lead to strong pseudopotentials for states with f character.

For the higher lying unoccupied states the performance of the pseudopotential method is even worse. The necessary number N of plane waves increases as $E^{3/2}$ and the computational work to obtain the eigenvalues of the secular matrix varies as $N^3 \sim E^{9/2}$. Furthermore, it can be expected that the number of plane waves increases even more because at higher energies the electrons can penetrate more to the core region and the potential there cannot be smoothed away, but must be treated correctly. To the author's knowledge no first-principle pseudopotential calculations for states higher than 10 or 20 eV above the Fermi level exist and only one empirical pseudopotential calculation has appeared [2.111]. Moreover, it has never been investigated whether the pseudopotentials constructed from the occupied atomic states can be transferred into solids to calculate the higher states. Nevertheless, the pseudopotential method has its definite range of applicability and is well suited for treating the lower unoccupied conduction states in semiconductors when the GW approximation [2.112] is used to improve density-functional eigenstates (Chap. 3).

In pseudopotentials calculations it is important to reduce the enormous amount of computing time which would arise if the large secular matrices are directly diagonalized. Several techniques have been proposed for that purpose. The plane waves with large values of k_n can be treated by Löwdin's perturbation method [Ref. 2.36, pp. 139, 140]. The plane waves may be replaced by Bloch sums of local orbitals or a mixed basis of both types is used [Ref. 2.36, pp. 155–160]. Fast-Fourier-transform methods [2.109, 113] and fast matrix-diagonalization techniques are also applied, see discussion in [2.109]. Another recently active field of research is to optimize the pseudopotentials with respect to the required number of plane waves by exploiting the freedom in the pseudopotential construction [2.114–117]. For example, it has been shown [2.117] that the number of plane waves for calculating the occupied states of Cu can be reduced to 500 compared to the 1700 necessary for the pseudopotential of [2.40].

2.4.2 Free-Electron Singularities, Spurious and Close Roots

The problems of free-electron singularities and spurious roots particularly affect the KKR method where sign changes of the determinant (2.19) must be detected numerically to obtain the band structure. But the signs also change for unphysical energies, as discussed in Sect. 2.3.1. This leads to spurious solutions which must be discarded.

For the small energy range of the occupied states, spurious solutions are normally easily detected. For example, in FCC crystals only two reciprocal lattice vectors, $\boldsymbol{K}_n=(0,0,0)$ and $\boldsymbol{K}_n=(-2\pi/a,\ -2\pi/a,0)$, lead to free-electron singularities $E=k_n^2$ in 1/48th of the Brillouin zone if the energy remains below $E=0.75(2\pi/a)^2$. For Cu with lattice constant $a=6.76$ a.u. this critical energy $E=0.648$ Ry is already higher than the Fermi level $E_{\mathrm{F}}=0.628$ Ry [2.31]. For higher energies, the detection of spurious solutions is more difficult because the number of dangerous reciprocal lattice vectors and the number of bands increase as $E^{3/2}$. An estimate for Pd is given in [2.98] stating that 107 bands are necessary below 10 Ry. Consequently, real and spurious roots become more densely spaced. Root-searching methods can easily fail if bands are nearly degenerate by symmetry or accidentally, or if there are nearby spurious roots. Therefore, instead of looking for sign changes of the secular determinant, it is safer to determine sign changes of the eigenvalues of the secular matrix. Both occur for the same energies because the determinant is equal to the product of the eigenvalues. An advantage arises because the eigenvalues vary less rapidly with energy [2.78] and resolve the degeneracies. The actual evaluation of the eigenvalues is not needed; instead it is enough to know the numbers of positive, negative and zero ones. For real symmetric and complex Hermitian matrices the effort to calculate these numbers is no larger than to calculate the determinants [2.82, 83]. Nearly degenerate bands and bands close to spurious roots also make the calculation of the expansion coefficients by Gaussian elimination (Sect. 2.3.1) problematic because linearly independent sets of coefficients are probably not obtained. Here, it is safer, but more time consuming, to calculate the coefficients from the eigenvectors of the secular matrix [2.84]. The root searching becomes considerably easier when the KKR matrix is transformed into a mixed representation of angular momenta and plane waves as described by *Williams* et al. [2.27]. Then all matrix elements and hence all eigenvalues are smooth functions of energy.

Close roots also lead to inaccurate results in the conventional tetrahedron-integration method (Sect. 2.3.2) when bands are interpolated from the corners of the tetrahedrons. Bands which cross (see Fig. 2.3) are incorrectly replaced by flat parallel bands and spurious peaks appear in the density of states (2.33). Extrapolation from the centers of the tetrahedrons is better [Ref. 2.98, Fig. 15] but the gradient of $E_\nu(\boldsymbol{k})$ with respect to $\boldsymbol{k}$ must be calculated additionally.

2.4.3 Basis Set and Linearization Problems

No particular basis set problems occur in the KKR method, where the angular-momentum cut-off $l_{\max}$ can be increased, or in the pseudopotential method with a plane-wave expansion. However, care is needed when the pseudopotential method is used with a mixed basis set of plane waves and local atomic-like wavefunctions. These may not be appropriate for the energy range under consideration. In linear methods, the atomic Schrödinger equation is solved at pivot energies, but away from the pivot energies the linearization introduces errors, particularly when states of two principal quantum numbers are close in energy, as are the 4d core and 5d valence states of Cd. The radial functions in (2.30) can describe either 4d or 5d states accurately, but not both. Similar difficulties are more frequent for the unoccupied states. A careful discussion of how the problem can be avoided is given by *Müller* and *Wilkins* [2.98]. They propose a variant of the LAPW method where the fixed pivot energies are replaced by energies $E_l(\boldsymbol{k}_n)$ which are different for different plane waves. These energies are determined by the condition that the logarithmic derivatives of $j_l(\kappa r)$ and $R_l(r, E)$ agree at the muffin-tin radius. In this LAPW variant, the usual introduction of energy panels with several values of E_l is not needed.

2.5 Non-Periodic Systems

Nowadays interest has been extended from perfect periodic crystals to real crystals containing imperfections such as impurities, surfaces, etc. In fact, some experiments themselves perturb periodicity—X-ray absorption and emission processes involve core holes in the excited atoms. The periodicity of the ideal, infinite crystal allows us to introduce the Bloch wave vector $\boldsymbol{k}$ by lattice Fourier transformation (2.16–17). As we have seen, this leads to a considerable simplification of the calculations, because the simultaneous treatment of an infinite number of sites, as in (2.13), is replaced by a separate treatment of each wave vector $\boldsymbol{k}$, as in (2.18). In non-periodic situations this simplification is lost.

In principle, it is always possible to introduce artificially a periodically repeating supercell with many atoms, so that the standard band-structure methods of Sect. 2.2 can be applied. In practice, this approach is problematic. It leads to secular determinants of large dimensions and to many closely lying energy bands. The KKR method is not well suited for supercell calculations because the close bands are difficult to find (Sect. 2.4.2) and even in the pseudopotential and linear methods, the number of atoms in the supercell is severely restricted by computer resources and spurious, uncontrolled interactions among the supercells are hardly avoided.

Instead of the supercell approach special multiple-scattering methods (Sects. 2.5.1–4) are more appropriate to treat non-periodic situations. These methods can be derived by working with Green functions instead of wavefunctions and with the Dyson equation

$$G(\boldsymbol{r},\boldsymbol{r}',E)=g(\boldsymbol{r},\boldsymbol{r}',E)+\int_{\infty} g(\boldsymbol{r},\boldsymbol{r}'',E)V(\boldsymbol{r}'')G(\boldsymbol{r}'',\boldsymbol{r}',E)d\boldsymbol{r}'' \tag{2.37}$$

instead of the Lippmann–Schwinger equation (2.8). Similarly to (2.10), the Green function $G(\boldsymbol{r},\boldsymbol{r}',E)$ is defined by

$$[-\nabla_r^2+V(\boldsymbol{r})-E]G(\boldsymbol{r},\boldsymbol{r}',E)=-\delta(\boldsymbol{r}-\boldsymbol{r}'). \tag{2.38}$$

The local density of states introduced in (2.33) is then given by $n(\boldsymbol{r},E)=-\frac{2}{\pi}\operatorname{Im}\{G(\boldsymbol{r},\boldsymbol{r},E)\}$. From (2.37) an algebraic system of equations

$$G_{LL'}^{nn'}(E)=g_{LL'}^{nn'}(E)+\sum_{n''}\sum_{L''}g_{LL''}^{nn''}(E)t_{l''}^{n''}(E)G_{L''L'}^{n''n'}(E) \tag{2.39}$$

can be derived [2.118] in a similar way to the derivation in Sect. 2.2.1. The coefficients $G_{LL'}^{nn'}(E)$ can be used to calculate the Green function $G(\boldsymbol{r},\boldsymbol{r}',E)$ by a double expansion [2.119].

Compared to supercell methods, where available standard band-structure programs can be used, the Green function techniques require the development of new computer programs. This is a field of active current research.

2.5.1 Local Perturbations

For local perturbations such as impurities or excited atoms, for example with core holes as produced in X-ray absorption experiments, the t matrices in (2.39) differ from the ones of the unperturbed crystal only for the atoms with perturbed potentials. It is convenient to solve (2.39) in two steps by first calculating the Green function $\mathring{G}$ of the periodic unperturbed crystal by Fourier transformation and then adding the perturbation. In matrix notation (2.39) can be written as $G=g+gtG$ and formally equivalently as $G^{-1}=g^{-1}-t$. Subtracting the similar result $\mathring{G}^{-1}=g^{-1}-\mathring{t}$ for the unperturbed crystal with Green function $\mathring{G}$ and t matrices $\mathring{t}$ leads to $G^{-1}=\mathring{G}^{-1}-(t-\mathring{t})$ or to

$$G_{LL'}^{nn'}(E)=\mathring{G}_{LL'}^{nn'}(E)+\sum_{n''}\sum_{L''}\mathring{G}_{LL''}^{nn''}(E)(t_{l''}^{n''}(E)-\mathring{t}_{l''}^{n''}(E))G_{L''L'}^{n''n'}(E). \tag{2.40}$$

In metals it seems [2.100] that densities of states can be calculated accurately enough if the potential perturbations are restricted to the impurity and its nearest neighbour atoms. Then (2.40) requires to invert matrices of moderate size. The LMTO method as the $\kappa \to 0$ limit of the KKR method is also straightforwardly extended to treat local perturbations [2.65, 120–121]. For an extension of the pseudopotential method I refer to a recent review [2.122].

2.5.2 Clusters

Equation (2.40) describes the embedding of a cluster of perturbed atoms in an otherwise unperturbed host crystal and (2.39) describes a cluster of atoms in free space similarly to the scattered-wave method [2.123]. For more details of cluster techniques, the reader is referred to in Chap. 5 by Vvedensky, but here we note that (2.39) can be solved iteratively

$$G^{nn'}_{LL'}(E) = g^{nn'}_{LL'}(E) + \sum_{n''}\sum_{L''} g^{nn''}_{LL''}(E) t^{n''}_{l''}(E) g^{n''n'}_{L''L'}(E) + \cdots . \tag{2.41}$$

At high energies, or if self-energy effects produce a damping imaginary part in the potential, this iteration rapidly converges, and higher terms in (2.41) can be neglected. As shown in [2.124] the standard formula of extended-X-ray-absorption-fine-structure (EXAFS) calculations follows from (2.41) when the Hankel functions in the expression (2.15) for $g^{nn'}_{LL'}(E)$ are approximated by their asymptotic expansions [2.125, (10.116)] $\mathrm{i}^{-l''-1}(\kappa R^{nn'})^{-1}\exp(-\mathrm{i}\kappa R^{nn'})$. The cluster methods are the real-space equivalent of the band-structure approach. They are better suited at high energies because the band structure contains an unnecessarily high degree of information which is smeared out anyway by various broadening mechanisms and life-time effects.

2.5.3 Surfaces

Equation (2.39), used with $t^{n''}_{l''}(E) = 0$ in the half-space outside the surface, is well suited to determine electronic states at surfaces. The periodicity parallel to the surface allows to apply two-dimensional Fourier transformation. Then the site representation in (2.39) is transformed into one with wave vectors $\boldsymbol{k}_{\parallel}$ parallel and layers perpendicular to the surface. The remaining problem in the resulting layer KKR method [2.126–129] is the solution of an infinite number of coupled layer equations. This can be done in different ways, by the layer-doubling technique [2.127–129], by an embedding procedure [2.130], by removal invariance [2.131] or by the slab approximation in which a finite number of layers is used as an approximation [2.58]. Alternatively, the problem can be treated as a local perturbation [2.132] in perpendicular direction to the surface by removing enough adjacent layers of atoms from the infinite crystal so that the resulting two inner surfaces do not interact.

2.5.4 Disordered Alloys

The electronic states of substitutionally disordered alloys can be described by configurationally averaged Green functions $\langle G \rangle$. When correlations between sites are neglected the approximation $\langle G \rangle = \mathring{G} + \mathring{G}(\langle t \rangle - \mathring{t})\langle G \rangle$ follows from (2.40) by averaging. The average t matrix approximation (ATA) [2.133] and the coherent potential approximation (CPA) [2.134, 135] differ in the choice of the reference system. In the ATA the pure crystals determine $\mathring{G}$, whereas in the CPA the reference system is chosen to give no scattering on the average and must be calculated self-consistently by the condition $\mathring{t} = \langle t \rangle$. An efficient implementation of the KKR-CPA method [2.136] combines both self-consistency iterations for the coherent potential and the density (Sect. 2.3.4) and uses only one $\boldsymbol{k}$ point in each iteration to refine the BZ integration.

References

2.1 P.C. Hohenberg, W. Kohn: Phys. Rev. **136**, B864–871 (1964)

2.2 W. Kohn, L.J. Sham: Phys. Rev. **140**, A1130–1138 (1965)

2.3 A.K. Rajagopal: "Theory of Inhomogeneous Electron Systems: Spin-Density-Functional Formalism", in *Advan. Chem. Phys.*, ed. by I. Prigogine, S.A. Rice, Vol. 41 (Addison-Wesley, New York 1980) pp. 59–193

2.4 D.D. Koelling: Rep. Prog. Phys. **48**, 139–212 (1981)

2.5 S. Lundqvist, N.H. March (eds.): *The Theory of the Inhomogeneous Electron Gas* (Plenum, New York 1983)

2.6 J. Callaway, N.H. March: "Density Functional Methods: Theory and Applications", in *Solid State Physics*, ed. by H. Ehrenreich, F. Seitz, D. Turnbull, Vol. 38 (Academic, New York 1984) pp. 135–221

2.7 U. von Barth: "Density Functional Theory for Solids", in *The Electronic Structure of Complex Systems*, ed. by P. Phariseau, W.M. Temmerman, NATO ASI Series B, Vol. 113 (Plenum, New York, 1984) pp. 67–140

2.8 R.M. Dreizler, J. da Providencia (eds.): *Density Functional Methods in Physics*, NATO ASI Series B, Vol. 123 (Plenum, New York 1985)

2.9 W.E. Pickett: Comments on Solid State Phys. **12**, 1–15 (1985)

2.10 W.E. Pickett: Comments on Solid State Phys. **12**, 57–68 (1985)

2.11 R.O. Jones, O. Gunnarsson: Rev. Mod. Phys. **61**, 689–746 (1989)

2.12 O. Gunnarsson, B.I. Lundqvist: Phys. Rev. **B 13**, 4274–4298 (1976)

2.13 L. Hedin, S. Lundqvist: J. Phys. **C 4**, 2064–2083 (1971)

2.14 U. von Barth, L. Hedin: J. Phys. **C 5**, 1629–1642 (1972)

2.15 D.M. Ceperley, B.J. Alder: Phys. Rev. Lett. **49**, 566–569 (1980)

2.16 S.H. Vosko, L. Wilk, M. Nusair: Can. J. Phys. **58**, 1200–1211 (1980)

2.17 Lord Rayleigh: Philos. Mag. **34**, 481–502 (1892)

2.18 P.P. Ewald: Ann. Phys. (Leipzig) **53**, 117–143 (1916)

2.19 J. Korringa: Physica **13**, 392–400 (1947)

2.20 W. Kohn, N. Rostocker: Phys. Rev. **94**, 1111–1120 (1954)

2.21 P.M. Morse: Proc. Nat. Acad. Sci. US **46**, 276–286 (1956)

2.22 F.S. Ham, B. Segall: Phys. Rev. **124**, 1786–1796 (1961)

2.23 B. Segall, F.S. Ham: "The Green's Function Method of Korringa, Kohn, and Rostocker for the Calculation of the Electronic Band Structure of Solids", in *Methods in Computational Physics*, ed. B. Alder, S. Fernbach, M. Rotenberg, Vol. 8 (Academic, New York 1968) pp. 251–294

2.24 A.R. Williams, S.M. Hu, D.W. Jepsen: "Recent Developments in KKR Theory", in: *Computational Methods in Band Theory*, ed. by P.M. Marcus, J.F. Janak, A.R. Williams (Plenum, New York 1971) pp. 157–177

2.25 H.L. Davis: "Efficient Numerical Techniques for the Calculation of KKR Structure Constants", in: *Computational Methods in Band Theory*, ed. by P.M. Marcus, J.F. Janak, A.R. Williams (Plenum, New York 1971) pp. 183–199
2.26 J.M. Ziman: "The Calculation of Bloch Functions", in *Solid State Physics*, ed. by H. Ehrenreich, F. Seitz, D. Turnbull, Vol. 26 (Academic, New York 1971) pp. 1–101
2.27 A.R. Williams, J.F. Janak, V.L. Moruzzi: Phys. Rev. **B6**, 4509–4517 (1972)
2.28 J.S. Faulkner, Phys. Rev. **B19**, 6186–6206 (1979)
2.29 J.M. Ziman: Proc. Phys. Soc. **86**, 337–353 (1965)
2.30 P.P. Ewald: Ann. Phys. (Leipzig) **64**, 253–287 (1921)
2.31 V.L. Moruzzi, J.F. Janak, A.R. Williams: *Calculated Electronic Properties of Metals* (Pergamon, New York 1978)
2.32 J.C. Phillips, L. Kleinman: Phys. Rev. **116**, 287–294 (1959)
2.33 W.A. Harrison: *Pseudopotentials in the Theory of Metals* (Benjamin, New York 1966)
2.34 V. Heine: "Electronic Structure of Metals", in *The Physics of Metals*, ed. by J.M. Ziman, (Cambridge University Press, Cambridge 1969) pp. 1–61
2.35 V. Heine: "The Pseudopotential Concept", in *Solid State Physics*, ed. by H. Ehrenreich, F. Seitz, D. Turnbull, Vol. 24 (Academic, New York 1970) pp. 1–36
2.36 W.E. Pickett: Comp. Phys. Rep. **9**, 115–198 (1989)
2.37 M.L. Cohen, V. Heine: "The Fitting of Pseudopotentials to Experimental Data and Their Subsequent Applications", in *Solid State Physics*, ed. by H. Ehrenreich, F. Seitz, D. Turnbull, Vol. 24 (Academic, New York 1970) pp. 37–248
2.38 W.C. Topp, J.J. Hopfield: Phys. Rev. **B7**, 1295–1303 (1974)
2.39 T. Starkloff, J.D. Joannopoulos: Phys. Rev. **B16**, 5212–5215 (1977)
2.40 G.B. Bachelet, D.R. Hamann, M. Schlüter: Phys. Rev. **B26**, 4199–4228 (1982)
2.41 E.L. Shirley, D.C. Allan, R.M. Martin, J.D. Joannopoulus: Phys. Rev. **B40**, 3652–3660 (1989)
2.42 V. Heine: Phys. Rev. **153**, 673–682 (1967)
2.43 J. Hubbard: Proc. Phys. Soc. **92**, 921–937 (1967)
2.44 R.L. Jacobs: J. Phys. **C1**, 492–506 (1968)
2.45 J. Hubbard, N.W. Dalton: J. Phys. **C1**, 1637–1649 (1968)
2.46 D.M. Hum, K.C. Wong: J. Phys. **C2**, 833–840 (1969)
2.47 D.G. Pettifor: J. Phys. **C2**, 1051–1069 (1969)
2.48 J. Hubbard: J. Phys. **C2**, 1222–1229 (1969)
2.49 N.W. Dalton: "Approximate KKR Band-Structure Schemes for Transition Metals", in: *Computational Methods in Band Theory*, ed. by P.M. Marcus, J.F. Janak, A.R. Williams (Plenum, New York 1971) pp. 225–259
2.50 P. Ziesche, G. Lehmann (eds.): *Ergebnisse in der Elektronentheorie der Metalle* (Akademie-Verlag, Berlin 1983)
2.51 J.C. Slater: Phys. Rev. **55**, 846–851 (1937)
2.52 J.C. Slater: Phys. Rev. **92**, 603–608 (1953)
2.53 M.M. Saffren, J.C. Slater: Phys. Rev. **92**, 1126–1128 (1953)
2.54 D.D. Koelling: Phys. Rev. **B2**, 290–298 (1970)
2.55 H. Bross, G. Bohn, G. Meister, W. Schubö, H. Stöhr: Phys. Rev. **B2**, 3098–3103 (1970)
2.56 O.K. Andersen: Phys. Rev. **B12**, 3060–3083 (1975)
2.57 D.D. Koelling, G. Arbman: J. Phys. **F5**, 2041–2054 (1975)
2.58 O. Jepsen, J. Madsen, O.K. Andersen: Phys. Rev. **B18**, 605–615 (1978)
2.59 E. Wimmer, H. Krakauer, M. Weinert, A.J. Freeman: Phys. Rev. **B24**, 864–875 (1981)
260 O.K. Andersen: "Linear Methods in Band Theory", in *The Electronic Structure of Complex Systems*, ed. by P. Phariseau, W.M. Temmerman, NATO ASI Series B, Vol. 113 (Plenum, New York, 1984) pp. 11–66
2.61 L.F. Mattheiss, D.R. Hamann: Phys. Rev. **B33**, 823–840 (1986)
2.62 P. Blaha, K. Schwarz, P. Sorantin, S.B. Trickey: Comp. Phys. Commun. **59**, 399–415 (1990)
2.63 O.K. Andersen: "Linearized Band Structure Methods", in *Lect. Notes Phys.*, Vol. 283 (Springer, Berlin, Heidelberg 1987) pp. 1–57
2.64 H.L. Skriver: *The LMTO Method* (Springer, Berlin, Heidelberg 1984)
2.65 O. Gunnarsson, O. Jepsen, O.K. Andersen: Phys. Rev. **B27**, 7144–7168 (1979)
2.66 A.R. Williams, J. Kübler, C.D. Gelatt: Phys. Rev. **B19**, 6094–6118 (1979)
2.67 J.W. Davenport: Phys. Rev. **B29**, 2896–2904 (1984)
2.68 G.W. Fernando, J.W. Davenport, R.E. Watson, M. Weinert: Phys. Rev. **B40**, 2757–2766 (1989)
2.69 H. Bross, G.M. Fehrenbach: Z. Phys. **B81**, 233–243 (1990)
2.70 A.R. Williams, J. van W. Morgan: J. Phys. **C7**, 37–60 (1974)

2.71 M. Weinert, E. Wimmer, A.J. Freeman: Phys. Rev. **B 26**, 4571–4578 (1982)
2.72 H.J.F. Jansen, A.J. Freeman: Phys. Rev. **B 30**, 561–569 (1984)
2.73 M. Methfessel: Phys. Rev. **B 38**, 1537–1540 (1988)
2.74 H. Bross, K.H. Anthony: Phys. Status Solidi **22**, 667–682 (1967)
2.75 F. Beleznay, M.J. Lawrence: J. Phys. **C 1**, 1288–1295 (1968)
2.76 P. Ziesche: J. Phys. **C 7**, 1085–1097 (1974)
2.77 R.G. Newton: Phys. Rev. Lett. **65**, 2031–2034 (1990)
2.78 R.H. Lasseter, P. Soven: Phys. Rev. **B 8**, 2476–2484 (1973)
2.79 B. Segall, A.-B. Chen: Phys. Rev. **B 16**, 2556–2563 (1977)
2.80 J.H. Wilkinson: *The Algebraic Eigenvalue Problem* (Oxford University Press, London 1965)
2.81 B.T. Smith, J.M. Boyle, J.J. Dongarra, B.S. Garbow, Y. Ikebe, V.C. Klema, C.B. Moler: *Matrix Eigensystem Routines—EISPACK Guide*, in *Lect. Notes Comp. Sci.*, ed. by G. Goos, J. Hartmanis, Vol. 6 (Springer, Berlin, Heidelberg 1976)
2.82 B.N. Parlett: *The Symmetric Eigenvalue Problem* (Prentice-Hall, Englewood Cliffs 1980)
2.83 J.J. Dongarra, C.B. Moler, J.R. Bunch, G.W. Stewart: *LINPACK User's Guide* (SIAM, Philadelphia 1979)
2.84 L.F. Mattheiss, J.H. Wood, A.C. Switendick: "A Procedure for Calculating Electronic Energy Bands Using Symmetrized Augmented Plane Waves", in *Methods in Computational Physics*, ed. B. Alder, S. Fernbach, M. Rotenberg, Vol. 8 (Academic, New York 1968) pp. 63–147
2.85 J.F. Janak, A.R. Williams, V.L. Moruzzi: Phys. Rev. **B 11**, 1522–1536 (1975)
2.86 M. Methfessel, A.T. Paxton: Phys. Rev. **B 40**, 3616–3621 (1989)
2.87 D.J. Chadi, M.L. Cohen: Phys. Rev. **B 8**, 5747–5753 (1973)
2.88 H.J. Monkhorst, J.D. Pack: Phys. Rev. **B 13**, 5188–5192 (1976)
2.89 S. Froyen: Phys. Rev. **B 39**, 3168–3172 (1989)
2.90 L. Macot, B. Frank: Phys. Rev. **B 41**, 4469–4474 (1990)
2.91 A. Bansil: Sol. State Commun. **16**, 885–889 (1975)
2.92 W.R. Fehlner, S.B. Nickerson, S.H. Vosko: Sol. State Commun. **19**, 83–86 (1976)
2.93 W.R. Fehlner, S.H. Vosko: Can. J. Phys. **54**, 2159–2169 (1976)
2.94 G. Gilat, L.J. Raubenheimer: Phys. Rev. **144**, 390–395 (1966)
2.95 O. Jepsen, O.K. Andersen: Solid State Commun. **9**, 1763–1767 (1971)
2.96 G. Lehmann, M. Taut: Phys. Status Solidi **B 57**, 469–477 (1972)
2.97 P.T. Coleridge, J. Molenaar, A. Lodder: J. Phys. **C 15**, 6943–6953 (1982)
2.98 J.E. Müller, J.W. Wilkins: Phys. Rev. **B 29**, 4331–4348 (1984)
2.99 M.S. Methfessel, M.H. Boon, F.M. Mueller: J. Phys. **C 20**, 1069–1077 (1986)
2.100 R. Zeller: Z. Phys. **B 72**, 79–85 (1988)
2.101 J.F. Janak: Phys. Rev. **B 9**, 3985–3988 (1974)
2.102 J. van W. Morgan: J. Phys. **C 10**, 1181–1202 (1977)
2.103 M. Weinert: J. Math. Phys. **22**, 2433–2439 (1981)
2.104 J.E. Müller, R.O. Jones, J. Harris: J. Chem. Phys. **79**, 1874–1884 (1983)
2.105 P.J. Feibelman: Phys. Rev. **B 33**, 719–725 (1986)
2.106 N. Stefanou, R. Zeller, P.H. Dederichs: Phys. Rev. **B 36**, 6372–6382 (1987)
2.107 P.H. Dederichs, R. Zeller: Phys. Rev. **B 28**, 5462–5472
2.108 J. Treusch, R. Sandrock: Phys. Status Solidi **16**, 487–497 (1966)
2.109 X. Gonze, J.P. Vigneron and J.-P. Michenaud: J. Phys.: Condens. Matter **1**, 525–540 (1989)
2.110 A. Zunger, M.L. Cohen: Phys. Rev. **B 19**, 568–582 (1979)
2.111 J.R. Chelikowsky, T.J. Wagener, J.H. Weaver, A. Jin: Phys. Rev. **B 44**, 9644–9651 (1989)
2.112 M.S. Hybertsen, S.G. Louie: Phys. Rev. **B 34**, 5390–5413 (1986)
2.113 J.L. Martins, M.L. Cohen: Phys. Rev. **B 37**, 6134–6138 (1988)
2.114 D.R. Hamann: Phys. Rev. **B 40**, 2980–2987 (1989)
2.115 D. Vanderbilt: Phys. Rev. **B 41**, 7892–7895 (1990)
2.116 N. Troullier, J.L. Martins: Solid State Commun. **74**, 613–616 (1990)
2.117 A.M. Rappe, K.M. Rabe, E. Kaxiras, J.D. Joannopoulus: Phys. Rev. **B 41**, 1227–1230 (1990)
2.118 R. Podloucky, R. Zeller, P.H. Dederichs: Phys. Rev. **B 22**, 5777–5790 (1980)
2.119 R. Zeller: J. Phys. **C 20**, 2347–2360 (1987)
2.120 C. Koenig, E. Daniel: J. Physique Lett. **42**, L193–L195 (1981)
2.121 F. Beeler, O.K. Andersen, M. Scheffler: Phys. Rev. **B 41**, 1603–1624 (1990)
2.122 A. Zunger: "Electronic Structure of 3d Transition-Atom Impurities in Semiconductors", in *Solid State Physics*, ed. by H. Ehrenreich, D. Turnbull, Vol. 39 (Academic, New York 1986) pp. 275–464
2.123 K.H. Johnson: J. Chem. Phys. **49**, 3085–3095 (1966)

2.124 W.L. Schaich: Phys. Rev. **B 29**, 6513–6519 (1984)
2.125 M. Abramowitz, I.A. Stegun (eds.): *Handbook of Mathematical Functions*, 9th printing (Dover, New York 1972)
2.126 J.L. Beeby: J. Phys. **C 1**, 82–87 (1968)
2.127 J.B. Pendry: *Low Energy Electron Diffraction* (Academic, London 1974)
2.128 J.M. MacLaren, S. Crampin, D.D. Vvedensky, R.C. Albers, J.B. Pendry: Comp. Phys. Commun. **60**, 365–389 (1990)
2.129 F. Maca, M. Scheffler: Comp. Phys. Commun. **51**, 381–390 (1988)
2.130 J.E. Inglesfield, G.A. Benesh: Phys. Rev. **B 37**, 6682–6700 (1988)
2.131 J.M. MacLaren, X.-G. Zhang, A. Gonis, S. Crampin: Phys. Rev. **B 40**, 9955–9958 (1989)
2.132 P. Krüger, J. Pollmann: Phys. Rev. **B 38**, 10578–10599 (1988)
2.133 H. Ehrenreich, L.M. Schwartz: "The Electronic Structure of Alloys", in *Solid State Physics*, ed. by H. Ehrenreich, F. Seitz, D. Turnbull, Vol. 31 (Academic, New York 1982) pp. 149–286
2.134 B.L. Gyorffy, G.M. Stocks: "Electronic States in Random Substitutional Alloys: the CPA and beyond", in *Electrons in Disordered Metals and at Metallic Surfaces*, ed. by P. Phariseau, B.L. Gyorffy, L. Scheire, NATO ASI Series B, Vol. 42 (Plenum, New York, 1979) pp. 89–192
2.135 W.M. Temmerman, Z. Szotek: Comp. Phys. Rep. **5**, 173–220 (1987)
2.136 H. Akai: J. Phys.: Condens. Matter **1**, 8045–8063 (1989)

3. Exchange and Correlation in Solids

R.W. Godby

With 23 Figures

In this chapter theories of the effects of the electron–electron interaction on optical and electronic properties of solids are described, with particular focus on the new first-principles approaches to many-body phenomena. First, the elementary theory of non-interacting electrons moving in a periodic potential is reviewed, and then it is shown how the use of Green's functions enables the many-body problem to be developed along analogous lines. Hedin's coupled equations for the self-energy operator and Green's function are described, and used to introduce the GW approximation for the self-energy operator. Density-functional theory, in which the electron density plays the central role, is also explained, and its relationship to many-body theory and the theoretical limitations of various quantities calculated in density-functional theory are examined. Then the application of many-body theory to actual systems is described, starting with the uniform electron gas, and continuing with real metals, semiconductors and insulators, including a discussion of the so-called band-gap problem.

The elementary theory of electrons in solids is familiar to anyone who has followed an undergraduate course on solid state physics. In such a theory, non-interacting electrons move in a periodic potential. The one-electron eigenstates are the well-known Bloch functions, which may be labelled with their wavevector $\boldsymbol{k}$ and the band index n, as discussed in Chap. 2. Because the electrons are non-interacting, their spectroscopic properties are particularly simple: the energy required to excite an electron from one state $n\boldsymbol{k}$ to another $n'\boldsymbol{k}'$ is the difference between the energies of the final and initial one-electron states, and the energies to remove an electron from, or add an electron to, a state are both equal to the energy of that state.

Real electrons are, of course, not non-interacting. At the simplest level, the electron–electron interaction may be taken into account using the Hartree theory: the Hartree potential (the electrostatic potential due to the average electron density) is added to the external potential. (Since the electrons still move in a periodic potential with no explicit interaction, the Hartree theory is in practice still an independent-electron theory, although one in which the effective potential depends on the electron density.) Effects of the electron–electron interaction beyond the Hartree theory are termed *exchange and correlation.* In modern theories of the electron–electron interaction, exchange and correlation are normally treated together. Historically, however, the term "exchange" was applied to Hartree–Fock theory. (We shall discuss Hartree–Fock theory briefly in the next section.) The relationships between the

Table 3.1. A summary of the relationship between different approaches to the electron–electron interaction. h is the single-electron Hamiltonian $-\frac{1}{2}\nabla^2 + V_{ext}$. The "Hamiltonian" and "potential" shown refer to the terms that appear in the Schrödinger-like equation obeyed by the relevant one-electron eigenvalues (see text). Note that in exact DFT, the potential is *local* (i.e. is a function only of $\boldsymbol{r}$) as shown, but not *local-density* (i.e. has a functional dependence on the electron density, $n(\boldsymbol{r})$, at all points). Dashes indicate that the properties mentioned bear no rigorous relationship to physical properties of the real system of interacting electrons

	Non-interacting electrons	Hartree theory	Hartree–Fock theory	Many-body theory	Exact DFT	DFT in the LDA
"Hamiltonian"	h	$h + V_H$	$h + V_H + \Sigma_x$	$h + V_H + \Sigma$	$h + V_H + V_{xc}$	$h + V_H + V_{xc}^{LDA}$
Nature of "potential"	Local, energy-independent	Local, energy-independent	Non-local, energy-independent	Non-local, energy-dependent	Local, energy-independent	Local, energy-independent
Meaning of one-electron eigenvalues	—	—	—	Quasiparticle energies	None in general	—
Meaning of highest occupied eigenvalue	—	—	—	Ionisation potential or chemical potential	Ionisation potential or chemical potential	—
Meaning of ground-state electron density	—	—	—	Exact ground-state electron density	Exact ground-state electron density	—
Meaning of Fermi surface	—	—	—	True Fermi surface	—	—

different approaches to exchange and correlation discussed here are summarised in Table 3.1.

This chapter is concerned with the effects of exchange and correlation on the spectroscopic properties of solids. It is not intended to be an exhaustive survey of many-body theories or of many-body effects in experiments; rather, the focus will be on the progress that is now being made in the practical treatment of exchange and correlation in real solids, and, in particular, the computational applications of many-body theory. In the remainder of this introduction (Sect. 3.1) the theoretical ideas that underlie such work are described.

3.1 Many-Body Theory of Electronic States in Solids

3.1.1 Introduction to Many-Body Theory

In this section we introduce the main formal ideas of many-body theory that we require. For more details the reader is referred to a textbook on many-body theory, for example [3.1] or the review article [3.2].

The key to formulating a tractable many-body theory of solids is the avoidance of explicit reference to the many-body wavefunction $\Psi(\boldsymbol{r}_1, \boldsymbol{r}_2, \boldsymbol{r}_3, \ldots, \boldsymbol{r}_N)$, where N is of the order of 10^{23}. In the initial stages of the development of the theory, however, it is necessary to have a way of describing a general many-body wavefunction. We shall use the fact that if ϕ_k are a complete set of one-electron wavefunctions, then the (infinite) set of Slater determinants

$$|\boldsymbol{k}_1\boldsymbol{k}_2\boldsymbol{k}_3\cdots\boldsymbol{k}_N\rangle = \begin{vmatrix} \phi_{\boldsymbol{k}_1}(\boldsymbol{r}_1)\phi_{\boldsymbol{k}_1}(\boldsymbol{r}_2)\cdots\phi_{\boldsymbol{k}_1}(\boldsymbol{r}_N) \\ \phi_{\boldsymbol{k}_2}(\boldsymbol{r}_1)\phi_{\boldsymbol{k}_2}(\boldsymbol{r}_2)\cdots\phi_{\boldsymbol{k}_2}(\boldsymbol{r}_N) \\ \vdots \\ \phi_{\boldsymbol{k}_N}(\boldsymbol{r}_1)\phi_{\boldsymbol{k}_N}(\boldsymbol{r}_2)\cdots\phi_{\boldsymbol{k}_N}(\boldsymbol{r}_N) \end{vmatrix} \tag{3.1}$$

(where the indices $\boldsymbol{k}_1\cdots\boldsymbol{k}_N$ run over all possible combinations of N positive integers as one moves from determinant to determinant within the set) forms a *complete* set for representing the general antisymmetric N-electron wavefunction $\Psi(\boldsymbol{r}_1, \boldsymbol{r}_2, \boldsymbol{r}_3, \ldots, \boldsymbol{r}_N)$. (For older work on the uniform electron gas, $\boldsymbol{k}$ labelled the wavevector of the set of plane waves $e^{i\boldsymbol{k}\cdot\boldsymbol{r}}$. However, $\boldsymbol{k}$ need not necessarily refer to a wavevector; it may be thought of as the means of labelling *any* complete set of one-electron orbitals.)

The first step in many-body theory is the definition of creation and annihilation operators $\hat{a}_k{}^\dagger$ and $\hat{a}_k$. The effect of $\hat{a}_k{}^\dagger$ on a Slater determinant is to add an extra row and column corresponding to an additional electron and the additional one-particle state $\boldsymbol{k}$; the effect of $\hat{a}_k$ on the same Slater determinant is to remove the row and column corresponding to one of the electrons and the one-particle state $\boldsymbol{k}$. Thus

$$\begin{aligned} \hat{a}_k{}^\dagger|\boldsymbol{k}_1\boldsymbol{k}_2\cdots\boldsymbol{k}_N\rangle &= |\boldsymbol{k}\boldsymbol{k}_1\boldsymbol{k}_2\cdots\boldsymbol{k}_N\rangle \\ \hat{a}_k|\boldsymbol{k}\boldsymbol{k}_1\boldsymbol{k}_2\cdots\boldsymbol{k}_N\rangle &= |\boldsymbol{k}_1\boldsymbol{k}_2\cdots\boldsymbol{k}_N\rangle. \end{aligned} \tag{3.2}$$

The operators $\hat{a}_k^\dagger$ and $\hat{a}_k$ are also declared to be linear, so that their action on linear combinations of Slater determinants is now defined. Roughly speaking, $\hat{a}_k^\dagger$ and $\hat{a}_k$ may be thought of as adding and subtracting an electron in the one-electron state ϕ_k.

The *field operators* $\hat{\Psi}^\dagger(\boldsymbol{r})$ and $\hat{\Psi}(\boldsymbol{r})$ are then defined by

$$\hat{\Psi}(\boldsymbol{r}) = \sum_k \phi_k(\boldsymbol{r})\hat{a}_k; \quad \hat{\Psi}^\dagger(\boldsymbol{r}) = \sum_k \phi_k{}^*(\boldsymbol{r})\hat{a}_k{}^\dagger \tag{3.3}$$

and may correspondingly be thought of as adding and subtracting an electron at the point $\boldsymbol{r}$.[1]

For time dependence, one uses the Heisenberg picture, in which the time dependence is placed in the operators rather than the many-body-wavefunction, e.g.

$$\Psi(\boldsymbol{r}, t) = \mathrm{e}^{\mathrm{i}Ht}\hat{\Psi}(\boldsymbol{r})\mathrm{e}^{-\mathrm{i}Ht} \tag{3.4}$$

where H is the full many-body Hamiltonian.

The central quantity in spectroscopy is the correlation function of an electron with itself at another position and time. This is the one-particle Green's function, G, defined as

$$G(\boldsymbol{r}, t, \boldsymbol{r}', t') = -\mathrm{i}\langle N | T[\hat{\Psi}(\boldsymbol{r}, t)\hat{\Psi}^\dagger(\boldsymbol{r}', t')] | N\rangle \tag{3.5}$$

where $|N\rangle$ is the many-body ground state, and T denotes "time ordering", that is, a re-ordering of the operators in order of their time arguments with the latest on the left, with a minus sign for each transposition. The interpretation of G is simple: using the definition of the field operators it may be shown to describe the probability amplitude for an electron, added at $\boldsymbol{r}t$ to the N-electron ground state, to have propagated to $\boldsymbol{r}'$ at a later time t'. (Or, if $t > t'$, the propagation is that of a *hole*.) This illustrates the Green's function as a *propagator*.

(In the case of non-interacting electrons, and after taking a Fourier transform from the time domain to the frequency domain, G reverts to the familiar form

$$G_{\mathrm{nonint}}(\boldsymbol{r}, \boldsymbol{r}', \omega) = 2\sum_i \frac{\psi_i(\boldsymbol{r})\psi_i^*(\boldsymbol{r}')}{\omega - E_i + \mathrm{i}\delta\,\mathrm{sgn}(E_i - \mu)} \tag{3.6}$$

where the sum is over the one-electron eigenfunctions ψ_i, and δ is an infinitesimal.)

---

[1] To see this, it is helpful to recall that

$$\sum_k \phi_k^*(\boldsymbol{r})\phi_k(\boldsymbol{r}') = \delta(\boldsymbol{r} - \boldsymbol{r}');$$

in other words, the amplitude of each state ϕ_k that one needs to sum in order to obtain a delta function centered at $\boldsymbol{r}$ is $\phi_k^*(\boldsymbol{r})$.

By analogy with the non-interacting Green's function above, which in the time domain obeys the equation[2]

$$\left[\mathrm{i}\frac{\partial}{\partial t} - h(\boldsymbol{r})\right] G^{\text{nonint}}(\boldsymbol{r}, t, \boldsymbol{r}', t') = \delta(\boldsymbol{r} - \boldsymbol{r}')\delta(t - t') \tag{3.7}$$

(where h is the one-electron Hamiltonian $-\frac{1}{2}\nabla^2 + V_{\text{ext}}$), it is conventional to define the *self-energy operator*, Σ, so that the effective "Hamiltonian" $h + V_{\text{H}} + \Sigma$ (where V_{H} is the Hartree potential—see Sect. 3.1.4) and the Green's function G obeys the analogous equation

$$\left[\mathrm{i}\frac{\partial}{\partial t} - h(\boldsymbol{r}) - V_{\text{H}}(\boldsymbol{r}, t)\right] G(\boldsymbol{r}, t, \boldsymbol{r}', t')$$
$$- \int d^3r'' dt'' \Sigma(\boldsymbol{r}, t, \boldsymbol{r}'', t'') G(\boldsymbol{r}'', t'', \boldsymbol{r}', t') = \delta(\boldsymbol{r} - \boldsymbol{r}')\delta(t - t'). \tag{3.8}$$

Σ, like G, is a non-local, time-dependent (or frequency-dependent) quantity: $\Sigma(\boldsymbol{r}, \boldsymbol{r}', t, t')$ or $\Sigma(\boldsymbol{r}, \boldsymbol{r}', \omega)$. We note that setting $\Sigma = 0$ results in the elementary Hartree theory; Σ therefore represents the effects of *exchange and correlation* as defined in Sect. 3.1.2 above. Using the Heisenberg equation of motion of the field operators, the equation of motion of G, and thus an implicit expression for Σ, may be derived.

So far we have defined the quantity of interest (the Green's function) in terms of the operators of many-body theory. However, we do not yet have a practical scheme, as the expression for the Green's function still contains the many-body ground-state wavefunction, $|N\rangle$, which is, of course, not known. The breakthrough was obtained by *Hedin* [3.2], who obtained equations involving G and Σ, but not $|N\rangle$. By the mathematical device of taking the functional derivative with respect to an artificial change $\delta\phi$ in the external potential, which is set equal to zero at the end, he derived the set of coupled equations[3]

$$\Sigma(12) = \mathrm{i}\int W(1^+3) G(14) \Gamma(42;3) d(34) \tag{3.9}$$

$$W(12) = v(12) + \int W(13) P(34) v(42) d(34) \tag{3.10}$$

$$P(12) = -\mathrm{i}\int G(23) G(42) \Gamma(34;1) d(34) \tag{3.11}$$

$$\Gamma(12;3) = \delta(12)\delta(13) + \int \frac{\delta\Sigma(12)}{\delta G(45)} G(46) G(75) \Gamma(67;3) d(4567) \tag{3.12}$$

where G and Σ are the Green's function and self-energy operator defined above, P is the *polarization propagator*

$$P(12) = \mathrm{i}\int G(23) G(42) \frac{\delta G^{-1}(34)}{\delta V(1)} d(34) \tag{3.13}$$

[2] We use Hartree atomic units ($\hbar = e = m_e = 4\pi\varepsilon_0 = 1$) throughout.
[3] The standard abbreviated notation $(1) = (\boldsymbol{r}_1, t_1); (1^+) = (\boldsymbol{r}_1, t_1 + \delta)$, etc. has been used.

(where V is the external potential plus the instantaneous Hartree potential), v is the Coulomb interaction $1/|\boldsymbol{r}-\boldsymbol{r}'|$, W is termed the *screened Coulomb interaction* and Γ is the *vertex function*

$$\Gamma(12;3) = \delta(12)\delta(13) + \frac{\delta\Sigma(12)}{\delta V(3)}. \tag{3.14}$$

The four equations (3.9–12) together with the equation of motion of the Green's function (3.8) form a coupled set of five equations which may in principle be solved exactly for a given external potential.[4]

Hedin [3.2] showed that an iterative solution of these equations is possible, starting with $\Sigma = 0$; successive iterations give an expansion of the various many-body quantities in powers of the screened Coulomb interaction, W. In particular, the first-order expression for the self-energy operator is

$$\Sigma(12) = \mathrm{i}W(1^{+}2)G(12) \tag{3.15}$$

or equivalently, in the frequency domain,

$$\Sigma(\boldsymbol{r},\boldsymbol{r}',\omega) = \frac{\mathrm{i}}{4\pi}\int_{-\infty}^{\infty} \mathrm{e}^{\mathrm{i}\omega\delta} W(\boldsymbol{r},\boldsymbol{r}',\omega')G(\boldsymbol{r},\boldsymbol{r}',\omega+\omega')\,d\omega'. \tag{3.16}$$

This is known as the *GW* approximation. It is the key expression in much computational many-body theory. The important point is that because screening in solids is relatively strong, a first-order expansion of Σ in powers of W should be much more successful than an expansion in powers of the *unscreened* Coulomb interaction, v. Indeed, in later sections we shall see that the *GW* approximation provides a good description of the self-energy operator in many real solids.

The identification of the *quasiparticle energies* (the energies of the single-particle-like approximate eigenstates of the many-body system that are measured in spectroscopic experiments[5]) then follows through consideration of the imaginary part of the Green's function, $\mathrm{Im}\,G$ (also known as the spectral function, A). In a non-interacting system, the Green's function obeys (3.7) and may be written in the form shown in (3.6). Taking the imaginary part shows that $\mathrm{Im}\{G\}$ has a delta function centred at each one-electron energy, E_i. In the interacting system the quasiparticle energies are similarly identified from peaks

---

[4] By considering interacting electrons moving in a fixed external potential, we are of course excluding any effects of phonons on the electronic structure. However, these may in principle be treated in a very similar way, leading to a formally similar set of coupled equations [3.2], in which the definitions of the self-energy, screened Coulomb interaction and vertex function are modified to account for the nuclear motion.

[5] The quasiparticle energies refer to the energies to add or subtract an electron to or from the system. Spectroscopies in which an electron is excited from one state to another are formally described using a two-electron Green's function, although if the electron and hole state can be considered to be independent of each other (as in the case of non-excitonic transitions across the band gap of a semiconductor, for example) the quasiparticle picture remains valid. In other cases (such as core-level excitations in X-ray absorption), it is possible to use a modified one-electron Green's function formalism, but this is not discussed further here.

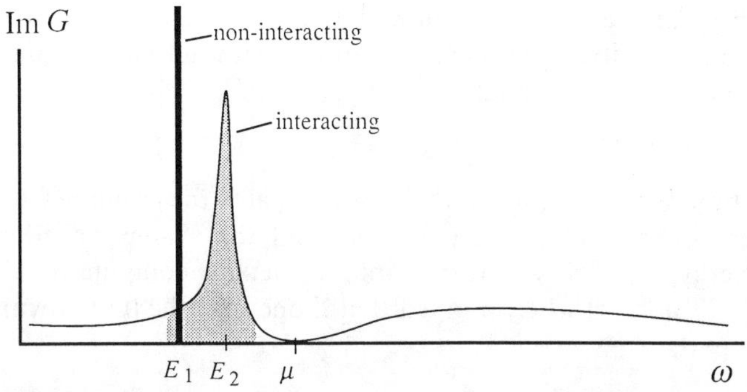

Fig. 3.1. Schematic illustration of the spectral function, Im G, in systems of non-interacting and interacting electrons. In each case the single-particle(-like) energies are identified from the energies of peaks in Im G

in Im $\{G\}$, which in general are Lorentzian peaks rather than delta functions, resulting from the fact that the quasiparticle states have a finite lifetime[6] (see Fig. 3.1). The relationship between the self-energy operator and the Green's function (3.8) then leads to the Schrödinger-like equation obeyed by the quasiparticle energies and quasiparticle "wave functions":

$$[-\tfrac{1}{2}\nabla^2 + V_{\text{ext}}(\boldsymbol{r}) + V_{\text{H}}(\boldsymbol{r}) - E_i]\psi_i(\boldsymbol{r}) + \int \Sigma(\boldsymbol{r}, \boldsymbol{r}', E_i)\psi_i(\boldsymbol{r}')d^3\boldsymbol{r}' = 0. \tag{3.17}$$

3.1.2 Density-Functional Theory and Connections with Many-Body Theory

The many-body problem of finding the electronic-ground-state electron density and total energy of a system of interacting electrons moving in an external potential is often encountered in the physics of solids. In this case the external potential is the electrostatic potential provided by the atomic nuclei or ion cores, which are to be regarded as fixed. In density-functional theory (DFT), this problem is rigorously transformed into a parallel problem of a system of non-interacting electrons moving in a different external potential (which is itself a functional of the electron density). In practice, density-functional theory calculations are so much easier to perform than other forms of many-body theory that nearly all so-called "electronic structure" calculations are in fact within the framework of DFT, and the one-electron band structure which appears in DFT as a mathematical artifact is frequently interpreted as the quasiparticle band structure. In this section we shall describe DFT and also discuss the validity of such an interpretation.

---

[6] An exponential decay $e^{-\gamma t}$ in the time domain becomes a Lorentzian of width γ in the frequency domain.

In 1964 *Hohenberg* and *Kohn* [3.3] showed that there exists a *unique* functional of the electron density, $F[n]$, such that for a given external potential V_{ext} the minimum value of the functional

$$E[n] = F[n] + \int V_{\text{ext}}(\boldsymbol{r})n(\boldsymbol{r})d^3\boldsymbol{r} \tag{3.18}$$

(with respect to variations in $n(\boldsymbol{r})$) occurred when n was equal to the ground-state density of the system of interacting electrons, n_0, and that value was the ground-state total energy, E_0. This was turned into a tractable computational scheme by *Kohn* and *Sham* [3.4] who observed that if one split the (unknown) functional $F[n]$ into parts

$$F[n] = T_s[n] + E_{\text{H}}[n] + E_{\text{xc}}[n] \tag{3.19}$$

where T_S is the kinetic energy of a system of *non*-interacting electrons with density n and E_{H} is the Hartree interaction energy

$$E_{\text{H}} = \frac{1}{2}\int d^3\boldsymbol{r}\int d^3\boldsymbol{r}' \frac{n(\boldsymbol{r})n(\boldsymbol{r}')}{|\boldsymbol{r}-\boldsymbol{r}'|} \tag{3.20}$$

(thereby *defining* the *exchange-correlation energy* functional E_{xc}), the Euler equation obtained when $E[n]$ was varied with respect to n was the same as that of a system of *non*-interacting electrons moving in the effective potential

$$V_{\text{eff}}(\boldsymbol{r}) = V_{\text{ext}}(\boldsymbol{r}) + V_{\text{H}}(\boldsymbol{r}) + V_{\text{xc}}(\boldsymbol{r}) \tag{3.21}$$

(itself a functional of the density), where the Hartree potential is the electrostatic potential due to the total electron density,

$$V_{\text{H}}(\boldsymbol{r}) = \int d^3\boldsymbol{r}' \frac{n(\boldsymbol{r}')}{|\boldsymbol{r}-\boldsymbol{r}'|}, \tag{3.22}$$

and the exchange-correlation potential is

$$V_{\text{xc}}(\boldsymbol{r}) = \frac{\delta E_{\text{xc}}[n]}{\delta n(\boldsymbol{r})}, \tag{3.23}$$

the functional derivative of the exchange-correlation energy.

Thus (assuming for the moment that the functional E_{xc} and therefore V_{xc} are known) the ground-state density n_0 may be obtained by solving the non-interacting Schrödinger equation

$$[-\tfrac{1}{2}\nabla^2 + V_{\text{eff}}(\boldsymbol{r}) - E_{i,\text{DFT}}]\psi_{i,\text{DFT}}(\boldsymbol{r}) = 0, \tag{3.24}$$

constructing the density from the wave functions of the Kohn-Sham electrons

$$n(\boldsymbol{r}) = \sum_{\text{occ}} |\psi_{i,\text{DFT}}(\boldsymbol{r})|^2 \tag{3.25}$$

and iterating until the density and effective potential are consistent with one another. The ground-state electronic total energy may then be obtained from (3.19).

In principle this gives a method for finding the exact ground-state electron density and total energy of a system of interacting electrons, without recourse to conventional many-body theory. However, in practice the exchange-correlation energy functional E_{xc} and its functional derivative V_{xc} are not known exactly, and it is necessary to approximate them. A very successful approximation has been the *local density approximation* (LDA), in which the contribution from the small region of space around each point $\boldsymbol{r}$ to E_{xc} is taken to be the same as it would be in a *uniform* electron gas with the same local electron density $n(\boldsymbol{r})$:

$$E_{xc} = \int n(\boldsymbol{r})\varepsilon_{xc}(n(\boldsymbol{r}))d^3\boldsymbol{r} \tag{3.26}$$

where $\varepsilon_{xc}(n)$ is the exchange-correlation energy per electron of a uniform electron gas with density n. This is known from accurate quantum Monte Carlo calculations [3.5] for the total energy of uniform electron gases of various densities, which may be thought of as effectively dealing with the many-electron wave function. Using the LDA, total energy calculations have been performed for many real atoms, molecules, clusters, surfaces, solids and defects [3.6]. These have usually resulted in good agreement with experiment where the structures etc. are known experimentally, and therefore it has been possible to perform "computer experiments" with some confidence to explore possibilities inaccessible in real experiments, to develop an understanding of the physical principles involved in bonding and the adiabatic energy surfaces for atomic motion in complex structures.

However, many workers have attempted to extend the use of DFT and LDA calculations beyond their true applicability, by interpreting the one-electron eigenvalues $E_{i,\mathrm{DFT}}$ that appear in the Schrödinger equation of the fictitious system of non-interacting electrons (3.24) as if they were the quasiparticle energies E_i of (3.17). This procedure is certainly not valid for all the eigenvalues, but it was once thought that it would be correct for eigenvalues sufficiently close to the Fermi energy. This argument was based on the DFT analogue of Koopmans' theorem [3.7] which states (correctly) that *one* DFT eigenvalue *is* the same as its quasiparticle counterpart: the highest occupied DFT eigenvalue ($E^{(N)}_{N,\mathrm{DFT}}$ in a (N)-electron system) is equal to the ionization potential of the system, $I^{(N)}$; that is, the energy to remove an electron to infinity, leaving the ($N-1$)-electron system in its ground state:

$$I^{(N)} = E^{(N)}_{N,\mathrm{DFT}}. \tag{3.27}$$

Using "continuity" arguments it was thought that this could be extended to the eigenvalues in the neighbourhood of the Fermi energy, so that the part of a metal band near the Fermi energy would be exactly correct in an exact DFT calculation, and similarly the band gap of a semiconductor or insulator. This will be addressed further in Sect. 3.4.2.

Certain rigorous connections are, however, known between exact DFT and the true quasiparticle energies. Firstly, the DFT analogue of Koopmans' theorem [3.7] has been proved, as stated above. (For an extended system with a well-defined chemical potential, this theorem is equivalent to stating that the DFT chemical potential is the same as the true chemical potential.) The second known connection, derived by *Sham* and *Schlüter* [3.8], is an equation that directly relates the self-energy operator Σ to the exchange-correlation potential V_{xc}. The origin of the relationship is simply the fact that, by construction, the true ground-state electron density of the real system, $n(\boldsymbol{r})$, is reproduced by that of the Kohn–Sham electrons. It may be recalled from Green's function theory that the imaginary part of the Green's function, taken at $\boldsymbol{r}' = \boldsymbol{r}$, and integrated over frequency, is proportional to the electron density. Furthermore, this result applies both to the true single-particle Green's function of the interacting system, G, and to the non-interacting Kohn–Sham electrons' Green's function, G^{DFT}. Thus we have

$$n(\boldsymbol{r}) = \frac{1}{\pi} \operatorname{Im} \left\{ \int G(\boldsymbol{r}, \boldsymbol{r}' = \boldsymbol{r}, \omega) d\omega \right\} \tag{3.28}$$

$$n(\boldsymbol{r}) = \frac{1}{\pi} \operatorname{Im} \left\{ \int G^{\mathrm{DFT}}(\boldsymbol{r}, \boldsymbol{r}' = \boldsymbol{r}, \omega) d\omega \right\}. \tag{3.29}$$

Then, using the fact that G and G^{DFT} are related by the Dyson equation

$$G = G^{\mathrm{DFT}} + G^{\mathrm{DFT}} \cdot (\Sigma - V_{xc}) \cdot G \tag{3.30}$$

(which follows from (3.8) and its DFT counterpart in which Σ is replaced by V_{xc}), the result

$$\operatorname{Im} \left\{ \int [G^{\mathrm{DFT}}(\omega) \cdot (\Sigma(\omega) - V_{xc}) \cdot G(\omega)]_{\boldsymbol{r}=\boldsymbol{r}'} d\omega \right\} = 0 \tag{3.31}$$

is obtained. In Sect. 3.4.2 we will describe how this equation has been used in practice to calculate an exchange-correlation potential from a self-energy operator.

The third known connection is related directly to the band-gap problem, and was proved by *Sham* and *Schlüter* [3.8], and by *Perdew* and *Levy* [3.9]. It states that the minimum band gap calculated in a DFT calculation, E_g^{DFT}, differs from the true minimum band gap, E_g, by a discontinuity in the exchange-correlation potential that occurs upon going from the (N)-electron system (with the valence band exactly filled) to the $(N+1)$-electron system:

$$E_g = E_{g,\mathrm{DFT}} + \Delta \tag{3.32}$$

where

$$\Delta = V_{xc}^{(N+1)}(\boldsymbol{r}) - V_{xc}^{(N)}(\boldsymbol{r}). \tag{3.33}$$

To prove this, we consider the two DFT "band structures" shown schematically in Fig. 3.2, which correspond to the (N) and the $(N+1)$-electron systems. Because the Kohn–Sham electrons are non-interacting, any difference between the two band structures must reflect a change in the DFT effective potential in which they move (3.21). Of the three parts, the external potential

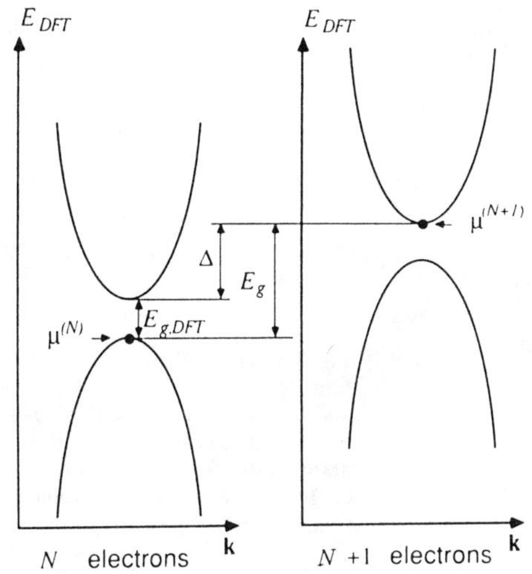

Fig. 3.2. Illustration of the significance of Δ, the discontinuity in the exchange-correlation potential. The exact Kohn–Sham one-electron energies are shown schematically in the form of a band structure for the N and $N+1$-electron systems. The two differ in a uniform increase of the eigenvalues by Δ, as explained in the text. The quasiparticle band gap E_g is the difference between the two eigenvalues indicated: $E_g = E^{(N+1)}_{N+1,\mathrm{DFT}} - E^{(N)}_{N,\mathrm{DFT}}$. It is evident that $E_g = E_{g,\mathrm{DFT}} + \Delta$. From [3.10]

is independent of the number of electrons, and the Hartree potential is a known analytical functional of the electron density $n(\mathbf{r})$, and so, like n, will change only by an amount of order $1/N$ when an electron is added. The remaining part of the potential, however, the exchange-correlation potential V_{xc}, is not necessarily an *analytical* functional of n, and may change by an amount of order 1 when an electron is added. However, any change of order 1 can only be a position-independent constant, Δ, since a position-dependent extra potential of order 1 would give rise to a change in n of order 1 when the Schrödinger equation is then solved. Thus the total effective potential can change by a constant, Δ, on going from the (N) to the ($N+1$)-electron system. This causes the Kohn–Sham energy bands in the ($N+1$)-electron system to be higher in energy by Δ (Fig. 3.2).

The significance of this for the minimum band gap is that the true minimum band gap, E_g, can be written in terms of the DFT eigenvalues shown in Fig. 3.2. The energy to add an electron in the conduction band is the chemical potential of the ($N+1$)-electron system, which by the eigenvalue theorem above is equal to $E^{(N+1)}_{N+1,\mathrm{DFT}}$. Likewise, the energy to subtract an electron from the valence band is the chemical potential of the (N)-electron system, $E^{(N)}_{N,\mathrm{DFT}}$. Provided that the electron and hole are in extended states (i.e. if we are excluding excitonic effects from our definition of the minimum band gap), the minimum band gap is the difference between these: $E_g = E^{(N+1)}_{N+1,\mathrm{DFT}} - E^{(N)}_{N,\mathrm{DFT}}$. It is clear from Fig. 3.2 that this is equal to $E_{g,\mathrm{DFT}} + \Delta$, which proves the theorem.

Finally, a comment is in order at this point on the question of what exactly is "a ground-state property" of a system of interacting electrons. This is a source of much confusion, partly because of a misunderstanding about what the term "ground-state property" means, and has led to inaccurate statements being made about the interpretation of quantities derived from DFT or LDA

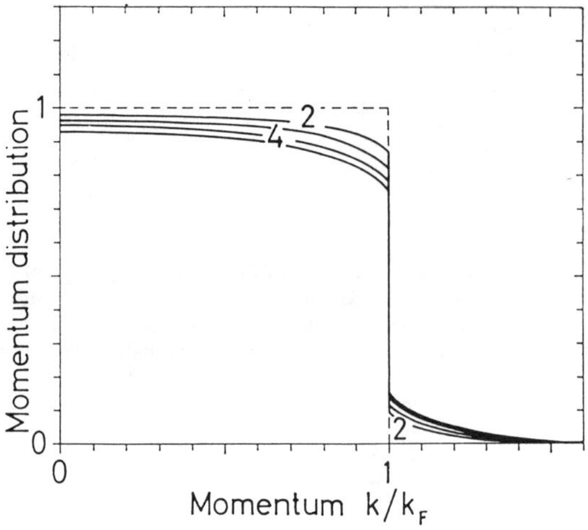

Fig. 3.3. The calculated momentum density of jellium ($r_s = 2$, 3, 4 and 5). (The extension of $n(\boldsymbol{p})$ beyond the Fermi momentum is a many-body effect, the consequence of the wavefunction renormalsation.) From [3.2]

calculations. Certainly, the original paper by *Hohenberg* and *Kohn* [3.3] showed that no two external potentials (except, trivially, two which differ by only a constant) generate the same ground-state electron density. Thus the external potential V_{ext} is in principle a functional of $n(\boldsymbol{r})$. But if the external potential is known, the complete many-electron Hamiltonian, H, is known, and of course a knowledge of H may in principle be used to compute any desired property of the system. In this sense, then, *any* property of the system is a ground-state property, i.e. a functional of the ground-state electron density. However, that does *not* mean that a particular property of the system is the same as the corresponding property of the fictitious system of non-interacting electrons that appear in DFT. For example, the true momentum density, $n(\boldsymbol{p})$, of jellium (Fig. 3.3) shows a "renormalisation tail" above k_F. However, the momentum density of the Kohn–Sham electron, $n_{\text{DFT}}(\boldsymbol{p})$, is simply a step function, with no renormalisation tail, since the Kohn–Sham electrons in jellium are simply non-interacting electrons moving in a constant effective potential, and therefore, by symmetry, have a spherical Fermi surface with occupation 1 inside and 0 outside. Thus $n(\boldsymbol{p})$ and $n_{\text{DFT}}(\boldsymbol{p})$ are not the same.

3.1.3 Practical Calculation of Many-Body Quantities

Since W and G themselves depend on Σ, the practical use of the GW approximation relies on making a good initial approximation for W and G. What is done in practice is to note that LDA calculations, although certainly not perfect, go a considerable way towards inclusion of the exchange and correlation effects, and therefore W and G are constructed from LDA wavefunctions and eigenvalues. For W, use is made of the fact that the exact *static* response function χ of a system of interacting electrons to an external potential may be written down using DFT, as[7] [3.11]

$$\chi = (1 - \chi_0 v - \chi_0 K_{xc})^{-1} \chi_0 \tag{3.34}$$

where

$$K_{xc}(\boldsymbol{r}, \boldsymbol{r}') = \frac{\delta^2 E_{xc}}{\delta n(\boldsymbol{r}) \delta n(\boldsymbol{r}')} \tag{3.35}$$

and χ_0 is the Alder–Wiser independent-electron response function of the Kohn–Sham electrons, calculated using elementary perturbation theory [3.12]. The inverse static *dielectric* function $\varepsilon^{-1} = 1 + v\chi$ may then be formed. For the quasiparticle calculations, the screened Coulomb interaction $W = \varepsilon^{-1} v$ is required at all frequencies, not just $\omega = 0$, and the same formula is used, with χ_0 replaced by the corresponding non-zero-frequency Adler–Wiser response function. (If the K_{xc} term is neglected, this procedure reduces to the familiar RPA expression for the inverse dielectric function [3.13].) Within the LDA, K_{xc} may be calculated straightforwardly. It is essential to include two aspects of the screening: *local field effects* and *dynamic screening*. Local field effects refer to the fact that the unit cell is not homogeneous, so that W is not merely a function of $|\boldsymbol{r} - \boldsymbol{r}'|$, and therefore becomes (in reciprocal space) a series of matrices $W_{GG'}(\boldsymbol{k}, \omega)$ in a periodic solid. (In the case of simple metals, however, the solid is sufficiently homogeneous that these local field effects are much less important than in semiconductors and insulators [3.14, 15].) Dynamic screening simply means that the frequency dependence of W must be taken into account, for example using a plasmon pole-like model (see below).

G is also obtained from the LDA eigenfunctions and eigenvalues. It is initially approximated by its equivalent in the LDA one-electron theory,

$$G^{\text{LDA}}(\boldsymbol{r}, \boldsymbol{r}', \omega) = 2 \sum_{n,\boldsymbol{k}} \frac{\psi_{n\boldsymbol{k}}^{\text{LDA}}(\boldsymbol{r}) \psi_{n\boldsymbol{k}}^{\text{LDA}*}(\boldsymbol{r}')}{\omega - E_{n\boldsymbol{k}}^{\text{LDA}} + \mathrm{i}\delta \,\mathrm{sgn}\,(E_{n\boldsymbol{k}}^{\text{LDA}} - \mu^{\text{LDA}})} \tag{3.36}$$

(which is a special case of (3.6) for the Green's function of non-interacting electrons).

In the spirit of Hedin's coupled equations, these starting points for W and G should be viewed as the initial guesses in an iterative procedure: W and G will be used to calculate a self-energy Σ, which will in turn be used to update W and G. However, in practice the LDA exchange-correlation potential is sufficiently close in its effect to that of the true self-energy operator that in many cases only a single calculation of Σ is necessary, with no updating of W and G. Limited updating of the energy denominators in G has been explicitly shown to alter the quasiparticle energies in semiconductors by less than 0.1 eV [3.16, 17], although in work on simple metals [3.14, 15], in which smaller energy scales were significant, such updating was important in achieving agreement with experiment (see Sect. 3.3.1 below).

---

[7] In this and other equations, matrix-like multiplication of non-local quantities is implied.

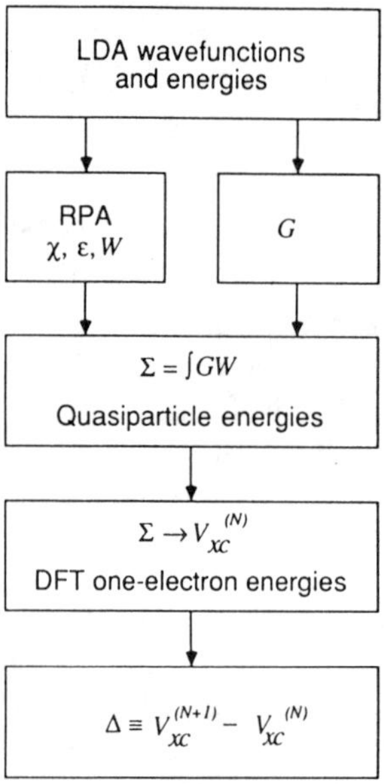

Fig. 3.4. Flowchart of a typical GW calculation. The Green's function, G, and screened Coulomb interaction, W, are combined to form the self-energy operator, Σ. This is used to calculate quasiparticle energies, and also (in the case shown) to calculate an exchange-correlation potential, V_{xc}, from which the discontinuity Δ may be determined. From [3.10]

The whole quasiparticle calculation is shown in Fig. 3.4. W and G are calculated as just described, and are represented in a convenient form for the basis set being used. (For example, the majority of calculations are performed within a plane wave basis set throughout, and therefore non-local quantities within a periodic system are represented by matrices at each $\boldsymbol{k}$-point on the chosen grid within the Brillouin zone and each value of frequency ω, such as $W_{GG'}(\boldsymbol{k}, \omega)$, with the rows and columns of the matrices corresponding to the plane waves $e^{i(\boldsymbol{k}+G)\cdot \boldsymbol{r}}$.)

Performing the GW frequency integration requires a knowledge of G and W at all frequencies. *Godby* et al. [3.18–20, 10] used an explicit knowledge of G and W, together with analytic continuation and contour integration techniques, to evaluate the frequency integral in full. However, the computational effort is large, and it is more common to make some model assumption about the frequency dependence of the elements of $W_{GG'}(\boldsymbol{k}, \omega)$, such that the frequency integrals can be done analytically for each term. For example, several works have used plasmon-pole-like models, based on the success of plasmon-pole models for the inverse dielectric function of jellium [3.2]. *Hybertsen* and *Louie* [3.14–17, 21–23] calculated plasmon-pole-like parameters for each element of the W matrices by constraining the model to have the calculated $\omega = 0$ dependence and also to obey the Kramers–Kronig and f-sum rules; *von der Linden* and *Horsch* [3.24] developed a more concise parameterisation by using the representation in which the inverse dielectric matrix is diagonal; and *Godby* and *Needs* [3.25] have calculated plasmon-pole-like parameters by fitting the full, frequency-dependent inverse dielectric matrix elements to the chosen model form at a few *imaginary* frequencies in the range important for the self-energy.

The calculation of the quasiparticle energies then proceeds as follows. Comparison of the "Schrödinger-like" quasiparticle equation (3.17) with (3.14), and recalling our belief that the LDA wavefunctions and energies already contains much of the truth about exchange and correlation, shows that the quasiparticle energies E_i may be formulated in terms of the LDA eigenvalues E_i^{LDA} in perturbation theory:

$$E_i = E_i^{\mathrm{LDA}} + \langle \psi_i^{\mathrm{LDA}} | \Sigma(E_i) - V_{\mathrm{xc}}^{\mathrm{LDA}} | \psi_i^{\mathrm{LDA}} \rangle + O((\Sigma(E_i) - V_{\mathrm{xc}}^{\mathrm{LDA}})^2) \tag{3.37}$$

where, additionally, the need to compute Σ self-consistently at the energy E_i requires a Taylor expansion about E_i^{LDA}:

$$\langle \Sigma(E_i) \rangle = \langle \Sigma(E_i^{\mathrm{LDA}}) \rangle + (E_i - E_i^{\mathrm{LDA}}) \left\langle \left(\frac{\partial \Sigma}{\partial \omega} \right)_{E_i^{\mathrm{LDA}}} \right\rangle + O((E_i - E_i^{\mathrm{LDA}})^2). \tag{3.38}$$

In practice, it is found to be sufficient to include only the first-order terms in (3.37) and (3.38).

3.2 The Self-Energy and Green's Function of Jellium

The uniform electron gas (or *jellium* as we shall refer to it), in which interacting electrons (of total electron density n, independent of position) move in the constant electrostatic potential provided by a fixed positive "background charge" of the same density n, is a widely used prototype of metallic systems. Its main appeal is its computational simplicity: the existence of complete homogeneity rather than simple translational symmetry (as exists in real solids) means that elementary quantities such as the electrostatic potential and electron density are simply numbers rather than functions of position, while the non-local many-body quantities (e.g. $\Sigma(\boldsymbol{r}, \boldsymbol{r}', \omega)$) depend only on $|\boldsymbol{r} - \boldsymbol{r}'|$ and therefore may be represented in reciprocal space simply as $\Sigma(|\boldsymbol{k}|, \omega)$, etc. Furthermore, varying the density of the jellium, n (often described by the "density parameter" r_s, equal to the radius of the sphere containing one electron: $(4\pi/3)r_s^3 = 1/n$), allows different régimes of exchange and correlation to be explored.

For these reasons the uniform electron gas has been a popular test bed for the ideas of many-body theory. Indeed, from the start of many-body perturbation theory in the 1950s to the mid-1980s, computational restrictions prevented investigation of many-body theory of real materials, and jellium was the only "system" to have been exhaustively investigated without drastic approximations.

We shall concentrate particularly on the *GW* approximation for the self-energy operator, since that is the formulation of many-body theory which has been most fruitful for real materials. In 1965 *Hedin* [3.26] introduced the *GW* approximation (3.16) (whose main novelty, as mentioned above, was the expansion of the self-energy operator in terms of the screened Coulomb interaction rather than the bare Coulomb interaction), and showed how the formalism could be applied to jellium, with most of the work being done analytically and just a few integrals left to be done numerically. Since the self-energy operator of jellium exhibits most of the key features of the self-energies of real materials (in particular, features which are absent from LDA calculations) we shall devote the next section to a discussion of the main results. A more detailed review is given by *Hedin* and *Lundqvist* [3.2].

Figure 3.5 shows the key effects of the self-energy operator as far as the quasiparticles are concerned. For each of three values of k, the upper panel shows the real and imaginary parts of the self-energy, and the lower panel shows the spectral function ($\mathrm{Im}\{G\}$). All these quantities are plotted as a function of ω. The self-energy has been calculated using the GW approximation, using the Lindhard RPA for W. $\mathrm{Re}\{\Sigma\}$ shows structure on the scale of the plasmon energy, ω_p, (10–30 eV in typical metals) but in the neighbourhood of the quasiparticle energy (the intersection of the curve $\mathrm{Re}\{\Sigma\}$ with the straight line $\omega - E(k)$ (where $E(k)$ is the non-interacting eigenvalue $k^2/2$)) it is relatively

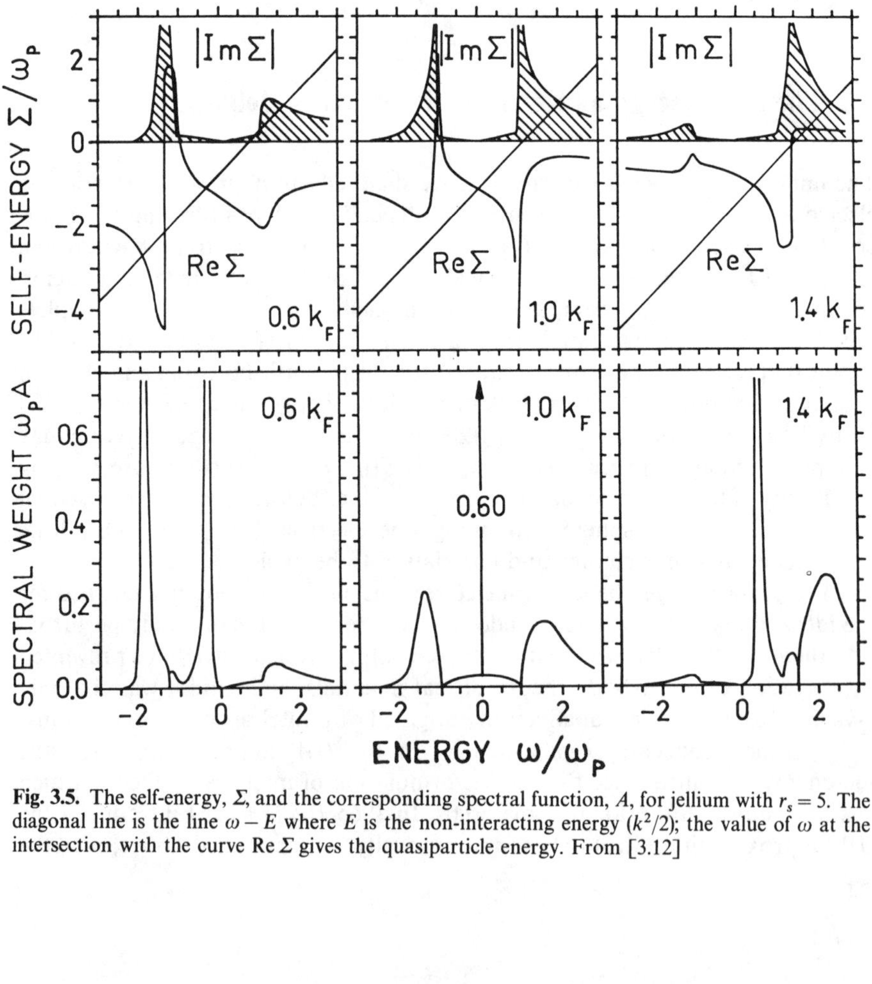

Fig. 3.5. The self-energy, Σ, and the corresponding spectral function, A, for jellium with $r_s = 5$. The diagonal line is the line $\omega - E$ where E is the non-interacting energy ($k^2/2$); the value of ω at the intersection with the curve $\mathrm{Re}\,\Sigma$ gives the quasiparticle energy. From [3.12]

smooth. Im$\{\Sigma\}$ characteristically goes to zero at the quasiparticle energy at $k = k_F$, resulting in the quasiparticle lifetime becoming infinite at the Fermi surface. This is caused by the limited amount of phase space available for electron–hole excitations across the Fermi surface. Although the peak in the spectral function corresponding to the quasiparticle is evident, not all the spectral weight is contained in this peak. The proportion of the spectral weight under the quasiparticle peak, also known as the renormalisation constant Z, is one measure of the strength of exchange and correlation effects, and is related to the energy derivative of the self-energy operator:

$$Z_k = \left(1 - \left(\frac{\partial \Sigma_k(\omega)}{\partial \omega}\right)_{\omega = E_k}\right)^{-1}. \tag{3.39}$$

Figure 3.6 shows the components of the quasiparticle energies in jellium as a function of k. The upper curves show the simple kinetic energy of the non-interacting electrons, $k^2/2$. The lower curves show the remainder of the quasiparticle energy, Σ_k. The relatively flat curves are the result of a full GW calculation for Σ_k, while the rapidly varying curves are what is termed "exchange only"; that is, the bare Coulomb interaction v rather than the screened Coulomb interaction W has been used to calculate Σ_k (corresponding to Hartree–Fock theory). This illustrates the importance of using the screened interaction, and

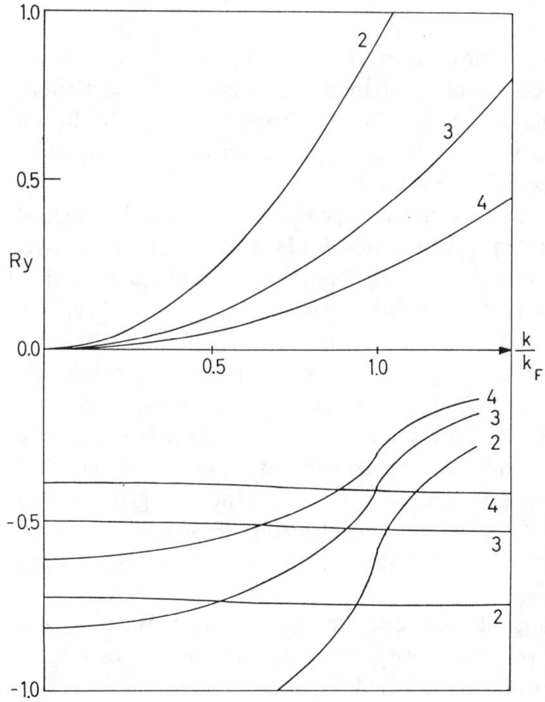

Fig. 3.6. Upper curves: the kinetic (i.e. non-interacting) energy of electrons in jellium with $r_s = 2, 3$ and 4. Lower curves: the quasiparticle energy corrections, in (i) bare exchange (steep curves) and (ii) the GW approxi-mation (almost flat curves). From [3.26]

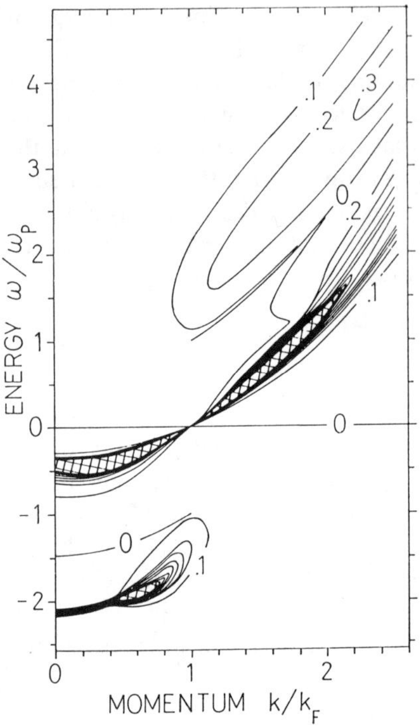

Fig. 3.7. Contour plot in the $k\omega$ plane of the spectral function $A(k,\omega)$, for jellium with $r_s = 4$. From [3.12]

shows that exchange and correlation terms have a significant amount of cancellation as far as the dispersion of the bands is concerned. In particular, the artificial logarithmic singularity in the Hartree–Fock bands at the Fermi energy (which results in the density of states falling to zero there) is completely removed by the use of the screened interaction.

In Fig. 3.7 the ingredients of the quasiparticle peaks may be seen: the spectral function is plotted as a contour map in the $\omega - k$ plane. As discussed above, the width of the quasiparticle peak (inversely proportional to the quasiparticle lifetime) goes to zero at the Fermi surface, but elsewhere is non-zero. Much of the spectral weight not associated with the quasiparticle peak can be seen to be in two broad areas. One, for $k < k_F$, at energies lower than ω_p below the quasiparticle band, is known as the *plasmaron* peak, and corresponds to a hole–plasmon resonance. The other, for $k > k_F$, at energies ground ω_p above the quasiparticle band, corresponds to an electron–plasmon resonance. The redistribution of weight from the quasiparticle peaks into the background, and the corresponding renormalisation of the one-electron density of states, is reflected in the quasiparticle density of states (the $\boldsymbol{k}$-integral of the spectral function) (Fig. 3.8).

Figure 3.9 shows the non-locality of the self-energy operator. With ω fixed at the Fermi surface, $4\pi r^2 r_s^2 \Sigma(r)$ (a quantity which, as a function of r/r_s, is independent of r_s for bare exchange (dashed line)) is plotted against r/r_s.

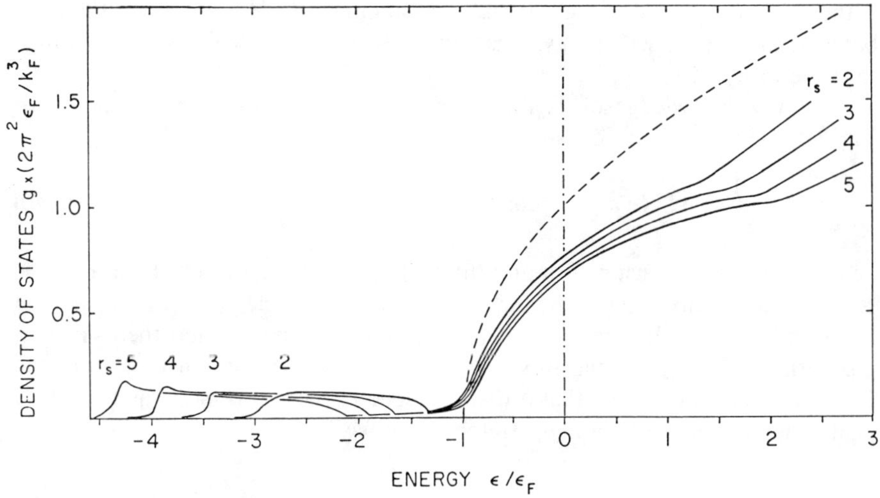

Fig. 3.8. The *GW* quasiparticle density of states of jellium with $r_s = 2-5$. The result in the simple Hartree approximation is also shown (dashed line). The vertical broken line denotes the Fermi level. From [3.2]

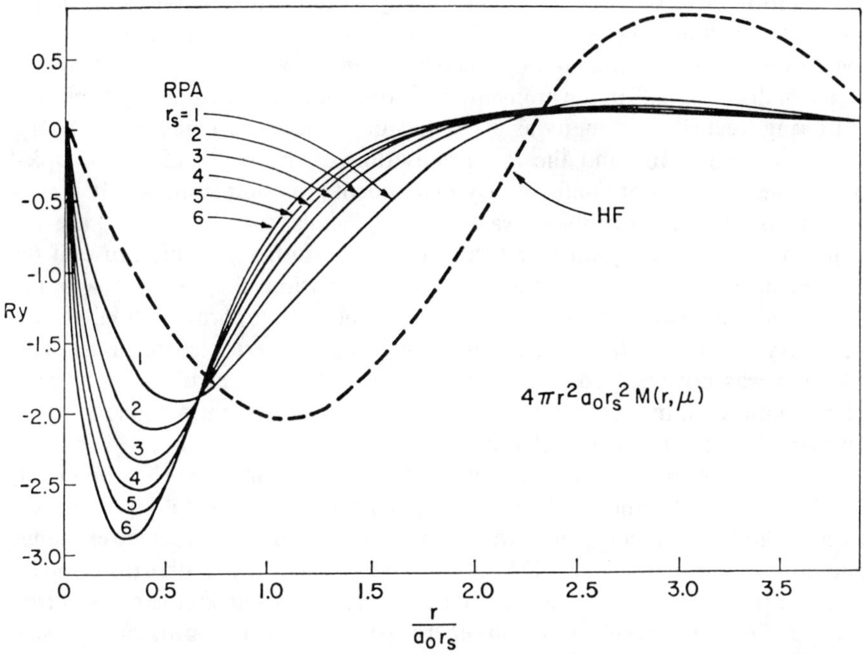

Fig. 3.9. The non-locality of the *GW* self-energy operator for jellium with $r_s = 1-6$. The bare exchange self-energy operator ("HF") is also shown, which in the combination plotted is independent of r_s. From [3.26]

Although the GW self-energy operator is significantly less non-local than the bare exchange (Gv) self-energy, it is still far from being local: its typical range is of the order of r_s.

From a knowledge of Σ, the Green's function G may be calculated. In Fig. 3.3, the momentum density

$$n(\boldsymbol{p}) = \frac{1}{8\pi^4} \mathrm{Im}\{\textstyle\int G(\boldsymbol{p},\boldsymbol{p}'=\boldsymbol{p},\omega)d\omega\} \tag{3.40}$$

has been plotted against $\boldsymbol{p}$. (For non-interacting electrons, this would be a simple step function from 1 to 0.) The fact that the quasiparticle peaks in the spectral function become delta functions at the Fermi energy means that there is still a discontinuity in $n(\boldsymbol{p})$ (equal in size to the renormalisation constant Z) at the Fermi surface, but the fact that not all the spectral weight is in the quasiparticle peak leads to the "tail" in $n(\boldsymbol{p})$ stretching above k_F.

3.3 Real Metals

3.3.1 Simple Metals: Effect of the Crystal Structure

The uniform electron gas (Sect. 3.2) is clearly an inadequate model of a simple metal. The seriousness of this for the many-body effects was known early in the development of many-body theory when it was recognised that the "renormalisation tail" in the momentum distribution, as measured by Compton scattering (regarded by many as a key test of the ideas of many-body theory) [3.27] (see Fig. 3.10), and the renormalisation of the width of the occupied bands [3.28] were not confirmed by experiments on simple metals. It is now known that the main problem was that real metals, such as sodium, are not uniform but have a crystal structure. However, another possible cause of the discrepancies was that the many-body perturbation theory used was not converged (for example, in the context of the GW approximation, it was necessary to include the next terms—the so-called vertex corrections). This question was not resolved for many years, until the advent of the techniques of computational many-body theory discussed above made it possible to apply many-body theory to real materials.

The GW approximation has been applied to the nearly-free-electron metals Li, Na, Al [3.14] and K [3.15], using pseudopotentials to represent the electron–ion core interaction. It was found to be important to include exchange and correlation (within the LDA) in the inverse dielectric matrix (as opposed to using the RPA inverse dielectric matrix), and also to make the one-electron energies in the Green's function energy denominator approximately self-consistent with the self-energy operator (as opposed to using the unperturbed LDA eigenvalues). However, the local field effects in the dielectric matrix (which are crucial for self-energies in semiconductors) were found to be unimportant,

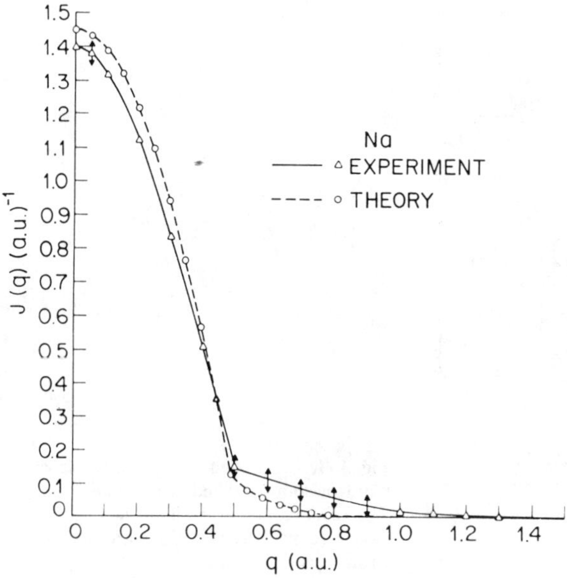

Fig. 3.10. Points: the measured Compton profile (the projection of the momentum density onto a given direction in $\boldsymbol{k}$-space) of sodium. Solid line: the calculated Compton profile of jellium of the corresponding density ($r_s = 3.96$). From [3.27]

presumably because of the relative homogeneity of the electron density within the unit cell compared with diamond-structure semiconductors. With two minor exceptions, good agreement with photoemission and soft X-ray emission experiments is obtained, within about 0.1 eV. Most notably, in Na the free-electron and LDA valence band widths are about 3.2 eV, while the measured band width is 2.5 eV. The *GW* calculations give 2.5 eV (Fig. 3.11). Previous attempts to model this many-body band narrowing using jellium had been unsuccessful (giving at most a 0.3 eV narrowing), which shows the importance of including the effect of the crystal structure, even in these free-electron-like metals. However, in Al the calculations gave a band width of 10.0 eV, 0.6 eV smaller than the experimental value from angle-resolved photoemission [3.29]. The authors speculate [3.14] that this may reflect a limitation of their plasmon-pole model for the frequency dependence of W, or of the use of the LDA to incorporate exchange and correlation into the dielectric function.

While computational many-body theory has allowed the many-body techniques that were once restricted to jellium to be applied to real materials, work on jellium has not stopped. Indeed, an exchange of ideas between model systems and real systems will probably continue to make a significant contribution to progress in the field. For example, recent work by *Mahan* and *Sernelius* [3.30] examines the role of vertex corrections (i.e. going beyond the first-order terms in the expressions for Σ and W and not simply setting the vertex function, Γ, to a delta function; see (3.9–14)). Their conclusion is that while the vertex corrections in jellium make a substantial difference to W and Γ individually, the net effect on Σ (which is $GW\Gamma$) is small. This is an encouraging result for practical many-body calculations for real materials. However, it does

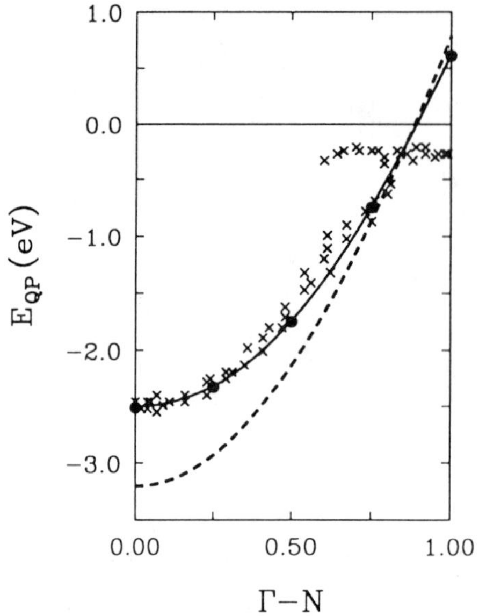

Fig. 3.11. Calculated *GW* quasiparticle valence band of sodium (filled circles and solid line). The LDA band (dashed line) and experimental data (crosses) (from [3.28]) are also shown. From [3.14]

cast doubt on whether attempts to go beyond the RPA in calculating W (such as the K_{xc} term discussed in Sect. 3.1.3) are useful, if W is to be used in a *GW* calculation.

3.3.2 Transition and Noble Metals and Rare Earths

The presence of the compact d-orbitals in transition and noble metals, and f-orbitals in rare earths, gives rise to stronger many-body effects than are found in simple metals and semiconductors. At present, it has not proved possible to apply first-principles many-body perturbation theory (such as the *GW* approximation) systematically to such materials; instead, progress has been made using simple model Hamiltonians that describe the electron–electron interaction in terms of a local orbital picture, with parameters such as the Hubbard U (the strength of the interaction between a spin-up electron and a spin-down electron on the same lattice site). Much elaboration of such model Hamiltonians is possible. For example, interactions between electrons in the localised orbitals (the d- or f-orbitals) and other ("band") electrons can be included. The various parameters are often fitted to experimental data, or (a recent innovation) calculated using constrained density-functional theory [3.31].

This empirical application of model Hamiltonians to describe strongly correlated materials has, of course, become a very large field in the last 20 years. Although it takes us away from our focus on the application of first-principles many-body perturbation theory to real materials, we will give one example: a recent model calculation for the temperature dependence of the photoemission spectrum of Ni [3.32]. The two parameters in the model (the Hubbard parameter

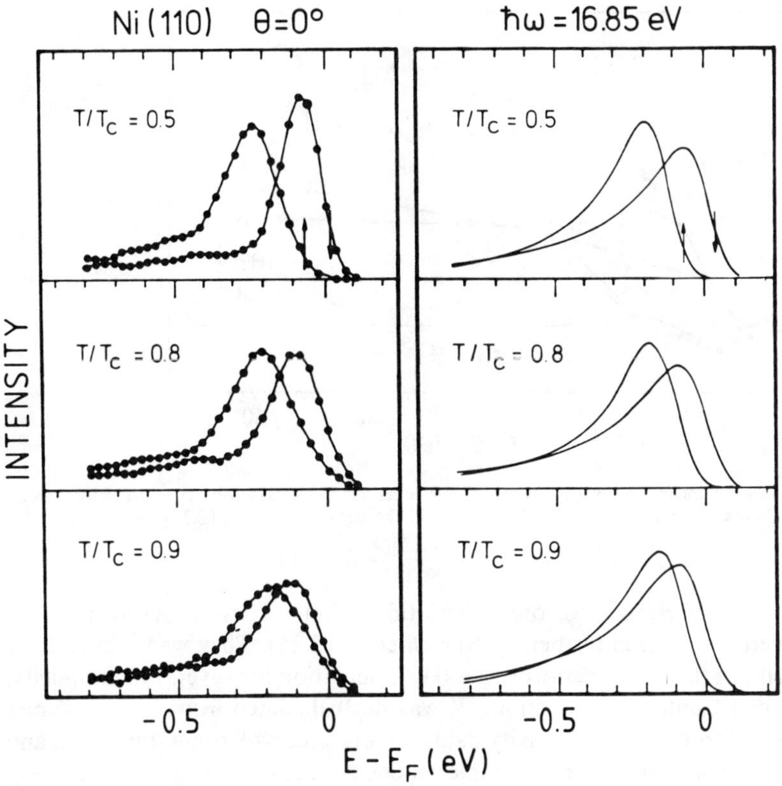

Fig. 3.12. Experimental (left) and calculated (right) spin-resolved normal emission photoemission spectra for ferromagnetic Ni(110) ($A \parallel [110]$) as a function of temperature. From [3.32]

U and the exchange parameter J) have been fitted to the experimental magnetic moment of the ferromagnetic ground state. Figure 3.12 shows the success of the model in describing the experimental photoemission spectrum.

Unoccupied bands in these metals also exhibit strong many-body effects. Figure 3.13 shows the discrepancies in peak energies[8] between bremsstrahlung isochromat spectroscopy (BIS) (or inverse photoemission) spectra [3.33] and LDA calculations [3.34] for four d-band metals (see also Chap. 10.) The quasiparticle energy corrections for a uniform electron gas (also shown) are quite different, which must be due to a combination of the crystal structure and the failure of the many-body perturbation theory used for the uniform electron gas

[8] Such discrepancies are sometimes loosely referred to as "self-energy corrections". However, strictly they are differences between quasiparticle energies and LDA eigenvalues, which, as (3.28) shows, is (to first order) equal to $\langle \Sigma(E) \rangle - \langle V_{xc}^{LDA} \rangle$. The confusion has arisen because of the fact that in the uniform electron gas $\langle V_{xc}^{LDA} \rangle$ is independent of the state and may be ignored; this is not true in real materials.

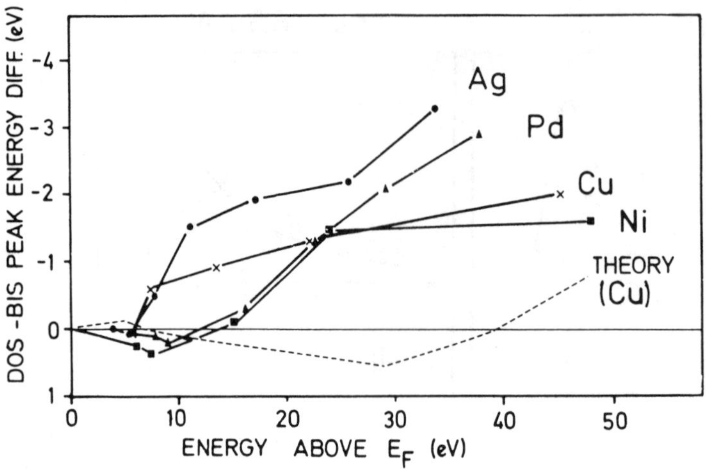

Fig. 3.13. Discrepancies in energy between the energies of peaks observed in BIS spectra of Ag, Pd, Cu and Ni and the corresponding peaks in LDA densities of states [3.34]. From [3.33]

to describe the relatively strong correlation effects. These experimental BIS results were considered further by *Horsch* et al. [3.25]. They went beyond the simple jellium theory by performing a GW calculation for silver. For simplicity, the screened Coulomb interaction, W, was not calculated from first principles but obtained from $q = 0$ reflectivity data, via a Kramers–Kronig transform, and extended to non-zero q using a simple model. Moreover, the Green's function, G, was modelled by the sum of a jellium term (to represent the s electrons) and other terms from the filled core states. The screened exchange contribution to the GW self-energy was then evaluated, which consequently consists of a free-electron-like term plus a core screened-exchange term. The results are shown in Fig. 3.14. It was found that the results with the core-screened exchange term included (curve b) were in good agreement with the experimental results. (Curve a, which omits the screening of the core exchange, is not very different from b, showing the relatively small effect of electron–electron screening on the length scale of the size of the atomic core.) The lesson is clear: once one gets more than a few eV from the Fermi level the self-energy (and therefore the quasi-particle energies) start to reflect the structure in the Green's function on an energy scale where deviations from free-electron behaviour are significant; in this case the energies corresponding to the energy differences between the core levels and the valence states.

This question of the state- and energy-dependence of the self-energy operator makes a comment on the "inner potential" of electron diffraction appropriate. In models of the diffraction of electrons by the crystal lattice in photoemission spectroscopy and low-energy electron diffraction, the effects of exchange and correlation are often described by part of an energy-independent, local, effective potential [3.36], in much the same way as in density-functional theory (3.21).

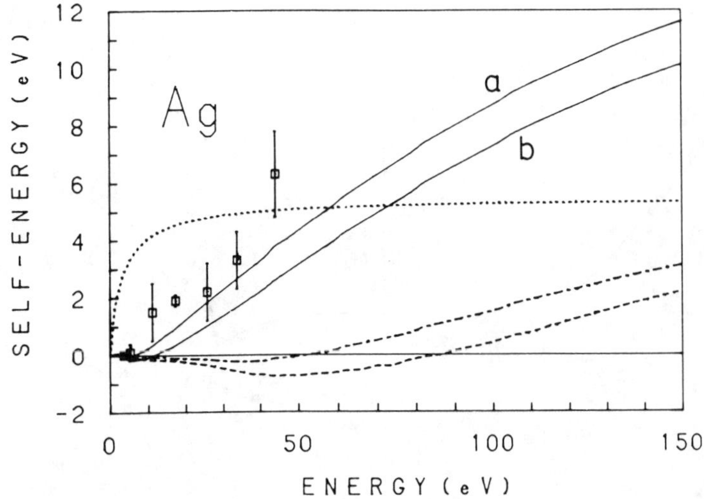

Fig. 3.14. The quasiparticle energy correction (to the LDA value) for unoccupied states in Ag. Points: experimental BIS data [3.33]. Other curves: various levels of approximation of the screened exchange self-energy operator, as follows. (i) Dotted curve: free-electron bare exchange, (ii) Dashed curve: as (i) but screened exchange, (iii) Dot-dash curve: as (ii) but including lifetime effects in the self-energy's energy denominators, (iv) Curve a: as (iii) but with unscreened core exchange added, (v) Curve b: as (iii) but with screened core exchange added. The last is in principle the most accurate, and is indeed in good agreement with the experimental data. From [3.35]

This, of course, is an approximation: the electron being diffracted is in reality a quasiparticle, and, as such, obeys the quasiparticle equation (3.17) in which exchange and correlation are represented by the self-energy operator. The effective potential is sometimes modelled by a "muffin-tin" form, in which the value of the potential in the interstitial region between the atoms is taken to be a constant, with the value of the constant known as the "inner potential". Since the true self-energy operator is non-local and energy dependent, it cannot be represented by a local, energy-independent potential, and the introduction of energy dependence of the inner potential is in part a necessary attempt to compensate for this. (In addition, it is also to compensate for the constraint of the muffin-tin representation.)

3.3.3 Collective Excitations and Lifetime Effects

The interaction between single–particle-like excitations and collective excitations such as plasmons causes satellites in many spectra (see, for example, Chap. 7 and Sect. 10.4), and is in principle contained within many-body theory. For example, at a plasmon energy above the Fermi energy an electron is able to decay by exciting a plasmon, and so is heavily damped: $\mathrm{Im}\{\Sigma\}$ becomes large, and the quasiparticle peak in the spectral function, $\mathrm{Im}\{G\}$, becomes greatly broadened. Consequently $\mathrm{Re}\{\Sigma\}$ also shows structure at this energy. Figure 3.15

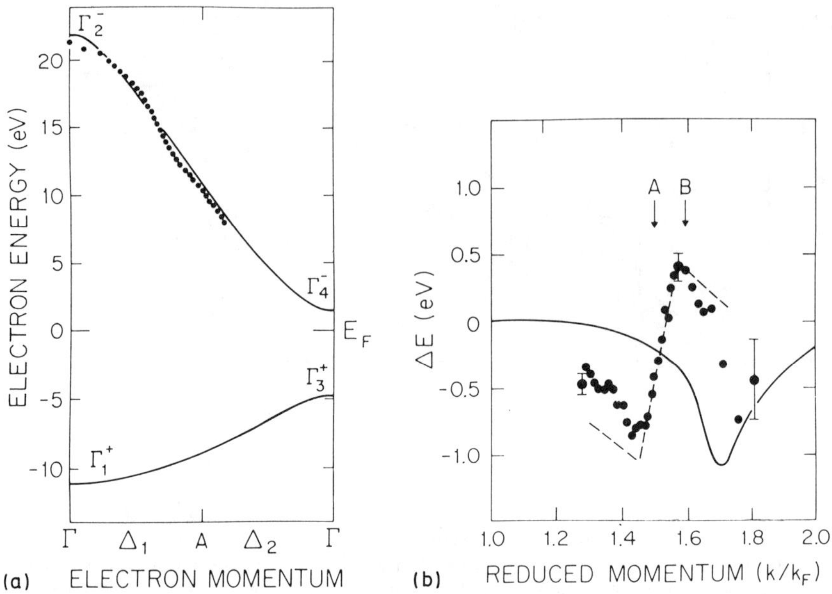

Fig. 3.15. Experimental and theoretical band structure of Be (from [3.37]). (**a**) The lower (occupied) band is empirically determined, and the upper (unoccupied) band is calculated using the local density approximation [3.38]. (**b**) The energy difference between the experimental and theoretical unoccupied bands (points): the structure is due to plasmon emission at the energies shown. The solid curve shows the corresponding structure in the many-body quasiparticle energy corrections for a uniform electron gas of the appropriate density [3.39]; this is unable to reproduce even the qualitative form of the feature in the real metal

illustrates the failure of calculations for a uniform electron gas to reproduce this structure in the quasiparticle bands correctly (see also Fig. 3.7), and again shows the necessity of including the effect of the crystal structure in many-body calculations.

The inelastic mean free path for the quasiparticle state in band n and wavevector $\boldsymbol{k}$ is proportional to the lifetime of the quasiparticle and is defined by

$$\frac{1}{l_{nk}} = -\frac{2}{v_{nk}} \operatorname{Im}\{E_{nk}\} \tag{3.41}$$

where v_{nk} is the usual quasiparticle group velocity and E_{nk} the quasiparticle energy. In general it depends on the direction of $\boldsymbol{k}$ as well as its magnitude. Within the GW approximation, this expression leads to an inverse mean free path proportional to an integral of the inverse dielectric function. The calculation may easily be done for jellium, and results for $r_s = 2$ are compared with experimental results for aluminium in Fig. 3.16. It is clear that the predicted mean free path is in good agreement with the experimental measurements in this case, and the same has been found for a variety of free-electron-like metals. First-principles calculations of the mean free path in other metals, semiconductors

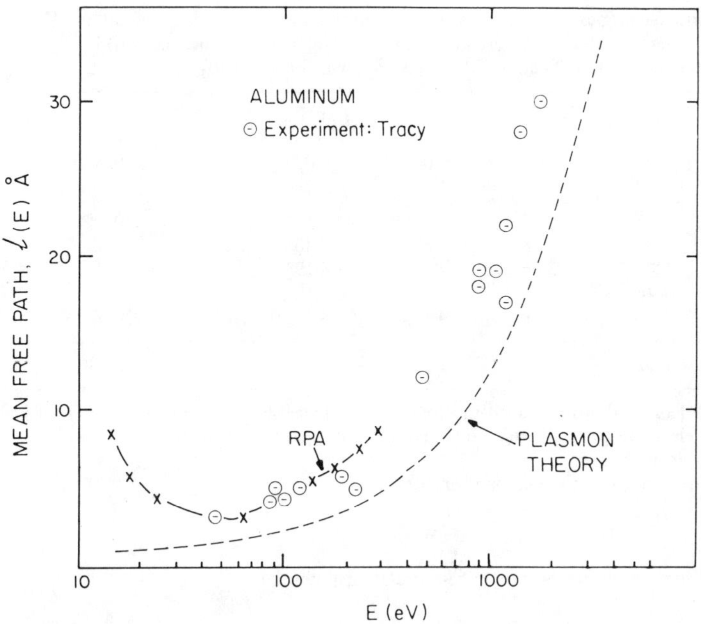

Fig. 3.16. Experimental measurements of the inelastic mean free path in aluminium (points; from [3.40]) compared with *GW* calculations of the inelastic mean free path in jellium of the relevant density ("RPA"; from [3.41]). (A simpler calculation, which includes only plasmons in the inverse dielectric function, is also shown.) From [3.1]

and insulators have not yet been performed, but *Tanuma* et al. [3.42] have calculated inelastic mean free paths for a variety of materials by combining an inverse dielectric function obtained from optical measurements (and extended to non-zero q using a LDA-like expression) with a free-electron-like Green's function to form a *GW* self-energy operator, much in the same spirit as *Horsch* et al. above [3.35].

3.4 Real Semiconductors and Insulators

3.4.1 Quasiparticle Energies in Semiconductors and Insulators

Most applications of the *GW* approximation to real materials have so far been to semiconductors and insulators. *Hybertsen* and *Louie* [3.16, 17, 21, 23] have studied silicon, LiCl, Ge and diamond. *Godby* et al. [3.10, 18–20] calculated the quasiparticle properties of Si, GaAs, AlAs and diamond. *Godby* and *Needs* [3.25] calculated the quasiparticle energies in compressed silicon as a function of volume. *Von der Linden* and *Horsch* [3.43] have performed calculations for Ge.

Table 3.2. Quasiparticle energies (GW) at Γ, X and L, and the minimum gap, in silicon, in eV. The experimental (Expt.) energies (at $T=0$ where known) and the DFT eigenvalues using (i) V_{xc}^{LDA}(LDA) and (ii) the V_{xc} calculated from Σ (Sect. 3.4.2) are also shown. (From [3.10])

	LDA[a]	DFT[b]	GW	Expt.	$GW-$DFT
Γ'_{25v}	−0.07	0.00	0.00	0.00[c]	0.00
L'_{3v}	−1.29	−1.21	−1.19	-1.2 ± 0.2[d], −1.5[e]	0.02
c-band min.	0.45	0.66	1.24	1.17[d]	0.58
Γ_{15c}	2.50	2.68	3.30	3.40[d]	0.62
Γ_{2c}	3.49	3.66	4.27	4.19[d]	0.61
L_{1c}	1.46	1.62	2.30	2.1[f], 2.4[e]	0.68
L_{3c}	3.30	3.49	4.11	4.3 ± 0.2[d], 4.0[d,e]	0.62
Min. gap	0.52	0.66	1.24	1.17[d]	0.58[g]

[a] Hedin RPA LDA [3.26]; the more accurate (but less comparable) Ceperley–Alder LDA [3.5] would increase the eigenvalues by 0.49 eV and leave the band gaps unaffected
[b] The V_{xc} calculated from Σ_{GW}
[c] Aligned with the quasiparticle valence-band maximum
[d] [3.44]
[e] [3.45]
[f] [3.46]
[g] Equal to Δ, the discontinuity in the exchange-correlation potential

Typical results, obtained for silicon, are shown in Table 3.2. (The column headed "DFT" will be discussed in the next section.) Most quasiparticle energies, whether near or far from the band gap, are within 0.1 eV of the experimental values. In contrast, the LDA energies are seen to be very poor. The features of the self-energy operator which give rise to these quasiparticle energy corrections are discussed in Sect. 3.4.3. A summary of the results obtained for four semiconductors (Si, GaAs, AlAs and diamond) is presented graphically in Fig. 3.17.

Systems other than bulk semiconductors have also now been studied: *Hybertsen* and *Louie* [3.23] have calculated the dispersion of quasiparticle surface states on As-terminated Ge, *Zhu* et al. [3.47] have calculated the quasiparticle surface band structure of relaxed GaAs (110) and *Hybertsen* and *Schlüter* [3.48] have studied quasiparticle bands in short-period Si/Ge (001) superlattices. Again, good agreement with experimental data was obtained.

All the above calculations were performed using first-principles norm-conserving pseudopotentials to represent the effect of the ion cores, which permits a simple plane-wave basis set to be used. An alternative approach has recently been demonstrated by *Hamada* et al. [3.49], who performed a GW calculation for Si using a linearised-augmented-plane-wave (LAPW) basis set, which enabled them to treat the true electron–ion potential: an "all-electron" calculation.

Among many proposed simple models which offer the possibility of achieving much of the accuracy of the full quasiparticle calculations without the computational effort are those of *Gygi* and *Baldereschi* [3.50] who made

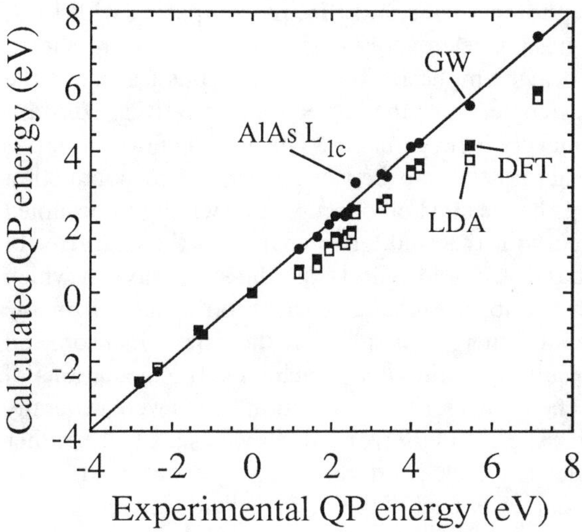

Fig. 3.17. The calculated quasiparticle energies of Si, GaAs, AlAs and diamond plotted against the experimental values. The 45° line is to guide the eye. For each experimental energy three calculated energies are shown: (i) the *GW* quasiparticle energy (filled circle), (ii) the LDA eigenvalue (open square) and (III) the calculated DFT eigen value (filled square) (see text). The *GW* quasiparticle energies are in good agreement with experiment, except for the L_{1c} conduction band minimum in AlAs, for which a revision of the accepted experimental value is proposed [3.20]. Data from [3.10]

use of the LDA and approximated the non-local-density terms, and *Hybertsen* and *Louie* [3.22] who proposed model dielectric matrices, obtained using a LDA-like procedure, which could subsequently be used in a *GW* calculation.

Quasiparticle energies in Si, Ge, GaAs and ZnSe over a large range of energies in both valence and conduction bands have been measured by *Jackson*

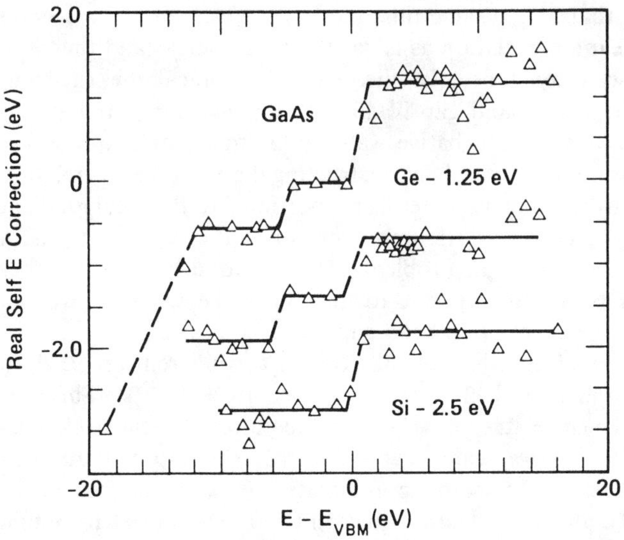

Fig. 3.18. The real part of the quasiparticle energy correction as a function of quasiparticle energy in three semiconductors, as deduced from XPS and BIS measurements. (For clarity, the curves have been offset vertically by the amounts indicated.) From [3.51]

and *Allen* [3.51] using X-ray photoemission and BIS. By comparing with LDA calculations, they deduced quasiparticle energy corrections, which are shown in Fig. 3.18. The guide lines drawn emphasize the most obvious feature of the measured quasiparticle energy corrections: the correction is relatively constant within a band, but changes sharply *between* bands. This was mentioned above in connection with changes across the gap between valence and conduction bands, but also applies to gaps between wholly occupied or wholly unoccupied states. This effect is associated with the qualitative change in the character of the wave function across a band gap, and reflects the different ways in which wave functions interact with the local exchange-correlation potential on the one hand, and the non-local self-energy operator on the other. Jackson and Allen also compared their results with the first-principles *GW* calculations of *Hybertsen* and *Louie* [3.21]. The results for the conduction bands were generally roughly within the estimated experimental error (which was said to be rather better than 0.2 eV). In the valence bands, good agreement was obtained for Si, but deviations were found in the lower part of the valence band for Ge. Jackson and Allen suggest that this may be caused by the breakdown of the plasmon-pole approximation for the dielectric function used in [3.21].

3.4.2 The Band Gap Problem

We have referred above, in Sects. 3.1.2 and 3.4.1, to the fact that band gaps of semiconductors and insulators calculated using conventional "electronic structure" methods (i.e. performing a LDA calculation and interpreting the DFT eigenvalues as quasiparticle energies) gives band gaps which are much too small. It was puzzling, then, that at the same time the LDA appeared to be a good treatment of exchange and correlation as far as the total energy (and therefore structural properties) were concerned. (In silicon, which is typical, the true band gap is 1.2 eV, while the LDA band gap is 0.52 eV.) Furthermore, attempts to improve upon the LDA in a perturbative way (by taking account of the fact that the real system was not uniform when constructing the exchange-correlation potential, for instance by making use of terms proportional to the local gradient of the electron density) gave very little improvement to the band gap [3.52]. This became known as the "band gap problem". The basic question was whether the error in the LDA band gap was due to the use of the LDA, or was an inherent error in even the exact DFT band gap.

The problem was resolved when calculations of the inherent error, Δ, in the DFT minimum band gap (3.32), were performed [3.10, 18, 19]. This was done by using (3.30), which relates the self-energy operator Σ to the exchange-correlation potential V_{xc}. Σ was calculated using the *GW* approximation as described above, and G and G^{DFT} were approximated by G^{LDA} (as done in the calculation of Σ itself), and the integral equation (3.30) was solved to obtain an exchange-correlation potential $V_{xc}(\boldsymbol{r})$. This potential, for silicon, is shown in Fig. 3.19 together with its LDA counterpart. The key to the band gap problem lies in the surprising similarity of these two potentials, which have been calculated in quite different ways.

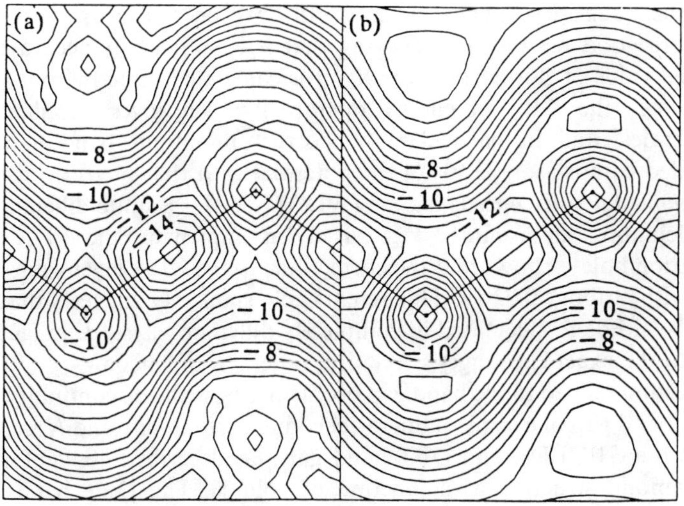

Fig. 3.19. The exchange-correlation potential of silicon calculated (**a**) from the self-energy operator, and (**b**) in the local-density-approximation. In each case the potential is shown in eV in the (110) plane containing the bond chains as indicated. From [3.10]

In Table 3.2, the column headed "DFT" shows a selection of the Kohn–Sham eigenvalues that would be obtained in a DFT calculation for silicon if the calculated $V_{xc}(\boldsymbol{r})$ were used instead of V_{xc}^{LDA} [3.10]. It may be seen that in each case these eigenvalues are very close to the LDA eigenvalues—as might be expected from the similarity of the potentials shown in Fig. 3.19. In particular, the minimum band gap is only very slightly greater than the LDA minimum band gap. Thus the calculated value of Δ, obtained by subtracting the "DFT" band gap from the GW band gap, is substantial: 0.72 eV. The same picture emerges for GaAs, AlAs and diamond: in each case Δ is about 80% of the LDA band gap error (see [3.10] for more details). The values of Δ obtained are summarised in Table 3.3.

It should be emphasized that the calculated exchange-correlation potential, although labelled DFT for brevity, is not the *exact* exchange-correlation

Table 3.3. Δ, the discontinuity in the exchange-correlation potential; Eg, the calculated minimum quasiparticle band gap; and the LDA gap error $E_g - E_g^{LDA}$ for four materials. (From [3.10])

Material	Δ(eV)	E_g(eV)	LDA gap error (eV)	Δ/LDA gap error
Si	0.58	1.24	0.72	81%
GaAs	0.67	1.58	0.91	74%
AlAs	0.65	2.18	0.81	80%
Diamond	1.12	5.33	1.43	78%

potential, since the quantities apppearing in (3.30) have been approximated. However, there are good reasons for believing that it is sufficiently accurate that the calculated value of Δ is correct. Firstly, it is calculated directly from a self-energy operator (the *GW* self-energy operator) which gives excellent quasiparticle energies, and in particular the correct minimum band gap. Secondly, a detailed comparison [3.10] of the potential with that provided for silicon by the *weighted density approximation* (WDA) [3.52]—a quite different approach to calculating an accurate V_{xc}, based on improving the LDA's exchange-correlation hole—shows that the corrections to the LDA eigenvalues are identical to within about 0.02 eV. This remarkable similarity suggests that the two potentials, although calculated in completely different ways, are in fact both very close to the exact exchange-correlation potential of silicon.

A further question related to the band gap problem is how the discontinuity, Δ, behaves as the band gap of a semiconductòr becomes zero. *Godby* and *Needs* [3.25] calculated the *GW*, LDA and DFT band gaps (as described above) for silicon in the diamond structure as a function of volume. The DFT bands, within the aproximations used, became metallic at a volume of 0.8 of the equilibrium volume (Fig. 3.20), while at the same volume the *GW* quasiparticle bands were non-metallic. This result is inconsistent with the previously asserted view that the DFT bands would become metallic at the same volume as the quasiparticle bands. Furthermore, it also shows that the DFT and quasiparticle Fermi surfaces are, in general, not the same[9].

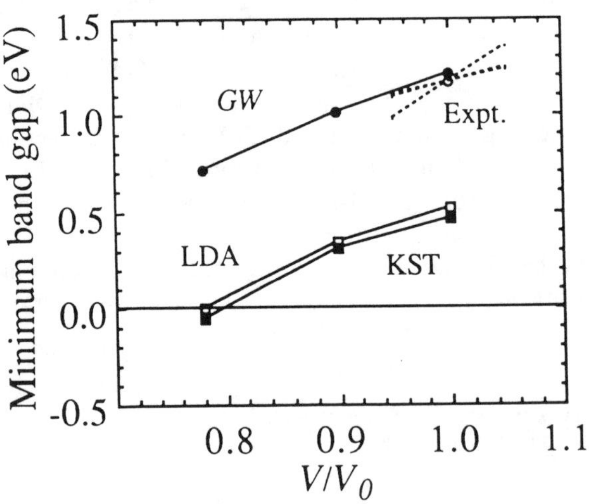

Fig. 3.20. The minimum band gap of silicon in the diamond structure plotted against the unit cell volume. Three calculated band gaps are shown: (i) the quasiparticle band gap (*GW*), which is in good agreement with the experimental gap and volume derivative (also shown), (ii) the LDA band gap (which, as is well known, is much too small) and (iii) the calculated DFT band gap (KST; see text), which is only slightly different from the LDA. From [3.25]

[9] The two are, however, identical for the special case of the uniform electron gas, being constrained by the isotropic symmetry to be spherical and by *Luttinger's* theorem [3.53] to have the same volume.

3.4.3 The Self-Energy Operator in Semiconductors and Insulators

Figure 3.21 shows the calculated self-energy operator in real space, for several semiconductors. In each case, $\Sigma(\boldsymbol{r}, \boldsymbol{r}', \omega = \text{mid-gap})$ is shown as a function of $\boldsymbol{r}'$ (in the (110) plane), with $\boldsymbol{r}$ fixed at the point shown. This should be interpreted as the "non-locality" of the operator felt by electrons near the band gap; for comparison purposes, if Σ were completely local, the plot would show a delta function centred at $\boldsymbol{r} = \boldsymbol{r}'$. It can be seen that the degree of non-locality is large: for instance, the range of Σ from $\boldsymbol{r} = \boldsymbol{r}'$ to the first zero is roughly the same size as the wavelength of the wave functions in the valence and conduction bands.

Non-locality is one way in which the self-energy operator differs from the local exchange-correlation potential of DFT. Another is energy dependence. This is illustrated in Fig. 3.22. Figure 3.22a shows the matrix elements (with respect to the LDA wave functions) of Σ and $V_{\text{xc}}^{\text{LDA}}$ individually as a function of the quasiparticle energy: there is no obvious pattern, since the matrix elements are strongly affected by the particular wave function in question. In contrast, Fig. 3.22b shows the matrix elements of $\Sigma(E) - V_{\text{xc}}^{\text{LDA}}$: the first-order-perturbation-theory quasiparticle energy corrections. For silicon, this is very similar to a step function, with the height of the step corresponding to the correction to the LDA

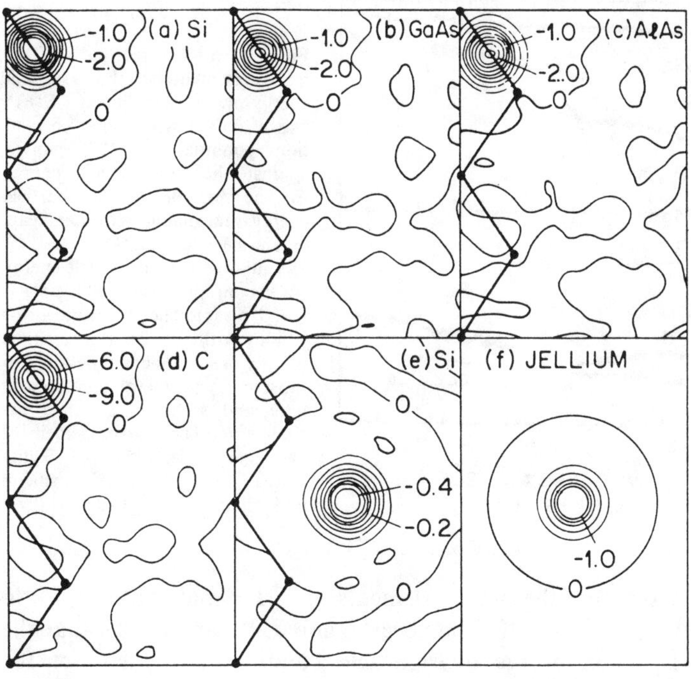

Fig. 3.21. Contour plots of the self-energy operator $\Sigma(\boldsymbol{r}, \boldsymbol{r}', \omega)$ in eV a.u.$^{-3}$ for $\boldsymbol{r}$ fixed at the point in the bond centre ((**a**)–(**d**)) and interstitial site (**e**). For comparison, the self-energy operator of jellium with $r_s = 2.0$ (the average density of silicon) (data from [3.26]) is shown in (**f**). From [3.10]

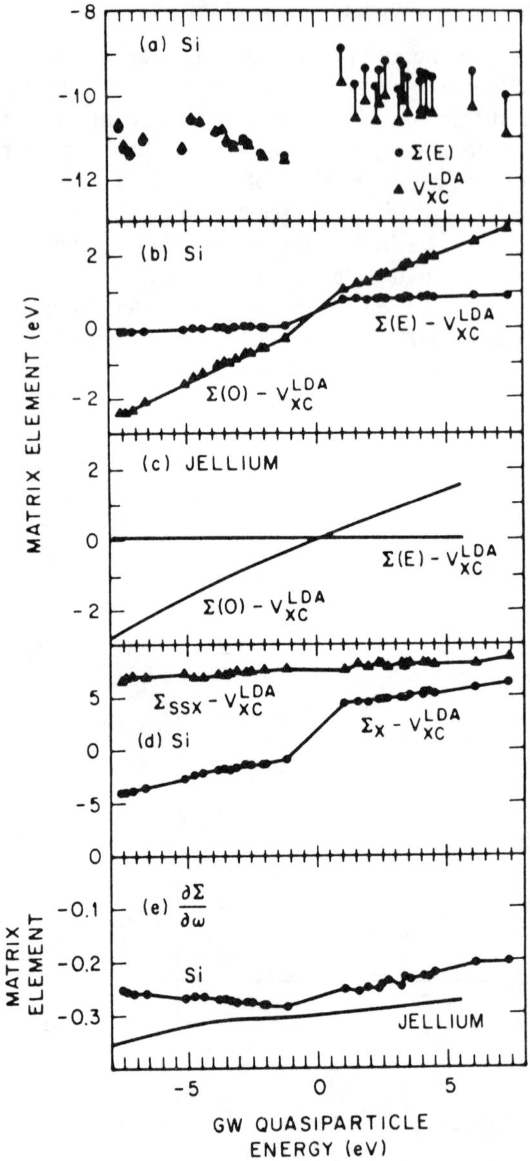

Fig. 3.22. (**a**) The real parts of the matrix elements of the *GW* self-energy operator in silicon $\langle \Sigma(E) \rangle$ and of the LDA exchange-correlation potential $\langle V_{xc}^{LDA} \rangle$, plotted against the quasiparticle energy E. The zero of the quasiparticle energy is the middle of the band gap. (**b**) The differences $\langle \Sigma(E) \rangle - \langle V_{xc}^{LDA} \rangle$ (essentially the quasiparticle energy correction) and also $\langle \Sigma(\text{mid-gap}) \rangle - \langle V_{xc}^{LDA} \rangle$. (**c**) The quantitites for jellium with $r_s = 2.0$ (the average density of silicon) corresponding to those in (a). (**d**) As (b), but showing bare and statically screened exchange. (**e**) The matrix element of the energy derivative of Sigma $\langle \psi(E) | \partial\Sigma(E)/\partial E | \psi(E) \rangle$ in silicon and jellium. From [3.10]

band gap. Also shown are the matrix elements of $\Sigma(E = \text{mid-gap}) - V_{xc}^{LDA}$, in which the energy dependence of Σ has been ignored. It can be seen that the effect on the dispersion of the quasiparticle bands, and the size of the quasiparticle band gaps, is great: the bands are much too wide, and the band gap too large. Thus the energy dependence of Σ is also crucial.

For comparison, Fig. 3.22c shows the corresponding quantities for jellium with the average density of silicon ($r_s = 2.0$). Of course, the abrupt step in the

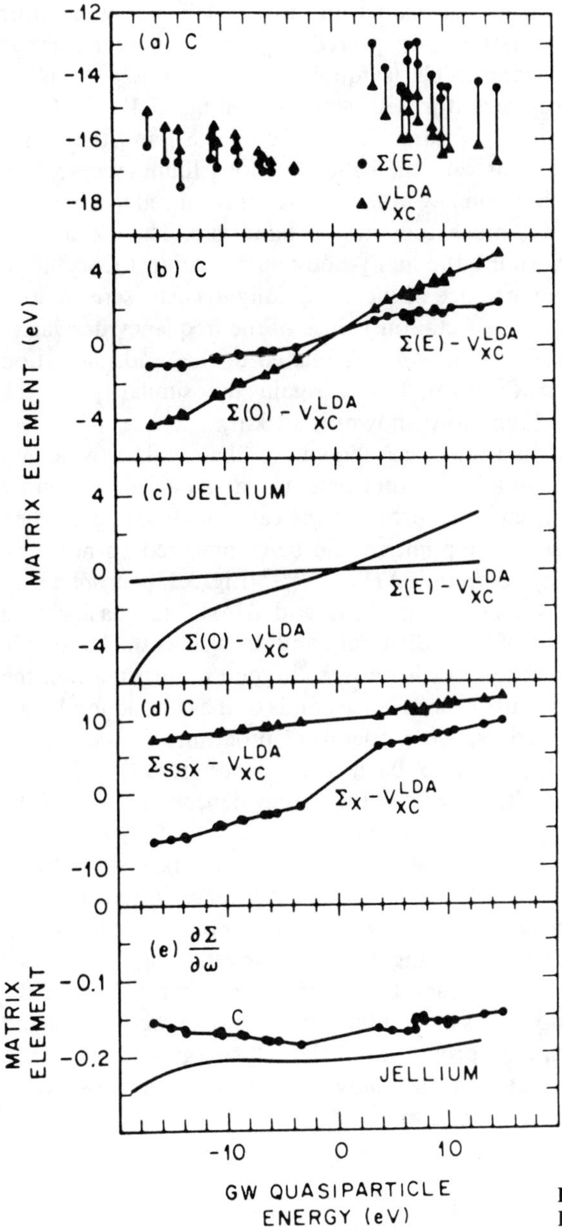

Fig. 3.23. As 3.22, but for diamond. From [3.10]

matrix elements across the band gap of silicon, which arises mainly from the qualitatively different nature of the wave functions at the top of the valence band and the bottom of the conduction band, is absent from the jellium case, but the underlying behaviour of $\Sigma(E) - V_{xc}^{LDA}$ and $\Sigma(E = \text{mid-gap}) - V_{xc}^{LDA}$ are very similar to silicon.

In Fig. 3.22d the effect of using approximations for the self-energy operator other than the *GW* approximation is explored. Σ_X is the bare exchange (Hartree–Fock) self-energy operator *Gv* (symbolically, as opposed to *GW*), whereas Σ_{SSX} is the self-energy operator with static screening, $G \cdot W(\omega = 0)$ Σ_X causes the band gap to be much too large (about 5 eV for Si) and the bands much too wide. (This is consistent with the behaviour for jellium (Sect. 3.2) in which a band gap (albeit of infinitesimal width) was even introduced in a metallic system.) In contrast, Σ_{SSX} gives much too small a band gap. These examples show the importance of performing the many-body perturbation theory in *W* rather than *v*, and also the importance of including fully *dynamic* screening.

Finally, Fig. 3.22e gives the expectation values of the frequency derivative of the self-energy operator, $\langle \partial\Sigma/\partial\omega \rangle$, which is related to the renormalisation constant for electrons at that energy (3.39). Again the similarity to the corresponding quantity for jellium (also shown) is striking.

In Fig. 3.23, the same quantities are shown for diamond, a wide-gap semiconductor. The same qualitative picture emerges, and indeed, the wider band gap is accompanied by even larger errors in the calculated band gap when the LDA, Σ_X or Σ_{SSX} are used. One point should be commented on here: the calculated quasiparticle energy correction $\Sigma(E) - V_{xc}^{LDA}$ (Fig. 3.23b) is *not* a step function in diamond (as it is also not in GaAs and AlAs): the quasiparticle correction alters the dispersion of the individual bands as well as the band gap. The fact that the energy correction in silicon is, to a good approximation, the same for different states within either the valence bands or the conduction bands (which has led to the empirical "scissors operator" procedure for correcting LDA band structures [3.54] seems to be a coincidence, resulting from a fortuitous cancellation between the effects of the energy dependence and of the non-locality of the self-energy operator.

In conclusion, the self-energy operator in semiconductors is seen to be similar in many ways to that in jellium. The main difference in the operator itself is a variation in the strength of the self-energy from point to point within the unit cell (though not its non-local range), due so the inhomogeneity of the electron distribution. However, the qualitative difference in the nature of the quasiparticle energy corrections in semiconductors—the opening of the band gap—is largely not due to special properties of the self-energy operator itself, but rather to the interaction between the change in the wave functions across the band gap with the self-energy operator.

3.5 Concluding Remarks

The calculations described in this chapter have shown that it is now possible to treat exchange and correlation from first principles for a range of solid state systems. By combining the mathematical results of formal many-body theory with the techniques of electronic structure calculations, it is possible to calculate quasiparticle energies and other properties of the Green's function, such as the ground-state total energy and the momentum density. Of the many challenges that lie ahead in this field, two are paramount. Firstly, the extension of the methods to transition metals, oxides, and other more strongly correlated materials, which are currently understood only at the level of simple model Hamiltonians such as the Hubbard model (Sect. 3.3.2). This will require the developments of new computational techniques and probably the extension of the current many-body perturbation theory. Secondly, the investigation of many-body phenomena in more complex systems, such as the structures of metals and semiconductors that can be made using modern device fabrication techniques.

References

3.1 G.D. Mahan: *Many-Particle Physics* (Plenum, New York 1981)
3.2 L. Hedin, S. Lundqvist: In *Solid State Physics*, Vol. 23, ed. by H. Ehrenreich, F. Seitz, D. Turnbull (Academic, New York 1969), p. 1
3.3 P. Hohenberg, W. Kohn: Phys. Rev. **136**, B864 (1964)
3.4 W. Kohn, L.J. Sham: Phys. Rev. **140**, A1133 (1965)
3.5 D.M. Ceperley, B.I. Alder: Phys. Rev. Lett. **45**, 566 (1980); parameterised in J.P. Perdew, A. Zunger: Phys. Rev. **B 23**, 5048 (1981)
3.6 See, for example, *Theory of the Inhomogeneous Electron Gas*, ed. by S. Lundqvist, N.H. March (Plenum, New York 1983) and references therein
3.7 M. Levy, J.P. Perdew, V. Sahni: Phys. Rev. **A 30**, 2745 (1984); C.-O. Almbladh, U. von Barth: Phys. Rev. **B 31**, 3231 (1985)
3.8 L.J. Sham, M. Schlüter: Phys. Rev. Lett. **51**, 1888 (1983); Phys. Rev. **B 32**, 3883 (1985)
3.9 J.P. Perdew, M. Levy: Phys. Rev. Lett. **51**, 1884 (1983)
3.10 R.W. Godby, M. Schlüter, L.J. Sham: Phys. Rev. **B 37**, 10159 (1988)
3.11 e.g. P. Singhal, J. Callaway: Phys. Rev. **B 14**, 2347 (1976)
3.12 B.I. Lundqvist: Physik Kondensierten Materie **7**, 117 (1968)
3.13 J. Lindhard: Dan. Math. Phys. Medd. **28**, No. 8 (1954)
3.14 J.E. Northrup, M.S. Hybertsen, S.G. Louie: Phys. Rev. Lett. **59**, 819 (1987); Phys. Rev. **B 39**, 8198 (1989)
3.15 M.P. Surh, J.E. Northrup, S.G. Louie: Phys. Rev. **B 38** 5976 (1988)
3.16 M.S. Hybertsen, S.G. Louie: Phys. Rev. Lett. **55**, 1418 (1985)
3.17 M.S. Hybertsen, S.G. Louie: Phys. Rev. **B 32**, 7005 (1985)
3.18 R.W. Godby, M. Schlüter, L.J. Sham: Phys. Rev. Lett. **56**, 2415 (1986)
3.19 R.W. Godby, M. Schlüter, L.J. Sham: Phys. Rev. **B 36**, 6497 (1987)
3.20 R.W. Godby, M. Schlüter, L.J. Sham: Phys. Rev. B (Rap. Com.) **35**, 4170 (1987)
3.21 M.S. Hybertsen, S.G. Louie: Phys. Rev. **B 34**, 5390 (1986)
3.22 M.S. Hybertsen, S.G. Louie: Phys. Rev. **B 38**, 4033 (1988)
3.23 M.S. Hybertsen, S.G. Louie: Phys Rev. Lett. **58**, 1551 (1987); Phys. Rev. **B 38**, 4033 (1988)
3.24 W. von der Linden, P. Horsch: Phys. Rev. **B 37**, 8351 (1988)
3.25 R.W. Godby, R.J. Needs: Phys. Rev. Lett. **62**, 1169 (1989)
3.26 L. Hedin: Phys. Rev. **139**, A796 (1965)
3.27 P. Eisenberger, L. Lam, P.M. Platzman, P. Schmidt: Phys. Rev. **B 6**, 3671 (1972)
3.28 E. Jensen, E.W. Plummer: Phys. Rev. Lett. **55**, 1912 (1985)
3.29 H.J. Levinson, F. Greuter, E.W. Plummer: Phys. Rev. **B 27**, 727 (1983)
3.30 G.D. Mahan, B.E. Sernelius: Phys. Rev. Lett. **62**, 2718 (1989)
3.31 O. Gunnarsson, A.V. Postnikov, O.K. Andersen: Phys. Rev. **B 40**, 10407 (1989)
3.32 W. Nolting, J. Braun, G. Borstel , W. Borgiel: Physica Scripta **41**, 601 (1990)
3.33 W. Speier, R. Zeller, J.C. Fuggle: Phys. Rev. **B 32**, 3597 (1985)

3.34 V.L. Moruzzi, A.R. Williams, J.F. Janak: Phys. Rev. **B 15**, 2584 (1977)
3.35 P. Horsch, W. von der Linden, W.-D. Lukas: Solid State Comm. **62**, 359 (1987)
3.36 M.S. Woolfson, S.J. Gurman, B.W. Holland: Surf. Sci. **117**, 450 (1982)
3.37 E. Jensen, R.A. Bartynski, T. Gustafsson, E.W. Plummer: Phys. Rev. Lett. **52**, 2172 (1984)
3.38 F.J. Himpsel: Adv. Phys. **32**, 1 (1983)
3.39 L. Hedin, B.I. Lundqvist: J. Phys. **C 4**, 2064 (1971)
3.40 J.C. Tracy: J. Vac. Sci. Technol. **11**, 280 (1974)
3.41 B.I. Lundqvist: Phys. Stat. Sol. **32**, 273 (1969)
3.42 S. Tanuma, C.J. Powell, D.R. Penn: Surf. Interf. Anal. **11**, 577 (1988)
3.43 W. Von der Linden, P. Horsch: Phys. Rev. **B 37**, 8351 (1988)
3.44 *Physics of Group IV Elements and III–V Compounds*, ed. by O. Madelung, M. Schulz, H. Weiss, *Landolt-Börnstein, Numerical Data and Functional Relationships in Science and Technology*, New Series Group III, Vol. 17a (Springer, New York 1982)
3.45 D. Straub, L. Ley, F.J. Himpsel: Phys. Rev. Lett. **54**, 142 (1985)
3.46 R. Hulthen, N.G. Nilsson: Solid State Commun. **18**, 1341 (1976)
3.47 X Zhu, S.B. Zhang, S.H. Louie, M.L. Cohen: Phys. Rev. Lett. **63**, 2112 (1989)
3.48 M.S. Hybertsen, M. Schlüter, R. People, S.A. Jackson, D.V. Lang. T.P. Pearsall, J.C. Bean, J.M. Vandenberg, J. Bevk: Phys. Rev. **B 37**, 10195 (1988); M.S. Hybertsen, M. Schlüter: Phys. Rev. **B 36**, 9683 (1987)
3.49 N. Hamada, M. Hwang, A.J. Freeman: Phys. Rev. **B 41**, 3620 (1990)
3.50 F. Gygi, A. Baldereschi: Phys. Rev. Lett. **62**, 2160 (1989)
3.51 W.B. Jackson, J.W. Allen: Phys. Rev. **B 37**, 4618 (1988)
3.52 M.S. Hybertsen, S.G. Louie: Phys. Rev. **B 30**, 5777 (1984)
3.53 J.M. Luttinger: Phys. Rev. **119**, 1153 (1960)
3.54 G.A. Baraff, M. Schlüter: Phys. Rev. **B 30**, 3460 (1984)

4. Theory of Correlated Holes (and Electrons): from Satellites to Luttinger Liquids

Jan Zaanen

With 32 Figures

The study of strongly correlated electron systems is still at a stage where the search is for the basic principles. Unoccupied state spectroscopies have started to play a significant role in this field, and on more than one occasion break-throughs occurred after new experimental data became available. However, it is the *combination* of high-energy spectroscopy for both occupied and unoccupied states which has had the most impact, and therefore we shall consider states below and above the Fermi energy in this chapter.

A fundamental property of the electronic state is the single electron Green's function. This Green's function, the starting point for most many-body descriptions of the interacting electron system, is at zero temperature defined as ($|N_0\rangle$ being the N-particle ground state with energy E_0)

$$G_1(z,\boldsymbol{k}) = \left\langle N_0 \left| c_{\boldsymbol{k}}^{\dagger} \frac{1}{z+H-E_0} c_{\boldsymbol{k}} \right| N_0 \right\rangle + \left\langle N_0 \left| c_{\boldsymbol{k}} \frac{1}{z-H+E_0} c_{\boldsymbol{k}}^{\dagger} \right| N_0 \right\rangle . \tag{4.1}$$

The object which we wish to interpret is its imaginary part, the single electron spectral function

$$\begin{aligned} A(\boldsymbol{k},\omega) &= \frac{1}{\pi} \lim_{0^+ \to 0} \operatorname{Im}\{G_1(\boldsymbol{k},\omega - \mathrm{i}0^+)\} \\ &= \sum_f |\langle N_-^f | c_{\boldsymbol{k}} | N_0 \rangle|^2 \delta(\omega + E_-^f - E_0) \\ &\quad + \sum_f |\langle N_+^f | c_{\boldsymbol{k}}^{\dagger} | N_0 \rangle|^2 \delta(\omega - E_+^f + E_0), \end{aligned} \tag{4.2}$$

where $|N_+^f\rangle$ and $|N_-^f\rangle$ are the fth eigenstates of the system with one electron added or removed, respectively, at energies E_+^f and E_-^f. In other words, this spectral function measures the probability of adding or removing an electron to the system at energy ω. When the sudden approximation holds, that is, when the kinetic energy of the outgoing/incoming electron is large, this is exactly the quantity which we measure in (angle-resolved) photoelectron spectroscopy (ARPES) and (angle-resolved) inverse photoemission (ARIPES). This spectral function is the main subject of this chapter. Although they are less direct because they involve a core hole, the spectral functions for core level photoemission (c-XPS) and X-ray absorption spectroscopy (XAS) offer useful information too—in the end it is the combination of all these spectroscopies which makes our interpretations trustworthy.

Traditionally, studies of correlated systems concentrated on macroscopic properties, such as transport, susceptibility, specific heat and so on. It was believed that away from the ground state and the low-lying excitations, the problem was too complicated to be understood in any detail. However, thanks to high-energy data, it became more and more clear that the physics on a scale of $\sim 20\,\text{eV}$ around the Fermi energy is much simpler than expected. In fact, simple models seem to tell us rather precisely what we see in the spectra. This works so well that a belief has emerged that a 'top-to-bottom' approach might work: starting out from the understanding of the physics on short time- and length-scales, one might attempt to deduce the proper model for macroscopic physics. The seminal work in this respect has been the demonstration by *Gunnarsson* and *Schönhammer* that all known high-energy data (core-XPS, XAS, valence band photoemission and inverse photoemission) of Ce intermetallics could be explained with a single underlying (Anderson impurity) model [4.1–3]. Moreover, the renormalization down to the lowest energy scales could be followed in detail, showing that this model could explain everything from the full valence band scale down to energies below the Kondo temperature [4.3, 4]. In the wake of this development, *Sawatzky*, *Fujimori* and coworkers repeated this exercise for the classic 3d Mott–Hubbard insulators, and found out that the approach is equally successful for these materials [4.5–7].

With non-interacting electrons, (4.2) corresponds directly to the band structure. The interactions lead to more complicated electron states, and some of the features in (4.2) might still resemble ones in the non-interacting system (quasiparticle (QP) states). However, the overall shape of (4.2) can be quite different from the non-interacting band structure, and new peaks show up which cannot be explained at all with single-particle theory ('satellites'). The reason is that the hard core atomic repulsions survive to some extent in the solid. If these interactions are larger than the bandwidth, the single-particle bands split up into Hubbard bands, and these are seen in the spectra as satellites. By neglecting the effects of hybridization on these bands altogether, one can already obtain an impression of some general features of the strongly correlated state, as will be discussed in Sect. 4.1. In order to arrive at a realistic description of the correlated state, this picture has to be augmented with multiplet splittings, band formation and, last but not least, the effects of hybridization. Using impurity models, one can account for at least the overall shape of the spectra of real materials, as shown in Sect. 4.2. Although the focus point of this section is PES-IPES, I will also discuss XAS and c-XPS in cases where they offer simple examples of principles of more general interest. This has been elucidated for the $3d$-insulators and the $4f$-systems. I have chosen the former as illustration material, because the effect of hybridization is more outspoken, and they offer in this respect more general examples. Furthermore, the $4fs$ have already been reviewed extensively [4.3, 8–10]. Although the $4fs$ have much in common with the $3d$ insulators, the $4f$ intermetallics are different in the respect that their low-energy physics is dominated by the Kondo effect, causing the Kondo resonance in PES-IPES. As I will show in Sect. 4.3, it is rather easy to understand this phenomenon using the $1/N$ expansion technique.

The near-ground-state electronic structure of a strongly correlated impurity is well understood, but this is not the case for a lattice of correlated ions. This problem has been under attack for a long time and recently received new impetus by the discovery of the heavy fermion systems and, especially, the high T_c superconductors. For the heavy fermion systems, firm evidence is present in favour of a Fermi liquid (FL) description [4.11]. On the other hand, it has been argued that in the high T_c cuprates, which are from a 'high energy' perspective strongly correlated (see e.g. [4.12]), a non-FL state is realized [4.13, 14]. Thanks to the improved experimental resolution, PES-IPES has recently started to play a role as a probe for the low-energy electronic structure [4.15]. In Fig. 4.1 I show the low-energy region of the ARPES data of the 80 K superconductor $Ba_2Bi_2Ca_1Cu_2O_8$, as measured by *Olsen* et al. [4.16] (for similar ARIPES data with lower resolution, see [4.17]). At $\theta = 12°$ (close to the LDA k_F) a peak

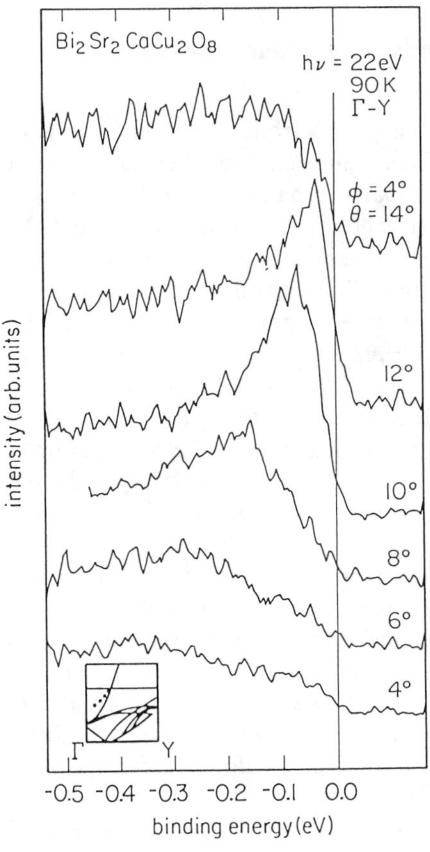

Fig. 4.1. Angle-resolved photoemission data for the 80 K high T_c superconductor $Bi_2Sr_2CaCu_2O_8$, for momentum directions parallel to the Cu—O planes [4.16]

crosses E_F and at smaller θs the peak moves away from E_F, at the same time picking up width. At first sight, this is what one expects for a Fermi liquid. On the other hand, one could imagine that the peak is actually a cusp in the incoherent background, i.e. below the cusp $A(\omega) \sim 1/\omega$ and above $\sim \omega$. Obviously, such a system would not be a Fermi liquid and this behaviour would be more reminescent of a so-called Luttinger liquid [4.18].

Also on the theoretical side, there has been recently substantial progress on the problem of band formation. Although the link with experiment remains to be established, these new results at least stretch our imagination about what can happen in the strongly correlated electronic state, and offer a perspective for looking at the PES-IPES data. In the second half of Sect. 4.3 I will touch on the renormalized band pictures based on the slave boson techniques, which are believed to make sense for the heavy fermion systems. In Sect. 4.4 I will discuss a number of rigorous results for the single-electron spectral function as deduced from the simplest $(t-J)$ model for correlated bands, including the recently discovered 'Landau–Luttinger liquid state', which is realized in one dimension.

4.1 The Principle of Hubbard Bands, and Their Systematics

In many materials the interactions between the electrons play a minor role. In the first instance, the electron–electron interactions can be taken into account in an average way, leading to the single-particle bands of the local density approximation (LDA). Comparison with experiment shows that these LDA electrons make a lot of sense, although there are still some minor imperfections left, and these 'shifts and broadenings' are well described by the lowest order correction to LDA (GW, see Chap. 3). This averaging out of the electron–electron interactions is not at all obvious, keeping in mind that these interactions are dominant in isolated atoms. *The essence of the strong correlation problem is that in some (mostly 3d, 4f, 5f) materials these atomic electron–electron interactions do not disappear completely.* In the 4fs the (quasi-) atomic nature of the electrons is most outspoken. The 4f PES-IPES of an isolated, say, Gd atom can be understood as follows: in the neutral state, the Gd atom has a $4f^7 6s^2$ configuration and, because of the exchange, the spins of the f-electrons would be aligned in a ^8S multiplet. Upon adding or removing an electron in the f-angular momentum channel, the atom would end up in a $4f^8 6s^2$ or a $4f^6 6s^2$ configuration, respectively. In both cases, one expects to see a large number of lines, because fractional parentage rules tell us that many of the f^6 and f^8 multiplets can be reached, starting from a $f^7(^8S)$ state. Furthermore, the f^6 and f^8 manifolds would lie well below and above the Fermi energy, respectively, because the f-ionization energy and the first f-electron affinity are of the order of Rydbergs.

In fact, this argument works as well for the rare earth *solids*. In Fig. 4.2 the PES–IPES spectra of the $4f$ metals are shown, as obtained by *Baer* and

Schneider [4.9]. The bars come from atomic multiplet calculations, and the drawn lines are obtained by broadening these with experimental Gaussian and life-time Lorentzians. The splitting of the f^{n-1} and f^{n+1} 'bands',

$$U = E(f^{n+1}) + E(f^{n-1}) - 2E(f^n), \tag{4.3}$$

is caused by the strong on-site electrostatic Coulomb interaction between the f-electrons: it costs U additional energy to change the f occupancy by two

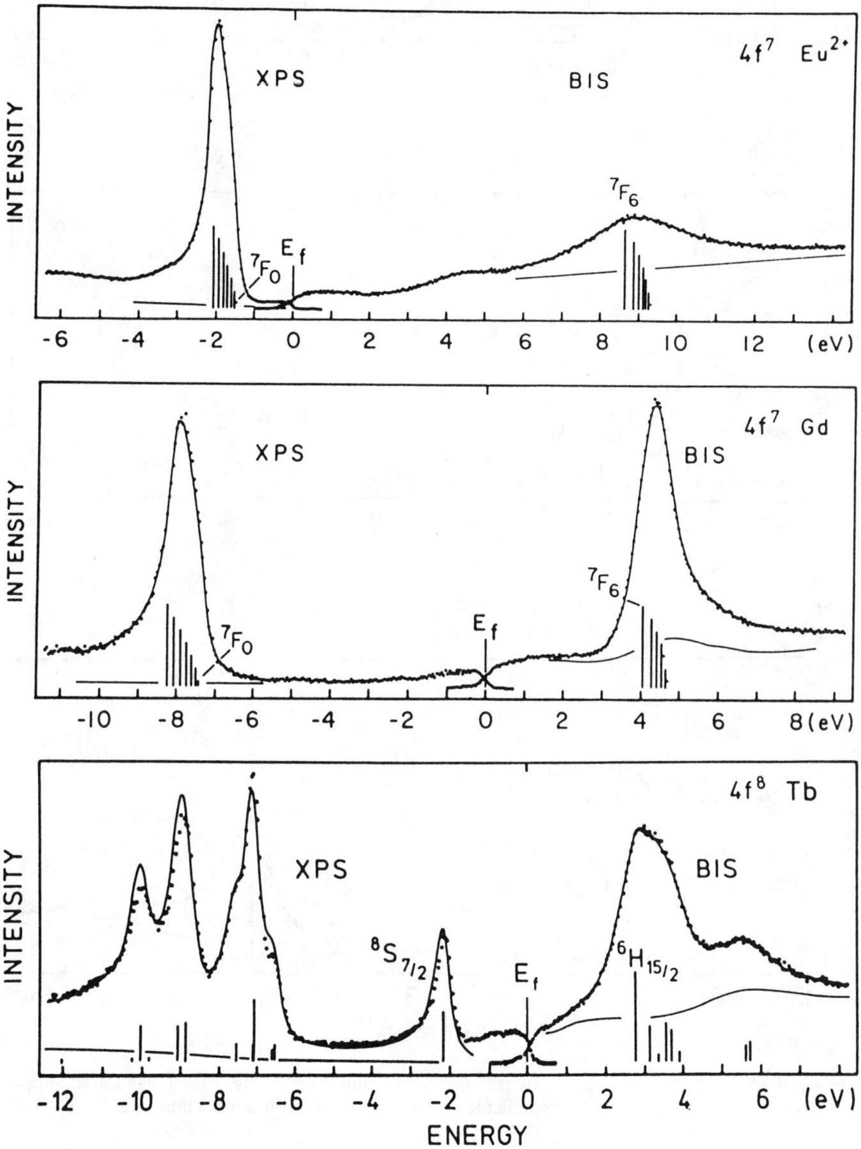

Fig. 4.2. (*Continued*)

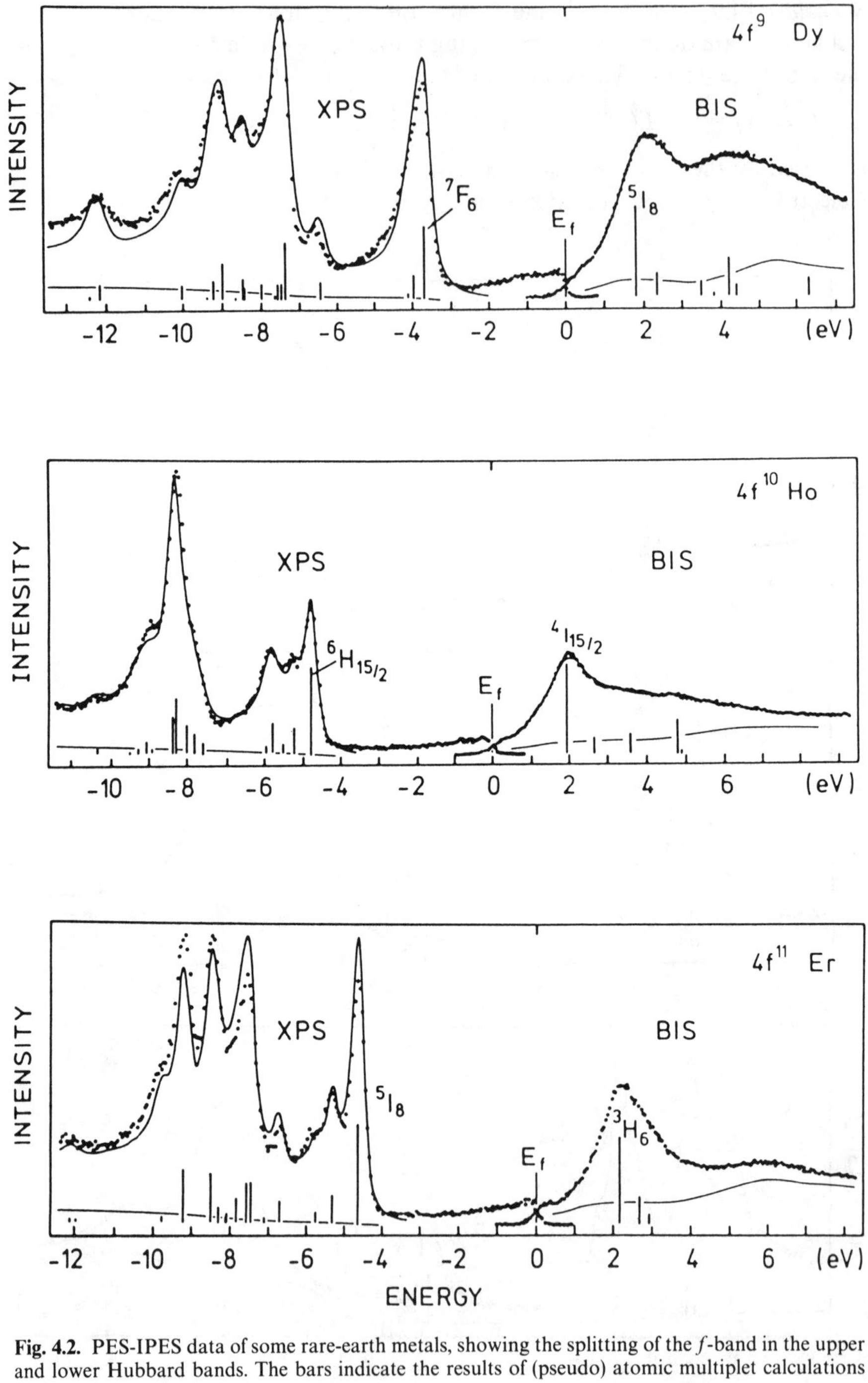

Fig. 4.2. PES-IPES data of some rare-earth metals, showing the splitting of the f-band in the upper and lower Hubbard bands. The bars indicate the results of (pseudo) atomic multiplet calculations [4.9]

electrons (e.g. $f^6 \to f^8$), compared to adding only a single electron (either $f^6 \to f^7$ or $f^7 \to f^8$). Apart from the broadenings and the broad background, coming from the $5d$–$6s$, p-derived LDA-like bands, the difference with the free atom is that U is reduced by a factor of ~ 3. The reason is that the bare U of (4.3) does not exist on the time scale we probe experimentally; it gets renormalized or 'screened'. This renormalization is of the simplest sort, as *Herring* pointed out a long time ago [4.19]. The typical time scale of the dynamics of the $6s$ (and $5d$) electrons is much shorter than that of the $4f$ electrons. Upon a change of the f-occupancy, the s-like electrons will react instantaneously [4.20]. Because of the small screening length in normal metals, the net effect is that the s-electrons exactly compensate the change in f-count within the Wigner–Seitz sphere. Therefore, instead of (4.3), a better way to calculate U from the atom is to use configurations in which the change in f-count is completely compensated by a change in $6s$ and/or $5d$ count ('metallic screening'). Schematically

$$U = E(f^{n+1}s^0) + E(f^{n-1}s^2) - 2E(f^n s^1). \tag{4.4}$$

One way of estimating the configurational energies in (4.4) is by using numbers derived from atomic spectroscopy. They can also be calculated using an LDA-based first-principle method ('constrained LDA' [4.21]). Ideally [4.22], one represents the LDA problem in an atomic-like single-particle basis in order to identify the 'f' orbitals, and the transfer integrals between these orbitals and the rest of the Hilbert space are removed (the hopping is treated explicitly in the model). If the screening is truly instantaneous, the U is then given by the second derivative of the total energy to the change in 'f' occupancy. In this way, excellent results are obtained which are even better than the empirical estimates, because one accounts for the fact that the s-orbitals change their shape in the solid ('renormalized atom' [4.23]). Finally, the exchange interactions responsible for the multiplet splittings are barely renormalized. This is consistent with the strong renormalization of the U. The presence of s-electrons etc. has only a minor influence on the value of exchange type integrals.

The essence of the previous paragraphs can be formulated as follows: *the electronic eigenstates in strongly correlated systems are grouped according to the charge distribution they represent* (the Hubbard band principle). The simplest model containing this physics is named after Hubbard [4.24]

$$\begin{aligned} H &= H_0 + H_1, \\ H_0 &= U \sum_i n_{i\uparrow} n_{i\downarrow}, \\ H_1 &= t \sum_{\langle ij \rangle \sigma} c^\dagger_{i\sigma} c_{j\sigma}, \end{aligned} \tag{4.5}$$

representing a tight binding lattice of s atoms (H_1) with an on-site electron–electron repulsion (H_0). If $U \ll W$ (single-particle bandwidth), the kinetic energy wins and the problem can be treated using conventional diagrammatic perturbation theory around the non-interacting limit (weak coupling). However, 'strongly correlated' means that the system is closer to $U/W \gg 1$ and in this case it is more convenient to take the H_0 term as the zero order, and the hopping as the perturbation (strong coupling). The most difficult case is $U \sim W$ where one does not know where to start. Furthermore, $U/W > 1$ is a necessary, but not sufficient condition for 'strong correlations'. Especially near the beginning or the end of the series, the number of d (or f) electrons or holes present in the ground state is equally important. Consider, for instance, a d^{10} system such as Cu metal. Upon removing a d-electron, one ends up with a single d^9 site, in a background of d^{10} sites. The probability that in a hopping process a d^8 site is created is identically zero, and therefore the photoemission spectrum is well described by LDA. This is different if one looks at the Auger spectrum. Here one creates at $t = 0$ two d-holes on a single site, and one hits the d^8 Hubbard band of Cu. This spectrum looks perfectly atomic like, showing that $U > 2W$ [4.25]!

This reflects a general rule: *the experimental response of a strongly correlated system depends in the first instance on what one is doing with respect to the charge distribution.* Let us neglect the hopping, and consider the case of 'half filling' with one electron per site in the ground state (Fig. 4.3a). In the optical experiment the total number of electrons is not changed. Transferring an electron from one side to another will automatically produce a doubly occupied site (Fig. 4.3b) and this will cost an energy U. The energy level diagram is indicated in Fig. 4.3c. There is a manifold of states at 0 (everywhere s^1), at U (one electron–hole excitation), at $2U$ (two electron–hole excitations), etc. Modulating this with the optical dipole matrix elements yields the optical spectrum. The single particle

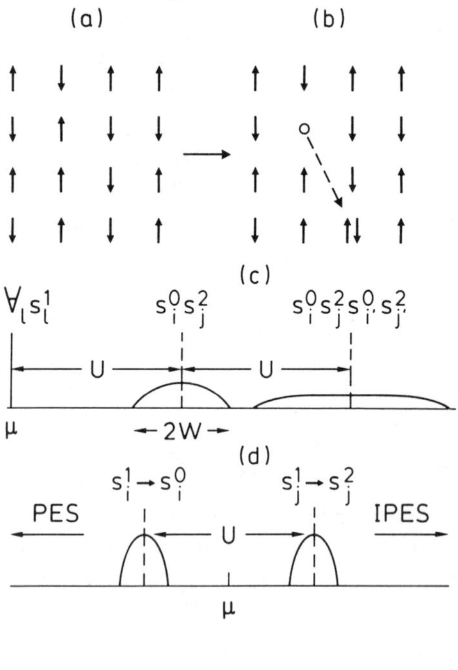

Fig. 4.3. The situation at half filling (**a**) for large U/W (arrows indicating the spins), and the lowest lying (at U) charge excitation, (**b**) in the single-band model. For the optical spectrum and the hopping the total energy diagram (**c**) is relevant, which is different from $A(\omega)$ (**d**). The states (dashed lines) will broaden in bands if the hopping is switched on

spectrum (4.2) is something quite different: the removal of an electron will produce an empty site, and this costs the chemical potential μ. Adding an electron produces a doubly occupied state, and this costs $U-\mu$. By convention, ionization energies are drawn to the left and affinities to the right, yielding the spectrum in Fig. 4.3d. The '$s^1 \rightarrow s^0$' band is called the lower and the '$s^1 \rightarrow s^2$' the upper Hubbard band. Both in the single-particle and the electron–hole spectrum a gap $\sim U$ is present. This is the correlation gap, and this system is the generic Mott–Hubbard insulator [4.24, 26].

Although less fundamental than PES-IPES, the core spectroscopies (c-XPS and XAS) yield quite interesting independent information about the Hubbard band structure. If the electrostatic interaction between the f (or d) electrons is large, the interaction of the core hole with such a valence electron is even larger. The core spectra give information about the strongly perturbed electronic structure in the presence of this core hole (see, however, [4.27]). The core hole–valence electron interaction is

$$Q = E(\underline{c}_i d_i^{n+1}) - E(\underline{c}_i d_i^n). \tag{4.6}$$

We gain Q if we put an extra electron on the core-ionized site, and typically $Q \sim 1.5\,U$. The Hubbard band structure for these experiments is sketched in Fig. 4.4. In c-XPS a localized core hole is created at site i. If $Q > U$, the 'singly occupied site everywhere' ($s^1 \leftrightarrow d^n$) state is no longer lowest in energy. Promoting

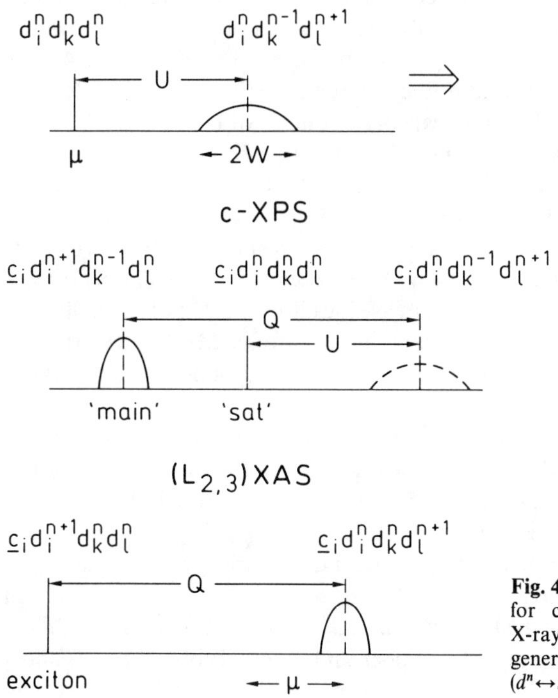

Fig. 4.4. Total energy diagrams relevant for core photoemission (c-XPS) and X-ray absorption spectroscopy (XAS), generalized to the degenerate case ($d^n \leftrightarrow s^1$)

an electron to the core-ionized site costs the (electron–hole pair) energy U, but yields Q (Fig. 4.4). Amusingly, this 'well-screened' state [4.28] broadens into a band because the newly made hole is free to hop, while the 'poorly screened' state (the core-hole satellite) remains sharp. In the XAS experiment, we put back the electron on the same site (Fig. 4.4). The lowest state is now directly reached in the experiment, and after investing $|Q|$, the additional valence electron is free to leave the core-ionized site, as in the familiar core-exciton picture (see Chap. 5).

The single-band Hubbard model is too oversimplified to account for the valence electronic structure of real materials. A first complication is the splitting of the Hubbard bands into multiplet subbands, as occurs in the rare earths. Furthermore, besides the d- or f-shells of the transition metal/rare earth atoms, other types of electrons are present, such as the broad $5d/6s$ bands in the rare earth example. In the absence of hopping, this band corresponds with $f^n\Phi \to f^n c_{k\sigma}\Phi$ below E_F and $f^n\Phi \to f^n c^\dagger_{k\sigma}\Phi$ above E_F, where $c^\dagger_{k\sigma}$ creates a $6s/5d$ LDA-like electron, and $\Phi = \prod_{k,\sigma}^{k_F} c^\dagger_{k\sigma}|\text{vac}\rangle$. Although this looks like a 'normal' band, it implicitly contains the constrained f-state. If the hybridization is switched on, very interesting phenomena occur at low energies: the Kondo effect, mixed-valency and heavy-fermion behaviour, which are discussed in Sect. 4.3. The d or f state of an impurity in a s/p-type metal would, if U was switched off, broaden into a virtual bound state, and the width of this resonance (Γ) is the appropriate kinetic scale to compare U with. Plotting the 'metallic screening' U^{eff} (which includes the Hund's rule interactions) against Γ, *van der Marel* and *Sawatzky* obtained a 'U/W' diagram for all $3d$, $4d$, $5d$ and $4f$, $5f$ impurities in a number of simple and noble metal hosts (Fig. 4.5) [4.29]. Region I, II and III (bounded by the drawn lines) correspond with $U > \Gamma$, $|U| < \Gamma$ and $-U > \Gamma$, respectively. The last possibility, an attractive U, is rather exotic and is based on the idea that if the monopole part of the Coulomb interactions is strongly screened, the (unscreened) exchange interactions could give rise to effective attractive interactions. The dashed lines correspond with the $|U| = \Gamma$ lines in other hosts. As can be seen, the $4f$'s are always $U \gg \Gamma$. The $5f$'s have a smaller U but are nevertheless, in most cases, well inside $U > \Gamma$. Finally, the d-impurities are usually borderline (except the d^5 atoms Mn(II), Mo(II) and Cr(I)), and whether they display strong or weak coupling depends critically on the host.

In materials with a more complicated single-particle band structure, the Hubbard band structure is also richer. Let us consider the divalent transition metal (TM) compounds, such as the oxides TM(*II*)O or halides TM(*II*)X_2 with X = F, Cl, Br, I [4.5, 6]. Besides the partly filled d-shells, these materials are characterized by occupied bands of largely ligand p-character, separated by a rather large gap ('itinerant gap', Δ_{it}) from the unoccupied TM s-like conduction band. In the ground state, the TM ions are in a d^n configuration (Cu(*II*) $\sim d^9 \cdots$ Ti(*II*) $\sim d^1$) and in G_1 the d^{n-1} and d^{n+1} upper and lower Hubbard bands show up, besides the 'normal' $d^n\underline{p}$ (ligand p-) and $d^n s$ (metal s-) bands (underlining

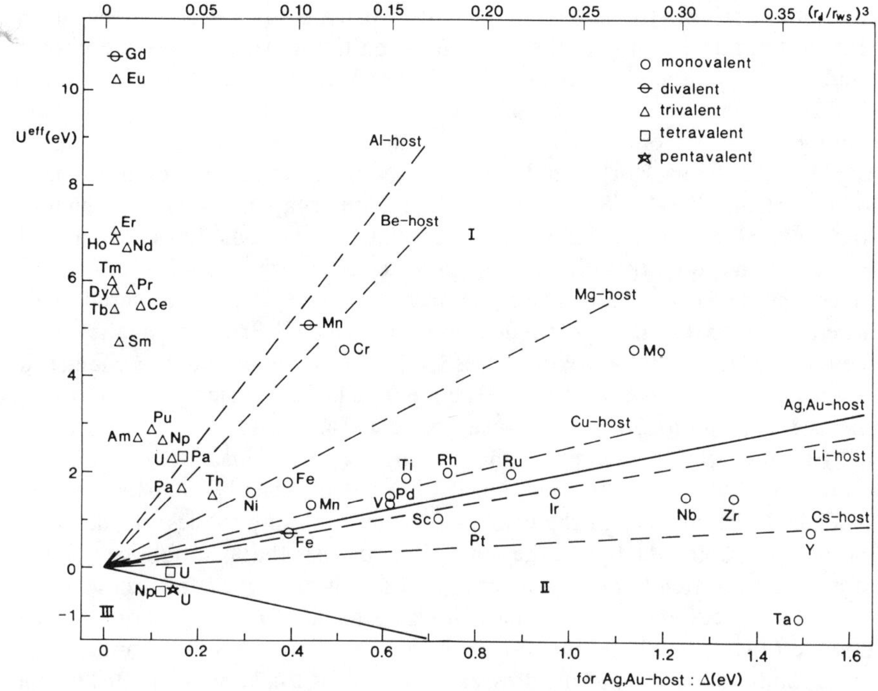

Fig. 4.5. Virtual bound-state width ($\Delta = \Gamma$) versus U^{eff} for a variety of d and f impurities in Ag and Au hosts [4.29]. The dashed lines are the dividing lines in other hosts

indicates hole). In addition to Δ_{it}, the d-band-splitting U enters, which is larger in insulators than in metals because the metallic screening gets frustrated [4.30]. Finally, a third energy is needed, telling us how to position the d-Hubbard bands with respect to the other bands (Fig. 4.6). The splitting between the upper Hubbard band and the p-band can be taken,

$$\Delta = E(d^{n+1}) + E(d^n \underline{p}) - 2E(d^n) \tag{4.7}$$

being the energy cost to transfer a hole from the TM ion to the ligand ($d^n \to d^{n+1}p$, 'charge-transfer energy'). A couple of possibilities arise for the nature of the

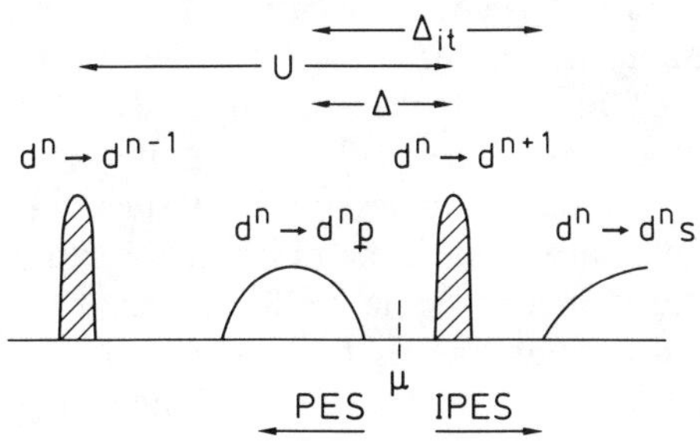

Fig. 4.6. The single-electron spectrum above and below the Fermi energy (μ) in a charge-transfer insulator. Indicated are the upper and lower Hubbard bands (dashed) and the filled ligand p- and empty metal s-bands

correlation gap: firstly, assuming that the *s*-band is always above the upper Hubbard band, the gap can be either between the upper and lower Hubbard band ($U < \Delta$, *d–d* gap), or between the upper Hubbard band and the *p*-band ($U > \Delta$, *p–d* gap, 'charge transfer insulator' [4.5]). Including the *s*-band we have also the possibility of a *d–s* gap or, eventually, a purely 'itinerant' *p–s* gap.

How does this work out in the 3*d* oxides? One of the significant facts following from PES-IPES is the observation that the gap in NiO is of the charge-transfer type [4.31] (Sect. 4.2). Clearly, Δ is related to the 'ionicity' of the material: the larger this quantity, the stronger the tendency towards '2+' and '2−'. Indeed, going to NiF_2 Δ increases, compared to NiO, and this material might have a *d–d* gap. On the other hand, in a series like Cl-Br-I (and at the end S) the ionicity, and therefore Δ, decreases [4.32, 33]. Assuming modest changes in U, the gap will become more and more a pure charge-transfer gap, as has been checked with photoconductivity measurements [4.34].

Although the 'screening' of the parameters in the insulators is a rather complicated affair (see the constrained LDA calculations by *Gunnarsson* et al. [4.22, 35]), the changes in the crystal structure are quite modest if one varies the 3*d* ion, keeping the ligand the same. This suggests that it should be possible to estimate the trends in such a series and it can be argued that these are in the first instance determined by the 3*d* ionization and Madelung energies [4.36]. Using estimates for these quantities, one arrives at the PES-IPES and optical absorption Hubbard band pictures, as shown in Figs. 4.7 and 4.8 [4.36] for the divalent chlorides (the oxides are very similar). As in rare earth, the multiplet splittings are quite important, and in the figures only the energies of the lowest multiplets are indicated. Looking at the figures, one can distinguish four classes of (divalent) compounds. The late 3*d*'s (Cu, Ni) are the candidate charge transfer insulators because of the small Δ's and large U's [4.37]. As expected, Δ increases strongly to the left of the series and the gaps of the early 3*d*'s (Ti, V) are of strongly mixed *d–d*, *d–s* character. Mn(II) is a class of its own because of the strongly exchange stabilized (high spin) d^5 ground state, leading to an increase of all excitation energies. Invariably, Mn(II) compounds have the largest gaps. Finally, the neighbours of Mn (Fe, Cr) are quite similar. These are special because one can hit the d^5 state by removing (Fe) or adding (Cr) an electron. This leads in both cases to a small U and the other consequence is the anomalously small Δ in Cr(II) and the low lying $d \rightarrow s$ threshold in Fe(II). The above picture compares quite nicely with optical data [4.36].

Finally, let us consider what happens if Δ becomes negative. This means that $d^{n+1}\underline{p}$ is at lower energy than d^n, and the former is therefore the ground state. The *p*-band is therefore partly filled, and the lower and upper Hubbard band are of d^n and d^{n+2} character, respectively (Fig. 4.9). This is an abundant phenomenon at the end of the 3*d*-series. Famous examples are the pyrites, with the formula $TM^{2+}(S_2)^{2-}$, where the sulfur atoms sit in pairs in the structure. For instance, in CuS_2, the Cu ion is clearly in a d^{10} state [4.38] and the pairs are formed because the hole goes in an antibonding S_2 band [4.39]. This is also the standard situation in the trivalent late 3*d* compounds. Looking at, e.g.,

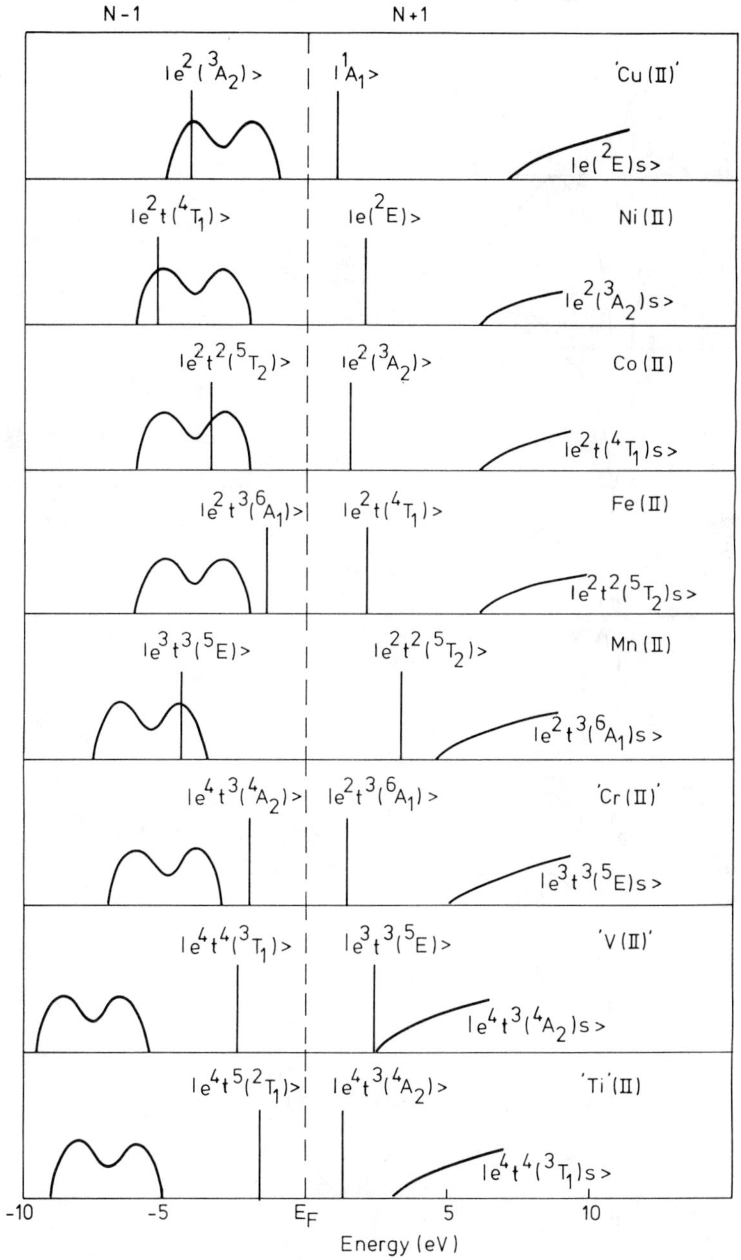

Fig. 4.7. The appearance of the single-electron spectrum in the transition metal dichloride series. The monoxides are very similar, except that the *s*-band is located at somewhat higher energy [4.36]

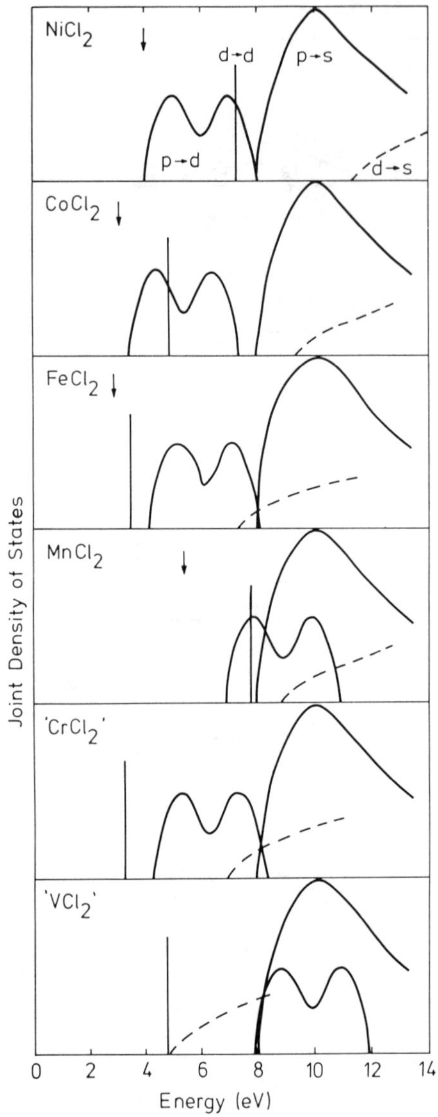

Fig. 4.8. As Fig. 4.7, but now for the optical spectra [4.36]

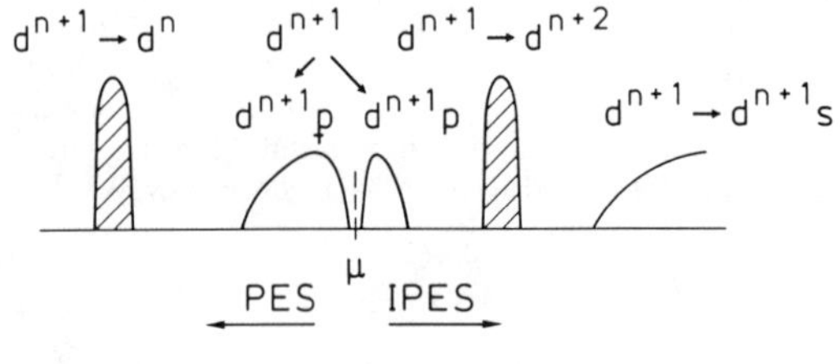

Fig. 4.9. The single electron spectral function in a *p*-type material

NiO one would expect that additional holes introduced by doping with a monovalent ion would go in the oxygen band. This is indeed what happens in $LiNiO_2$, where it has been shown that Ni stays locally d^8, and a new band appears above E_F which is largely of oxygen character [4.40]. The most famous examples of this behaviour are, of course, the high T_c cuprates [4.41].

4.2 Large Energy Spectroscopies (Mostly Insulators)

In many cases the Hubbard bands do not have such clear signatures in the spectra as in the rare earth. This is caused by delocalization effects, having two consequences: the different Hubbard bands start to hybridize and each Hubbard band acquires a dispersional width. If the bands are reasonably well separated, the former processes dictate the short-time behaviour (involving a small number of hops) and determine therefore the overall shape of the spectra. The d- (or f-) band formation effects involve a scale which is typically an order of magnitude smaller. For instance, in the 3d compounds the valence band extends ~ 20 eV around E_F, and the width of the upper or lower Hubbard band $\sim 1-2$ eV. This suggests that the latter can be in the first instance neglected, and one ends up with the problem of a single (d or f) impurity in an uncorrelated host (Anderson [4.42] impurity model).

One can simplify the problem further by considering only the correlated ion and its nearest neighbours (cluster approximation), neglecting the widths of the non-d bands also. Here the most important 'principles' governing the overall line shapes are already active. The simplest example is the cluster model description of the 2p c-XPS of Cu(II) compounds [4.43]. The Cu is coordinated by four ligands in a square planar geometry and this is taken as the cluster. In the ground state a single hole is present, being either in the Cu $3d_{x^2-y^2}$('d') orbital or in a linear combination of ligand p-orbitals ('p') with x^2-y^2 symmetry, and these are hybridized with $T=\langle p|H|d\rangle$. The splitting between the d^9 and $d^{10}p$ configurations is Δ and we end up with the two-level problem indicated in Fig. 4.10. The ground state is the 'bonding' combination $|\Phi_0\rangle=\cos(\theta)|d^9\rangle-\sin(\theta)|d^{10}p\rangle$ $(\tan(2\theta)=2T/\Delta)$ and the ground state energy is $E_0=\frac{1}{2}\Delta-\frac{1}{2}\sqrt{\Delta^2+4T^2}$. In the presence of the core hole, the $|\underline{c}d^{10}\underline{p}\rangle$ state comes down with Q, relative to $|cd^9\rangle$ (Fig. 4.10), and the eigenstates are now

$$
\begin{aligned}
|\Psi_c^1\rangle &= \sin\theta'|cd^9\rangle+\cos\theta'|cd^{10}p\rangle,\\
|\Psi_c^2\rangle &= \cos\theta'|cd^9\rangle-\sin\theta'|cd^{10}p\rangle,
\end{aligned}
\tag{4.8}
$$

at

$$
E_c^{1,2}\sim\tfrac{1}{2}(\Delta-Q)\pm\tfrac{1}{2}\sqrt{(\Delta-Q)^2+4T^2},
\tag{4.9}
$$

with $\tan(2\theta')=2T/(\Delta-Q)$. In analogy to Eq. (4.2), the c-XPS spectral function

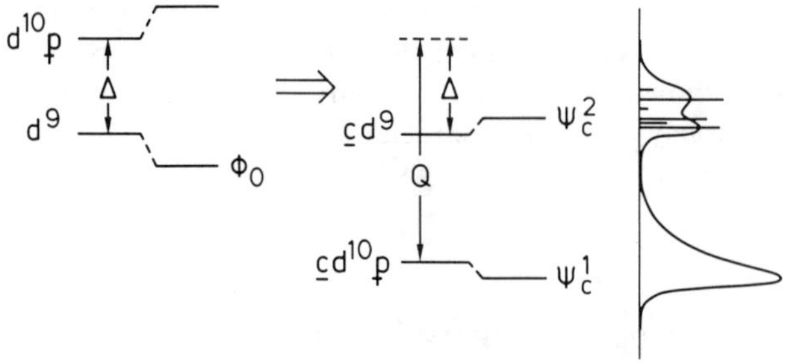

Fig. 4.10. c-XPS of divalent Cu compounds. Due to the core-hole d-electron Coulomb interaction (Q), $d^{10}\underline{p}$ band jumps below the d^9 band. This gives rise to the multiplet-split satellite in the (artist's impression) spectrum

is given by (c_i creates a core hole at site i)

$$\rho^c(\omega) = \sum_{f=1}^{2} |\langle \Psi_c^f | c_i | \Phi_0 \rangle|^2 \, \delta(\omega - E^f + E^0), \tag{4.10}$$

and it follows that *both states (4.8) acquire intensity in the spectrum.* The ratio of the 'main line' (I_1) to 'satellite' (I_2) intensity is given by

$$\frac{I_2}{I_1} = \left(\frac{\sin\theta' \cos\theta - \cos\theta' \sin\theta}{\cos\theta' \cos\theta + \sin\theta' \sin\theta} \right)^2 = \tan^2(\theta' - \theta). \tag{4.11}$$

If the Coulomb interaction (Q) is zero, $\theta = \theta'$ and only the lowest lying eigenstate (Ψ_c^1) is seen. The rule 'no satellites—no correlations' is quite general. It also applies to the low hole density limit (one line if $\Delta/T \ll 0$, independent of Q), and also to, e.g., PES-IPES. The relevant limit for 'ionic' cuprates is $\Delta/T \geqq 1$ and $Q > \Delta$ and both lines now show up. In the limit $Q \gg \Delta, T$ the satellite-to-main-line ratio samples directly the d^9 and d^{10} occupancies in the ground state ($I_2/I_1 = I(\underline{c}d^9)/I(\underline{c}d^{10}\underline{p}) = \cos^2(\theta)/\sin^2(\theta)$, (4.11)). This is not too far from reality; there are two (about equally as large) peaks present in the experimental spectra (Fig. 4.10). The higher lying satellite has a peculiar line shape which maps very well on the expected multiplet structure of a $d^9 \to \underline{2p}d^9$ transition [4.43, 44]. What happens with the intensity ratio if $(Q - \Delta)/T$ gets smaller, such that the $\underline{c}d^9$ and $\underline{c}d^{10}\underline{p}$ states start to hybridize? Here is a subtlety: the intensities are not simply given by the sum of the products of the weights of the basis states in the initial and final state, but also contain the cross term $\sim 2\sin\theta\cos\theta\sin\theta'\cos\theta'$, and this is added to the main line and subtracted from the satellite intensity. This term is relatively large, and the satellite is therefore much smaller than expected intuitively. This is a general rule; the reason is that the phasing of the ground-state wave function is more similar to the phasing of low-lying states than that of high-lying states, which means that the interference terms are optimally constructive at the low end of the spectrum, and optimally destructive at the high end. For instance, the Cu ground-state configuration is rather close to d^9 and in PES one would thus naively expect that most of the spectral weight would be in the $d^9 \to d^8$ lower Hubbard band (Fig. 4.6). However, this band can be unambigously [4.45] identified with the -9 to -12 eV satellite in the PES, and this is actually quite a week feature in the spectrum compared to the $\sim d^9\underline{p}$ main band. This is no longer a paradox if one accounts for the interference terms.

By considering only the (dominant) hybridization between the d-shell of a single Cu and the filled ligand p-band, we ended up with an effective single-hole problem for the ground state of the Cu cluster. Similarly, going away from the end of the 3d series, we find *few*-particle problems which can also be solved exactly. In fact, the Ni (II) cluster [4.7, 32] is fairly representative for the rest of the series. This cluster contains two holes. If both are localized on the Ni (II), the exchange and crystal field stabilized $d^8(^3A_2)$ state ($|d^8\rangle = |\varepsilon\uparrow\theta\uparrow\rangle$; $\varepsilon, \theta \sim 3d_{x^2-y^2}, 3d_{3z^2-1}$) is lowest in energy. In analogy to the Cu(II) case, $|\varepsilon\rangle$ and $|\theta\rangle$ orbitals from the ligand p-states can be constructed (p_ε, p_θ with $\langle\varepsilon|H|p_\varepsilon\rangle = \langle\theta|H|p_\theta\rangle = T$). We are now forced to work in a two-hole basis: $|d^8\rangle, |d^9\underline{p}; 1\rangle = |\varepsilon\uparrow p_\theta\uparrow\rangle, |d^9\underline{p}; 2\rangle = |p_\varepsilon\uparrow\theta\uparrow\rangle, |d^{10}\underline{p}_2\rangle = |p_\varepsilon\uparrow p_\theta\uparrow\rangle$. These two-hole states are of the Hubbard band sort: taking as a reference $\langle d^8|H|d^8\rangle = 0$, by definition $\langle d^9\underline{p}; i|H|d^9\underline{p}; i\rangle = \Delta$, and then $\langle d^{10}\underline{p}^2|H|d^{10}\underline{p}^2\rangle = 2\Delta + U$, because there are *two* additional electrons on the Ni compared to d^8. Hopping is only possible between configurations differing by one hole, and the two d^9 states can be combined in a 'bonding' and 'non-bonding' configuration and only the former ($|d^9\underline{p}\rangle = (1/\sqrt{2})(|d^9\underline{p}; 1\rangle + |d^9\underline{p}; 2\rangle)$) matters. This introduces an innocent-looking factor $\sqrt{2}$ in the hopping ($\langle d^8|H|d^9\underline{p}\rangle = \langle d^9\underline{p}|H|d^{10}\underline{p}^2\rangle = \sqrt{2}T$), which will play an important role in the theory of Ce systems (1/N expansion, next section). Taken together, the ground state of the cluster is the lowest eigenstate of the three level problem

$$\begin{bmatrix} -\omega & \sqrt{2}T & 0 \\ \sqrt{2}T & \Delta - \omega & \sqrt{2}T \\ 0 & \sqrt{2}T & 2\Delta + U - \omega \end{bmatrix} \begin{bmatrix} u(d^8) \\ u(d^9\underline{p}) \\ u(d^{10}\underline{p}^2) \end{bmatrix} = 0. \tag{4.12}$$

It is easily checked that for $U = 0$ (4.12) reduces to two single-hole ('molecular orbital') problems. However, for finite U the molecular orbital solution would overestimate the double occupancy, and in order to project out d^{10} one has to solve the three-level problem. This example is representative for many other correlated systems, because for large U the other Hubbard bands are at much higher energies, i.e. d^{n+3} at $3U$, d^{n+4} at $6U$, etc. As a consequence, only three Hubbard bands are important for the ground state. The 'three levels on a short time scale' bears physical significance. Consider again c-XPS; with respect to the cd^n reference state, the $cd^{n+1}\underline{p}$ state will be pulled down with Q by the core hole, but the $cd^{n+2}\underline{p}^2$ state by $2Q$. For not too large Δ it can be imagined that the d^{n+2} state ends up below the d^n, or even below the screening d^{n+1} state ('overscreening') and one will see three peaks in the spectrum. This happens in all Ce systems [4.1, 46] and in many 3d compounds [4.32, 47].

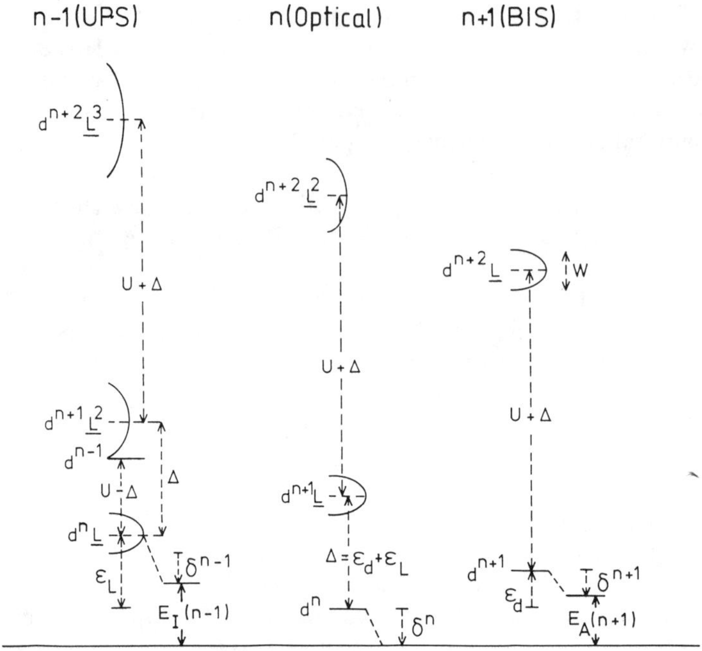

Fig. 4.11. Diagram showing the various states and parameters, used in the calculation of PES-IPES of $3d$ compounds

In order to calculate PES-IPES, one has to repeat the above calculation with one electron removed ($N-1$) or added ($N+1$). For instance, (4.12) relates also to either the PES of Cu(II) or the IPES of Co(II). The calculation of (4.2) is then similar to (4.8–11), except that three states come into play. The typical energetics of the important bands (i.e. up to d^{n+2}) of the $3d$ charge transfer insulators is indicated in Fig. 4.11 [4.5]. In the IPES only the upper Hubbard band is seen ($d^{n+1}\underline{p} \to d^{n+2}\underline{p}^2$ is very weak). In PES not only d^{n-1} and $d^n\underline{p}$ come into play, but also $d^{n+1}\underline{p}^2$, and the PES is therefore harder to calculate and to interpret. In Ce systems the situation is in this respect reversed (next section). The spectra calculated in this way would not compare very well with experiment. Two further extensions are necessary: inclusion of multiplet (and crystal field) splittings, and the width of the non-d bands.

A consequence of the interplay of the multiplet splitting and mixing of Hubbard bands is the phenomenon of (covalent) reduction of the exchange. Although this is a general effect, it can be nicely illustrated with the $L_{2,3}$ XAS spectra of Ni compounds [4.33] (see [4.48] for the more complicated rare-earth spectra). These are shown in Fig. 4.12, in order of increasing ionicity. In the ionic

Fig. 4.12. Calculated $L_{2,3}$ XAS spectra of Ni halogenides (dashed line) using the impurity model, compared to the experimental spectra [4.33]

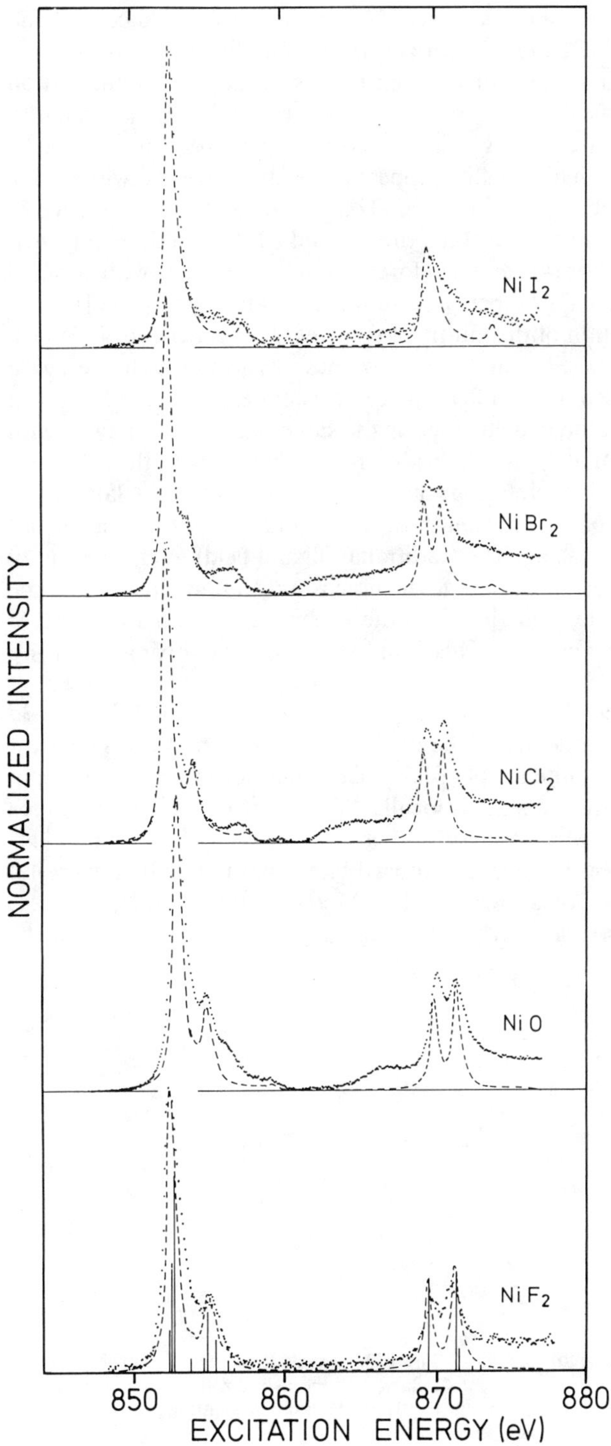
NORMALIZED INTENSITY
Ni I2
Ni Br2
Ni Cl2
Ni O
Ni F2
850
860
870
880
EXCITATION ENERGY (eV)

limit, the ground state would be d^8 and the XAS 'white lines' would correspond with the transition $d^8 \rightarrow \underline{2p}d^9$, and this works very well for NiF_2 (see also Chap. 5). The bars in its spectrum are given by the multiplets of the $\underline{2p}d^9$ configuration in a (octahedral) crystal field of $10D_q = 1.5\,\mathrm{eV}$, weighted by the dipole strengths. If the covalency is increased, the spread of the multiplet spectrum decreases, and at the same time a small 'satellite' appears slightly above the white lines, as is seen most clearly in the $2p^{3/2}$ region. This is caused by the presence of $d^9\underline{p} \rightarrow \underline{c}d^{10}\underline{p}$ transitions. In the final state, the $\underline{c}d^9$ and $\underline{c}d^{10}\underline{p}$ states are only $\Delta + U - Q$ apart, and they hybridize. Therefore, a satellite is seen, with a small intensity because $(U-Q)/T$ (the perturbation) is small (see also [4.49]). How do we explain the reduction of the multiplet splittings? Consider Fig. 4.13: pick out two $\underline{c}d^9$ multiplets (Γ_1, Γ_2) and ask how these hybridize with the $\underline{c}d^{10}\underline{p}$ states. The Γ_1, Γ_2 representations of $\underline{c}d^{10}\underline{p}$ are degenerate, and it can be shown that the hybridization in the two channels is the same [4.33]. Starting out with a splitting δE, it is clear that Γ_2 will hybridize more strongly with the d^{10} states than Γ_1, and the former is therefore pushed down further than the latter. The net result is a decrease of the splitting $(\delta E' < \delta E)$, and this is the reason that the $\underline{c}d^9$ splittings are decreasing. The traditional (ligand field) interpretation of this reduction is less general: on average the hole is only a fraction δ of its time in the multiplet split state, and therefore the exchange constants are only δ times the bare exchange constants. This is only correct if the mixing is not too strong [4.50].

What is the influence of the multiplet splittings on the PES-IPES of the $3d$ insulators? The internal structure of the d^{n+1} band as seen in IPES is (at least in high-spin compounds) quite simple [4.51]. Consider, for example, high-spin Co(II) compounds; the ground state is locally ${}^4T_2(\varepsilon\uparrow\theta\uparrow t\uparrow)$. ($t$ being a t_{2g} hole). Upon adding a hole, one ends up in either a ${}^3A_2(\varepsilon\uparrow\theta\uparrow)$ or a ${}^3T_2(\varepsilon\uparrow t\uparrow)d^8$ configuration and one sees the upper Hubbard band split in two by the crystal field splitting. The same holds in Fe(II) and Mn(II), while in Ni(II) one sees only one (e_g) peak. This single-particle-like signature in IPES comes from the

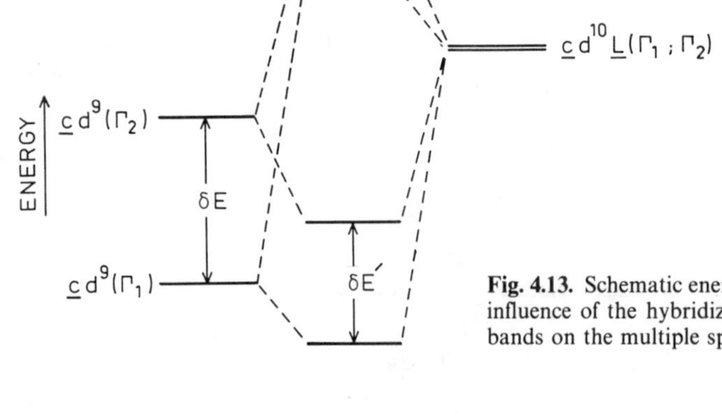

Fig. 4.13. Schematic energy diagram showing the influence of the hybridization between Hubbard bands on the multiple splitting

fact that only the largest total spin states are involved [4.52]. Again, in PES this is much more complicated [4.7]. Consider NiO; the d^7 lower Hubbard band splits in many multiplet sub-bands, and three of the irreducible representations can be reached by emitting a d-electron from the $^3A_2$ ground state: $^4T_1$, $^2T_1$ and ^2E, having 2, 5 and 4 d^7 components, respectively. These d^7 states mix with $d^8\underline{p}$ states which are also split up by exchange interactions, and in Fig. 4.14 the transfer integrals between the most important d^7 and $d^8\underline{p}$ are indicated, derived from fractional parentage [4.6]. Finally, the $d^8\underline{p}$ states in their turn hybridize with the $d^9\underline{p}^2$ states and the latter are split by the crystal field.

Usually, the ligand bandwidths are rather large, ~ 4 eV, and these should also be taken into account. This leads to the Anderson impurity Hamiltonian

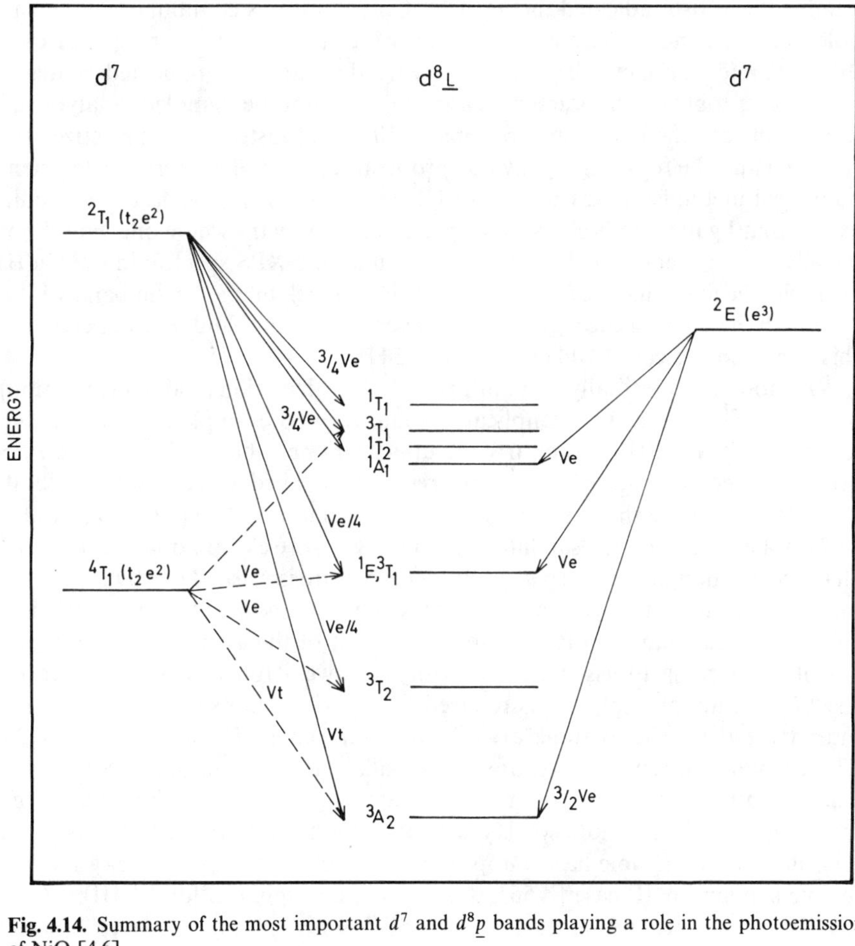

Fig. 4.14. Summary of the most important d^7 and $d^8\underline{p}$ bands playing a role in the photoemission of NiO [4.6]

[4.42]

$$
\begin{aligned}
H^{\mathrm{imp}} = & \sum_n \int d\varepsilon \varepsilon c^\dagger_{n\varepsilon} c_{n\varepsilon} + \sum_n \varepsilon_{nd} d^\dagger_n d_n + \sum_{n\sigma} \int d\varepsilon [V_n(\varepsilon) d^\dagger_n c_{n\varepsilon} + h.c.] \\
& + \sum_{n_1 n_2 n_3 n_4} U(n_1 n_2 n_3 n_4) d^\dagger_{n_1} d_{n_2} d^\dagger_{n_3} d_{n_4},
\end{aligned}
\tag{4.13}
$$

where $d^\dagger_n$ creates an electron in the impurity d- (or f-) orbital ($n = m_l, \sigma$), ε_{nd} are the d-level positions (including crystal fields, spin-orbit coupling) and in the U term all exchange terms have to be included (for simple approximations see e.g. [4.53]). Instead of a single-ligand level per d single-hole state, there is now a band of non-d states ($c^\dagger_{n\varepsilon}$) and it is convenient to label these by energy (ε) and point group label, instead of k and band label. All the band detail and the k-dependence of the hopping are now collected in the energy and symmetry dependence of the hopping function $V_n(\varepsilon)$ [4.1]. The effect of the ligand bandwidth is also indicated in Fig. 4.11. Configurations containing one ligand hole become bands with once the p-bandwidth (W), the $d^x\underline{p}^2$ are spread over $2W$, etc., and one faces the task of solving this rather complicated 'states + continua' problem. In practical cases this cannot be handled analytically (exceptions are the two-hole problems [4.50]), and instead one discretizes the continua and the resulting eigenvalue problems are solved numerically (efficient numerical methods are discussed in [4.2, 3]). Looking at Fig. 4.11, one would expect that ligand bands show up at places in the spectra where one would not naively expect them. Good examples are found in c-XPS spectra; in the Cu(II) example, the main line is $\underline{c}d^{10}\underline{p}$-like and it should pick up the ligand band width. Although the phasing effect gives rise to a peaking of this band to lower energies, this is indeed the case [4.44] (see also [4.54]).

Although conceptually straightforward, the calculation (and interpretation) of spectra can be a rather complicated affair. A good example is NiO. The big low-energy peak in IPES is clearly the upper Hubbard band. However, in PES everything comes together and this requires detailed calculations. A 'best' theoretical fit, using the impurity model, is shown in Fig. 4.15 [4.6, 55] (see also [4.77]). The states in the 'satellite' region ($-8 \to -16$ eV) are dominantly of d^7 character, although the sharp d^7 multiplet structure is washed out by the strong mixing with high-energy d^8 and d^9 states. The two peak structure at the low end of the spectrum is in fact more reminiscent of the d^7 multiplet signature, and it is tempting to associate this with the lower Hubbard band. However, the ${}^4T$–${}^2E$ multiplet splitting is reduced by $\sim 50\%$ from its free ion value and, indeed, these low energy states are of strongly mixed $d^7, d^8\underline{p}$ character ($U \sim \Delta$). The 'exchange screening' is a more delicate affair here than in the XAS example. The ${}^4T$ state is lower lying than the ${}^2E$ state in the d^7 manifold; however, the former is hybridizing more weakly with the $d^8\underline{p}$ states than the latter, because a t_{2g} instead of a e_g hole has to hop (Fig. 4.14). As *Eskes* and *Sawatzky* showed for the similar Cu(II) case [4.56], this favours low spin (${}^2E$ for Ni(III), ${}^1A$ for Cu(III)) $N-1$ ground states for large $d^{n-1} - d^n\underline{p}$ splitting $(U - \Delta)/T$, while high-spin states (such as ${}^4T$) are more favourable for smaller $(U - \Delta)/T$. In fact,

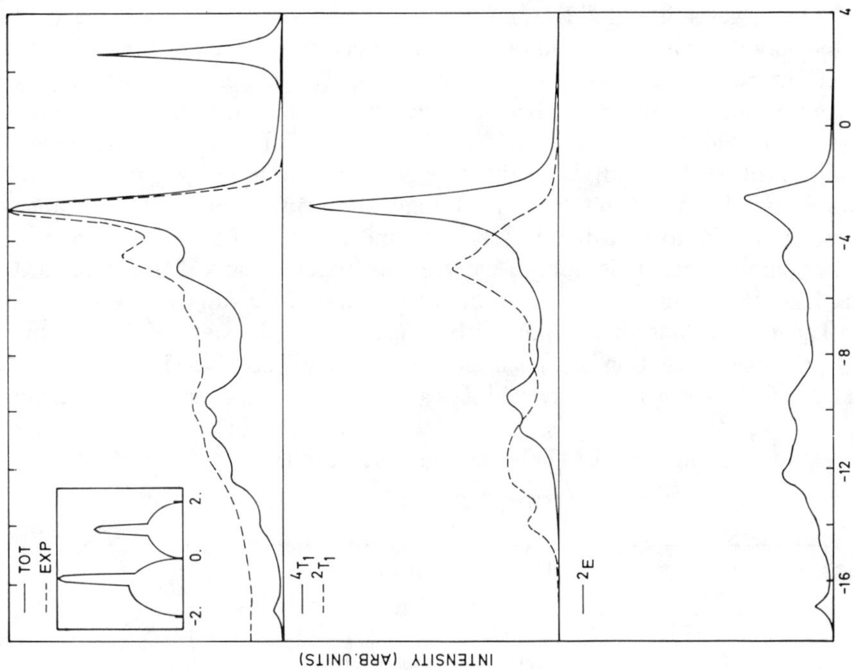

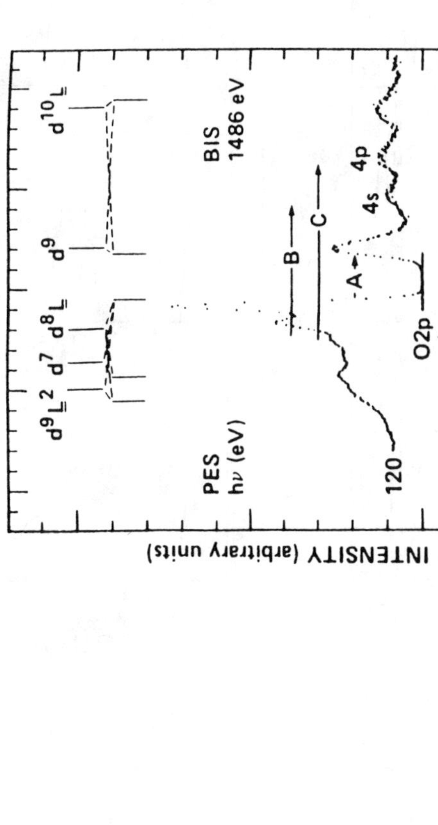

Fig. 4.15. Complete experimental valence band structure of NiO [4.31] (left) and the comparison of an impurity model fit (drawn line) with the experimental PES (dashed) in the right figure. Also indicated are how the different symmetries contribute [4.6, 55]

Fig. 4.15 suggests that the $^4T-{}^2E$ splitting is an optical effect: ^2E states extend all the way down to threshold, and the shoulder at ~ -5 eV is according to the calculation a *resonance* in the ^2E continuum. There might be some truth in this, according to a recent study of the trivalent compound $NiLiO_2$ by *Kuiper* et al. [4.40]. The facts are as follows: the $L_{2,3}$ XAS data indicate that the Ni-ion is still locally $\sim {}^3A_2$, indicating that the ground state is $\sim d^8\underline{p}$. According to the susceptibility, the spin state is $S=1/2$ and apparently the (Kondo) exchange is strong enough to bind the Ni $^3A_2$ state and the spin of the oxygen hole in a low-spin ^2E state. It is not yet clear if this material has a Mott–Hubbard gap. If so, this would be a realization of the Zhang–Rice gap (Sect. 4.4).

Let me conclude this section with an example of the state of the art: the first-principle calculation of *Gunnarsson* et al. of the PES of a Mn impurity in CdTe [4.22]. Besides the screened U's (and Δ's etc.), the hopping function $\Delta_n(\varepsilon)=\pi|V_n(\varepsilon)|^2$ can also be determined from the LDA band-structure calculation. Using the LMTO method, the localized (d,f) level can be

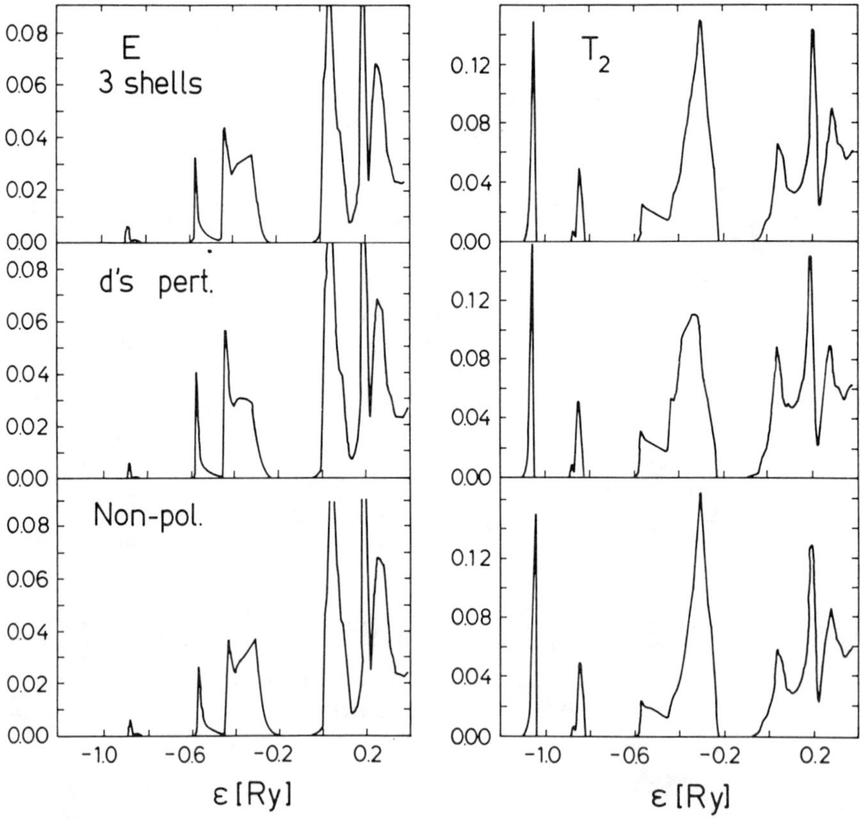

Fig. 4.16. The hopping function $\Delta(\varepsilon)$ for Mn in CdTe as a function of energy for E and T_2 symmetries [4.22]

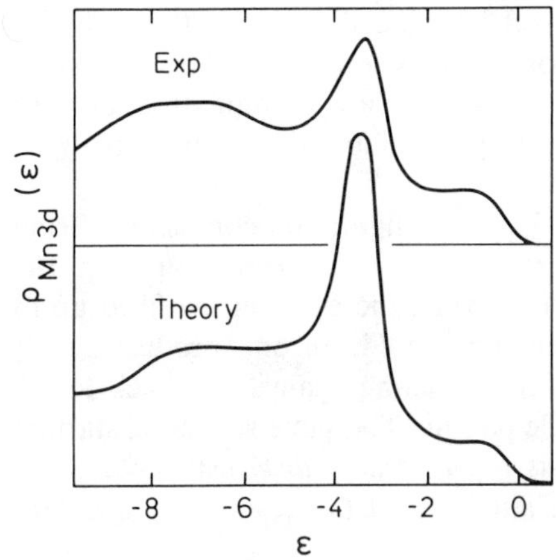

Fig. 4.17. Comparison of the d PES of Mn in CdTe according to theory [4.22] and experiment [4.128]

represented by orthogonal, short-range tight-binding orbitals, as in the models [4.57]. The LDA impurity Green's function is then given by

$$G_{nn}(z) = \frac{1}{z - \varepsilon_{dn} - \int d\varepsilon \frac{|V_n(\varepsilon)|^2}{z - \varepsilon}}, \tag{4.14}$$

and it follows that

$$\Delta_n(\varepsilon) = -\operatorname{Im}\left\{\frac{1}{G_{nn}(\varepsilon - \mathrm{i}0)}\right\}. \tag{4.15}$$

In Fig. 4.16, the 'hopping functions' Δ of e_g and t_{2g} symmetry for a substitutional impurity in CdTe are shown. The host valence band extends up to -0.2 Ry, and the conduction band starts at ~ 0. A surprise is the strong hybridization with the conduction band states, because one would expect this hopping to be less important, involving next-nearest-neighbour overlaps with Cd s-states. The PES can now be calculated by using these first-principle parameters in the impurity model. The result is shown in Fig. 4.17 and appears to be in good agreement with experiment. The Mn(II) PES is the worst case with respect to the number of multiplets to be taken into account ($\sim 10^2$ for d^4, d^5, d^6) and there is no simple interpretation of the spectrum, except that the lowest states are of $d^5(^6A)$ character, with a (nearly?) unbound valence band hole.

4.3 The Kondo Resonance and Renormalized Bands

The problem of a correlated (model) impurity in an insulator is straightforward to solve, as shown in the last section. In fact, I exploited implicitly the presence of the itinerant gap; quite generally, upon going to longer and longer times there will be a point where the (effective) hopping divided by this gap will become a small number, and the expansion becomes convergent. This is different for a correlated impurity in a metal. In Fig. 4.18 I show a (limited) set of basis states for a Ce impurity in a metallic host [4.3]. Invariably, Ce fluctuates between the f^0 and f^1 configurations and the f^2 sits at U larger energy. Because $U/\pi V^2 \gg 1$, it is safe to first project out the f^2 configuration completely. Being interested in the $S=0$ states, one can start (rather arbitrary) with the unhybridized state $f^0\Phi$ (0 in the figure), Φ being the electron sea filled up to E_F (Sect. 4.1). Acting with the hopping term of (4.13), one produces a f^1 configuration with a hole in the band (a) and acting again leads back to f^0, but now with an additional electron–hole pair (c). The same sequence, starting with c, leads to state f, and so on. *This is an infrared divergent problem*: all states $0, c, f$, etc., with the band electrons and holes at E_F are equally important in the ground state.

In general, this problem goes under the name 'valence fluctuations' or 'mixed valency'. If $n_f \to 1$ the f-electron can be replaced by its spin (next section) and then it is called the Kondo problem. A remarkable many-body effect occurs in such system ('Kondo effect'), which is fully understood for a single impurity. Let me summarize its most important ingredients (for an exhaustive discussion, see e.g. [4.4, 11]). At high temperatures, the impurity spin is free, as manifested by e.g. the Curie–Weiss susceptibility. Upon lowering the temperature a cross over occurs at a specific temperature, the Kondo temperature [4.58]

$$T_K \sim e^{-\pi|\varepsilon_f|/N\Gamma}. \tag{4.16}$$

Below T_K, the impurity spin binds to a spin polarization cloud in the electron sea, forming a singlet. As conjectured by *Anderson* et al. [4.59], and later confirmed by *Wilson* [4.58], the Kondo effect is a scaling phenomenon: the physical properties of all Kondo systems are universal functions of a single parameter (T_K). At temperatures well below T_K a (local) Fermi liquid is realized [4.60], characterized by universal FL relations such as the Wilson ratio relating the linear coefficient of the specific heat (γ) to the magnetic susceptibility (χ) by $\chi/\gamma = 1 + 1/N$ (N is the degeneracy).

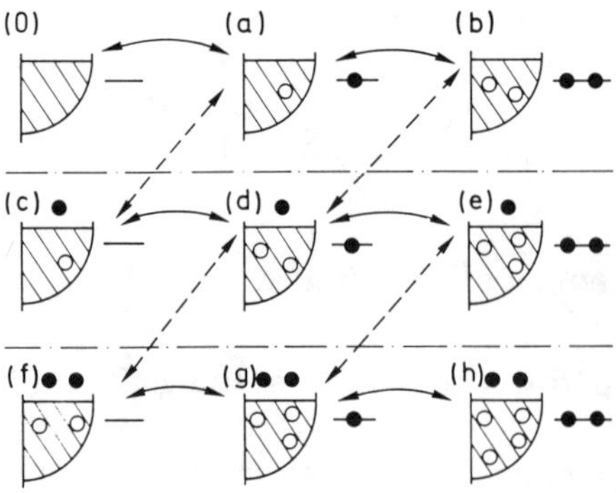

Fig. 4.18. Schematic representation of the basis states in the variational $1/N$ expansion for the ground state. Solid circles show electrons, open circles holes, the hatched part the filled conduction band and the horizontal line the f-level. The arrows show which states hybridize: solid line $\sim N$ and dashed line ~ 1 [4.3]

One of the universal features is the Kondo (or Abrikosov–Suhl [4.61, 62]) resonance (KR), showing up both in G_1 and the QP density of states. This peak is centered at T_K, and has a width of the same order. It has a weight ~ 1 in the QP DOS and it is responsible for the large γ's in the Kondo limit. However, in G_1 it is a weak feature with weight $\sim 1 - n_f$. In Fig. 4.19 examples are shown [4.10]. Both in $CeNi_2$ and CeAl one sees the $f^1 \to f^0$ (at ~ -3 eV) and $f^1 \to f^2$ (~ 5 eV) peaks in $A(\omega)$. However, an additional peak is seen close to E_F, which is the KR. $CeNi_2$ has a $T_K \sim 0.1$ eV, and accordingly we see a big peak, showing up both in PES and IPES. However, in CeAl $T_K \sim 10$ K and it is very hard to resolve the KR. Gunnarsson and Schönhammer demonstrated that these line shapes could be calculated in detail (Fig. 4.19), using the expansion of Fig. 4.18 for the Anderson impurity model [4.3]. The crucial point is that the infrared divergency, although essential for the ground state, is rather harmless if one calculates local properties such as G_1 and if one can live with finite energy resolution or temperature. The 'finite resolution' convergence of the expansion Fig. 4.18 is very much helped by the use of the inverse of the degeneracy (N) as expansion parameter. In Ce one has typically $N = 6$, because the crystal field effects on the $f^{5/2}$ state can be neglected. The degeneracy enters in the following way: there is only one f^0 state $|0\rangle$(0), but N degenerate f^1 states (a). Only the symmetric combination $|f^1\varepsilon\rangle = (1/\sqrt{N}) \sum_{n=1}^{N} f_n^\dagger c_{n\varepsilon}|0\rangle$ of the latter is hybridizing

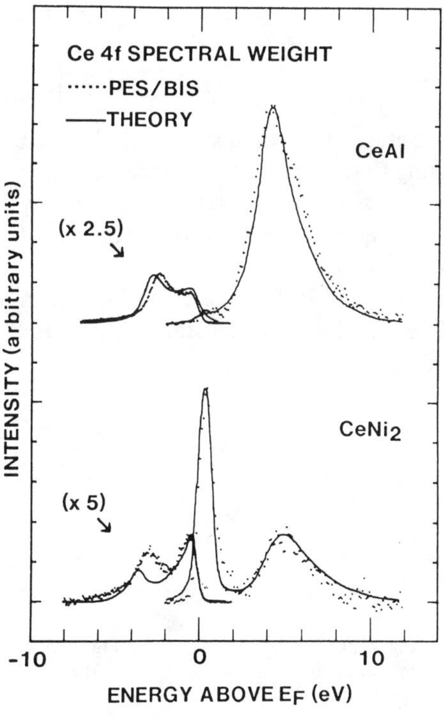

Fig. 4.19. Combined theoretical and experimental $4f$ PES-IPES spectra for CeAl and $CeNi_2$ [4.10]. The PES portion is increased by a scale factor for clarity. In $CeNi_2$ the Kondo temperature is ~ 0.1 eV and the large peak centred slightly above E_F is the Kondo resonance in this material

with the former, and $\langle 0|H|f^1\varepsilon\rangle = \sqrt{N}V(\varepsilon)$. On the other hand, the relevant states of type c are $|f^0\varepsilon E\rangle = (1/\sqrt{N})\sum_{n=1}^{N} c_{nE}^\dagger c_{n\varepsilon}|0\rangle$ and it follows that $\langle f^1\varepsilon|H|f^0\varepsilon E\rangle = V(E)$. The mixing $0\to a$, $c\to d$, etc. is thus N times stronger than $a\to c, d\to f$ etc., and this has the effect of speeding up the downward renormalization by a factor $\sim N$. The qualitative aspects of the problem are already present in $N\to\infty$ and it is instructive to follow in detail the derivation of G_1 in this limit [4.1].

For $N\to\infty$, only the states (0) and (a) (Fig. 4.18) are relevant for the ground state and we write for the ground-state wave function [4.63]

$$|N_0\rangle = A\left[|0\rangle + \int_{-W}^{0} d\varepsilon a(\varepsilon)|f^1\varepsilon\rangle\right], \tag{4.17}$$

corresponding to the problem of a discrete state ($|0\rangle$) mixing with a continuum. Because the continuum has a sharp (Fermi) cut off, a bound state is always formed, with binding energy

$$\delta = \int_{-W}^{0} d\varepsilon \frac{NV^2(\varepsilon)}{\delta+\varepsilon} - \varepsilon_f, \tag{4.18}$$

with respect to the bottom of the f^1 continuum (f-level position $\varepsilon_f < 0$). Furthermore,

$$\alpha(\varepsilon) = \frac{\sqrt{N}V(\varepsilon)}{\delta+\varepsilon}, \tag{4.19a}$$

$$A^2 = 1 - n_f, \tag{4.19b}$$

where n_f is the f (electron) count. The magnetic states would start at the bottom of the unhybridized f^1 continuum suggesting that $|\delta|$ is the Kondo temperature. In fact, one finds in the Kondo limit ($\Gamma \ll \varepsilon_f, n_f\to 1$) for the static susceptibility $\chi \sim n_f/|\delta|$ and the scale is set by $\delta = -W\exp(\pi\varepsilon_f/N\Gamma)$, i.e. the exponential form for T_K, with the correct exponent for large N.

The relevant basis states for PES are indicated in Fig. 4.20. The f-electron can only be removed from the $|f^1\varepsilon\rangle$ component of the ground state. This means

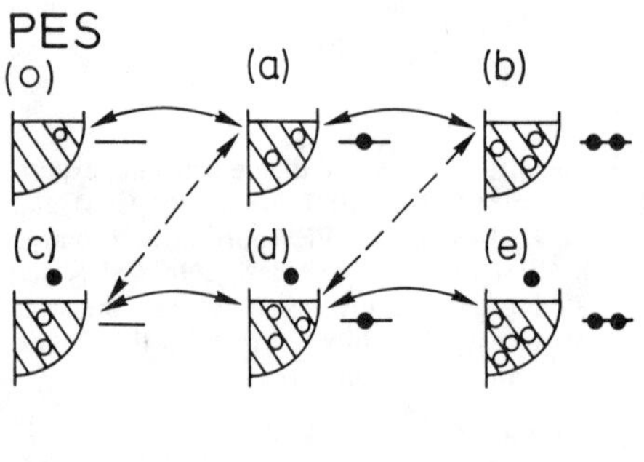

Fig. 4.20. The basis states as in Fig. 4.18 of relevance for the calculation of the PES spectrum [4.3]

for the PES part of (4.1)

$$G_1^<(z) = \langle N_0 | f_n^\dagger \frac{1}{z - E_0 + H} f_n | N_0 \rangle = A^2 \int_{-W}^{0} d\varepsilon a^2(\varepsilon) \tilde{g}(z - \varepsilon), \tag{4.20}$$

where

$$\tilde{g}(z - \varepsilon) = \langle f^0 \varepsilon | \frac{1}{z - E_0 + H} | f^0 \varepsilon \rangle, \tag{4.21}$$

where $|f^0 \varepsilon\rangle$ is the f^0 state with a spectator hole at ε, 'picked up' in the ground state. The state $|f^0 \varepsilon\rangle$ hybridizes strongly ($\sim N$) with the continuum (a), and except for the spectator hole ε, this problem is identical to the ground-state problem. $\tilde{g}$ is thus the Green's function of the state plus continuum problem having as ground state (4.17),

$$\tilde{g}(z) = \frac{1}{z - \varepsilon_f - \delta - \tilde{\Gamma}(z - \delta)}, \tag{4.22}$$

with

$$\tilde{\Gamma}(z) = N \int_{-W}^{0} d\varepsilon \frac{V^2(\varepsilon)}{z - \varepsilon}. \tag{4.23}$$

In Fig. 4.22a I have sketched the construction of the PES spectrum. According to (4.20), the PES is the convolution of $a^2(\varepsilon)$ (4.19) and the imaginary part of $\tilde{g}$. In the Kondo limit, the former is strongly peaking at $\varepsilon \to 0$. In the low-energy region ($\delta < \omega < 0$), the PES is given by the convolution of this function and the bound-state delta function. Hence, the PES is strongly peaking towards E_F on a scale $\sim \delta$. *This is the occupied part of the Kondo resonance.* In the region $-W < \omega < -\delta$ the continuum part of $\mathrm{Im}\{\tilde{g}\}$ is picked up in the convolution and this gives rise to the '$f^1 \to f^0$' peak with spectral weight $\sim n_f$. Consistent with experiment (Fig. 4.19), this peak has a large width $\sim N\Gamma$, reflecting the spread of f^0 weight in the f^1 continuum.

In IPES one can either add an f electron to the $|f^0\rangle$ or the $|f^1 \varepsilon\rangle$ states in $|N_0\rangle$, ending up in $|f^1\rangle$ and $|f^2 \varepsilon\rangle$ states, respectively (Fig. 4.21a, b). In this order, things are very similar to e.g. Cu(II) photoemission, as discussed in Sect. 4.2. An important difference is in the magnitudes of the prefactors for the mixing. Besides the N-times enhanced $f^0 \to f^1$ mixing, the $f^1 \to f^2$ mixing is larger by a factor $N - 1$. Although $\Gamma / U \ll 1, N\Gamma$ and U are not that different, and the interference effects are not unimportant, forcing one to include f^2 explicitly in the calculation of realistic spectra [4.2].

What is gong on near E_F? Taking $U \to \infty$ and infinite N, the only relevant state is Fig. 4.21a and *the* IPES *consists of a single peak with weight* $1 - n_f$ *times the number of channels* (N), *located at the Kondo temperature* $|\delta|$. This is the unoccupied part of the Kondo resonance. In order to recover the (finite N) width of the KR, consider the first order $1/N$ corrections. The state (a) can decay in continuum (c), and the latter plays again 'ground state' with (d). This

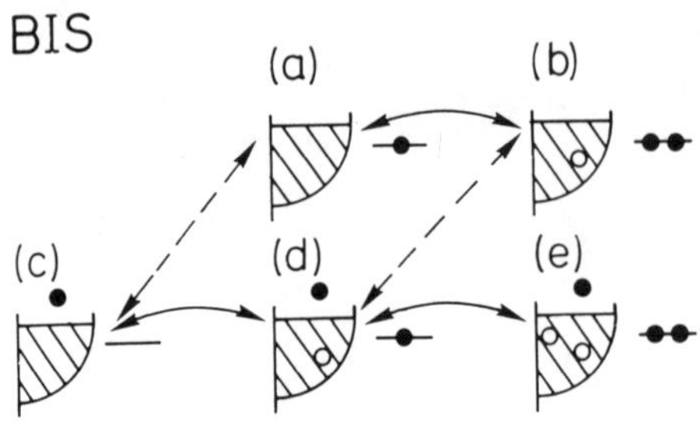

Fig. 4.21. The basis states as in Fig. 4.18 for the IPES spectrum [4.3]

leads to

$$G_1^>(z) = \frac{(1-n_f)N}{z+\delta-\mu(z)}, \tag{4.24}$$

where

$$\mu(z) = -\int_0^W dE V^2(E)\tilde{g}(-z+E). \tag{4.25}$$

The IPES spectrum is constructed in Fig. 4.22*b*. The convolution of the 'ground state' $\tilde{g}$ with the unoccupied band-hopping function $V^2(E)$ yields the continuum in which the $|f^1\rangle$ state (*a*) decays. At low energy ($\omega < |\delta|$), $\mathrm{Im}\{u\} = (1-n_f)\pi V^2$, and accordingly the KR is narrowed by a factor $1-n_f$, compared to the bare virtual bound state. Apparently, thanks to the $1/N$ principle, the derivation of the KR has become a very simple exercise. For $N=6$, one has to go to higher order to describe the near-ground-state properties quantitatively. However, using numerical methods this is not difficult and convergency is achieved in $O(1/N^2)$ for the local properties [4.2, 64]. For detailed comparisons between

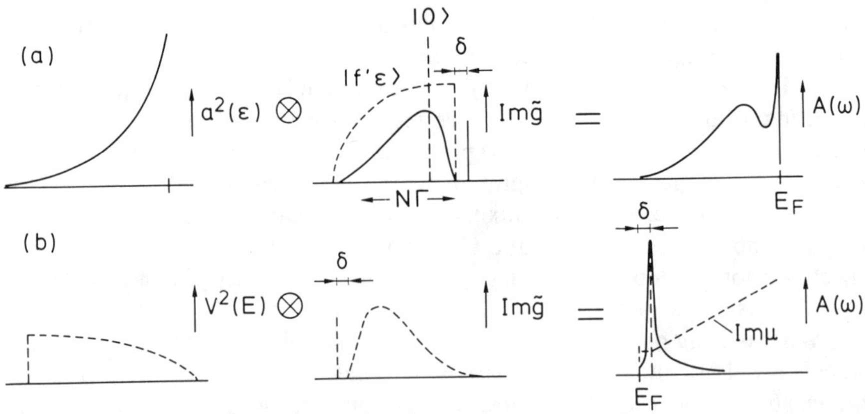

Fig. 4.22. Construction of the PES ($A^<(\omega)$), (**a**) and IPES ($A^>(\omega)$), (**b**) in the large N limit, as described in the text

the results obtained with the Gunnarsson and Schönhammer method and experiment, see [4.1, 10].

Up to now, I have used 'brute force' expansions of the type in Fig. 4.18. However, there are alternative formulations which are mathematically more elegant (although not obviously more powerful) and conceptually interesting because they offer a different way of looking at the problem. The key idea is to exploit a freedom of representation, inherent to Hubbard bands. For $U \to \infty$, the doubly occupied states (f^2, etc.) are projected out from the low-energy region. One can absorb this projection in the hopping term,

$$U \sum_{klmn} f_k^\dagger f_l f_m^\dagger f_n + \sum_n \int d\varepsilon [V(\varepsilon) c_{n\varepsilon}^\dagger f_n + h.c] \to \sum_n \int d\varepsilon [V(\varepsilon) c_{n\varepsilon}^\dagger X_{0n} + h.c.], \tag{4.26}$$

introducing the Hubbard operators X which have the meaning $X_{0n} = P_0 f_n$ and $X_{n0} = f_n^\dagger P_0$, where the operator P_0 projects on states with no f electrons. The Hubbard operators therefore express the fact that the impurity fluctuates exclusively between f^0 and f^1 charge states. The problem is that these Hubbard operators obey a difficult (in fact, supersymmetric [4.65–67]) algebra and, consequently, there is no Wick's theorem, precluding the application of standard diagrammatic techniques.

The trick is to represent the Hubbard operators as products of fermion- ($f^\dagger$) and boson ($b^\dagger$) operators [4.68, 69]

$$X_{0n} = b^\dagger f_n, \quad X_{n0} = f_n^\dagger b, \tag{4.27}$$

where the boson and the fermion represent the f^0 and (spin-bearing) f^1 impurity states, respectively. The substitution (4.27) only makes sense if there is either a single boson or fermion present. This can be handled as follows: the number operator

$$Q = b^\dagger b + \sum_n f_n^\dagger f_n \tag{4.28}$$

commutes with the Hamiltonian, implying that the subspaces with definite Q are disjoint, and one has to take care not to leave the physical subspace with $Q = 1$.

For the impurity, the resulting fermion + boson problem can be handled using standard field theoretical techniques [4.4, 69, 70]. The crucial observation is that the vertex corrections are of order $1/N^2$, and the neglect of these for large N makes the theory managable. Let me quote the final result, the so called non-crossing approximation (NCA; for a derivation see [4.69]). The single f-electron spectral function relates to the convolution of the fermion (A_{fn}) and boson (B) spectral functions,

$$A_n(\omega) \sim \int_{-\infty}^{\infty} d\nu \, A_{fn}(\omega + \nu) B(\nu) (e^{-\beta(\omega+\nu)} + e^{-\beta\nu}), \tag{4.29}$$

where β is the inverse temperature. A_{fn} and B are derived from the fermion and boson Greens function,

$$G_n(\tau) = \langle T f_n(\tau) f_n^\dagger(0) \rangle, \tag{4.30a}$$

$$D(\tau) = \langle T b(\tau) b^\dagger(0) \rangle. \tag{4.30b}$$

These Green's functions are solutions of the self-consistency equations ($f(\omega) = (1 + \exp(\beta\omega))^{-1}$)

$$\begin{aligned}
G_n(z) &= [z - \varepsilon_f - \Sigma_n(z)]^{-1}, \\
D(z) &= [z - \Pi(z)]^{-1}, \\
\Sigma_n(z) &= \int d\varepsilon\, V^2(\varepsilon)(1 - f(\varepsilon)) D(z - \varepsilon), \\
\Pi(z) &= \sum_n \int d\varepsilon\, V^2(\varepsilon) f(\varepsilon) G_n(z + \varepsilon),
\end{aligned} \tag{4.31}$$

i.e. the dressed fermion Green's function is used to calculate the boson self-energy, and vice versa. In terms of the variational basis, this amounts to walking down the 'ladder' $0 \to a \to c \to d$, etc. (Fig. 4.18) to infinite order, taking the additional hole in e.g. $c \to d$ as a spectator. In fact, there are phase space constraints such as $\varepsilon < \varepsilon'$ and these are neglected in the NCA. Equation (4.31) can be solved numerically by iteration [4.4] and in Fig. 4.23 I show some representative results for the f-electron spectral function [4.69]. Although the NCA is not well behaved in the infrared limit [4.71], it is remarkably accurate for a number of response functions on a scale ranging from photoemission etc. to temperatures far below the Kondo temperature [4.4]. Only if vertex corrections become important (small N, low density [4.72]), does the scheme seriously break down.

In the NCA the slave boson appears as an auxiliary device, introduced to get the diagrams working. However, there might be more in it, as suggested by a functional integral approach invented by *Newns* and *Read* [4.73]. Although it applies equally well to the impurity limit, let me illustrate the idea in the lattice case, where it leads to the concept of renormalized bands. The lattice generalization of (4.13) in slave boson representation is written as ($i = f$-site index)

$$\begin{aligned}
h = H^{\text{band}} &+ \sum_{in} \varepsilon_f f_{in}^\dagger f_{in} \\
&+ \sum_{ink} [V_{in}(\boldsymbol{k}) b_i f_{in}^\dagger c_{kn} + h.c.] + \sum_i \lambda_i \left(b_i^\dagger b_i + \sum_n f_n^\dagger f_n - 1 \right),
\end{aligned} \tag{4.32}$$

collecting the non-f bands in H^{band} and the constraint is imposed by the Lagrange multiplier λ_i. The trick is now to construct a mean-field approximation for (4.32), by assuming a site-independent Lagrange multiplier and a macroscopic

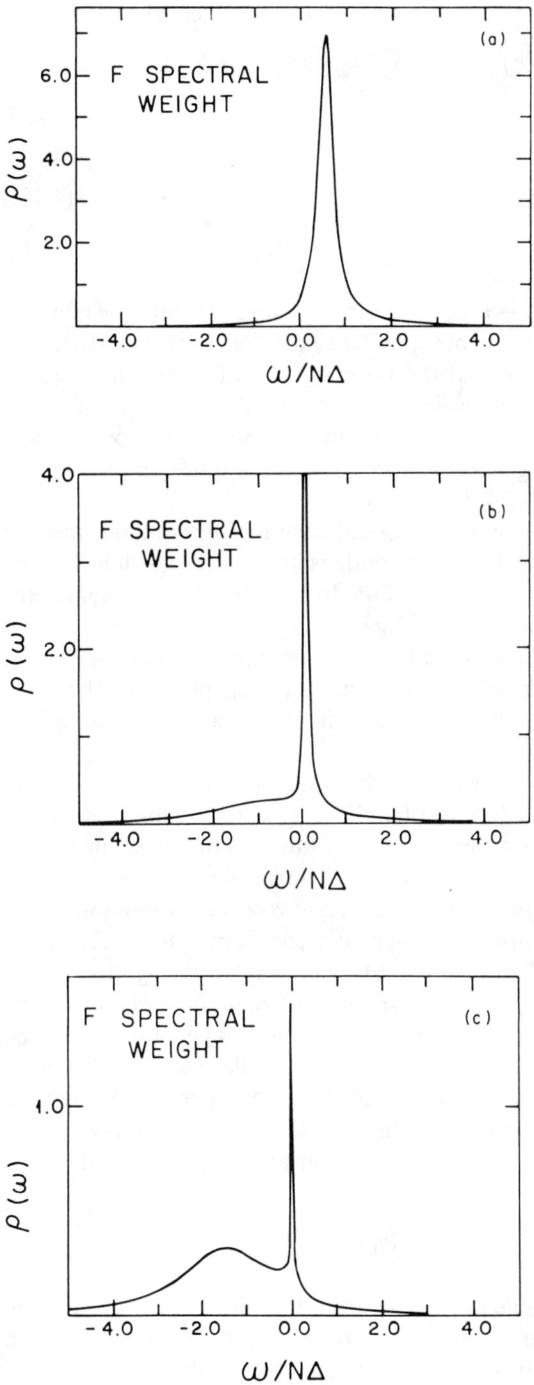

Fig. 4.23. Theoretical f-spectrum for $U \to \infty$, $N = 6$ and bandwidth $= 10N\Delta(\Delta = \Gamma)$ calculated using the non-crossing approximation [4.69]. (a) $\varepsilon_f = 0$, empty impurity regime, (**b**) $\varepsilon_f/N\Delta = -1.1$, mixed-valence regime, (c) $\varepsilon_f = -2.0N\Delta$, Kondo regime

occupation of the slave boson $z^{1/2} = \langle b \rangle = \langle b^\dagger \rangle$. The mean-field Hamiltonian is

$$H^{\text{MF}} = H^{\text{band}} + \sum_{in} (\varepsilon_f + \lambda) f^\dagger_{in} f_{in} + z^{1/2} \sum_{ink} [V_{in}(\boldsymbol{k}) f^\dagger_{in} c_{\boldsymbol{k}n} + h.c.] + \lambda \sum_i (z-1), \tag{4.33}$$

leading to the saddle-point equations

$$z = 1 - \langle n_f \rangle, \tag{4.34a}$$

$$z^{1/2} \sum_{nk} V_{in}(\boldsymbol{k}) \langle f^\dagger_{in} c_{\boldsymbol{k}n} \rangle = \lambda, \tag{4.34b}$$

One can see what happens if $n_f \to 1$: according to (4.34a) $z \to 0$, and the effective hopping tends to zero, expressing they physical fact that a quasiparticle can only hop if it finds an empty f-site. If the hopping tends to zero, the effective f-level has to approach the Fermi level in order to fulfill $n_f \to 1$, and one thus obtains a narrowed band, pinned to the Fermi level (in the impurity case $T_{\text{K}} \to W \exp(\pi \varepsilon_f / N V^2)$). This is generally believed to be the type of physics going on in heavy fermion systems [4.74, 75].

In this mean-field approximation, the bold assumption is made that the slave boson has a physical meaning: its condensation is responsible for the symmetry breaking and coherence of the QP. In fact, this kind of symmetry breaking is rigorously forbidden by local gauge invariance. However, it has been asserted that this symmetry is nearly, but not quite, broken [4.73, 76]: the boson autocorrelation falls off as the inverse $(1/N)$th power of the time, indicating that the slave bosons fluctuate very slowly, eventually freezing in if $N \to \infty$.

A serious drawback of the above approach is the unsymmetric way in which charge- and spin-degrees of freedom are handled. By construction, the charge fluctuations are treated in zero order, while spin fluctuations show up for the first time in order $1/N^2$, and this is a rather inconvenient starting point for studying cases where magnetism plays an important role (heavy fermions, high T_c). Recently *Kotliar* and *Ruckenstein* [4.77] proposed an approach which seems to cure this problem. Instead of a single slave boson, counting the charge fluctuation, also slave (vector [4.78]) bosons are introduced to represent the spin of e.g. the f^1 state. From this formulation, a mean-field theory can be obtained which turns out to be identical to the Gutzwiller mean-field (GMF) theory. From the conventional derivation of the GMF, one learns that an approximation is involved of another type than in the Newns/Read slave boson theory. The Gutzwiller ansatz is a Jastrow-type variational ground-state wave function [4.79].

$$|\Psi^{\text{GA}}\rangle = \prod_i [1 - (1-g) d^\dagger_{i\uparrow} d_{i\uparrow} d^\dagger_{i\downarrow} d_{i\downarrow}] |\Phi_0\rangle, \tag{4.35}$$

where one projects out the double occupancy from the single determinant state $|\Phi_0\rangle$. In order to simplify the evaluation of the expectation values, a mean field approximation is introduced amounting to the neglect of the spin-configuration dependence of the matrix elements (GMF) [4.79, 80]. Applied

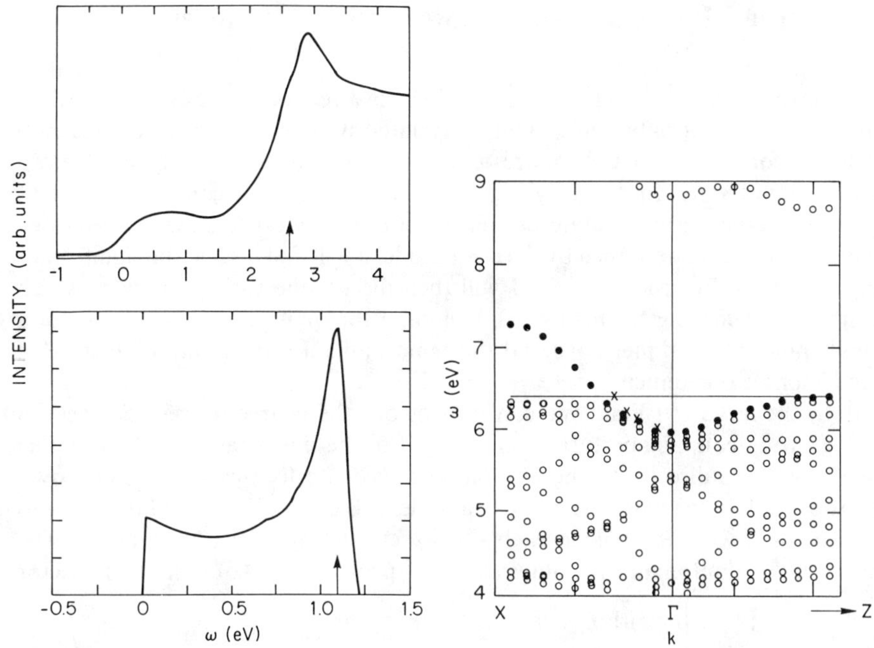

Fig. 4.24. Left: experimental IPES spectrum of the high-Tc superconductor $Bi_2Sr_2CaCu_2O_8$ [4.129] compared to a theoretical result for the d-IPES obtained from slave bosson renormalized band theory [4.87]. Right: dispersion of quasiparticle bands according to slave boson theory [4.87], compared to ARPES data (crosses) [4.130]

to (4.32) this leads to a mean field theory, similar to (4.33) except that the z factors and the Lagrange multipliers become spin dependent [4.81–83]. The GMF predicts the magnetic instability in e.g. the high-T_c cuprates correctly [4.84, 85], and it has been shown that the local quantities as derived from GMT are (also for small N) in excellent agreement with quantum Monte Carlo results [4.86]. It is still waiting for a calculation of G_1, starting from the GMT.

What has this all to do with PES-IPES? These strong-coupling mean-field theories are quasiparticle theories. In lowest order the QP bands are obtained and the large-energy scale properties show up if the fluctuations are included. In Fig. 4.24 the IPES is shown, obtained in this way from the Newns/Read theory [4.87]. The model is thought to be representative of the (unoccupied) Cu $3d$–O $2p$ states of the perovskite planes in the high-T_c superconductors. The peak at ~ 1 eV corresponds, roughly speaking, with the $d^9 \rightarrow d^{10}$ Hubbard band. However, instead of the Kondo peak onset, as in the analogous Ce-PES case, a broad continuum is seen between F_F and the d^{10} peak. *This reflects the presence of a quasiparticle band with a modest* (~ 2) *mass enhancement, and a pole strength reduced by about the same factor*. This quasiparticle band of course persists below E_F, and in Fig. 4.24 the AR-PES dispersion (e.g. Fig. 4.1) is compared with the calculated QP dispersion [4.87, 88].

4.4 Doped Mott–Hubbard Insulators: the $t-J$ Model

The slave boson mean-field approach of the last section suggests an appealing picture of band formation in strongly correlated systems. However, very recently it has become clear that there are other options than the slave boson 'heavy' Fermi liquid state. Inspired by high T_c, there has been a concerted effort to unravel the near-ground-state properties of doped Mott–Hubbard insulators, which is in essence described by the $t-J$ model. I will first discuss the significance of this model. Focussing on G_1, I will then discuss the variety of states which can occur according to this model, including the quasiparticle behaviour of a single hole in two dimensions and the remarkable 'Luttinger liquid' picture of Anderson in one dimension.

In the large U/t limit one can project out the charge degrees of freedom using strong-coupling perturbation theory. For the single band Hubbard model (4.5) this leads directly to the $t-J$ model. Technically this works as follows: consider half filling with one additional hole. If $U/t \to \infty$ the only allowed states are the ones with one hole at each site, except the site which is empty (i). The effective Hamiltonian is obtained by projecting (4.5) on this sector $\left(P_1 = \sum_i \left[\prod_{i' \neq i} \hat{n}_{i'}\right] \bar{n}_i \text{ with } \hat{n}_i = n_{i\uparrow} + n_{i\downarrow} - 2n_{i\uparrow}n_{i\downarrow} \text{ and } \bar{n}_i = (1-n_{i\uparrow})(1-n_{i\downarrow})\right)$

$$H^0_{\text{eff}} = P_1 H^{\text{Hub}} P_1 = \sum_{\langle ij\rangle\sigma} t_{ij} \tilde{c}^\dagger_{i\sigma} \tilde{c}_{j\sigma}, \tag{4.36a}$$

where $\tilde{c}^\dagger_{i\sigma} = (1-n_{i-\sigma})c^\dagger_{i\sigma}$. One finds (trivially) that the additional hole is free to hop, as long as it does not find a site which is already empty. For finite U, fluctuations occur to other Hubbard bands and perturbation theory [4.89] tells us that the second order contribution is given by (see Fig. 4.25) $H^2_{\text{eff}} = -\frac{1}{U} P_1 H^{\text{Hub}} P_2 H^{\text{Hub}} P_1$, where $P_2 = \sum_{i \neq k \neq l} \left[\prod_{i' \neq i,k,l} \hat{n}_{i'}\right] \bar{n}_i \bar{n}_k \tilde{n}_l$ (with $\tilde{n}_l = n_{l\uparrow}n_{l\downarrow}$) projects on the sector with one additional electron–hole excitation. Omitting some terms which are irrelevant in practice, this yields the Heisenberg Hamiltonian (see Fig. 4.25)

$$H^2_{\text{eff}} = 2J \sum_{\langle ij\rangle} \boldsymbol{S}_i \cdot \boldsymbol{S}_j, \tag{4.36b}$$

where $J = 2t^2/U$. Thus, the spins on the occupied sites are antiferromagnetically coupled and this is Anderson's superexchange mechanism [4.90]. Equations (4.36a) and (4.36b) define the $t-J$ model.

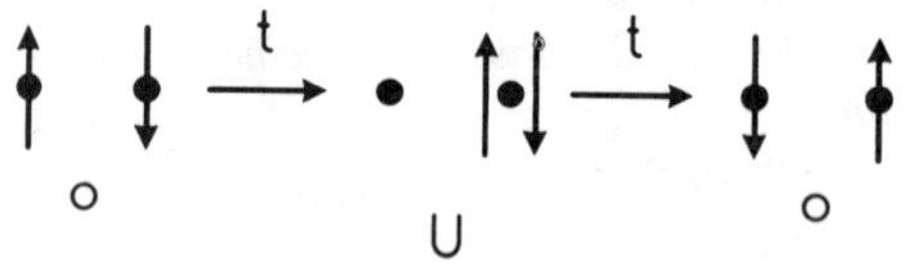

Fig. 4.25. Lowest order fluctuations in the strong-coupling limit, giving rise to superexchange

It is often argued that the $t-J$ model is in fact more general than the single-band Hubbard model. For the Heisenberg part this is unproblematic. It reflects the remarkable (and perfectly general) phenomenon that the charge and the spin of the electron are separated at energies lower than the gap. The reason is that the constraint works only on the charge and not on the spin of the electron. The description of the hole is not so obvious in cases when, for example, the holes are predominantly O $2p$-like, as in the high-T_c cuprates [4.41]. However, it can be shown that at least in one limit the $t-J$ model is also relevant for this case. This involves a non-trivial mapping, discovered by *Zhang* and *Rice* [4.91], which might be of more general applicability. The basic structural unit, common to all cuprate high-T_c superconductors, are the Cu-O perovskite planes (Fig. 4.26). The simple model proposed by *Emery* [4.92] is the lattice generalization of the impurity model, specified to these planes, including only the Cu $d_{x^2-y^2}$- and two O p orbitals in each unit cell (Fig. 4.26). At half filling, the holes are predominantly localized on the Cu sites and the statement of Zhang and Rice is that, upon adding holes by doping, a low-energy sector splits off which can be described by the $t-J$ model, also *if the holes have predominantly oxygen character* ($U \gg \Delta$). This is straightforward to show, assuming that the width of the oxygen band is zero, together with $U, \Delta, U-\Delta \gg t$ (strong coupling) [4.91]. The oxygen orbitals are combined in Cu-centered Wannier states, which are bonding or non-bonding with respect to the $d_{x^2-y^2}$ orbital. The spin of the oxygen hole in the bonding orbital has an antiferromagnetic Kondo-type interaction with the Cu spin [4.93]. One can construct a basis by combining the p-hole and a Cu spin in 'local' singlets and higher lying local triplets, etc. The former are at lower energy (ΔE_s) with respect to all other states and if one restores the translational symmetry, one finds that these local singlet (and other) states broaden into bands. The surprise is that *the widths of these bands are much smaller than the local singlet binding energy* (ΔE_s) *and therefore a gap separates the local singlet sector from all other states*. This is remarkable, because there was only a single scale in the original Hamiltonian (the Kondo exchange),

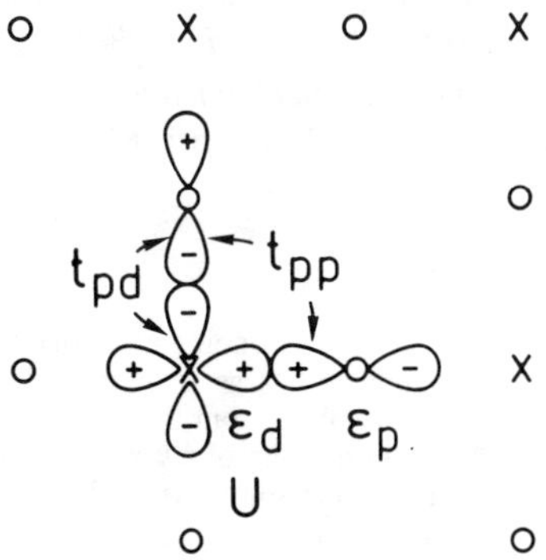

Fig. 4.26. The orbitals included in the Emery model [4.92] for the perovskite planes of the cuprate superconductors (crosses represent Cu and circles oxygens)

and the gap has a purely dynamical origin. The local singlet cannot be broken up if $\omega < \Delta E_s$ and we can think of it as the hole in the $t-J$ model. Finally, the most important [4.94] interactions appearing in the effective low-energy Hamiltonian are the ones of the $t-J$ model, although t and J have a different relationship than in the single-band model.

Let us now turn to the motion of a single hole according to the $t-J$ model. A priori, one cannot follow the perturbative procedure developed for the electron–lattice problem. The coupling between hole and spin system is essentially non-perturbative; if the hole hops from one site to the next, a spin is transferred in the opposite direction, leading to a strong distortion surrounding the hole. Things are further complicated by the quantum nature of the spins (assuming $S = 1/2$). Before going to the more conventional situation in two and higher space dimensions, let me first highlight the bizarre picture which holds in 1D: *Anderson's* [4.13, 18] (*Carmelo* and *Ovchinnikov's* [4.95] and *Haldane's* [4.96]) 'Landau–Luttinger liquid'.

Imagine a chain with non-interacting, antiferromagnetically oriented, spins. *The movement of a hole in such a chain is completely incoherent*, as made clear in the seminal work of *Brinkman* and *Rice* [4.97]. This is easy to see, using a real space basis. Let us start with the spin system and take away an up spin residing at site i (the hole). Acting repeatedly with the hopping operator on this state will produce a Hamiltonian which is identical to that of a free particle moving in the chain, as is easily checked. However, *this does not mean that the wave function can be written as a Bloch wave.* Every time the hole hops in one direction, a spin is transported in the other direction (Fig. 4.27). Therefore, for a given position of the hole the spin content of the total wave function is orthogonal to that corresponding with the hole at any other position. This implies for the single-hole spectral function [4.97]

$$A(k,\omega) = \frac{1}{\pi}\,\mathrm{Im}\left\{\frac{1}{\sqrt{\omega^2 - 4t^2}}\right\}. \tag{4.37}$$

The spectrum is independent of the wavenumber of the hole, and for all wavenumbers one measures the k-integrated density of states of the corresponding free system!

How may we understand this result? The clue is that for any finite interaction strength or concentration [4.95, 96], *only the charge* ('*holon*') *and the spin* ('*spinon*') *of the electron exist in one dimension, but not the electron itself.* This is trivial in the Brinkman–Rice example. Comparing Fig. (4.27b), with the hole

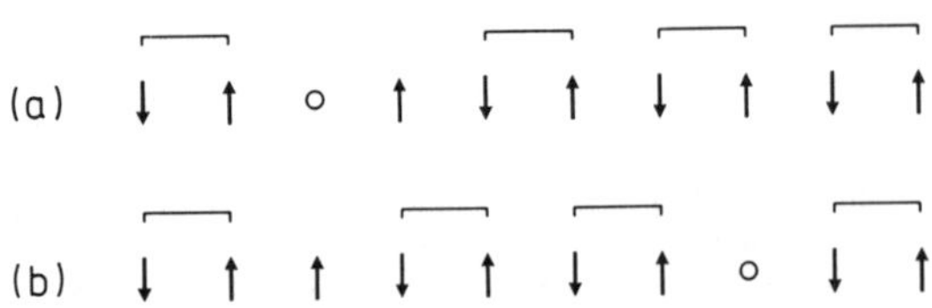

Fig. 4.27. The hopping of a hole in a chain in the large U limit. If the hole hops away from the origin (**a**) it creates a string of reversed spins (**b**) and a spin soliton is left behind at the origin

infinitely far from the origin, with Fig. (4.27a) with the down spin put back at the origin, the difference is clear: in Fig. (4.27b) a domain wall (or spin soliton) has been created at the origin and this carries the spin quantum number of the hole ($M_s = 1/2$, the hole being a missing down spin). Clearly, the charge of the hole is the (previously unidentified) freely hopping object (with dispersion $\omega^h = 2t\cos(k^h a)$) and the spectrum equation (4.37) is given by the convolution of the holon and (localized, $\omega^s(q) = 0$) spinon Green's functions $\sim \int dq d\omega' \delta(\omega' - \omega^h(k+q))\delta(\omega + \omega' - \omega^s(-q))$, because the hole is a composite particle.

What happens if the spin–spin interactions are turned on? Figure 4.28 shows where the states, having weight in the single-particle spectral function, are located in momentum-energy space, according to the exact diagonalization of a 16-site chain with one hole and periodic boundaries [4.98]. This result suggests that the spectrum is completely incoherent in the thermodynamic limit, as turns out to be the case. In Fig. 4.29 results are shown, obtained by *Becker* and *Fulde* using the Zwanzig–Mori formalism [4.99] (see also [4.100]), and we conclude that the hole has to be a composite particle, as in the Brinkman–Rice case. However, the new aspect is, that the *upper and lower bounds of the spectra are k-dependent.* The reason is that not only the Brinkman–Rice holon survives, but the spinon turns also into a propagating fermion-like QP. It is essential that the ground state of the spin system is different from the Néel state, which would not allow for such an elementary excitation. In 1D it is better to think of a resonating valence bond (RVB) state. One first binds pairs of spins in singlets, and the ground-state wave function is a linear superposition of all possible pairings on the lattice. In analogy with polyacetelene [4.101], the 'lone spin' $M_s = 1/2$ soliton (Fig. 4.27b) is then a good quasiparticle [4.102]. Indeed, if one calculates numerically the excitation spectrum of a uneven (15-site)

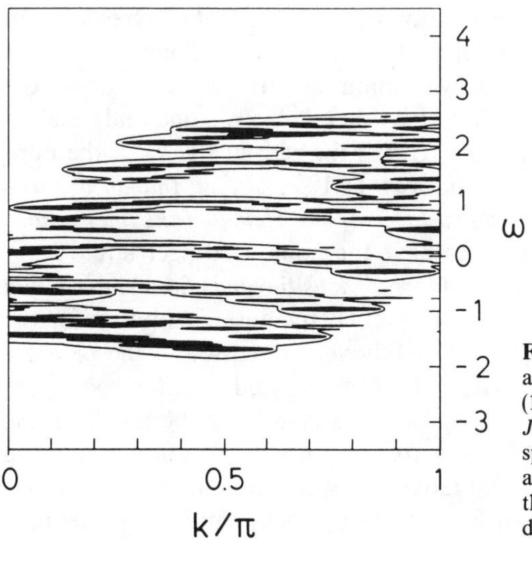

Fig. 4.28. Contour plot of $A(k,\omega)$ for a single hole in the one-dimensional (16-site) Heisenberg spin chain ($t = 1$, $J = 0.2$). The energy minimum of the spectrum is at the single particle $k_F \sim \pi/2$ and for $k < k_F(k) > k_F$) the lower bound of the spectrum is given by the spinon (holon) dispersion [4.98]

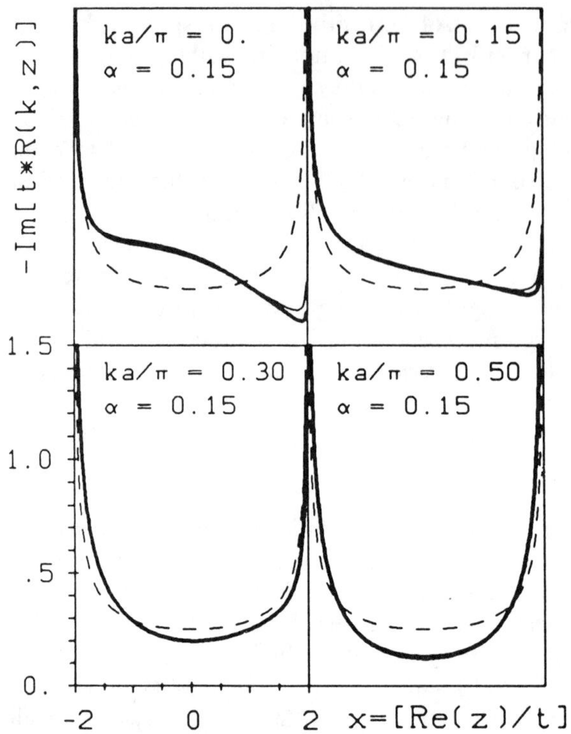

Fig. 4.29. The shape of $A(k, \omega)$ of a single hole in the one-dimensional Heisenberg spin chain with infinite length calculated using a mode coupling approximation (thick solid line, α = spin asymmetry parameter and $t/J = 0.1$), compared to the Brinkman–Rice spectrum (dashed line). Because the spin fluctuations are not treated correctly, the spinon dispersion is not picked up [4.99]

Heisenberg chain, one finds such an excitation which is like that of a free fermion in a half-filled chain ($\omega^2 \sim 2J\cos(k_s a)$ with $k_s < k_F \sim \pi/2$). Convoluting this spinon with the Brinkman-Rice holon (with minimum at $\sim \pi = 2k_F$), under the constraint that they are colinear (i.e. if $k_s < k_F$, then $k_h > 2k_F$), one finds exactly the distribution of Fig. 4.28. Amusingly, G_1 hits the zero at the k_F of the non-interacting system, as if *the Luttinger theorem* [4.103], *stating that in a Fermi liquid the volume enclosed by the Fermi surface will not change if the interactions are switched on, is obeyed.* Therefore, we have the same 'Luttinger liquid'.

The above spinon–holon picture is valid for all densities and (finite) interaction strength. Already since the late sixties, an exact (Bethe ansatz) solution for the Hubbard chain has been around [4.104]. Only very recently did the physical content of this solution become clear, when *Carmelo* and *Ovchinnikov* [4.95] showed that it boils down to a two-component Landau Fermi liquid. As in the single-hole case, the spinons and holons are the correct QP obeying a Pauli exclusion principle, having their Fermi momenta at the free particle k_F and $2k_F$, respectively, and the hole Fermi surface therefore obeys Luttinger's theorem.

Although the holons and spinons interact, their lifetimes go to zero upon approaching their Fermi momentum, as in a Fermi liquid (therefore 'Landau'). All near-ground-state properties can therefore be expressed in terms of spinon/holon properties and their interaction parameters, for which explicit expressions can be deduced from the Bethe ansatz solution.

The question arises of whether the same charge–spin separation could happen in higher dimensions. Anderson's assertion about the cusp-like nature of the spectra in Fig. 4.1 is in fact based on an extrapolation of the 1D convolution argument to two dimensions, where one expects a cusp at the holon energy [4.13]. However, the case of a single hole in a two-dimensional Heisenberg system is now cleared up and here it does not happen. If the spins are non-interacting, the problem bears similaritics to thc 1D case. One can divide the hopping pathways into the ones with or without closed loops. The former are much less abundant than the latter (Fig. 4.30b) and can be neglected in first instance. In this approximation, one has only to take into account the hopping walks on the corresponding Bethe lattice (retractable path approximation [4.97]), which is similar to the 1D problem, and the spectra are also incoherent. However, this is not completely true. As *Trugman* argued [4.105], by going 3/2 times around a plaquette, the hole can restore the spin background (Fig. 4.30b), and this will lead to small but finite dispersions.

A new aspect is introduced if the spins are interacting. This is most easily seen if one considers classical (Ising) spins. In two and more dimensions (near) self-localization occurs, by a mechanism which is unique for the problem. As *Bulaevskii* et al. pointed out [4.106], a hole moving in an interacting spin background will experience a potential which is proportional to the length of the hopping path ('string potential'). This is easily seen (Fig. 4.30a): every time the hole hops, a spin is transported in the opposite direction, and it ends up with the wrong orientation with respect to its neighbours. From a holon–spinon

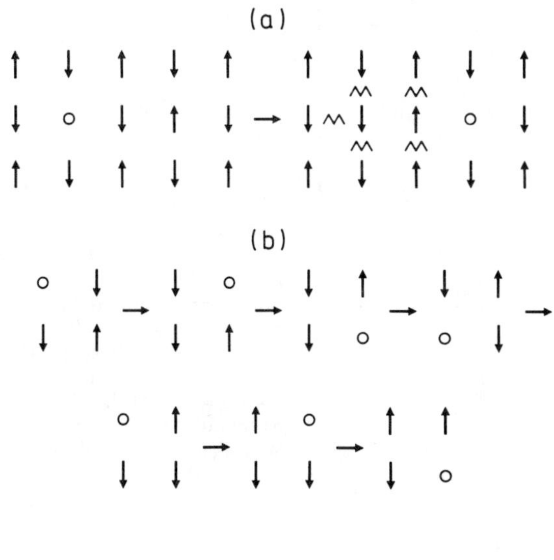

Fig. 4.30. In $D > 1$, the reversed spins in the string have the incorrect orientation with respect to their neighbours (**a**). By going 3/2 times around a plaquette, the hole can repair the damage in the spin system, giving rise to coherent transport [4.105]

perspective, this is similar to quark confinement. The charge and the spin of the hole experience a force increasing with their separation and they are therefore bound. In the (small J_z) continuum limit, the hole experiences a linear potential [4.107]

$$H = -\sqrt{z-1}\,t\frac{\delta^2}{\delta x^2} + J_z x - 2\sqrt{z-1}\,t, \tag{4.38}$$

and this Hamiltonian has a *completely discrete spectrum*, with eigenvalues

$$E_n = (J_z/t)^{2/3}\,t a_n - 2\sqrt{z-1}\,t, \tag{4.39}$$

i.e. spaced apart by $\sim t(J_z t)^{2/3}(a_n \cong 1)$. These discrete eigenstates correspond with the internal excitations of the Bulaevskii polarons. By including the Trugman walks (Fig. 4.30b), these spikes will get a small dispersion [4.108].

Finite-size diagonalization studies have played a key role in clarifying the role of the spin quantum fluctuations in this problem. In Fig. 4.31 $A(\boldsymbol{k},\omega)$ is shown for momentum $\boldsymbol{k} = \left(\frac{\pi}{2},\frac{\pi}{2}\right)$, calculated from a 4×4 Heisenberg cluster by *Horsch* and coworkers [4.98] (see also [4.109]). This spectrum is characterized by a large, isolated peak at threshold, separated by a gap of the order of several J from the rest of the spectrum. The spectral functions at other points in k-space look similar, and the position of the isolated low-energy peak can be monitored as a function of $\boldsymbol{k}$. This yields the quasiparticle dispersion as shown in Fig. 4.32 [4.98]. This dispersion is in fact close to the form (drawn line in the figure)

$$\omega_{\boldsymbol{k}} = J[\tfrac{1}{2}(\cos k_x + \cos k_y)^2 - 1], \tag{4.40}$$

and this is the dispersion of a free-particle hopping on only one of the two AF

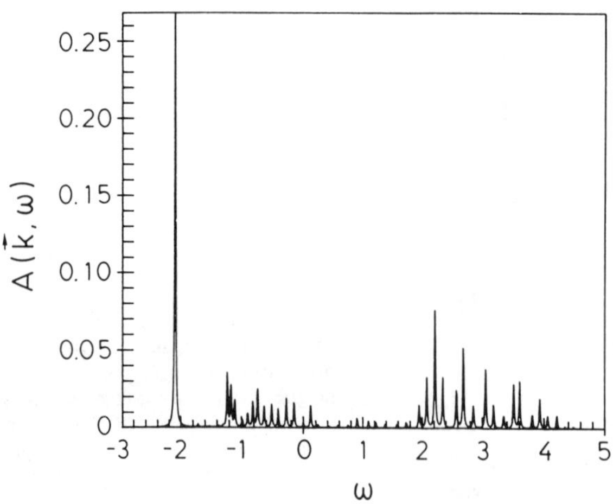

Fig. 4.31. Spectral function $A(\boldsymbol{k},\omega)$ for $\boldsymbol{k} = (\pi/2,\pi/2)$ of a single hole in a two-dimensional (16-site) Heisenberg spin system ($t=1, J=0.2$). At the bottom of the spectrum a large quasiparticle peak is seen [4.98]

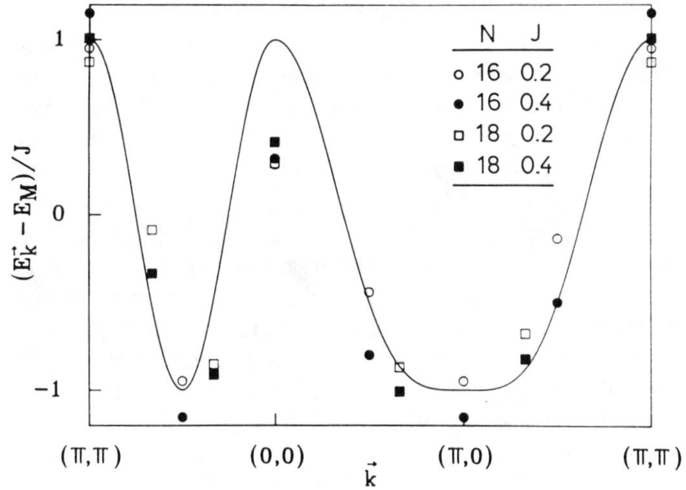

Fig. 4.32. The dispersion of the quasiparticle peak (Fig. 4.29) along the principle axis of the Brillioun zone after subtracting a constant energy (E_M) and normalization by J. The numerical results for 16 sites, open (solid) circles, and 18 sites, open (solid) squares are given for $J = 0.2$ (0.4). The drawn line correspond with the 'on sublattice' dispersion (4.40) [4.98]

sublattices, with a (on-sublattice) hopping matrix element J! Finally, the relative weight in this low-energy peak, averaged over the Brillioun zone, is $\sim J/t$ [4.98], suggesting that the conventional relation $Z_k \sim m/m^*$ between QP pole strength and mass enhancement holds.

The above result can be understood from the approach developed by *Schmitt-Rink* et al. [4.110] and applied to the present problem by *Kane* et al. [4.111]. In contrast to 1D, the Néel state is a good starting point, and its elementary excitations (magnons) should play a key role. After applying a Holstein–Primakoff transformation, the hole–spin coupling takes the fimiliar electron–phonon form $\sim f^\dagger_k f_{k-q}(b^\dagger_{-q} + b_q)$. Also after linearizing the spin waves, the problem cannot be solved exactly and it is further approximated by neglecting the hole–spin interaction vertex corrections, as in GW or NCA. In the present case, the boson propagator is not renormalized because of the vanishing hole concentration and the self-consistency equation takes the simple form

$$G(\boldsymbol{k}, \omega) = \frac{1}{\omega - \sum_{\boldsymbol{q}} f(\boldsymbol{k}, \boldsymbol{q}) G(\boldsymbol{k} - \boldsymbol{q}, \omega - E_{\boldsymbol{q}})}, \tag{4.41}$$

where $E_{\boldsymbol{q}}$ is the spin wave energy, $f(\boldsymbol{k}, \boldsymbol{q})$ the bare hole–magnon interaction vertex and G the hole propagator.

Applying this approach to the Heisenberg case yields spectral functions in good agreement with the numerical results [4.111, 112]. *This is a quite significant result because it indicates that we can interpret these results in terms of small polaron physics.* The QP is thus the hole itself, surrounded by a cloud of distorted spins. Besides the Fermi-liquid relation between pole strength and mass enhancement, this also explains why the QP does not interact with low-lying spin excitations. The hole will have quadratic dispersion (e.g. Fig. 4.32) and the magnons have a relativistic spectrum and then conservation of energy and momentum forbids scattering of the (heavily dressed) holes to lower energy by single- or multiple-spin excitations [4.111]. The remaining question is about the origin of the 'on-sublattice' shape of the QP dispersion. This can be understood qualitatively as follows [4.98]: if the hole has hopped twice, it is on the same sublattice as where it started from, and at the same time one has only to act once with the transverse part of the Heisenberg Hamiltonian to flip the spins back to their original positions. Therefore, the combination of two hops and one spin flip allows the hole to propagate coherently, and J is the limiting factor for the quasiparticle bandwidth.

4.5 Concluding Remarks

It might be clear from this chapter that the field of 'strongly correlated electrons' is, despite the rapid progress in understanding, still characterized by many open questions. Let me finish this chapter with what I feel are some main lines extending in the future from a theoretician's perspective.

Although there is little doubt that the Hubbard model approach to the large-energy-scale electronic structure makes at least sense from a phenomenological point of view, the success of this approach is not completely understood. A central issue is how to rationalize the use of effective parameters, such as the screened U. Although the adiabaticity argument of Sect. 4.1 can be shown to be sound by explicit calculation [4.113], it is not always clear if the time scales are indeed so different [4.114, 38]. Even if the adiabaticity argument holds, there are intricacies. For instance, upon changing the charge state of the atom, the orbitals will breath and this changes the overlap to the neighbours, leading to a charge-state dependence of the hopping [4.32, 115, 116]. Alternatively, if one changes the charge states of the correlated ion, the uncorrelated electrons will screen these changes. However, this implies that the hopping matrix elements have to be affected because the uncorrelated electron orbitals before and after the hopping are no longer orthogonal. This leads to an X-ray singularity problem, and it can be shown that the effective hopping becomes quite energy dependent [4.117].

In the same context, the Hubbard model approach is bound to break down at some point if the electrons become less localized. Where and how this happens is far from clear. There might exist a region where many interactions are relevant, instead of the few taken into account in contemporary models. If so, this calls for a first-principle theory, even if one is after qualitative understanding. In this respect, I mention the spectroscopy of uranium compounds which up to now has escaped any reasonable interpretation.

The final goal should, of course, be a quantitative ab initio theory for strongly correlated materials. Although this goal is still remote, some recent progress is worthwhile mentioning. The traditional quantum chemist's approach is hampered by the shortcomings of the (ab initio) Hartree–Fock approximation, yielding much too localized solutions. However, it was shown recently that a relatively limited amount of configuration interaction strongly improves the situation [4.118]. The same is found from the conceptually more transparent local ansatz of *Stollhof* and *Fulde* [4.119, 120].

Straightforward application of LSDA to Mott–Hubbard insulators leads to disasters such as the non-magnetism of La_2CuO_4 [4.121], the crystal field gap of NiO and the metallicity of CoO or FeO [4.122]. The problem is that the interaction, linking the magnetization (and orbital polarization) to the effective single–particle potentials, is of the order of the Hund's rule exchange, according to the homogeneous electron gas exchange-correlation potentials. To get a sensible theory, this interaction should be of order U, which is typically an order of magnitude larger. Based on this notion, very recently *Anisimov* et al. [4.123] proposed a modified functional, marrying the LDA-parameter approach with the band calculation. In this approach, at least the large-scale LSDA problems seem to be cured. One can also look from a different perspective to this 'U versus J' problem. In LSDA, the Kohn–Sham QP interact with themselves, and in principle this unphysical interaction should be subtracted. This so-called self-interaction correction (SIC) leads to a theory with a structure which is clearly different from the (LSDA) weak-coupling form. *Svane* and *Gunnarsson* showed very recently that SIC gives a much improved description of the electronic structure of the $3d$ monoxides [4.124]. Also, in this case the U seems to be responsible, although it is now interpreted as a self-interaction effect.

As might be clear from the last two sections, the knowledge about band formation in strongly correlated systems is still highly fragmented, and the little we know inspires further study. Definitely, large-energy-scale spectroscopies will play an increasingly important role here. Besides Fig. 4.1, there is another important piece of data: the ARPES spectra of NiO, showing quasiparticle peaks of predominantly O $2p$ character, which disperse very much like the O $2p$ bands in the LDA calculation [4.125] (see also [4.126]). At first sight, this seems to contradict the impurity model picture, which suggests a strong O $2p$-Ni $3d$ hybridization, which would tend to destroy the O $2p$ single–particle momentum. This puzzle can be resolved by exploiting the recent advancement in the theory [4.127]. The first step is to project out the Ni charge degrees of freedom by assuming that $U - \Delta, \Delta \gg t$, thereby exaggerating the problem a little. The linear spin wave approach of Sect. 4.4 is then applied to the resulting 'spin-fermion' model and the result shows exactly what happens in the experiment. The Kondo interactions give rise to a local singlet-like spectrum with d-like features at the top of the valence band. The O $2p$ bands stay close to what they are in the absence of the Ni spins, except that they acquire a life-time width, which is still significantly smaller than the total p-bandwidth.

References

4.1 O. Gunnarsson, K. Schönhammer: Phys. Rev. B **28**, 4315 (1983)
4.2 O. Gunnarsson, K. Schönhammer: Phys. Rev. B **31**, 4815 (1985)
4.3 O. Gunnarsson, K. Schönhammer: In *Handbook of the Physics and Chemistry of the Rare Earth*, Vol. 10, ed. by K.A. Schneider Jr., L. Eyring, S. Hüfner (Elsevier, Amsterdam 1987)
4.4 N.E. Bickers, D.L. Cox, J.W. Wilkins: Phys. Rev. B **36**, 2036 (1987)
4.5 J. Zaanen, J.W. Allen, G.A. Sawatzky: Phys. Rev. Lett. **55**, 418 (1985); J. Magn. Magn. Mater. **54–57**, 607 (1986)
4.6 J. Zaanen, G.A. Sawatzky: Can. J. Phys. **65**, 1262 (1987)
4.7 A. Fujimori, F. Minami, S. Sugano: Phys. Rev. B **29**, 5225 (1984); A. Fujimori, F. Minami: Phys. Rev. B **30**, 957 (1984)
4.8 A. Kotani, T. Jo, J.C. Parlebas: Adv. Phys. **37**, 37 (1988)
4.9 Y. Baer, W.D. Schneider: In *Handbook of the Physics and Chemistry of the Rare Earth*, Vol. 10, ed. by K.A. Schneider Jr., L. Eyring, S. Hüfner (Elsevier, Amsterdam 1987)
4.10 J.W. Allen, S.J. Oh, O. Gunnarsson, K. Schönhammer, M.B. Maple, M.S. Torikachvili, I. Lindau: Adv. Phys. **35**, 275 (1986)
4.11 P.A. Lee, T.M. Rice, J.W. Serene, L.J. Sham, J.W. Wilkins: Comm. Cond. Matt. Phys. **12**, 99 (1986)
4.12 K.C. Hass: Solid State Physics, Vol. 42, ed. by C.H. Eherenreich, D. Turnbull (Academic, New York 1989); W.E. Pickett: Rev. Mod. Phys. **61**, 433 (1989)
4.13 P.W. Anderson: Int J. Mod. Phys. B **4**, 181 (1990)
4.14 C.M. Varma, P.B. Littlewood, S. Schmitt-Rink, E. Abrahams, A. Ruckenstein: Phys. Rev. Lett. **63**, 1990 (1989)
4.15 F. Patthey, B. Delley, W.-D. Schneider, Y. Baer: Phys. Rev. Lett. **55**, 1518 (1985); J.M. Imer, F. Patthey, B. Dardel, W.-D. Schneider, Y. Baer, Y. Petroff, A. Zettl: Phys. Rev. Lett. **62**, 336 (1989); R. Manzke, T. Buslaps, R. Claessen, J. Fink: Europhys. Lett. **9**, 477 (1989)
4.16 C.G. Olson, R. Liu, D.W. Lynch, R.S. List, A.J. Arko, B.W. Veal, Y.C. Chang, P.Z. Jiang, A.P. Paulakis: Phys. Rev. B **42**, 381 (1990)
4.17 R. Claessen, R. Manzke, H. Carstensen, B. Burandt, T. Buslaps, M. Skibowski, J. Fink: Phys. Rev. B **39**, 7316 (1989)
4.18 P.W. Anderson, Y. Ren: In *High Temperature Superconductivity*, Proc. Int. Conf. on the Physics of Highly Correlated Electron Systems, (Addison-Wesley, Reading, MA 1990); P.W. Anderson: Phys. Rev. Lett. **64**, 1839 (1990); Phys. Rev. B **42**, 2625 (1190)
4.19 C. Herring: In *Magnetism*, ed. by G.T. Rado, H. Suhl (Academic, New York 1966)
4.20 O. Gunnarsson: Phys. Rev. B **41**, 514 (1990)
4.21 P.H. Dederichs, S. Blügel, R. Zeller, H. Akai: Phys. Rev. Lett. **53**, 2512 (1984)
4.22 O. Gunnarsson, O.K. Andersen, O. Jepsen, J. Zaanen: Phys. Rev. B **39**, 1708 (1989)
4.23 J.F. Herbst, R.E. Watson, J.W. Wilkins: Phys. Rev. B **13**, 1439 (1976); ibid. Phys. Rev. B **17**, 3089 (1978); J.F. Herbst, J.W. Wilkins: Phys. Rev. Lett. **43**, 1760 (1979)
4.24 J. Hubbard: Proc. R. Soc. London A **277**, 237 (1964); Proc. R. Soc. London A **281**, 401 (1964)
4.25 H. Haak, G.A. Sawatzky, T.D. Thomas: Phys. Rev. Lett. **41**, 1825 (1978); G.A. Sawatzky, A. Lenselink: Phys. Rev. B **21**, 1790 (1980)
4.26 N.F. Mott: Proc. Phys. Soc. Sect A **62**, 416 (1949)
4.27 J. Zaanen, M. Alouani, O. Jepsen: Phys. Rev. B **40**, 837 (1989)
4.28 J.C. Fuggle, M. Campagna, Z. Zolnierek, R. Lässer, A. Platau: Phys. Rev. Lett. **45**, 1597 (1980)
4.29 D. van der Marel, G.A. Sawatzky: Phys. Rev. B **37**, 10674 (1988)
4.30 M.R. Norman, A.J. Freeman: Phys. Rev. B **33**, 8896 (1986)
4.31 G.A. Sawatzky, J.W. Allen: Phys. Rev. Lett. **53**, 2239 (1984); S. Hüfner, J. Osterwalder, T. Riesterer, F. Hulliger: Solid State Commun. **52**, 793 (1984)
4.32 J. Zaanen, C. Westra, G.A. Sawatzky: Phys. Rev. B **33**, 8060 (1986)
4.33 G. van der Laan, J. Zaanen, G.A. Sawatzky, R. Karnatak, J.-M. Esteva: Phys. Rev. B **33**, 4253 (1986)
4.34 C.R. Ronda, G.J. Arends, C. Haas: Phys. Rev. B **35**, 4038 (1987)
4.35 O. Gunnarsson, A.V. Postnikov, O.K. Andersen: Phys. Rev. B **40**, 10407 (1989)
4.36 J. Zaanen, G.A. Sawatzky: J. Solid State Chem. **88**, 8 (1990)

4.37 Because we did not account for the strong Jahn-Teller distortion in Cu compounds, the U is underestimated
4.38 J. Folmer, D.K.G de Boer: Solid State Commun. **38**, 1135 (1981); D.K.G. de Boer, C. Haas, G.A. Sawatzky: Phys. Rev. B **29**, 4401 (1984)
4.39 F. Jellinek: In *Inorganic Sulphur Chemistry* (Elsevier, Amsterdam 1968) p. 669; MTP International Review of Science, Inorganic Chemistry Series I, Vol. 4 (Butterworths, London 1972) p. 339
4.40 P. Kuiper, G. Kruizinga, J. Ghijsen, G.A. Sawatzky, H. Verwey: Phys. Rev. Lett. **62**, 221 (1989); P. Kuiper, thesis, University of Groningen (1989)
4.41 N. Nücker, H. Romberg, X. X. Xi, J. Fink, B. Gegenheimer, Z.X. Zhao: Phys. Rev. B **39**, 6619 (1989); J.A. Yarmoff, D.R. Clarke, W. Drube, U.O. Karlssen, A. Talib-Ibrahimi, F.J. Himpsel: Phys. Rev. B **36**, 3967 (1987); P. Kuiper, G. Kruizinga, J. Ghijsen, M. Grioni, P.J.W. Weijs, F.M.F. de Groot, G.A. Sawatzky, H. Verweij, L.F. Feiner, H. Petersen: Phys. Rev. B **38**, 6483 (1988)
4.42 P.W. Anderson: Phys. Rev. **124**, 41 (1961)
4.43 G. van der Laan, C. Westra, C. Haas, G.A. Sawatzky: Phys. Rev. B **23**, 4369 (1981)
4.44 K. Okada, A. Kotani: J. Phys. Soc. Jpn. **58**, 2578 (1989)
4.45 O. Gunnarsson, J.W. Allen, O. Jepsen, T. Fujiwara, O.K. Andersen, C.G. Olson, M.B. Maple, J.-S. Kang, L.Z. Liu, J.H. Park, R.O. Andersen, W.P. Ellis, R. Liu, J.T. Markert, Y. Dalichaoux, Z.-X. Shen, P.A.P. Lindberg, B.O. Wells, D.S. Dessau, A. Borg, I. Lindau, W.E. Spicer: Phys. Rev. B **41**, 4811 (1990)
4.46 J.C. Fuggle, F.U. Hillebrecht, Z. Zolnierek, R. Lässer, Ch. Freiberg, O. Gunnarsson, K. Schönhammer: Phys. Rev. B **27**, 7330 (1983); E. Wuilloud, B. Delley, W.-D. Schneider, Y. Baer: Phys. Rev. Lett. **53**, 202 (1984)
4.47 S.J. Oh: In *Core-Level Spectroscopy in Condensed Systems*, Springer Series in Solid-State Sciences, Vol. 81, ed. by J. Kanamori, A. Kotani (Springer, Berlin, Heidelberg 1988), pp. 125
4.48 S. Imada, T. Jo: J. Phys. Soc. Jpn. **58**, 402 (1989); ibid. **58**, 2665 (1989)
4.49 A. Kotani, H. Ogasawara, K. Okada, B.T. Thole, G.A. Sawatzky: Phys. Rev. B **40**, 65 (1989)
4.50 J. Zaanen, G.A. Sawatzky, thesis, University of Groningen (1986)
4.51 F.M.F. de Groot, M. Grioni, J.C. Fuggle, J. Ghijsen, G.A. Sawatzky, H. Petersen: Phys. Rev. B **40**, 5715 (1989); J. van Elp, thesis, University of Groningen (1991)
4.52 In other respects things are not so simple, e.g. the $t-J$ model for electrons in CoO (J. Zaanen: Unpublished)
4.53 A.M. Oleś: Phys. Rev. B **28**, 327 (1983)
4.54 J.C. Fuggle, O. Gunnarsson, G.A. Sawatzky, Schönhammer: Phys. Rev. B **37**, 9919 (1988)
4.55 J. Zaanen, G.A. Sawatzky: Prog. Theor. Phys. Suppl. **101**, 231 (1990)
4.56 H. Eskes, G.A. Sawatzky: Phys. Rev. Lett. **61**, 1415 (1988)
4.57 O.K. Andersen, O. Jepsen: Phys. Rev. Lett. **53**, 2571 (1984); O.K. Andersen, O. Jepsen, K. Glötzel: In *Highlights of Condensed Matter Theory*, ed. by F. Basani, F. Fumi, M.P. Tosi (North Holland, New York 1985)
4.58 K.G. Wilson: Rev. Mod. Phys. **47**, 773 (1975); H.R. Krishna-Murthy, J.W. Wilkins, K.G. Wilson: Phys. Rev. B **21**, 1003 (1980)
4.59 P.W. Anderson: J. Phys. C **3**, 2436 (1970); P.W. Anderson, P.W. Yuval, D. Haman: Phys. Rev. **131**, 4464 (1970); F.D.M. Haldane: Phys. Rev. Lett. **40**, 489 (1978)
4.60 P. Nozières: J. Low Temp. Phys. **17**, 31 (1974)
4.61 A.A. Abrikosov: Phys **2**, 5 (1965)
4.62 H. Suhl: Phys. Rev. **138**, A515 (1965)
4.63 C.M. Varma, Y. Yafet: Phys. Rev. B **13**, 2950 (1976)
4.64 J.W. Rasul, A.C. Hewson: J. Phys. C **17**, 3337 (1984)
4.65 P.B. Wiegmann: Phys. Rev. Lett. **60**, 821 (1988)
4.66 W. Weller: to be published
4.67 D. Foerster: Phys. Rev. Lett. **63**, 2140 (1989)
4.68 S.E. Barnes: J. Phys. F **6**, 1375 (1976)
4.69 P. Coleman: Phys. Rev. B **29**, 3035 (1984)
4.70 Y. Kuramoto: Z. Phys. B **53**, 37 (1983); Y. Kuramoto, H. Kojima: Z. Phys. B **57**, 95 (1984)
4.71 E. Müller-Hartmann: Z. Phys. B **57**, 281 (1984)
4.72 J. Zaanen, O. Gunnarsson: Physica C **162–164**, 821 (1989)
4.73 N. Read, D.M. Newns: Solid State Commun. **52**, 993 (1984); D.M. Newns, N. Read: Adv. Phys. **36**, 799 (1987)

4.74 P. Fulde, J. Keller, G. Zwicknagel: Solid State Physics **41**, 1 (1988)
4.75 A. Auerbach, K. Levin: Phys. Rev. Lett. **57**, 877 (1986)
4.76 P. Coleman: Phys. Rev. B **29**, 3035 (1985)
4.77 G. Kotliar, A.E. Ruckenstein: Phys. Rev. Lett. **57**, 1362 (1986)
4.78 T. Li, P. Wölfle, P.J. Hirschfeld: Phys. Rev. B **40**, 6817, (1989); P. Wölfle, T. Li: Z. Phys. B **78**, 45 (1990)
4.79 M.C. Gutzwiller: Phys. Rev. **137**, A1726 (1965)
4.80 D. Vollhardt: Rev. Mod. Phys. **56**, 99 (1984)
4.81 T.M. Rice, K. Ueda: Phys. Rev. Lett. **55**, 995 (1985); Phys. Rev. B **34**, 6420 (1986)
4.82 P. Fazekas, B.H. Brandow: Phys. Scr. **36**, 809 (1987)
4.83 C.M. Varma, W. Weber, L.J. Randall: Phys. Rev. B **33**, 1015 (1985) B. Brandow: Adv. Phys. **26**, 651 (1977)
4.84 A.M. Oleś, J. Zaanen: Phys. Rev. B **39**, 9175 (1989)
4.85 C.A. Balseiro, M. Avignon, A.G. Rojo, B. Alascio: Phys. Rev. Lett. **62**, 2624 (1989)
4.86 L. Lilly, A. Maramatsu, W. Hanke: Phys. Rev. Lett. **65**, 1379 (1990)
4.87 P.C. Pattnaik, D.M. Newns: Phys. Rev. B **41**, 880 (1990); unpublished.
4.88 D.M. Newns, M. Rasolt, P.C. Pattnaik: Phys. Rev. B **38**, 6513 (1988); D.M. Newns, P.C. Pattnaik, M. Rasolt, D.A. Papaconstantopoulos: Phys. Rev. B **38**, 7033 (1988)
4.89 K.A. Chao, J. Spalek, A.M. Oles: Phys. Rev. B **18**, 3453 (1978)
4.90 P.W. Anderson: Phys. Rev. **115**, 2 (1959)
4.91 F.C. Zhang, T.M. Rice: Phys. Rev. B **37**, 3759 (1988)
4.92 V.J. Emery: Phys. Rev. Lett. **58**, 2794 (1987); C.M. Varma, S. Schmitt-Rink, E. Abrahams: Solid State Commun. **62**, 681 (1987)
4.93 J. Zaanen, A.M. Oleś: Phys Rev. B **37**, 9423 (1988)
4.94 G. Reiter, V.J. Emery: preprint, Brookhaven, 1990; V.J. Emery, G. Reiter: Phys. Rev. B **41**, 7247 (1990)
4.95 J.Carmelo, A.A. Ovchinnikov: To be published
4.96 F.D.M. Haldane: J. Phys. C **14**, 2585 (1981)
4.97 W.F. Brinkman, T.M. Rice: Phys. Rev. B **2**, 1324 (1970)
4.98 K.J. von Szczepanski, P. Horsch, W. Stephan, M. Ziegler: Phys. Rev. B **41**, 2017 (1990)
4.99 W. Brenig, K.W. Becker: Z. Phys. B **76**, 473 (1989)
4.100 M. Ziegler, P. Horsch: proc. of the NATO advanced research workshop on 'Dynamics of Magnetic Fluctuations in High Temperature Superconductors', ed. by G. Reiter, P. Horsch, G. Psaltakis (Plenum, New York 1990)
4.101 W.P. Su, J.R. Schrieffer, A.J. Heeger: Phys. Rev. B **22**, 2099 (1980)
4.102 S. Kivelson, D.S. Rokshar, J.P. Sethna: Phys. Rev. B **35**, 8865 (1987)
4.103 J.M. Luttinger: Phys. Rev. **121**, 942 (1961)
4.104 E.H. Lieb, F.Y. Wu: Phys. Rev. Lett. **20**, 1445 (1968)
4.105 S.A. Trugman: Phys. Rev. B **37**, 1597 (1988)
4.106 L.N. Bulaevskii, E.L. Nagaev, D.L. Khomskii: Zh. Eksp. Teor. Fiz. **54**, 1562 (1968) [Sov. Phys.-JETP **27**, 836 (1968)]
4.107 B.I. Shraiman, E.D. Siggia: Phys. Rev. Lett. **60**, 740 (1988)
4.108 M. Ziegler, W. Stephan, P. Horsch: unpublished.
4.109 E. Dagotto, R. Joynt, A. Moreo, S. Bacci, E. Gagliano: Phys. Rev. B **41**, 2585 (1990)
4.110 S. Schmitt-Rink, C.M. Varma, A.E. Ruckenstein: Phys. Rev. Lett. **60**, 2793 (1988)
4.111 C.L. Kane, P.A. Lee, N. Read: Phys. Rev. B **39**, 6880 (1989)
4.112 G. Martinez, P. Horsch: Phys. Rev. B **44**, 317 (1991)
4.113 O. Gunnarsson: Phys. Rev. B **41**, 514 (1990)
4.114 W. Folkerts, C. Haas: In *Narrow-Band Phenomena—Influence of Electrons with Both Band and Localized Character*, ed. by J.C. Fuggle et al. (Plenum, New York 1988) p. 183
4.115 O. Gunnarsson, O. Jepsen: Phys. Rev. B **38**, 3568 (1988)
4.116 J.E. Hirsch, S. Tang: Solid State Commun. **69**, 987 (1989)
4.117. O. Gunnarsson, K. Schönhammer: Phys. Rev. B **40**, 1989 (4160)
4.118 G.J.M. Janssen, W.C. Nieuwpoort: Phys. Rev. B **38**, 3449 (1988)
4.119 C.J. Mei, G. Stollhoff: Z. Phys. B **77**, 353 (1989)
4.120 G. Stollhoff, P. Fulde: Z. Phys. B **26**, 257 (1977); Z. Phys. B **29**, 231 (1978)
4.121 J. Zaanen, O. Jepsen, O. Gunnarsson, A.T. Paxton, O.K. Andersen, A. Svane: Physica C **153–155**, 1636 (1988)
4.122 K. Terakura, T. Oguchi, A.R. Williams, J. Kuebler: Phys. Rev. B **30**, 4734 (1984)

4.123 V.I. Anisimov, J. Zaanen, O.K. Andersen: Phys. Rev. B **44**, 943 (1991)
4.124 A. Svane, O. Gunnarsson: Phys. Rev. Lett. **65**, 1148 (1990)
4.125 Z.-X. Shen, C.K. Shih, O. Jepsen, W.E. Spicer, I. Lindau, J.W. Allen: Phys. Rev. Lett. **64**, 2442 (1990)
4.126 H.I. Starnberg, M.T. Johnson, H.P. Hughes: J. Phys. C **19**, 2689 (1986)
4.127 G. Martinez, J. Zaanen, P. Horsch: Unpublished.
4.128 L. Ley, M. Taniguchi, J. Ghijsen, R.L. Johnson, A. Fujimori: Phys. Rev. B **35**, 2839 (1987)
4.129 W. Drube, F.J. Himpsel, G.V. Chandrasekhar, M.W. Schafer: Unpublished
4.130 T. Takahashi, H. Matsuyama, H. Katayama-Yoshida, Y. Okabe, S. Hosoya, K. Seki, H. Fujimoto, M. Sato, H. Inokuchi: Nature **334**, 691 (1988)

5. Theory of X-Ray Absorption Fine Structure

D.D. Vvedensky

With 7 Figures

As suggested in Sect. 1.4.1, the modern interest in X-ray absorption spectroscopy initially centered around identifying the origin of the structure in the "extended" region of the absorption spectrum, i.e. oscillations with an amplitude of approximately 10% of the signal up to 1000 eV beyond the absorption edge. *Kronig* [5.1] recognized these oscillations as being due to the existence of allowed and forbidden bands of energy levels, and some forty years later *Sayers* et al. [5.2] suggested that these data could be inverted by Fourier-transform methods to obtain information about the local structure of the absorbing atom. With many subsequent developments, extended X-ray absorption fine structure (EXAFS) [5.3, 4], and the surface analogue, surface EXAFS (SEXAFS) [5.5], have evolved into standard techniques for determining local structure [5.6]. More recently, the analysis of X-ray absorption spectra has been extended into the near-edge regime (near-edge X-ray absorption fine structure, or NEXAFS) [5.7–9], where multiple scattering effects are strong and structural information supplementary to EXAFS can be obtained [5.10–13], but the extraction of local structural parameters must proceed by indirect methods, in analogy with the procedure used in low-energy-electron diffraction [5.14].

Many interpretations of X-ray absorption spectra are based upon either bare or modulated final densities of electronic states. The most successful approach for treating electrons in solids is based upon the observation that each electron may be viewed as moving in an average potential generated by all the other electrons. This potential must be found self-consistently, since the wavefunction of each electron affects the potential of all other electrons. *Hohenberg* and *Kohn* [5.15] proved that this intuitive mean-field argument is indeed correct and *Kohn* and *Sham* [5.16] showed that the total ground-state energy of a system is minimized by the ground-state charge density subject to the constraint of a fixed number of particles. The effective potential was shown to be a sum of a Hartree potential and an "exchange-correlation" potential, which is a quantum mechanical correction to the electrostatic Coulomb interaction.

While the Hohenberg–Kohn theorem establishes a unique correspondence between ground-state energy and the ground-state charge density, the functional form of the exchange-correlation potential is not determined by the theorem. This has led to many proposed approximate forms, of which the most common are local-density approximations, where the exchange-correlation potential of the interacting electron system is replaced by that corresponding to a homogeneous electron gas. However, none of these approaches has completely general applicability. For example, in applications where the final states are localized to the site of the excited atom and intra-atomic correlations dominate the hybridization with the conduction electrons, an approach based upon the interaction of individual atomic configurations is found to be more appropriate [5.17]. Another possibility is that the Coulomb attraction between the electron and hole pulls a state out of the band to form a bound electron–hole pair, i.e. an exciton (see [5.18, 19] for early reviews). The quantitative treatment of these complex many-body effects in the final state is beyond the scope of the local-density approximation. On a more fundamental level is the question of the applicability of the Hohenberg–Kohn theory to *excited* states, an issue which was touched upon in Chap. 3 by Godby.

The theoretical requirements for the interpretation of X-ray absorption measurements are thus seen to be quite diverse. In some applications, for example in the near-edge region of EXAFS and SEXAFS, the calculation of the underlying single-particle excitations is the first step in the direct interpretation of the absorption spectrum, and for assessing the requirements of a more refined theory. For systems where correlation effects are expected to be important, atomic-based multiplet calculations are used. There is yet no first-principles computational scheme that includes the effects of localized atomic correlations and delocalized band structure, with the appropriate mixing obtained as a natural result of the self-consistent competition of the two types of interaction. Calculations have been reported for specific systems [5.20] though, at present, each system much be studied on a case-by-case basis.

In this chapter, we review some of the theoretical approaches that are used for calculating X-ray absorption spectra. We begin with single-particle multiple scattering calculations, which have provided the basis of much of the development of EXAFS, NEXAFS and SEXAFS. This is followed by a discussion of single-particle selection rules for the matrix elements and the dependence of the absorption spectrum upon the polarization of the incident radiation. Symmetry considerations play an important role in several aspects of X-ray absorption spectroscopy, including SEXAFS, where the polarization dependence provides additional information about the local geometry of atomic adsorbates and orientation of molecular adsorbates, and in magnetic X-ray dichroism, where the selection rules are exploited to obtain information about the local magnetic structure. We then turn our attention to systems where correlations are important, focussing upon the multiplet effects in the $3d \rightarrow 4f$ spectra of rare-earth materials, and showing how X-ray absorption spectroscopy can be used to infer the ground state in highly correlated systems. This is followed by a discussion of magnetic X-ray dichroism, where many of the concepts and techniques developed in the preceding sections can be extended to obtain information about the local magnetic structure of materials in both the band and correlated limits. We conclude with a brief survey of related spectroscopic techniques to which the methods described in this section have been applied.

5.1 The Single-Particle Excitation Process

When the energy of an incident X-ray photon exceeds the binding energy of a particular core level, the photon can be absorbed, and the core electron is excited to an unoccupied state. In systems where intra-atomic correlation effects are dominated by band formation and exciton formation can be neglected, the final state is a delocalized scattering state similar to the states in photoemission and low-energy electron diffraction, and the propagation of the excited photoelectron can be calculated using the multiple scattering formalism [5.4, 7]. The scattering of the photoelectron by the neighboring atoms modifies the matrix element for the absorption coefficient due to the interference between the outgoing and backscattered wavefunctions. Since this interference depends upon interatomic distances and the phase shifts of the neighboring atoms, the absorption coefficient shows modulations as a function of the incident photon energy which contain information about the local electronic structure and the local arrangement of atoms with respect to the excited atom.

The interpretation of the absorption spectrum and the type of structural and electronic information that may be inferred from the data is determined by the nature of the photoelectron-scattering processes. In the EXAFS energy range, typically $\geqq 50$ eV above the absorption edge, the dominance of forward scattering facilitates an analysis of the data based upon a single back-scattering event of the photoelectron, with the possibility of multiple *forward* scattering taken into account with an effective refractive index of the medium through a renormalization of the photoelectron wavenumber [5.21]. The derivation of structural parameters in the form of a radial distribution function with respect to the excited atom then proceeds by standard Fourier methods [5.2]. Additionally, in the case of SEXAFS, the amplitude ratios of the signals with different polarizations can be used to establish the local geometry of atomic adsorbates [5.6].

Alternatively, within approximately 50 eV of the absorption edge, near-edge X-ray absorption fine structure (NEXAFS) shows large modulations as a function of the photon energy due to the relatively isotropic scattering factors of neighboring atoms and the relatively long mean free path of the photoelectron. A single-scattering picture is no longer valid and in general a full multiple scattering calculation is required. Thus, while the analysis in the near-edge region is more complicated than in the extended region, there is the prospect of determining structural information from NEXAFS that is inaccessible to EXAFS, such as multi-atom correlations and bond angles [5.10–13]. The ability to interpret NEXAFS data is also important in those cases with overlapping edges in the first 300 eV (e.g. N and C *K*-edges SEXAFS for NO and CO, and O *K*-edge SEXAFS on Ag due to the substrate *M* edges), or for submonolayer coverages, where there may be sensitivity problems in the extended region.

X-ray absorption by molecular adsorbates provides an interesting intermediate case between the delocalized scattering picture and the localized intra-atomic regime. The absorption spectrum is dominated by intra-molecular resonances, whose intensity depends upon the polarization of the incident radiation. The effect of the substrate is primarily to orient the molecule, usually with little or no effect upon the absorption spectrum. Thus, by examining the relative intensities of the molecular resonances as a function of polarization, the angular dependence of the dipole matrix elements (Sect. 5.3.1) can be used to determine the molecular orientation of simple molecules [5.6], as discussed in Chap. 6.

5.1.1 Fermi's Golden Rule

To calculate the transition probability per unit time Ω including the various excitation channels, we apply the Fermi Golden Rule [5.22]:

$$\Omega = \frac{2\pi}{\hbar}\rho(\omega)\sum_f |M_{i,f}|^2\delta(\hbar\omega + E_i - E_f) \tag{5.1}$$

where $\hbar\omega$ is the energy of the incident photons, $\rho(\omega)$ is the photon density of states, the subscripts "i" and "f" denote the initial and final N-electron states of the system, respectively, the delta function ensures energy conservation. The matrix element is given by

$$M_{i,f} = \int d\boldsymbol{r}_1 \cdots \int d\boldsymbol{r}_N \Psi_i^*(\boldsymbol{r}_1,\ldots,\boldsymbol{r}_N)\mathscr{H}'\Psi_f(\boldsymbol{r}_1,\ldots,\boldsymbol{r}_N) \tag{5.2}$$

where the initial- and final-state wave functions are given by Ψ_i and Ψ_f, and $\mathscr{H}'$ is the electron–photon interaction:

$$\mathscr{H}' = -i\hbar\frac{e}{c}A\sum_{j=1}^{N} e^{i\boldsymbol{k}\cdot\boldsymbol{r}_j}\boldsymbol{n}\cdot\nabla_j, \tag{5.3}$$

with $\boldsymbol{A}$ being the intensity of the radiation field, $\boldsymbol{n}$ is a unit vector in the direction of the polarization and $\boldsymbol{r}_j$ and $-i\hbar\nabla_j$ are the position and momentum operators of the jth electron. In the dipole approximation, which is effectively a long-wavelength approximation to the radiation field, the exponential factor in (5.3) is set equal to unity. Higher order multipole contributions can also be computed within the formalism described below [5.23, 24].

The initial- and final-state wavefunctions, Ψ_i and Ψ_f, appearing in (5.2) differ in the occupancy of two states: an electron in the core state ψ_i is annihilated and an electron in the final state ψ_f is created. However, Ψ_i and Ψ_f also differ in the single-particle bases from which they are constructed, which reflects the extent to which the core-hole potential affects the final state. In the widely used "sudden" approximation [5.22], the creation of the core-hole potential is treated as instantaneous, so Ψ_i and Ψ_f diagonalize different Hamiltonians. To see the implications of this in the simplest independent-particle approximation, we write Ψ_i and Ψ_f as Slater determinants for an N-particle system in the single-particle bases appropriate to the initial (ψ_k) and final (ϕ) states:

$$\begin{aligned}\Psi_i(\boldsymbol{r}_i,\boldsymbol{r}_1,\ldots,\boldsymbol{r}_{N-1}) &= (N!)^{-1/2}\sum_P(-1)^P[\psi_i(\boldsymbol{r}_i)\psi_1(\boldsymbol{r}_1)\cdots\psi_{N-1}(\boldsymbol{r}_{N-1})]\\ \Psi_f(\boldsymbol{r}_1,\ldots,\boldsymbol{r}_{N-1},\boldsymbol{r}_f) &= (N!)^{-1/2}\sum_P(-1)^P[\phi_1(\boldsymbol{r}_1)\cdots\phi_{N-1}(\boldsymbol{r}_{N-1})\phi_f(\boldsymbol{r}_f)]\end{aligned} \tag{5.4}$$

where P is the permutation operator and the initial core state and the final state in the presence of the core hole have been denoted by ψ_i and ϕ_f, respectively. *Stern* and *Rehr* [5.25] have suggested that the most natural single-particle states in (5.4) are those where the excited photoelectron state is orthogonal to the initially occupied states ψ_k and the core-hole state is orthogonal to all finally occupied states ϕ_k, which can be obtained from the original Hartree–Fock states by unitary transformations (thereby leaving the Slater determinants invariant). Adopting these basis functions and substituting (5.4) into (5.2), we obtain

$$M_{i,f} = \langle i|\mathscr{H}'|f\rangle \sum_P (-1)^P [\langle 1|i_1\rangle\langle 2|i_2\rangle \cdots \langle N-1|i_{N-1}\rangle] \tag{5.5}$$

where the permutation operator acts only on the i_k. We have introduced Dirac bracket notation for the inner products:

$$\langle k|i_k\rangle = \int d\boldsymbol{r}\psi_k^*(\boldsymbol{r})\phi_{i_k}(\boldsymbol{r}). \tag{5.6}$$

The matrix element is thus seen to factor into a single-particle matrix element $\langle i|\mathscr{H}'|f\rangle$ between the states involved directly in the transition, and an overlap integral between the Slater determinants of the "passive" electrons.

If the wavefunctions of the passive electrons are assumed to be unaffected by the core hole, then $\langle k|i_k\rangle = \delta_{k,i_k}$, and the matrix element (5.2) reduces to

$$M_{i,f} = \int d\boldsymbol{r}\psi_i^*(\boldsymbol{r})\mathscr{H}'\phi_f(\boldsymbol{r}). \tag{5.7}$$

In the more general case, the wavefunctions of the passive electrons after the excitation need not be adiabatically related to those before the excitation since, for example, the passive electrons can be left in excited states. As discussed in Sect. 5.4, most calculations of the absorption spectrum focus on how the core-hole potential affects the single-particle states involved in the transition, with relatively few studies [5.26] addressing the role of the passive electrons.

For further analysis and computation of the transition rate, (5.1) can be put into a more convenient form by first writing

$$\begin{aligned}\Omega &= \frac{2\pi}{\hbar}\rho(\omega)\sum_f \langle i|\mathscr{H}'|f\rangle\langle f|\mathscr{H}'^*|i\rangle\delta(E+E_i-E_f)\\ &= \frac{2\pi}{\hbar}\rho(\omega)\langle i|\mathscr{H}'|\boldsymbol{r}\rangle\left[\sum_f \langle \boldsymbol{r}|f\rangle\langle f|\boldsymbol{r}'\rangle\delta(E+E_i-E_f)\right]\langle \boldsymbol{r}'|\mathscr{H}'^*|i\rangle\end{aligned} \tag{5.8}$$

where an integration over $\boldsymbol{r}$ and $\boldsymbol{r}'$ is implied. Using the well-known expansion of the Green's function in terms of a complete set of states:

$$\begin{aligned}G(\boldsymbol{r},\boldsymbol{r}';E) &= \lim_{\varepsilon\to 0}\sum_j \frac{\psi_j(\boldsymbol{r})\psi_j^*(\boldsymbol{r}')}{E-E_j+\mathrm{i}\varepsilon}\\ &= P\sum_j \frac{\psi_f(\boldsymbol{r})\psi_j^*(\boldsymbol{r}')}{E-E_j} - \mathrm{i}\pi\sum_j \psi_j(\boldsymbol{r})\psi_j^*(\boldsymbol{r}')\delta(E-E_j)\end{aligned} \tag{5.9}$$

where "P" denotes the principal part, we have

$$\sum_f \langle \boldsymbol{r}|f\rangle\langle f|\boldsymbol{r}'\rangle\delta(E+E_i-E_f) = -\frac{1}{\pi}\mathrm{Im}\{G(\boldsymbol{r},\boldsymbol{r}';E+E_f)\} \tag{5.10}$$

where $G(\boldsymbol{r},\boldsymbol{r}';E)$ is the Green's function at energy E of the photoelectron, and "Im" denotes the imaginary part. Thus, the transition rate may be written in the form [5.7]

$$\begin{aligned}\Omega &= -\frac{2}{\hbar}\rho(\omega)\int \mathrm{d}\boldsymbol{r}\int \mathrm{d}\boldsymbol{r}'\psi_i^*(\boldsymbol{r})\mathscr{H}'[\mathrm{Im}\{G(\boldsymbol{r},\boldsymbol{r}';E+E_i)\}]\mathscr{H}'^*\psi_i(\boldsymbol{r}')\\ &= -\frac{2}{\hbar}\rho(\omega)\langle i|\mathscr{H}'\,\mathrm{Im}\{G\}\mathscr{H}'^*|i\rangle.\end{aligned} \tag{5.11}$$

In the real-space representation, the localized nature of the core wavefunction $\psi_i(\boldsymbol{r})$ confines both $\boldsymbol{r}$ and $\boldsymbol{r}'$ to the same atomic site. This has several convenient conceptual and computational consequences for the representation of the Green's function with respect to the site of the excited atom, which we explore in Sect. 5.2 within the framework of multiple scattering solutions of Schrödinger's equation.

5.1.2 Early Calculations

Müller et al. [5.27] performed the first calculations for solids with a one-electron potential for K-shell excitations for several $4d$ transition metals

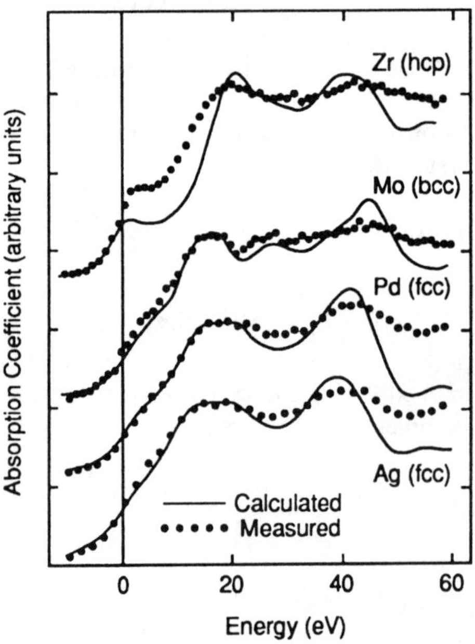

Fig. 5.1. Calculated (solid lines) and measured (dotted lines) X-ray absorption spectra for the $4d$ transition metals Zr, Mo, Pd, and Ag [5.27]. The energy scale is $(E-E_\mathrm{F})(S/S_\mathrm{Pd})$, where S is the radius of the Wigner–Seitz sphere on the indicated element and S_Pd is that for Pd. This scaling removes the atomic-volume dependence of the energy due to the predominantly kinetic character of the energy

(Fig. 5.1). The available Bloch states at each $\boldsymbol{k}$ point were calculated, a procedure which was made practical by the linear methods developed by *Andersen* [5.28] (see below), then the total cross section was obtained by summing over all points in $\boldsymbol{k}$-space. These calculations established that one-electron potentials could be used to describe the principal features of measured NEXAFS spectra for solids and, for K-shell excitations in particular, showed that the dominant systematic features can be reproduced by the projected density of states of the angular momentum channel $l = 1$.

By extending the quantum-chemical multiple-scattering approach of *Johnson* [5.29] to continuum states, *Dill* and *Dehmer* [5.30, 31] investigated the absorption spectra of CO and N_2 and also found that the single-particle theory reproduced the dominant features of the absorption spectrum. It is noteworthy that these early calculations by Müller et al. and Dehmer and Dill used ground-state potentials even for the excited atom. In contrast, *von Barth* and *Grossman* [5.32, 33] proposed the final-state rule whereby the single-particle approach can be applied provided the transition matrix elements are calculated in a basis of eigenstates of the final-state effective Hamiltonian. This was supported by numerical studies and empirical observations, but not proved in general.

5.2 Single-Particle Electronic Structure Calculations

As discussed in the introductory section, accurate single-particle calculations are a necessary first step in understanding various aspects of X-ray absorption spectra, including systematic trends in the Periodic Table, and in assessing the requirements of a more refined theory to include many-body effects such as excitons, multielectron excitations and corrections to the local-density approximation. There have been many methods developed for solving the one-electron Schrödinger equation from first principles, but it would digress too far from our aims here to discuss these in any detail. Instead, we focus on those methods that have been used most extensively for calculating X-ray absorption spectra. A more detailed discussion has been given by *Koelling* [5.34], and the discussion here is meant to complement that of Zeller in Chap. 2.

In close-packed solids, the high atomic coordination numbers have motivated the "muffin-tin" approximation to the potential, and forms the basis of the augmented plane wave (APW) [5.35] and the Korringa–Kohn–Rostoker [5.36, 37] methods. The solid is first divided into contiguous spheres centered around each atom. The potential is then spherically averaged within the spheres and volume averaged in the region between the spheres. The Schrödinger equation is numerically integrated within the sphere, where the wavefunction is strongly affected by the nuclear charge and retains many atomic characteristics, and matched onto the solution in the interstitial region. This matching over the boundary of the sphere produces a secular equation whose solution yields the one-electron energies and wavefunctions. Despite the reliance in the original formulation of the KKR method on the muffin-tin form of the potential, *Newton* [5.38] has recently formulated the theory without any assumption on the form of potential, and retains the separation of the crystal structure and the potential.

In the APW and KKR methods, solving the secular equation involves finding the solution of a large determinant as a function of energy. Since the energy dependence of the determinant is nonlinear and there are no algorithms for directly solving for all eigenfunctions and eigenvalues simultaneously, finding the zeroes involves many evaluations for trial energies. *Andersen* [5.28] proposed a modification of the APW and KKR methods whereby energy-independent basis functions are constructed for some reference energy and used throughout the energy range of interest. The determination of the band structure thus becomes a linear eigenvalue problem, so all eigenfunctions and eigenvalues for a given crystal momentum can be found simultaneously with standard diagonalization procedures. In the KKR and APW methods, this construction gives rise to the linear combination of muffin-tin orbitals (LMTO) [5.39] and linearized APW (LAPW) [5.40] methods, respectively.

For many practical reasons, the KKR, or Green's function, method has been used to the greatest extent in calculating X-ray absorption spectra. The formal properties of the Green's function, together with the intuitive association of the Green's function with the propagation of the photoelectron, facilitate many applications and approximations that would be less straightforward to implement with other methods. We now turn to a discussion of the Green's function approach to calculating X-ray absorption spectra.

5.2.1 Multiple Scattering Theory

Most of the formal structure of multiple scattering theory [5.36, 37] can be illustrated with a few operator equations derived from the Green's functions of Schrödinger's equation. The implementation of the multiple scattering equations of course requires choosing a particular representation, as discussed in Chap. 2. The primary advantages of this formulation are (i) the analytic properties of the Green's function, which permits analytic continuation to complex energies, which in turn facilitates many convenient simplifications in the solution of the Schrödinger equation, one of the most important of which is (ii) that the coupling of two sub-systems is much simpler and more natural within the Green's function formalism than with wavefunctions. In the context of X-ray absorption calculations, this means that single-particle calculations can be carried out self-consistently, including the embedding of the core hole, both in the bulk and at surfaces and interfaces. Details may again be found in Chap. 2.

The equations defining the free-space Green's function G_0 and the Green's function G in the presence of a potential V are:

$$(H_0 - E)G_0 = -1, \quad (H_0 + V - E)G = -1 \tag{5.12}$$

where H_0 is the kinetic energy operator. The eigenfunctions ψ of $H_0 + V$ are related to the free-space Green's function G_0 by the integral equation

$$\psi = G_0 V \psi \tag{5.13}$$

which is the Lippmann–Schwinger equation. G and G_0 are related by the Dyson equation:

$$G = G_0 + G_0 V G. \tag{5.14}$$

Solving (5.14) formally for $G, G = (1 - G_0 V)^{-1} G_0 = V^{-1}(1 - VG_0)^{-1} VG_0$, and substituting this expression for G on the right-hand side, produces the formally equivalent but more convenient solution

$$G = G_0 + G_0(1 - VG_0)^{-1} VG_0 \equiv G_0 + G_0 T G_0, \tag{5.15}$$

which defines the scattering matrix T of the system:

$$T = V + VG_0 T. \tag{5.16}$$

To obtain a coordinate representation of the Green's function, we have that for a single-scattering potential, $V = v(\boldsymbol{r})$, (5.16) reduces to $t = v + vG_0 t$. In the more general case of an arrangement of scattering centers, $V = \sum_i v_i$, we begin with a coordinate representation of (5.16):

$$\langle \boldsymbol{r}|T|\boldsymbol{r}'\rangle = \langle \boldsymbol{r}|V|\boldsymbol{r}'\rangle + \langle \boldsymbol{r}|V|\boldsymbol{r}''\rangle\langle \boldsymbol{r}''|G_0|\boldsymbol{r}'''\rangle\langle \boldsymbol{r}'''|T|\boldsymbol{r}'\rangle \tag{5.17}$$

with an implied integration over repeated indices. Equation (5.17) is suggestive of a representation in which the matrix elements of the operators are labelled by the site to which the coordinate labels are the nearest. This can be accomplished in practice for any collection of potentials by partitioning the system into contiguous cells; in the muffin-tin approximation these cells are spheres. In the site representation, the potential V is diagonal, $V_{ij} = v_i \delta_{ij}$, and (5.17) becomes

$$\tau^{ij} = v_i \delta_{ij} + v_i \sum_k G_0^{ik} \tau^{kj} \tag{5.18}$$

where, according to our construction, the quantities τ^{ij} are related to T by $T = \sum_{ij} \tau^{ij}$. By separating out the $k = i$ contribution in the summation, (5.18) may be rewritten as

$$\tau^{ij} = t^i \delta_{ij} + t^i \sum_{k \neq i} G_0^{ik} \tau^{kj} \tag{5.19}$$

where t^i is the scattering matrix of the ith potential alone: $t^i = v_i + v_i G_0^{ii} t^i$, and $G_0^{ii} \equiv G^i$ is the Green's function for the ith potential: $G^i = G_0 + G_0 t^i G_0$. The τ^{ij} are scattering path operators, first introduced by *Gyorffy* and *Stott* [5.41] and from (5.19) are seen to transform an incoming wave on site i to an outgoing wave from site j including all multiple scattering paths, with the propagation between successive scattering sites given by the free-space Green's function G_0.

5.2.2 The Single-Site Representation

The scattering path operators introduced in (5.18) and (5.19) can be used to obtain a representation of the Green's function G with respect to any site. We first define the quantity the T^{ii} by

$$T^{ii} = \sum_{j \neq i, k \neq i} \tau^{jk} \tag{5.20}$$

which, according to the definitions in the preceding section, sums all paths that neither begin nor end on the ith site. The matrix T can be expressed in terms of T^{ii} as

$$T = t^i + (1 + t^i G_0) T^{ii} (1 + G_0 t^i). \tag{5.21}$$

Thus, the Green's function G can be written entirely in terms of T^{ii} and the Green's function G^i

$$G = G^i + G^i T^{ii} G^i \tag{5.22}$$

which is valid for any site i. Denoting by "0" the site of the excited atom (5.22) may be written as

$$G = -(t^0)^{-1}(1 + t^0 G_0) + (t^0)^{-1}(1 + t^0 G_0)\tau^{00}(1 + G_0 t^0)(t^0)^{-1}, \tag{5.23}$$

which is formally equivalent to the expression derived originally by *Gyorffy* and *Stott* [5.41] with the identification $(v_0)^{-1} = (t^0)^{-1}(1 + t^0 G_0)$, which follows from (5.16) for a single scattering center. In fact,

$$\langle L | 1 + G_0 t^0 | \boldsymbol{r} \rangle = R_l^0(r) Y_{l,m}(\theta, \phi) \tag{5.24}$$

where $L = (l, m)$ is a composite angular momentum label and $R_l^0(r)$ is the regular solution to the Schrödinger equation for the isolated muffin-tin potential of the absorbing atom. Thus, taking matrix elements of the imaginary part of (5.23), and noting that the matrix t^0 is diagonal in angular momentum for a spherically symmetric potential, we obtain for real potentials

$$\begin{aligned} &\langle \boldsymbol{r} | \mathrm{Im}\{G\} | \boldsymbol{r}' \rangle \\ &\quad = \langle \boldsymbol{r} | (t^0)^{-1}(1 + t^0 G_0) | L \rangle \langle L | \mathrm{Im}\{\tau^{00}\} | L' \rangle \langle L' | (1 + G_0 t^0)(t^0)^{-1} | \boldsymbol{r}' \rangle. \end{aligned} \tag{5.25}$$

Introducing the quantities $\tilde{\phi}_L^0(\boldsymbol{r})$ by

$$\tilde{\phi}_L^0(\boldsymbol{r}) = \langle L | (1 + G_0 t^0)(t^0)^{-1} | \boldsymbol{r} \rangle, \tag{5.26}$$

the transition rate (5.7) may be written in the form [5.7]

$$\Omega = -\frac{2}{\hbar} \rho(\omega) \, \mathrm{Im} \left\{ \sum_{LL'} m_{i,L} \tau_{LL'}^{00} m_{L',i}^* \right\} \tag{5.27}$$

where $\tau_{LL'}^{00} = \langle L | \tau^{00} | L' \rangle$ and

$$m_{i,L} = \int \mathrm{d}\boldsymbol{r} \psi_i^*(\boldsymbol{r}) \mathscr{H}' \tilde{\phi}_L^0(\boldsymbol{r}). \tag{5.28}$$

The problem of calculating the transition rate has thus been reduced to calculating $\tilde{\phi}_L^0$, which is essentially an atomic quantity, and the scattering path operator τ^{00}, which sums all paths that begin and end on the excited atom and within which all the effects of the surrounding medium are contained.

We can obtain a formal expression for τ^{00} by again appealing to the scattering path operator formalism. Denoting by "m" quantities referring to the surrounding medium of the absorbing atom, the coupling of the excited atom to the surrounding medium is described by equations analogous to (5.14):

$$\tau^{00} = t^0 + t^0 G_0^{0\mathrm{m}} \tau^{\mathrm{m}0} \tag{5.29a}$$

$$\tau^{\mathrm{m}0} = T^{\mathrm{m}} G_0^{\mathrm{m}0} \tau^{00} \tag{5.29b}$$

where T^{m} is the scattering matrix for the surrounding medium. Substituting (5.29b) into (5.29a), we obtain [5.7]

$$\tau^{00} = [1 - t^0 G_0^{0\mathrm{m}} T^{\mathrm{m}} G_0^{\mathrm{m}0}]^{-1} t^0 \equiv [(t^0)^{-1} - R^{\mathrm{eff}}]^{-1} \tag{5.30}$$

where $R^{\mathrm{eff}} = G_0^{0\mathrm{m}} T^{\mathrm{m}} G_0^{\mathrm{m}0}$ is the effective reflectivity of the surrounding medium, i.e. R^{eff} transforms an outgoing wave from the excited atom to an incoming wave toward the excited atom including all multiple scattering within the surrounding medium.

5.2.3 The EXAFS Limit

Equation (5.30) represents an exact solution to the multiple scattering problem and, in conjunction with (5.27) and (5.28) can be used to calculate X-ray absorption spectra. However, with increasing energy, more phase shifts are needed to describe the increasingly forward-peaked scattering of the photoelectron. This necessitates the inversion of successively larger matrices. On the other hand, with increasing energy, the potential in (5.12) becomes of diminishing importance in comparison with the kinetic energy of the excited electron. In the extreme case, the Dyson equation (5.15) can be iterated and, if only the leading term containing the potential is retained, the expression for the Green's function reduces to

$$G = G_0 + G_0 T G_0 = G_0 + G_0 t_0 G_0 + \sum_{j \neq 0} G_0 t_j G_0 . \tag{5.31}$$

The first two terms on the right-hand side of (5.31) involve only the free-space Green's function and the scattering matrix of the excited atom. The second term describes the scattering by the surrounding medium in the single-scattering approximation, whereby the electron propagates from the excited atom to site j and is then scattered back to the excited atom. The coordinate representation, $G_0(\boldsymbol{r}, \boldsymbol{r}') = \langle \boldsymbol{r} | G_0 | \boldsymbol{r}' \rangle$, of the free-space propagator is [5.22]

$$G_0(\boldsymbol{r}, \boldsymbol{r}') \propto \frac{\mathrm{e}^{\mathrm{i}k \cdot |\boldsymbol{r} - \boldsymbol{r}'|}}{|\boldsymbol{r} - \boldsymbol{r}'|}, \tag{5.32}$$

where k is the magnitude of the photoelectron wavevector. Thus, denoting by

$\boldsymbol{R}_j$ the position vector of the jth atom with respect to the absorbing atom, the imaginary part of (5.31) is seen to contribute a sum of oscillatory terms proportional to $\sin(2k|\boldsymbol{R}_j|)$. These are the oscillations in the extended region discussed in the introduction, the measurement of which allows the determination of the positions of neighboring atoms by Fourier-transform methods.

5.2.4 Calculation of the Scattering Path Operator

The calculation of τ^{00} does not in itself impose any cut-off on the distance of the path of the photoelectron. However, the photoelectron does have a finite lifetime due to inelastic collisions that give rise to electron–hole pairs and plasmons, which is manifested by a finite path length of the photoelectron trajectory and by a broadening of the absorption spectrum [5.42]. These inelastic collisions destory the coherence of the photoelectron and inhibit further diffraction processes. In the case of photoemission, the mean free path as a function of energy has been found to exhibit remarkably universal behavior above $\approx 100\,\mathrm{eV}$ for a variety of metals [5.43].

The effect of the inelastic collisions can be taken into account phenomenologically by the introduction of an imaginary part to the energy in G, which

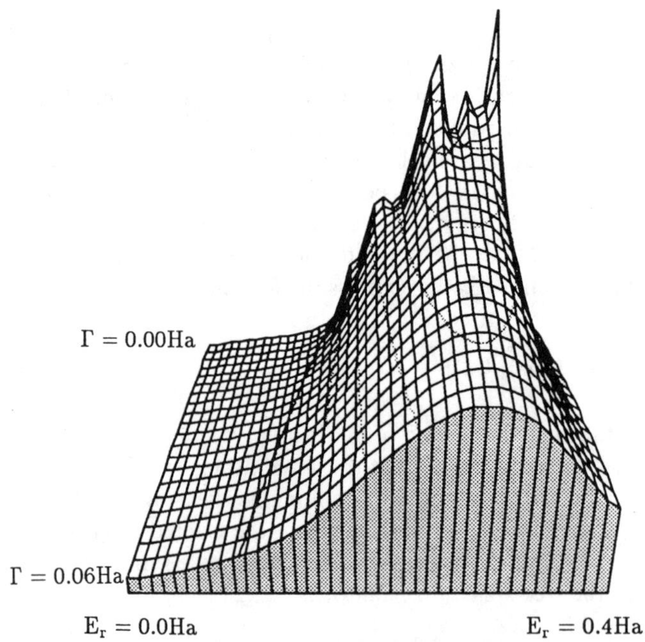

Fig. 5.2. Muffin-tin density of states of Ni obtained by integrating (5.25) within the muffin-tin volume, evaluated in the complex energy plane from $\Gamma = 0$ to $\Gamma = 0.06$ Hartree (1 Hartree = 27.2 eV) [5.45]. Only the occupied states are shown

may be viewed as an approximation to the self-energy and corresponds directly to the damping constant of the photoelectron propagator [5.44]. That this also results in a broadening of the absorption spectrum follows from the Kramers–Kronig relation, whereby the Green's function at imaginary energy Γ is given by the convolution of a Lorentzian with the Green's function evaluated at energy $\Gamma = 0$:

$$\begin{aligned}\rho(\boldsymbol{r}, E + \mathrm{i}\Gamma) &= -\frac{1}{\pi}\operatorname{Im}\{G(\boldsymbol{r}, \boldsymbol{r}; E + \mathrm{i}\Gamma)\} \\ &= \frac{1}{\pi}\int_{-\infty}^{\infty} \mathrm{d}E' \frac{\Gamma}{(E - E')^2 + \Gamma^2}\operatorname{Im}\{G(\boldsymbol{r}, \boldsymbol{r}; E')\},\end{aligned} \tag{5.33}$$

which leads to a broadening of the order of Γ. The effect of increasing the damping constant Γ upon the density of states of Ni is shown in Fig. 5.2. The sharp characteristic structure of the density of states on the real energy axis is seen to be gradually broadened into a smooth and featureless curve. As discussed in Chap. 2, this behavior can be exploited in performing energy integrations in band-structure calculations, since appreciably fewer energy points are required to accurately describe the form of the density of states with a modest damping, than the density of states at real energies. NEXAFS spectra are typically broadened by 1–10 eV, depending on the atomic number of the excited atom, the energy above the Fermi level and the absorption edge (see Appendix II on core-hole lifetimes).

The introduction of an imaginary part into the photoelectron Green's function has another beneficial consequence for the calculation of Ω. The finite path length suggests that an approach based upon surrounding the absorbing atom with a cluster representative of the local environment is more appropriate and flexible than using conventional band-structure methods, whose $\boldsymbol{k}$-space representation requires translational symmetry. In a real-space representation of (5.23), there is no requirement of long-range periodicity and, in the case of a crystalline solid, there is no requirement that the excited atom has the same potential as the surrounding symmetrically equivalent atoms, so effects of the core-hole potential can be studied very easily. This approach is therefore ideally suited to calculating the absorption spectra of amorphous materials [5.46] and biological systems [5.47], applications for which conventional band structure methods are not ideally suited.

For molecules the cluster approach is very natural [5.29–31]. However, for solids the cluster can become quite large and the inversion of R^{eff} can become a very time-consuming task. For a 1 eV imaginary part of the energy the size of cluster required to achieve convergence varies from approximately 30 to 150 atoms, depending upon the strength of scattering. Appealing to local point symmetry can reduce the computational task in some cases, but does not provide a sufficiently general scheme for treating systems with a large spatial extent. However, just as the total scattering matrix T can be decomposed into contributions from separate regions as in (5.19), so the surrounding medium can be partitioned into groups of atoms ("shells") whose T-matrices are computed separately before R^{eff} is calculated by successively incorporating the shells into the cluster. This results in a considerable economy of computer resources [5.7], both in terms of computation *time*, since the recursive construction of R^{eff} replaces the inversion of a single large matrix [cf. (5.11)] by a sequence of much smaller matrix inversions, and in *memory storage*, since the algorithm for assembling R^{eff} allows the scattering matrices of each shell to be calculated one at a time. Thus, to a great extent, the memory requirements are not determined by the cluster size. Since the contributions from the more distant shells diminish because of the longer path lengths involved, the calculation of ω converges to within some prescribed threshold as the size of the cluster is increased. This

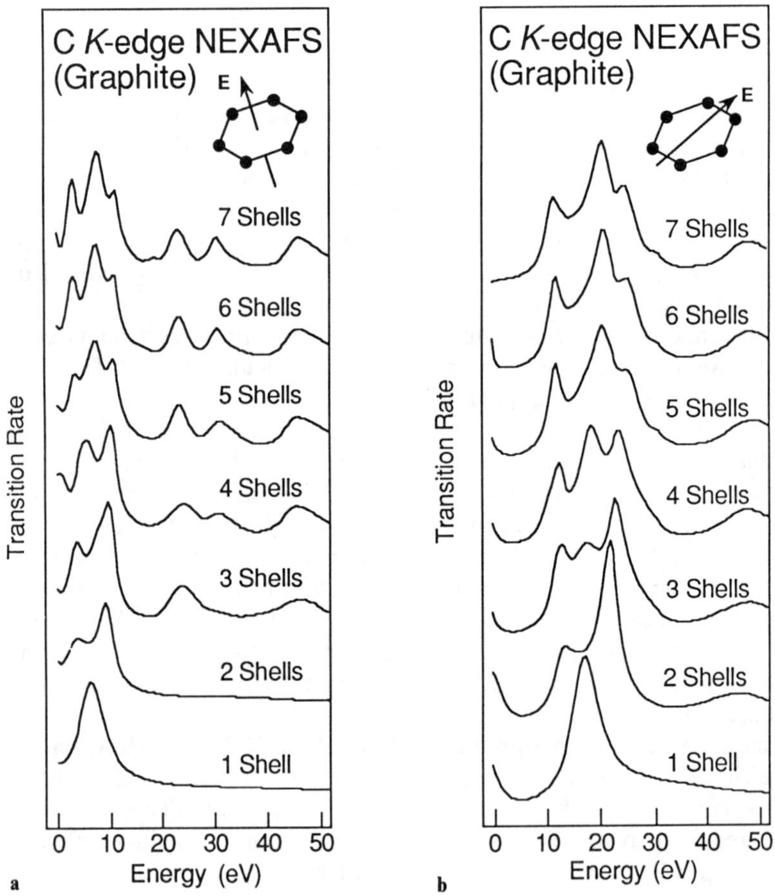

Fig. 5.3. Calculated C K-edge NEXAFS for graphite showing the convergence with an increasing number of shells for electric field polarization (**a**) normal to the basal planes and (**b**) along the basal planes. In several cases, neighboring coordination shells have been combined into a single shell in the calculation

recursive calculation of R^{eff} fails to converge for real energies except for regions in energy for which the system has no propagating states.

In Fig. 5.3 we show the convergence with increasing cluster size of the calculated C K-edge NEXAFS in graphite using the program ICXANES [5.48]. The imaginary part of the energy has been taken to have an energy-independent value of 1 eV. The potential was constructed by the superposition of neutral carbon potentials in the ground state atomic configuration, with no account taken of the change in potential due to the core hole. Nine coordination shells, involving more than 80 atoms surrounding the absorbing atom, are required to fully converge the absorption spectrum.

5.2.5 Relation to Band Structure

The general form of the manipulations carried out in (5.8–19) are not specific to the problem of calculating Ω. The same quantities are used for calculating the electronic structure using the theory of *Korringa*, *Kohn*, and *Rostoker* (KKR) [5.36, 37] and its extensions. In principle, calculations using either the single-center approach described above or one based upon the $\boldsymbol{k}$-space representation used in conventional KKR theory yield identical results for the same potentials [5.49, 50]. Other features common to both the single-center and band-structure approaches are the representation of the crystal potential by the muffin-tin approximation, the requirement of extended basis sets to describe the unoccupied states, and the modification of the unoccupied density of states due to the core hole and the excited electron. Many calculations have shown that the "muffin-tin" potential is adequate for most of the applications where the band picture is appropriate. Even for molecules, calculations based upon the muffin-tin approximation have proven to be very effective [5.6, 30, 31, 51, 52]. In some respects, the extended basis sets that are required to account for the enhanced forward scattering at higher energies are more conveniently carried out in the single-center framework because of the option to replace the matrix inversions required for a full multiple scattering treatment by various systematic approximations [5.3, 4, 14, 21, 48, 53]. The treatment of the effects of the core hole will be discussed in Sect. 5.4.

5.3 Single-Particle Selection Rules

The consequences of symmetry in determining the selection rules for the transition rate (5.27) may be assessed with a straightforward group-theoretic analysis [5.54, 55]. Selection rules for the matrix elements of the dipole operator determine the allowed final states for a given core state, while selection rules for the matrix elements of τ in the angular momentum representation determine the coupling among the excitation channels and lead to the polarization dependence of the X-ray absorption spectrum.

5.3.1 Dipole Selection Rules

The selection rules for the excitations are determined by the dipole matrix element $\langle L_i|\mathscr{H}'|L_f\rangle$. In spherical polar coordinates, the unit normal in (5.3) is given in terms of the polar and azimuthal angles θ and ϕ by $\boldsymbol{n}=(\sin\theta\cos\phi, \sin\theta\sin\phi, \cos\theta)$ and the perturbing Hamiltonian (5.3) can be written in the form

$$\begin{aligned}\mathscr{H}' &= -\mathrm{i}\hbar\frac{e}{c}A\left[\sin(\theta)\cos(\phi)\frac{\partial}{\partial x}+\sin(\theta)\sin(\phi)\frac{\partial}{\partial y}+\cos(\theta)\frac{\partial}{\partial z}\right]\\ &= -\mathrm{i}\hbar\frac{e}{c}A\left[\frac{1}{2}\sin(\theta)\mathrm{e}^{\mathrm{i}\phi}\left(\frac{\partial}{\partial x}-\mathrm{i}\frac{\partial}{\partial y}\right)\right.\\ &\quad\left.+\frac{1}{2}\sin(\theta)\mathrm{e}^{-\mathrm{i}\phi}\left(\frac{\partial}{\partial x}+\mathrm{i}\frac{\partial}{\partial y}\right)+\cos(\theta)\frac{\partial}{\partial z}\right].\end{aligned} \tag{5.34}$$

The quantities $Y_{1,m}(\nabla)$ defined by

$$Y_{1,1}(\nabla)=-\frac{1}{\sqrt{2}}\left[\frac{\partial}{\partial x}+\mathrm{i}\frac{\partial}{\partial y}\right],\quad Y_{1,-1}(\nabla)=\frac{1}{\sqrt{2}}\left[\frac{\partial}{\partial x}-\mathrm{i}\frac{\partial}{\partial y}\right],\quad Y_{1,0}(\nabla)=\frac{\partial}{\partial z} \tag{5.35}$$

are the components of a first-order irreducible tensor operator Y_1, in terms of which the matrix element (5.7) can be written as

$$M_{i,f}=-\mathrm{i}\hbar\frac{e}{c}\frac{4\pi}{3}A\sum_{m=-1}^{1}(-1)^m Y_{1,m}(\theta,\phi)\langle n_iL_i|Y_{1,-m}(\nabla)|n_fL_f\rangle, \tag{5.36}$$

where n_i and n_f are the principal quantum numbers of the initial and final states, respectively. The matrix element on the right-hand side of (5.36) can be evaluated by appealing to the Wigner–Eckart theorem [5.54, 55]. The inner product is broken up into angular and radial parts, with the angular part calculated according to the usual angular momentum coupling scheme in which the coupled angular momenta $(1,m)$ and L_f are projected onto L_i, which produces a Clebsch–Gordan coefficient $\langle l_im_i1m|l_fm_f\rangle$. The radial integral yields a *reduced matrix element* $\langle n_il_i\|Y_1\|n_fl_f\rangle$:

$$\langle n_iL_i|Y_{1,m}|n_fL_f\rangle=\frac{(-1)^{l_i-l_f-m_i-m_f}}{2l_f+1}\langle l_im_i;1,-m|l_fm_f\rangle\langle n_il_i\|Y_1\|n_fl_f\rangle. \tag{5.37}$$

The conditions under which the Clebsch–Gordan coefficient is nonvanishing thus yield selection rules for the matrix elements: $|l_f-l_i|=0, \pm 1$ and $m_i=m_f-m$. Note that this is a necessary but not sufficient condition for a matrix element to be nonvanishing. Since the dipole operator $\mathscr{H}'$ has odd spatial parity, transitions can only occur between states of different spatial parity which, in conjunction with the Clebsch–Gordan coefficient, yields the selection rules for dipole-induced transitions:

$$l_f = l_i \pm 1, \quad m_f = m_i + m. \tag{5.38}$$

Thus, using tabulated values of the Clebsch–Gordan coefficients [5.23, 54] we obtain the following forms for the nonvanishing matrix elements:

$$\begin{aligned}
\langle l_i m_i; 1, \pm 1 | l_i + 1 m_i \mp 1 \rangle &= \left[\frac{(l_i \pm m_i + 1)(l_i \pm m_i + 2)}{(2l_i + 1)(2l_i + 2)} \right]^{1/2} \\
\langle l_i m_i; 1, 0 | l_i + 1 m_i \rangle &= \left[\frac{(l_i + m_i + 1)(l_i - m_i + 1)}{(2l_i + 1)(l_i + 1)} \right]^{1/2} \\
\langle l_i m_i; 1, \pm 1 | l_i - 1 m_i \mp 1 \rangle &= \left[\frac{(l_i \mp m_i)(l_i \mp m_i - 1)}{2l_i(2l_i + 1)} \right]^{1/2} \\
\langle l_i m_i; 1, 0 | l_i - 1 m_i \rangle &= -\left[\frac{(l_i + m_i)(l_i - m_i)}{l_i(2l_i + 1)} \right]^{1/2} .
\end{aligned} \tag{5.39}$$

For later use, we include for angular momentum quantum numbers (j_i, m_i) and (j_f, m_f) for the case $j_f = j_i$:

$$\begin{aligned}
\langle l_i m_i; 1, \pm 1 | l_i m_i \mp 1 \rangle &= \pm \left[\frac{(l_i \pm m_i)(l_i + 1 \pm m_i)}{2l_i(2l_i + 1)} \right]^{1/2} \\
\langle l_i m_i; 1, 0 | l_i m_i \rangle &= -\frac{m_i}{\sqrt{l_i(l_i + 1)}} .
\end{aligned} \tag{5.40}$$

Table 5.1. Expressions for the matrix elements (5.36) with initial s and p states for the case of circular polarization

$\langle n_i s \| \mathscr{H}' \| n_f p_{-1} \rangle = \frac{1}{2}\sqrt{2}\bar{M}_{i,f} \sin(\theta) \mathrm{e}^{-\mathrm{i}\phi}$
$\langle n_i s \| \mathscr{H}' \| n_f p_0 \rangle = -\bar{M}_{i,f} \cos(\theta)$
$\langle n_i s \| \mathscr{H}' \| n_f p_{+1} \rangle = \frac{1}{2}\sqrt{2}\bar{M}_{i,f} \sin(\theta) \mathrm{e}^{\mathrm{i}\phi}$
$\langle n_i p_{-1} \| \mathscr{H}' \| n_f d_{-2} \rangle = \frac{1}{2}\sqrt{2}\bar{M}_{i,f} \sin(\theta) \mathrm{e}^{-\mathrm{i}\phi}$
$\langle n_i p_{-1} \| \mathscr{H}' \| n_f d_{-1} \rangle = -\frac{1}{2}\sqrt{2}\bar{M}_{i,f} \cos(\theta)$
$\langle n_i p_{-1} \| \mathscr{H}' \| n_f d_0 \rangle = \frac{1}{6}\sqrt{3}\bar{M}_{i,f} \sin(\theta) \mathrm{e}^{\mathrm{i}\phi}$
$\langle n_i p_0 \| \mathscr{H}' \| n_f d_{-1} \rangle = \frac{1}{2}\bar{M}_{i,f} \sin(\theta) \mathrm{e}^{-\mathrm{i}\phi}$
$\langle n_i p_0 \| \mathscr{H}' \| n_f d_0 \rangle = -\frac{1}{3}\sqrt{6}\bar{M}_{i,f} \cos(\theta)$
$\langle n_i p_0 \| \mathscr{H}' \| n_f d_{+1} \rangle = \frac{1}{2}\bar{M}_{i,f} \sin(\theta) \mathrm{e}^{\mathrm{i}\phi}$
$\langle n_i p_{+1} \| \mathscr{H}' \| n_f d_0 \rangle = \frac{1}{6}\sqrt{3}\bar{M}_{i,f} \sin(\theta) \mathrm{e}^{-\mathrm{i}\phi}$
$\langle n_i p_{+1} \| \mathscr{H}' \| n_f d_{+1} \rangle = -\frac{1}{2}\sqrt{2}\bar{M}_{i,f} \cos(\theta)$
$\langle n_i p_{+1} \| \mathscr{H}' \| n_f d_{+2} \rangle = \frac{1}{2}\sqrt{2}\bar{M}_{i,f} \sin(\theta) \mathrm{e}^{\mathrm{i}\phi}$

Table 5.2. Expressions for the matrix elements (5.36) with initial s and p states for the case of linear polarization

$$\begin{aligned}
\langle n_i s|\mathscr{H}'|n_f p_x\rangle &= -\bar{M}_{i,f}\sin\theta\cos\phi\\
\langle n_i s|\mathscr{H}'|n_f p_y\rangle &= -\bar{M}_{i,f}\sin\theta\sin\phi\\
\langle n_i s|\mathscr{H}'|n_f p_z\rangle &= -\bar{M}_{i,f}\cos\theta\\[6pt]
\langle n_i p_x|\mathscr{H}'|n_f d_{xy}\rangle &= -\tfrac{1}{2}\sqrt{2}\bar{M}_{i,f}\sin\theta\sin\phi\\
\langle n_i p_x|\mathscr{H}'|n_f d_{x^2-y^2}\rangle &= \tfrac{1}{2}\sqrt{2}\bar{M}_{i,f}\sin\theta\cos\phi\\
\langle n_i p_x|\mathscr{H}'|n_f d_{xz}\rangle &= -\tfrac{1}{2}\sqrt{2}\bar{M}_{i,f}\cos\theta\\
\langle n_i p_x|\mathscr{H}'|n_f d_{z^2}\rangle &= \tfrac{1}{6}\sqrt{6}\bar{M}_{i,f}\sin\theta\cos\phi\\[6pt]
\langle n_i p_y|\mathscr{H}'|n_f d_{xy}\rangle &= -\tfrac{1}{2}\sqrt{2}\bar{M}_{i,f}\sin\theta\cos\phi\\
\langle n_i p_y|\mathscr{H}'|n_f d_{x^2-y^2}\rangle &= \tfrac{1}{2}\sqrt{2}\bar{M}_{i,f}\sin\theta\sin\phi\\
\langle n_i p_y|\mathscr{H}'|n_f d_{yz}\rangle &= -\tfrac{1}{2}\sqrt{2}\bar{M}_{i,f}\cos\theta\\
\langle n_i p_y|\mathscr{H}'|n_f d_{z^2}\rangle &= \tfrac{1}{6}\sqrt{6}\bar{M}_{i,f}\sin\theta\sin\phi\\[6pt]
\langle n_i p_z|\mathscr{H}'|n_f d_{xz}\rangle &= -\bar{M}_{i,f}\cos\theta\cos\phi\\
\langle n_i p_z|\mathscr{H}'|n_f d_{yz}\rangle &= -\bar{M}_{i,f}\sin\theta\sin\phi\\
\langle n_i p_z|\mathscr{H}'|n_f d_{z^2}\rangle &= -\tfrac{1}{3}\sqrt{6}\bar{M}_{i,f}\cos\theta
\end{aligned}$$

Combining (5.36), (5.38), and (5.39) and introducing the quantity

$$\bar{M}_{if} = -\mathrm{i}\hbar\frac{e}{c}\frac{4\pi}{3}A\frac{1}{\sqrt{2l_f+1}}\langle n_i l_i \| Y_1 \| n_f l_f\rangle, \tag{5.41}$$

the matrix elements for K and L edges are compiled in Table 5.1 for the case of circular polarization, and in Table 5.2 for the case of linear polarization with the polarization vector in the direction $\boldsymbol{n}=(\sin\theta\cos\phi, \sin\theta\sin\phi, \cos\theta)$.

5.3.2 Selection Rules for the Tau-Matrix

To derive the selection rules for the angular momentum matrix elements of τ, suppose that the local environment of the absorbing atom has the symmetry of a point group P. Then all states of the system, when referred to the site of the absorbing atom, transform according to the irreducible representations of P. We denote the ith representation of P, whether irreducible or not, by Γ_i. Thus, the representations generated by Y_L, $Y_{L'}$, and τ are denoted by Γ_L, $\Gamma_{L'}$, and Γ_τ, respectively. The identical, or unit representation of P, i.e. the one-dimensional representation all of whose entries in the character table of P are unity, will be denoted by Γ_1. Quantities that transform according to Γ_1 under the action of operators in P are scalars of P. We see immediately that $\Gamma_\tau=\Gamma_1$, since any transformation that leaves the local environment invariant, also leaves the scattering paths invariant. Applying the standard procedure for determining selection rules [5.55], $\tau_{LL'}=\langle L|\tau|L'\rangle$ vanishes if the direct product $\Gamma_L\otimes\Gamma_{L'}$ does not include Γ_1, i.e. if the representation $\Gamma_L\otimes\Gamma_{L'}$ generated by

$Y_L^* Y_{L'}$ does not contain Γ_1. This has immediate implications for the polarization dependence of the absorption spectrum; for example, in the case of K-shell excitation, the polarization dependence vanishes for the tetrahedral point groups T, T_h, and T_d, and the octahedral point groups O and O_h. The operations contained in these groups may be found in [5.55].

The polarization dependence of the absorption spectrum is an important ingredient in SEXAFS, providing a sort of searchlight effect, since only bonds with an appreciable component along the electric field direction contribute to the signal [5.6]. In SEXAFS of atomic adsorbates on surfaces, the amplitude ratio is used to determine the numbers of nearest neighbors in given directions and so are useful in identifying adsorption sites [5.6]. Moreover, if the rotational symmetry in the surface plane is twofold or less, as is the case of (110) surfaces, there is an azimuthal-angle dependence in the SEXAFS [5.56], providing a total of three independent spectra for these systems. This facilitates the investigation of complex systems such as those involving absorbate-induced reconstructions on (110) surfaces [5.6, 56, 57].

For molecular adsorbates, the polar-angle dependence can be used to determine the orientation with respect to the surface plane of simple molecules, as we now show [5.6]. In the case of K-shell absorption, $L_i = (0,0)$, $l_f = 1$, and the allowed value(s) of m_f are determined by the polar angle of the electric field vector with respect to the surface normal, as shown in Table 5.1:

$$\langle 0,0|\mathscr{H}'|1,\pm 1\rangle = \frac{\bar{M}_{i,f}}{\sqrt{2}} \sin(\theta) \mathrm{e}^{\pm \mathrm{i}\phi}, \quad \langle 0,0|\mathscr{H}'|1,0\rangle = -\bar{M}_{i,f} \cos(\theta). \tag{5.42}$$

Consider the angular dependencies of the intensity of $\sigma(m_f = 0)$ and $\pi(m_f = \pm 1)$ resonances in the K-shell absorption of a molecule aligned normal to a surface. We see that the intensity of the σ resonance varies with the polar angle of the incident radiation as $\cos^2\theta$ and that the π resonance varies as $\sin^2\theta$. Thus, by measuring the polarization dependence of the σ and π intensities, the orientation of the molecule can be deduced. A more detailed discussion with specific examples may be found in Chap. 6.

5.3.3 Symmetry-Reduced Form of the Transition Rate

The discussion in this section suggests a more compact way of writing the transition rate (5.27) that directly incorporates the symmetry of the local environment by projecting the angular momentum basis onto the appropriate irreducible representations of the local point group:

$$\Omega = -\frac{2}{\hbar}\rho(\omega)\mathrm{Im}\left\{ \sum_{\Gamma,\Gamma'} m_{i,\Gamma} \tau^{00}_{\Gamma,\Gamma'} m^*_{\Gamma',i} \right\}. \tag{5.43}$$

In the single-particle case this notation is not particularly necessary. However, when the basis is composed of many-particle wavefunctions which interact with a crystal field, which is represented by $\tau^{00}_{\Gamma,\Gamma'}$, the form (5.43) is more appropriate for analyzing the coupling among different excitation channels. One feature of the single-particle case that emerges immediately from (5.43) is the factorization of contributions from different angular momentum channels for systems with sufficiently high local symmetry. Consider, for example, the *L*-edge absorption spectrum for a system with local O_h symmetry. Since $l_i = 1$, the allowed final states according to the dipole selection rule have $l_f = 0, 2$, i.e. *s* or *d* final states. The *s* states transform according to the A_{1g} irreducible representation of the O_h point group, while the *d* states corresponding to the T_{2g} and E_g irreducible representations. Since $\Gamma_\tau = \Gamma_1 = A_{1g}$, the transformation properties of the matrix elements of τ in the s, d subspace are determined by the direct product $A_{1g} \otimes [E_g \oplus T_{2g}] = E_g \oplus T_{2g}$, which clearly does not contain A_{1g}, and must therefore vanish. Thus, there are no matrix elements coupling the $l = 0$ and $l = 2$ channels in an O_h environment. In particular, if the energy dependence of the matrix elements and the effects of the core-hole potential can be neglected, the X-ray absorption spectrum is proportional to the sum of *s* and *d* partial densities of states. Evidently, for lower symmetries, such decompositions need not be valid.

5.4 Core-Hole Effects

As we have seen in Sect. 5.1.2, the early attempts at comparing X-ray absorption spectra with symmetry-projected densities of states proved successful despite no explicit account being taken of the presence of the core hole. In fact, the core hole is expected to modify the ground-state electronic structure in a variety of ways. In the simplest cases this can be seen as a pseudo-impurity effect with local relaxation of the electronic states in response to the attractive potential of the core hole, but there may also be very complicated multiplet effects, as described in Sect. 5.5. These perturbations may lead to distortions of the spectra due to modifications of the local density of states, and to core-exciton effects if the size of the perturbation is large in comparison with valence band widths. In addition, there are more complex dynamical effects due to the fact that the initial and final state Hamiltonians are not the same, and dynamical interactions between the core hole and the conduction electrons, which go beyond the local-density functional picture [5.20]. Some of these effects are treated within the approach based upon atomic configurations discussed in the next section. Calculations within the band-structure framework are described in Chap. 3.

5.4.1 Single-Particle Effects

The Green's function formalism presented in Sect. 5.2, can easily accommodate the core-hole potential to determine the self-consistent rearrangement of charge in the neighborhood of the core hole [5.58, 59]. Alternatively, if the extent of the electronic relaxation is confined to a few neighbor shells, a supercell calculation using any of the conventional band-structure methods can be used. For example, *Weijs* et al. [5.60] used this approach to investigate the effect of explicitly including the core-hole potential on the local projected density of states in K-edges of several transition-metal silicides. The authors found that including the core-hole potential alleviated most of the discrepancy between the measured absorption spectra and the partial density of p states by enhancing the spectrum near to threshold. However, the calculations were not able to account for the weight in the region far above threshold, which the authors attributed either to the energy dependence of the single-particle matrix elements, which was neglected in the calculation, or the presence of many-body excitations. Another important result to emerge from this work and that of *Speier* et al. [5.59] is the utility of the Clogston–Wolff model Green's function [5.61, 62] in reproducing the absorption spectrum given only the ground-state density of states and two parameters describing the depth of the impurity potential and the orbital hybridization.

Atomic and molecular systems, including molecular adsorbates, provide the greatest opportunity for systematic studies of the effects of the core hole, and effects originating in electronic correlations. For calculating single-particle excitation spectra of either free molecules or molecules adsorbed on surfaces, multiple scattering calculations based on a formalism similar to that described in Sect. 5.2, have been used to describe both bound and continuum states within a common framework [5.30, 31, 50, 51]. The presence of the core hole sometimes lowers the symmetry of the molecule compared with the symmetry of the ground state. Broken symmetry calculations have been known for some time to provide better results than full-symmetry calculations for excitations involving the removal of a core hole [5.63, 64]. This approach has been used for calculating the NEXAFS spectra of a number of complex adsorbates [5.51]. Chapter 6 gives a brief review of the various techniques used.

5.4.2 Edge Singularities

Mahan [5.65] first proposed the presence of a singularity in the X-ray absorption spectrum near threshold. Mahan argued that the core hole interacts both with the excited photoelectron and with the surrounding electron sea by increasing the charge density locally. This increases the electron–hole overlap integral and in Mahan's model leads to a divergence in the matrix element at threshold.

However, a competing effect to Mahan's picture is contained in a theorem due to *Anderson* [5.66, 67], which states that the overlap between the ground state $|N\rangle$ of a uniform N-electron gas and the ground state $|N^*\rangle$ of the system with a localized impurity potential (taken as a δ-function) is

$$\langle N^*|N\rangle = N^{-\delta^2/3\pi^2}, \tag{5.44}$$

where the bra and the ket refer to N-particle Slater determinants, and where δ is the $l=0$ phase shift of the impurity potential. The asymptotic orthogonality of the two states can be traced to the number of low-energy electron–hole pairs created by the impurity potential. This "orthogonality catastrophe" decreases the matrix element for the transition due to the diminished overlap between the ground-state and excited-stated wavefunctions.

Nozières and *De Dominicis* [5.68] placed Mahan's work on a firmer formal footing and calculated the transition rate of the absorption spectrum near threshold to be of the form

$$\Omega = \sum_{l_f} |M_{i,f}|^2 (1/E)\alpha_{l_f} \tag{5.45}$$

where the exponents α_l are given by

$$\alpha_l = \frac{2}{\pi}\delta_l - \frac{2}{\pi^2}\sum_{j=0}^{\infty}(2j+1)\delta_j^2 \tag{5.46}$$

and the phase shifts δ_l are evaluated at the Fermi energy in the presence of the core hole. Thus, if $\alpha < 0$, as is sometimes the case for p partial waves, then the contribution from that channel is suppressed at threshold, while if $\alpha > 0$, for example in the s partial wave, then the contribution from that channel is enhanced. For a core state of orbital angular momentum l_i, only the terms corresponding to $l_f = l_i \pm 1$ contribute to the sum in (5.45).

There have been many theoretical and experimental studies concerning the validity of the Mahan–Nozières–De Dominicis theory [5.33] (see also [5.69] and references therein). The final-state rule [5.32] described in Sect. 5.1 has been extended to include singular edge behavior [5.70] by multiplying the power-law and single-particle contributions from the corresponding angular momentum channels. *Swarts* et al. [5.71] questioned this form of the final-state rule by performing a series of model calculations which they claimed could be described by single-particle theory using wavefunctions in the presence of the core hole for both weak and moderately strong core-hole potentials. However, *von Barth* and *Grossman* [5.32] reinterpreted the results of Swarts et al. in terms of the Nozières–De Dominicis theory and found no inconsistency.

There have also been efforts centered around attempts to fit measured spectra to (5.44) for simple metals, e.g. Na, Mg, and Al, where the $l = 0$ and $l = 1$ phase shifts are presumed to be dominant. However, the results of these studies are far from unequivocal, with some providing firm support for the theory, while others provide results in clear contradiction to the predictions of the theory. In part this may be attributed to the large uncertainty in extracting the exponents α_l from measured absorption spectra, due to matrix-element and band-structure effects, and the finite energy resolution in the experiments and thermal and lifetime broadening effects. Reviews of this lively area of X-ray absorption spectroscopy may be found in [5.72, 73].

5.4.3 Excitons

The absorption of an X-ray and the excitation of a photoelectron can be considered as the creation of an electron–hole pair. In insulators and semiconductors, where the screening of the core hole is imperfect, the lowest energy excitation of core electrons is a bound hydrogen-like state, a core exciton. Because of the very small bandwidth of even the shallowest core states, the effective mass of the core hole is very large, so the photoelectron is bound to the site of the core-hold potential. By the same argument, the reduced mass of the exciton is nearly equal to the effective mass of the electron. The concept of excitons were introduced by *Frenkel* [5.74, 75] and *Wannier* [5.76] to explain the complicated structures near the optical absorption edges of insulators and semiconductors. However, one can also imagine "virtual" excitons in metals where, on account of the infinite static dielectric constant, the electron–hole pair has zero binding energy but a non-zero oscillator strength.

There have been few attempts to systematically assess the likely importance of excitons in X-ray absorption spectra outside the context of specific materials. A first-principles treatment must necessarily go beyond the local-density approximation [5.77], because the electron–hole bound state requires a single electron to have a much more localized interaction with the core-hole potential than the other electrons. On the other hand, for materials with relatively high dielectric constants, phenomenological approaches based upon effective-mass theory [5.78], in analogy with optical excitons in semiconductors [5.79], have produced absorption profiles in quite acceptable agreement with experiment. Similarly, calculations based upon model many-body Hamiltonians [5.80] can provide clues to mechanisms of exciton formation and give estimates of exciton satellite structures.

However, even in cases where the single-particle approach provides a good account of the measured spectrum, there remains the outstanding question of whether the core-hole potential can be neglected a priori. We are still far from having a good perspective of when core-hole effects are important in X-ray absorption spectroscopy and when not. At present, it is best to assume that these effects are important until proven otherwise.

5.5 Atomic Multiplet Structure and Crystal-Field Effects

The absorption spectra of narrow-band systems, including transition metals, actinides and the rare earths, present severe problems for single-particle calculations. There are strong indications that atomic correlations play an important role in determining the physical properties of these materials. For example, cerium appears to lie near the boundary of localized and delocalized behavior because of its single $4f$ electron and the rich variety of physical properties of the phases in the pressure temperature phase diagram. A theoretical explanation for this behavior requires considering the competition between the tendency toward electron localization, driven by the atomic correlation energy, and the tendency toward delocalization through the hybridization with the conduction electrons. Although band structure calculations are capable of reproducing many features of the electronic structure for such systems, and quasi-atomic theory can identify the appropriate many-electron configurations, the development of a general computational scheme that incorporates the competing tendencies of intra-atomic correlations and hybridization from first principles is proving much more difficult [5.81]. Nevertheless, the gross features of the absorption spectra of systems such as NiO, which show moderately strong correlations and are well known to be problematic for the local-density approximation, can often be calculated within the scattering framework described in Sect. 5.2 [5.82], with the effects of the correlations treated as an addendum [5.83]. An enlightening discussion of the relationship between local-density calculations and many-electron configurations may be found in [5.84].

In view of the preceding discussion, X-ray absorption spectroscopy acquires particular significance in narrow-band systems for probing the excited states and inferring the ground-state configuration and the term from selection rules, identifying the magnitude and orientation of the magnetic moment, and assessing the importance of the hybridization between the localized states and the conduction electrons. In many correlated systems, an approach that begins with the atomic many-electron picture is an appropriate starting point for interpreting the absorption spectrum. In the context of the discussion in Sects. 5.2 and 5.3, the atomic approach begins with many-electron wavefunctions constructed from single-electron wavefunctions using the appropriate angular momentum coupling scheme (see below). Thus, for systems where the effects of the neighboring environment can be neglected entirely, the calculation of the absorption spectrum reduces to calculating only the matrix elements, with the tau matrix in (5.27) set equal to the unit matrix. Any effects of interactions with the neighboring environment can then be taken into account using crystal-field theory, i.e. a reduction of symmetry from the group of all three-dimensional rotations, O_3, to that generated by the local atomic arrangement that leads to a coupling of different excitation channels as in the single-particle case discussed in Sect. 5.3. These interactions can be subsumed into parameters that specify the hybridization with the local wavefunctions; in this sense, the effects of the matrix τ are treated within the framework of crystal-field theory.

5.5.1 Many-Particle Matrix Elements

To gain some insight to the types of quantities that are required for calculating the X-ray absorption spectra for such highly correlated systems, we begin with the central-field Hamiltonian $\bar{\mathscr{H}}$ for an atom with atomic number Z:

$$\bar{\mathscr{H}} = \sum_{i=1}^{Z} \left[\frac{\boldsymbol{p}_i^2}{2m} + \bar{\mathscr{V}}(r_i) \right], \tag{5.47}$$

where $\bar{\mathscr{V}}$ is the sum of a central-field electron–electron and the electron–nuclear Coulomb interactions:

$$\bar{\mathscr{V}}(r_i) = \sum_{i=1}^{Z} \left[\bar{V}(r_i) + \frac{Ze^2}{r_i} \right]. \tag{5.48}$$

In the single-particle central-field approximation, each electron moves in an effective electrostatic field generated by the other electrons. $\bar{\mathscr{H}}$ can be regarded either as the Hamiltonian in the local-density approximation or the Hamiltonian in Hartree–Fock theory, though with applications to a configuration-based theory in mind, the latter choice is preferable. In the central-field approximation, the one-electron eigenfunctions are labelled by the angular momentum quantum numbers l_i and m_i and the principal quantum number n_i. These single-particle angular momentum quantum numbers must be combined to form total angular momentum states of the atom.

To describe the many-electron states of the atom by coupling the angular momenta of the single-particle eigenfunctions, we must add two additional terms to the Hamiltonian (5.47); those due to the residual Coulomb interaction and those due to magnetic forces within the atom in the form of spin–orbit coupling. The residual Coulomb interaction $\mathscr{H}^{(1)}$ is defined to be the difference between the actual Coulomb potential and that used in (5.48):

$$\mathscr{H}^{(1)} = \frac{1}{2}\sum_{i,j=1}^{Z} \frac{e^2}{|\boldsymbol{r}_i - \boldsymbol{r}_j|} - \sum_{i=1}^{Z}\left[\bar{V}(r_i) + \frac{Ze^2}{r_i}\right]. \tag{5.49}$$

The spin–orbit interaction $\mathscr{H}^{(2)}$ is given in terms of single-particle spin and orbital angular momenta $\boldsymbol{s}_i$ and $\boldsymbol{l}_i$ by

$$\mathscr{H}^{(2)} = \frac{1}{2m^2c^2}\sum_{i=1}^{Z} r_i^{-1}\frac{d\bar{V}}{dr_i}(\boldsymbol{l}_i \cdot \boldsymbol{s}_i). \tag{5.50}$$

The appropriate coupling scheme is determined by whether the Coulomb terms dominate (Russell–Saunders, or *LS* coupling), the spin–orbit terms dominates (*jj* coupling), or whether the two are comparable in magnitude (intermediate coupling) [5.85].

In the case that the Coulomb term dominates, which is appropriate for lighter atoms, the zeroth-order approximation neglects the spin–orbit term, and a series of levels results that correspond to well-defined pairs of L and S. The good quantum numbers of the individual configurations are $\boldsymbol{L}^2, \boldsymbol{S}^2, L_z$, and S_z which, upon introducing the spin–orbit coupling as a perturbation, yield states labelled by $\boldsymbol{L}^2, \boldsymbol{S}^2, \boldsymbol{J}^2$, and J_z, where $\boldsymbol{J} = \boldsymbol{L} + \boldsymbol{S}$. These quantities are related to single-particle angular momenta by $\boldsymbol{L} = \sum_i \boldsymbol{l}_i$ and $\boldsymbol{S} = \sum_i \boldsymbol{s}_i$. If the spin-orbit term is negligible in the Russell–Saunders coupling scheme, then $\boldsymbol{L}$ and $\boldsymbol{S}$ are separately good quantum numbers, and the dipole selection rules yield $\Delta L = \pm 1$ and $\Delta S = 0$, so that there are no transitions between states of different multiplicities.

The spin–orbit interaction is a relativistic effect that is magnetic in origin and whose contribution to the Hamiltonian becomes increasingly important for heavier atoms, as v/c increases. Thus, if we take the spin–orbit term to dominate the electrostatic term, the spin and orbital angular momenta of the individual particles must first be coupled to form single-particle total angular momenta: $\boldsymbol{j}_i = \boldsymbol{l}_i + \boldsymbol{s}_i$. Then, treating the electrostatic term as a perturbation, the corresponding many-electron states are eigenvectors of $\boldsymbol{j}_i^2, \boldsymbol{J}^2$, and J_z, where $\boldsymbol{J} = \sum_i \boldsymbol{j}_i$.

In intermediate coupling, the electrostatic and spin–orbit contributions are of comparable magnitude, so neither the LS nor jj coupling schemes are strictly appropriate, and the eigenstates are found by diagonalizing the full Hamiltonian comprised of the sum of (5.47), (5.49) and (5.50). The calculation proceeds by forming Slater determinants with the required values of angular momenta from central-field single-particle eigenfunctions of the Hartree–Fock equations in the LS basis, which is convenient for solving the secular equation because $\mathscr{H}^{(1)}$ is diagonal in this basis set, and $\mathscr{H}^{(2)}$ in diagonal in J and J_z (though not in L and S). The energy levels are then calculated using first-order degenerate perturbation theory. The splittings in the multiplet result from $\mathscr{H}^{(1)}$ and $\mathscr{H}^{(2)}$, with $\bar{\mathscr{H}}$ giving the average energy of the multiplet.

To calculate the matrix elements of the Coulomb potential, the quantity $v(\boldsymbol{x}, \boldsymbol{x}') = 1/|\boldsymbol{x} - \boldsymbol{x}'|$ is first expressed in terms of spherical harmonics [5.86]. Then, for example, the diagonal matrix element of states composed from central-field eigenfunctions of the form $\psi = R_l(r) Y_{l,m}(\theta, \phi)$, with total spin and orbital angular momentum quantum numbers L, M_L and S, M_S can be written as

$$\langle LSM_L M_S | v | LSM_L M_S \rangle = \sum_k (f_k F^k \pm g_k G^k), \tag{5.51}$$

where the upper (lower) sign on the right-hand side of (5.51) corresponds to $S = 0$ ($S = 1$). The f_k and g_k are angular integrals that can be computed by methods analogous to those used in deriving the Wigner–Eckart theorem [5.54, 55, 85, 87]. The Slater integrals F^k and G^k are the radial contributions arising from the direct and exchange contributions, respectively, from the Coulomb interaction. The selection rules on the f_k and g_k limit the range of k in the sum in (5.51). For example, for the direct Coulomb contributions, there are no odd-k terms, and the maximum value of k is twice the minimum l value.

The $M_{4,5}$ X-ray absorption spectra of Pr is shown in Fig. 5.4, together with theoretical curves based upon dipole excitations from the ground state determined by Hund's rules [5.88]. Also shown for comparison are spectra calculated with all levels in the $3d^9 4f^2$ configuration with each level given the weight $2J + 1$. The individual levels are represented by vertical bars, which have then been broadened by convolution with experimental Gaussian and Lorentzian lifetime broadening functions. The F and G integrals have been reduced to 80% of the Hartree–Fock values to account for configuration interaction. The spectral features are seen to occur in two groups separated approximately by the $3d_{3/2} - 3d_{1/2}$ spin–orbit splitting, though there is considerably structure within the individual groups due to the coupling of the angular momenta between the $3d$ and $4f$ electrons. The line shapes of these two groups in the full multiplet are much more similar than the dipole-selected counterparts because the dipole selection rules place appreciable weight to different parts of the multiplet. The correspondence between the calculation based upon the dipole-selection rules and the experimental absorption spectrum is seen to be quite close and is typical of the level of agreement found for other lanthanides [5.88].

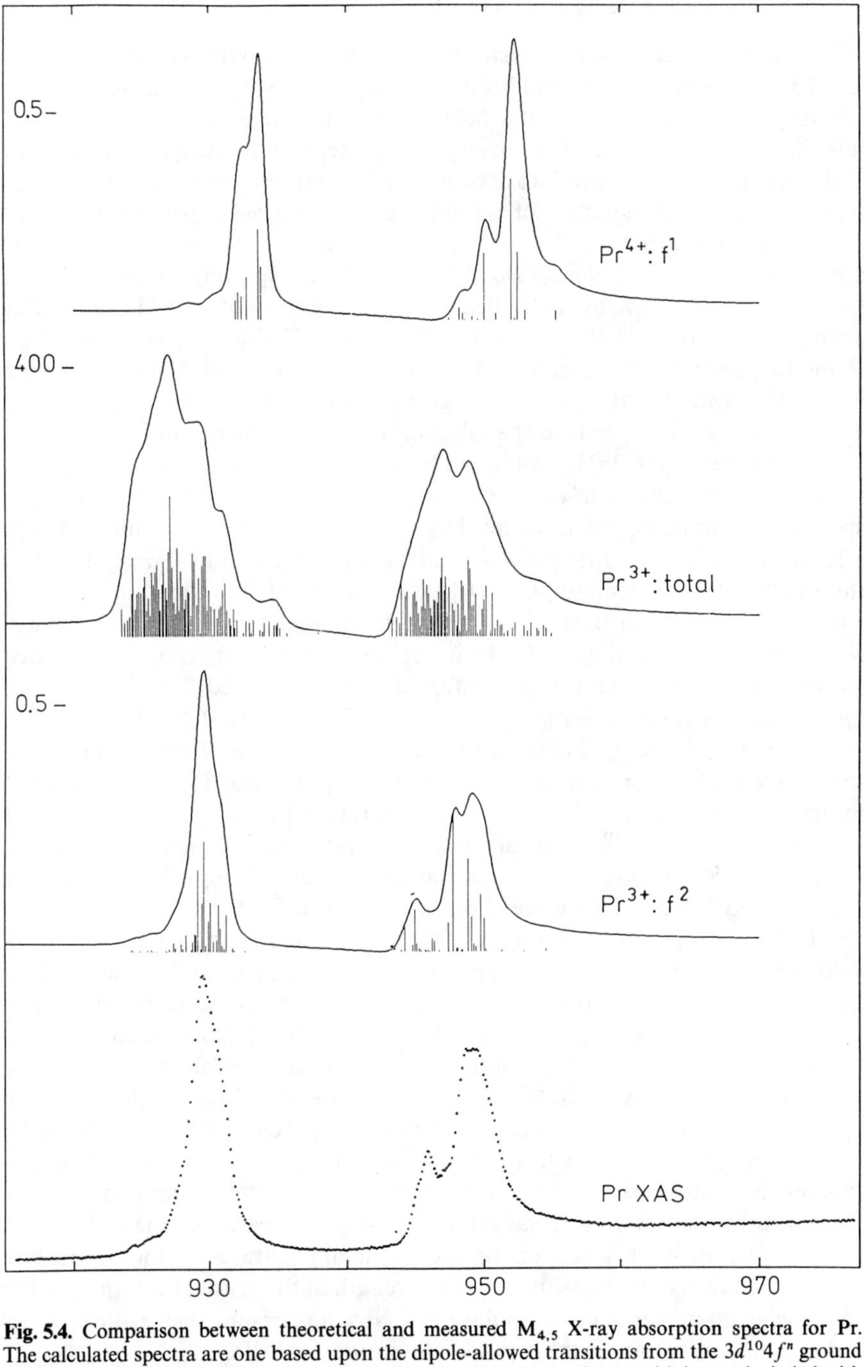

Fig. 5.4. Comparison between theoretical and measured $M_{4,5}$ X-ray absorption spectra for Pr. The calculated spectra are one based upon the dipole-allowed transitions from the $3d^{10}4f^n$ ground state determined by Hund's rules, and one where all of the levels of the multiplet are included with a weight factor of $2J+1$

5.5.2 Crystal-Field Effects

While an approach based exclusively upon atomic configurations is suitable for the $M_{4,5}$ edges of lanthanides, for the $L_{2,3}$ edges of the transition-metal elements, the effect of the crystal field must be included. As discussed above, this affects the calculation of the absorption spectrum in two ways: the symmetry of the system must be taken into account explicitly in the coupling of the excitation channels, and the effect of the interaction with the crystal environment must be parametrized. In the latter case, for high symmetries, this is usually done through a parameter, $10Dq$ that specifies the splitting of the pertinent levels, which is taken as an indication of the extent of hybridization. For simplicity, we consider the case of the $2p^6 3d^0 \rightarrow 2p^5 3d^1$ excitation appropriate to the $L_{2,3}$ edge of Ti^{4+} in an octahedral environment [5.89]. Using jj coupling in the O_3 point group, the $2p^6 3d^0$ ground state has $J_i = 0$, the perturbing Hamiltonian $\mathscr{H}'$ has $J = 1$, so the selection rule on the final state yields $J_f = 1$. The final states in the $2p^5 3d^1$ multiplet have $J = 0, 1, 2, 3, 4$ which form separate irreducible representations of the O_3 point group, and are thus uncoupled in a spherically symmetric environment. However, if these final states interact with a local environment whose point symmetry is a subgroup of O_3, say O_h, then the irreducible representations of O_3 are decomposed into direct sums of the irreducible representations of O_h through "branching" or "compatibility" relations [5.55]. Denoting by Γ_J the irreducible representation of O_3 generated by states with total angular momentum J, for the allowed final states of the $2p^5 3d^1$ multiplet, we have $\Gamma_0 = A_1$, $\Gamma_1 = T_1$, $\Gamma_2 = E \oplus T_2$, $\Gamma_3 = A_2 \oplus T_1 \oplus T_2$, and $\Gamma_4 = A_1 \oplus E \oplus T_1 \oplus T_2$. Thus, the final state $J = 1$ transforms according to the T_1 irreducible representation of O_h. Furthermore, since T_1 is also contained in the decompositions of Γ_3 and Γ_4, the crystal field couples all of these states together, leading to additional allowed final states, as indicated in (5.43), and so to additional features in the absorption spectrum. The peak positions can then be used to parametrize the effect of the crystal field.

In Fig. 5.5 we show the variation of the absorption spectrum with the crystal-field parameter $10Dq$ and a comparison between measured and calculated $L_{2,3}$ absorption spectra of two compounds: a high-spin Ni compound, Ni oxalate, and a low-spin Ni compound, K_2Ni dithio-oxalate [5.90]. Note the considerable difference in the characteristic features of the two sets of absorption spectra, which can be used as an indication of the spin state of Ni. In the high-spin spectrum the degree of interaction with the crystal field can be determined by the splitting of the peaks, with small values of $10Dq$ giving distorted structures that evolve into clearly resolved doublets as the interaction parameter is increased. In the low-spin spectrum, the crystal field parameter is deduced by the intensity of weak satellites, with the primary features of the absorption showing relatively little sensitivity to the strength of the crystal field. In addition to allowing an inference of the spin state of Ni, comparisons such as these allow a direct determination of the crystal-field parameter $10Dq$.

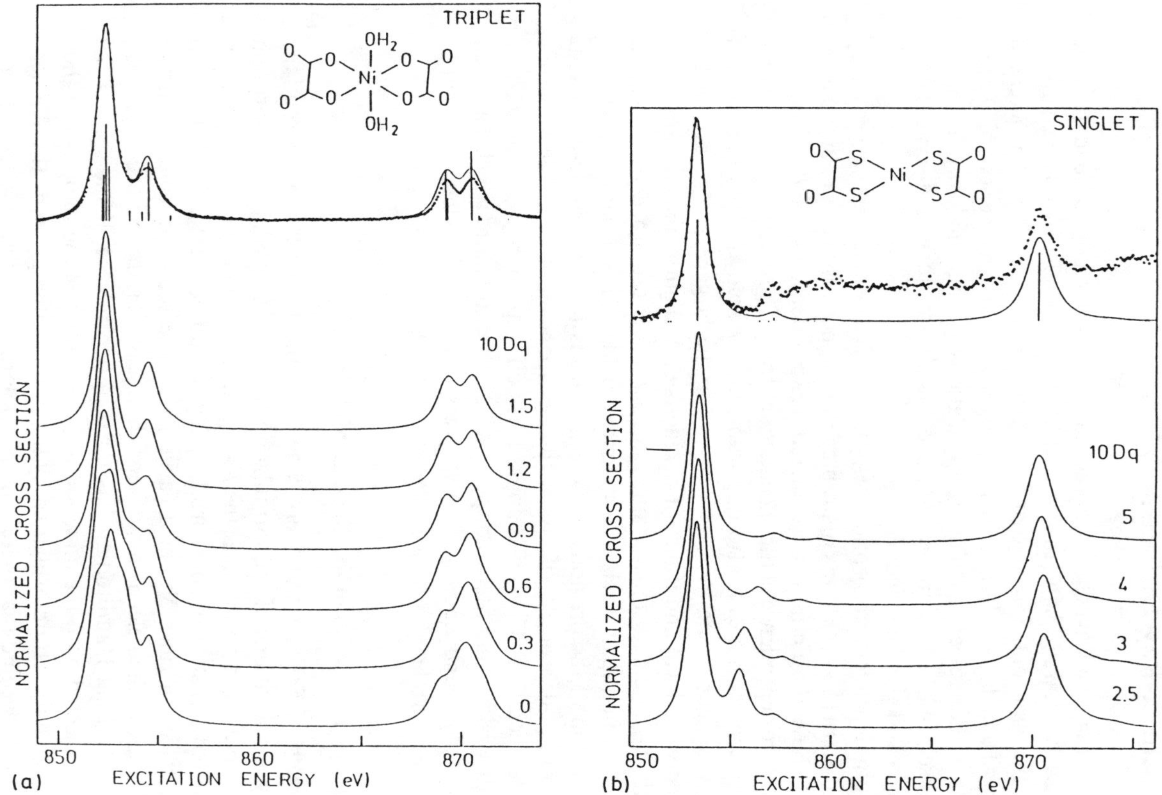

Fig. 5.5. Comparison of calculated and measured $L_{2,3}$ absorption spectra of Ni in two D_{4h} environments: (**a**) Ni oxalate, a high-spin Ni compound and (**b**) K_2Ni dithio-oxalate, a low-spin Ni compound. Included in the diagram are calculated spectra showing the sensitivity to the crystal-field parameter $10Dq$ and a direct comparison between the measured spectra and the optimized calculated spectra [5.87]. The atomic levels have been convoluted with a Gaussian and Lorentzian broadening functions

5.6 Magnetic X-Ray Dichroism

Although virtually all synchrotron radiation studies reported to date have utilized linearly polarized light, the radiation is in fact circularly polarized above and below the plane of the storage ring, as discussed in Sect. 1.3.2. The increased availability of circularly polarized light will open up and enhance several aspects of X-ray absorption research, including applications to magnetic X-ray dichrosim, which is discussed below, and to chiral molecules to allow a discrimination between left- and right-handedness.

The spin-dependent absorption of polarized X-ray was first suggested by *Erskine* and *Stern* [5.91] as a way of probing the magnetic structure of unoccupied bands just above the Fermi energy, including spin densities, band widths and the temperature dependence of the exchange splitting. The determination of spin-resolved unoccupied states is also important for the study of spin fluctuations and other magnetic excitations, as well as for the larger issue regarding the interplay between local and itinerant magnetism. A slightly different approach to magnetic X-ray dichrosim was suggested by *Thole* et al. [5.92] based upon the splitting of multiplet levels in a magnetic field, and is a natural extension of the discussion of the previous section, and leads to temperature-dependent and polarization-dependent effects as the occupancies of magnetic sublevels vary with temperature. These two approaches differ primarily in their starting points, with Erskine and Stern adopting a band formulation, which has been pursued by *Schütz* and *Ebert* and co-workers [5.24, 93–95], while Thole et al. begin with the correlated atomic limit.

5.6.1 Magnetic Dichrosim in the Band Limit

The approach of Erskine and Stern, and refinements thereto, are essentially magnetic generalizations of the formulae in Sect. 5.2. Consider the absorption of circularly polarized X-rays near the K-edge of Fe. The excitation occurs from an unpolarized $s_{1/2}$ state to spin–orbit-split $p_{1/2}$ and $p_{3/2}$ states. In addition to the dipole selection rules obtained earlier, we have the requirement that the spin state of the excited electron does not change during the excitation process, since the perturbing Hamiltonian has no spin dependence. If the spin of the electron is projected onto the direction of polarization of the photon and then summed over all final states weighted by the probabilities given by the Wigner–Eckart theorem, we find that there is no net polarization of the excited electron. However, this procedure assumes that the radial integral is the same for both the $p_{1/2}$ and $p_{3/2}$ final states. In fact the off-diagonal elements of the spin–orbit interaction couple the radial parts in the original basis of the p wavefunctions, as discussed originally by *Fermi* [5.96] and more recently by *Fano* [5.97]. This gives rise to deviations in the probabilities of excitations into the two final states. Thus, the excitation process creates partially polarized photoelectrons from the unpolarized initial state which then propagate in the exchange-split unoccupied states.

Let $n_{\uparrow\uparrow}$ and $n_{\uparrow\downarrow}$ denote the number of photoelectrons with spin parallel and antiparallel, respectively, to the photon polarization. Then, $P_e = (n_{\uparrow\uparrow} - n_{\uparrow\downarrow})/(n_{\uparrow\uparrow} + n_{\uparrow\downarrow})$ is the extent of polarization of the photoelectron. Thus, by decomposing the final density of p states $\rho(E)$ into states with spin parallel and antiparallel to the photon polarization, $\rho(E) = \rho_{\uparrow\uparrow} + \rho_{\uparrow\downarrow}$ and defining the difference of the two densities of states by $\Delta\rho(E) = \rho_{\uparrow\uparrow} - \rho_{\uparrow\downarrow}$, the transition rate for the K-edge may be written as [5.93]

$$\Omega = \hbar^{-1}|\tilde{M}_{i,f}|^2[\rho(E) + P_e\Delta\rho(E)] \equiv \Omega_0 + \Omega', \tag{5.52}$$

where $\tilde{M}_{i,f}$ is the matrix element averaged over final states, and Ω_0 and Ω' are the spin-independent and spin-dependent contributions to Ω, respectively. Measuring Ω when either $\Delta\rho$ or P_e vanishes yields Ω_0. Thus, measuring $\Omega'/\Omega_0 = P_e\Delta\rho/\rho$ for known electron polarization yields the relative difference $\Delta\rho/\rho$ in the spin densities at an energy E above the Fermi level.

In Fig. 5.6, are shown the measured absorption coefficients for Fe averaged over the two polarizations, and the difference for left and right circularly polarized radiation compared with a relativistic electronic structure calculation [5.24, 93]. The calculation clearly reproduces all of the main features of the experiment, though there is some discrepancy in the definition of the structure in the difference spectrum near 8 eV. Although the photoelectron polarization in quite small for the iron K-edge ($P_e = 8.4 \times 10^{-3}$[5.94]), which relies solely

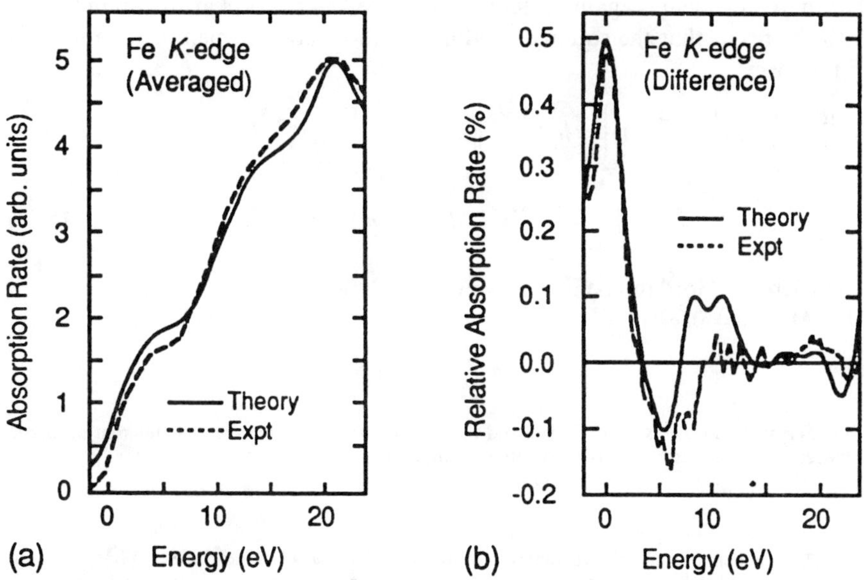

Fig. 5.6. Comparison between measured [5.93] and calculated [5.24] K-shell absorption coefficient of Fe, showing (**a**) the average and (**b**) the relative difference between right and left circularly polarized light

on the radial part of the dipole matrix element, a much larger polarization is seen for L edges (e.g. for Gd $P_e(L_2) \approx -0.45$ and $P_e(L_3) \approx 0.22$ [5.93]) due to the contribution of the angular part of the dipole matrix element (see below). This large polarization of the $L_{2,3}$ absorption spectra has led to applications of this technique to determining local magnetic order by measuring spin-dependent EXAFS [5.95].

5.6.2 Dichroism and Multiplets

We have seen in Sect. 5.5 that the $3d \to 4f$ transitions X-ray absorption spectra in rare-earth metals and compounds are explained quite well by atomic calculations for optical transitions from the $4f^n$ ground state determined by Hund's rules to the dipole-allowed manifold of $3d^9 4f^{n+1}$ multiplets. In the presence of a magnetic field H, the $2J+1$ degenerate states are split into states equally spaced in the energy, with an energy difference $-g\mu_B HM$ between successive sublevels (Fig. 5.7). Thus, at $T=0\,\mathrm{K}$, only the state corresponding to $M=-J$ is occupied. The matrix element for the transition from the state with total angular momentum quantum numbers (J_i, M_i) to the state with quantum numbers (J_f, M_f) is, in analogy with (5.36),

$$M_{i,f} = -\mathrm{i}\hbar \frac{e}{c}\frac{4\pi}{3} A \sum_{m=-1}^{1} (-1)^m Y_{1,m}(\theta,\phi) \langle n_i J_i M_i | Y_{1,-m}(\nabla) | n_f J_f M_f \rangle . \tag{5.53}$$

The matrix element is again evaluated using the Wigner–Eckart theorem [5.55], with the result that the squared modulus of the matrix element can be written in the form[1]

$$|M_{i,f}|^2 = \left[\hbar \frac{e}{c}\frac{4\pi}{3} A \right]^2 \langle n_i J_i \| Y_1 \| n_f J_f \rangle|^2 \cdot \sum_{m=-1}^{1} |Y_{1,m}(\theta,\phi)|^2 |\langle J_i M_i ; 1, -m | J_f M_f \rangle|^2 . \tag{5.54}$$

The Clebsch–Gordan coefficient is nonvanishing only if $J' - J = 0, \pm 1$ and if $M - M' = m = 0, \pm 1$.

[1] Since crystal-field effects are excluded, and there is no interference between the different magnetic sublevels, the cross terms involving different channels can be neglected

Fig. 5.7. (**a**) Schematic energy level diagram of the $3d^{10}f^{13} \to 3d^9 4f^{14}$ transition of Yb^{3+} with and without a magnetic field. The dipole-allowed transitions are indicated by vertical arrows [5.98]. (**b**) Comparison between measured (dotted lines) and calculated (solid lines) M_5 absorption spectra of a terbium iron garnet (TbIG) for various values of α, the direction between the X-ray polarization vector and the [111] magnetization direction [5.99]

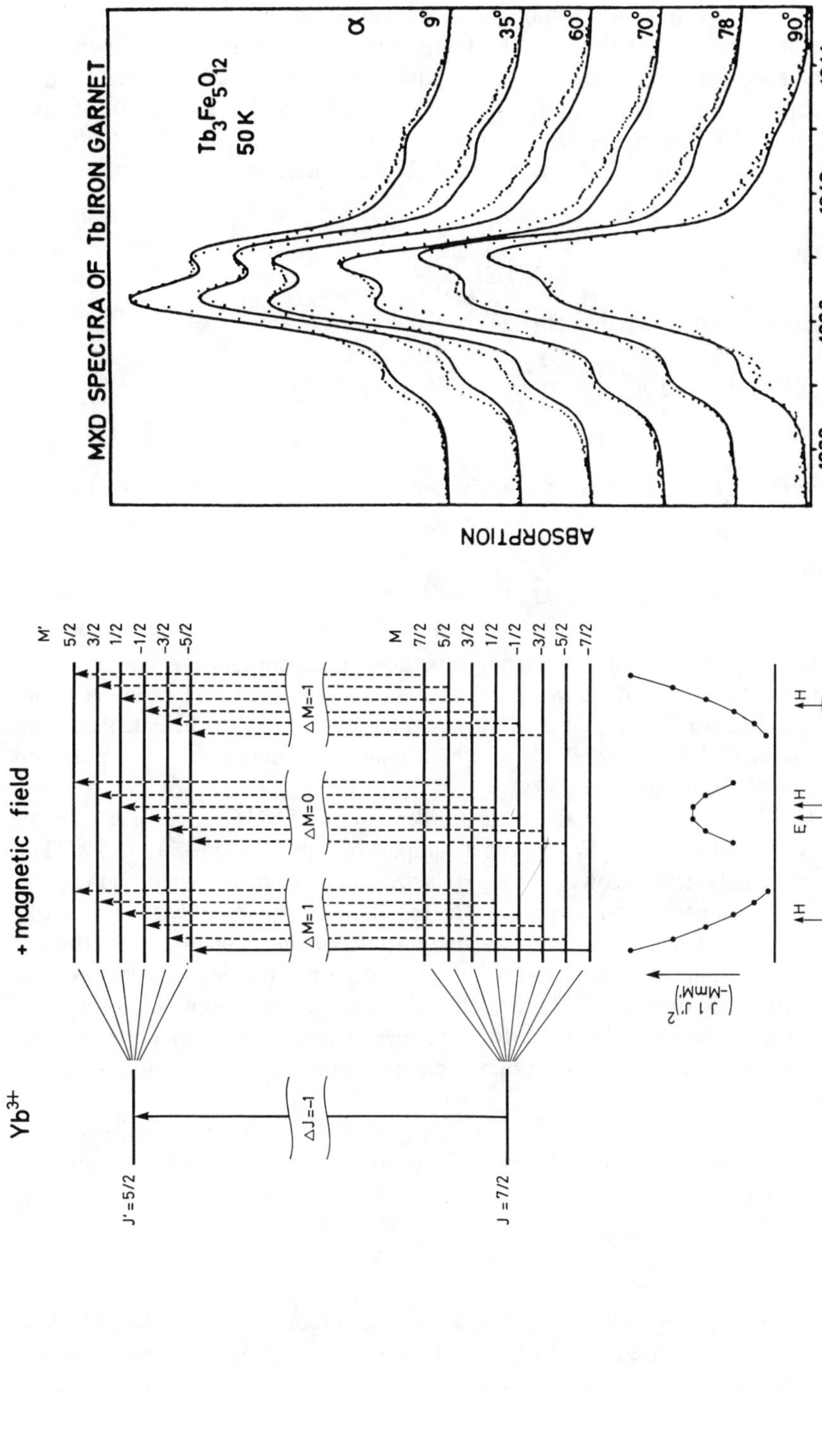

Yb3+
+ magnetic field
J' = 5/2
J = 7/2
ΔJ = -1
ΔM=1
ΔM=0
ΔM=-1
M'
M
(a)
MXD SPECTRA OF Tb IRON GARNET
Tb3Fe5O12
50 K
α
9°
35°
60°
70°
78°
90°
ABSORPTION
ENERGY (eV)
1232
1236
1240
1244
(b)

The observation of the splitting of the sublevels into components labelled by M will be evident through the temperature and polarization dependence of the absorption spectrum. The occupancy of the sublevels as a function of temperature and magnetic field is determined by the Boltzmann factor, $\exp(g\mu_B HM/kT) \equiv \exp(-M/t)$, where the reduced "temperature" is defined by $t = kT/g|\mu_B|H$. The partition function $\mathscr{Z}$ for the magnetic sublevels is thus given by

$$\mathscr{Z} = \sum_{M=-J}^{J} \mathrm{e}^{-M/\theta} = \frac{\sinh[(2J+1)/2t]}{\sinh(1/2t)}. \tag{5.55}$$

Thus, the thermal average of (5.54) is given by

$$\langle\langle |M_{i,f}|^2 \rangle\rangle = \left[\hbar \frac{e}{c} \frac{4\pi}{3} A\right]^2 |\langle n_i J_i \| Y_1 \| n_f J_f \rangle|^2 \sum_{m=-1}^{1} |Y_{1,m}(\theta,\phi)|^2 A^m_{J_i,J_f}, \tag{5.56}$$

where the thermal average of the Clebsch–Gordan coefficient is given by

$$\begin{aligned} A^m_{J_i,J_f} &\equiv \langle\langle |\langle J_i M_i; 1, -m | J_f M_f \rangle|^2 \rangle\rangle \\ &= \mathscr{Z}^{-1} \sum_{M_i=-J_i}^{J_i} |\langle J_i M_i; 1, -m | J_f M_f \rangle|^2 \mathrm{e}^{-M_i/t}. \end{aligned} \tag{5.57}$$

Explicit expressions for the thermal-averaged Clebsch–Gordan coefficents in terms of the averaged quantities $\langle\langle M \rangle\rangle$ and $\langle\langle M^2 \rangle\rangle$ can be obtained from (5.39) and (5.40). The only remaining quantities that need to be calculated are the averages $\langle\langle M \rangle\rangle$ and $\langle\langle M^2 \rangle\rangle$, which can be obtained from the partition function (5.54) in the usual way.

In Fig. 5.7b is shown the measurement of the Tb M_5 edge in a terbium iron garnet (TbIG), $Tb_3Fe_5O_{12}$ at a temperature of $T = 55 \pm 5$ K [5.99]. The absorption spectrum shows a strong polarization dependence, which is in agreement with the predictions of [5.92]. The results were consistent with magnetization measurements in terms of the magnitude of the magnetization, though there was insufficient resolution in the data to permit the determination of the magnetization direction. More detailed information about the magnetic structure can be obtained by performing measurements at lower temperatures, so only the lowest magnetic sublevel has an appreciable probability of being occupied.

For an arbitrary direction of polarization θ with respect to the z-axis, the effective thermal-averaged Clebsch–Gordan coefficient is, from (5.56) and Table 5.2,

$$\sum_{m=-1}^{1} |Y_{1,m}(\theta,\phi)|^2 A^m_{J_i,J_f} = A^0_{J_i,J_f} \cos^2\theta + \tfrac{1}{2}[A^1_{J_i,J_f} + A^{-1}_{J_i,J_f}] \sin^2\theta. \tag{5.58}$$

Thus, by combining (5.58) with (5.59), we see that a measurement of the dichroism with linearly polarized X-rays gives information only about the *magnitude* of the magnetization, not the *sign*, since the magnetization contributes terms proportional only to $\langle\langle M^2 \rangle\rangle$, and not to $\langle\langle M \rangle\rangle$. However, since the polarization dependence enters only as $\langle\langle M^2 \rangle\rangle$, the temperature dependence of the local magnetic moment can be determined for antiferromagnetic systems, where $\langle\langle M \rangle\rangle = 0$. Additionally, the polarization dependence can be used to determine the relative orientation of the sublattice magnetization directions in complex systems and, with the advent if circularly polarized sources, to determine the magnitude $\langle\langle M \rangle\rangle$. However, as discussed at the beginning, the increasing availability of circularly polarized X-rays will have an important impact in the study of magnetic X-ray dichroism.

5.7 Related Techniques

The theoretical techniques described in the preceding sections are not limited to applications in X-ray absorption spectroscopy. There are other techniques involving localized transitions where the type of information on both electronic structure and the local arrangement of atoms can in principle be extracted by comparison with calculations analogous to those described here.

5.7.1 X-Ray Bremsstrahlung Isochromat Spectroscopy

In X-ray bremsstrahlung isochromat spectroscopy (XBIS), an electron beam with an energy of several keV is injected into a sample. The electrons decelerate into lower-energy states, emitting an X-ray in the process. The intensity of X-rays at fixed energy are then measured as a function of the incident electron energy. Several experiments [5.100–102] have shown that the measured XBIS spectra show striking similarities to both the unoccupied states of the materials and to the corresponding EXAFS spectra. This opens up the possibility of XBIS being used as a probe of local structure complementary to EXAFS.

Šimunek et al. [5.103] calculated the matrix elements for XBIS in the region of energy 50–380 eV above the Fermi level for Cu and Pd and demonstrated that the decelerated final electron originates near the core region of an atom. Furthermore, these authors found that the relative ratios of s, p, and d final states varied slowly in the final-state energy range considered but depended strongly upon the X-ray energy. This provides a support for the similarity between XBIS and EXAFS spectra. *Šimunek* et al. [5.104] extended this model to include the multiple scattering of the final-state electron using the program ICXANES [5.48], and found excellent agreement between computed and measured spectra.

5.7.2 Electron-Energy-Loss Fine Structure

In electron-energy-loss core-level spectroscopy, a primary electron beam typically of energy of a few keV is directed onto a sample. A primary electron undergoes an inelastic collision during which energy is imparted to a core electron. The primary electron losses energy, which the core electron is excited into an unoccupied state. The subsequent scattering of the photoelectron is similar to that in an X-ray absorption measurement, and can be treated by analogous methods [5.105, 106]. Two important differences, however, are that the interaction between the primary electron after scattering and the excited photoelectron are neglected, and that the matrix element generating the transition does not have the dipole form, except in the limit of small momentum transfer.

The transition rate in an electron-energy-loss experiment can again be factored into atomic and solid-state contributions, with the multiple scattering treated as in [5.7, 48], with all angular momentum channels considered in the matrix element. The calculations [5.105, 106] show that different edges behave quite differently with respect to the dipole selection rules. Near some edges, dipole-allowed transitions dominate even at large momentum transfer, while near others, dipole-forbidden transitions appear even at small momentum transfer. As for other spectroscopies, such calculations are an important first step in identifying possible deviations from the simple single-particle picture.

References

5.1 R. de L. Kronig: Z. Physik **70**, 317 (1931)
5.2 D.E. Sayers, E.A. Stern, F.W. Lytle: Phys. Rev. Lett. **27**, 1204 (1971)
5.3 C.A. Ashley, S. Doniach: Phys. Rev. B **11**, 1279 (1975)
5.4 P.A. Lee, J.B. Pendry: Phys. Rev. B **11**, 2795 (1975)
5.5 P.A. Lee: Phys. Rev. B **13**, 5261 (1976)
5.6 J. Stöhr: "SEXAFS: Everything You Always Wanted to Know but Were Afraid to Ask," in *X-Ray Absorption: Principles, Applications and Techniques of EXAFS, SEXAFS and XANES*, ed. by R. Prinz, K. Koninsberger (Wiley, New York 1986)
5.7 P.J. Durham, J.B. Pendry, C.H. Hodges: Comp. Phys. Commun. **25**, 193 (1982)
5.8 D. Norman, J. Stöhr, R. Jaeger, P.J. Durham, J.B. Pendry: Phys. Rev. Lett. **51**, 2052 (1983)
5.9 D.D. Vvedensky, J.B. Pendry, U. Döbler, K. Baberschke: Phys. Rev. B **35**, 7756 (1987)
5.10 P.J. Durham, J.B. Pendry, C.H. Hodges: Solid State Commun. **38**, 159 (1981)
5.11 G. Bunker, E.A. Stern: Phys. Rev. Lett. **52**, 1990 (1984)
5.12 M. Benfatto, C.R. Natoli, A. Bianconi, J. Garcia, A. Marcelli, M. Fanfoni, I. Davoli: Phys. Rev. B **34**, 5774 (1986)
5.13 A. Bianconi, A. Di Cicco, N.V. Patel, M. Benfatto, A. Marcelli, C.R. Natoli, P. Pianetta, J. Woicik: Phys. Rev. B **36**, 6426 (1987)
5.14 J.B. Pendry: *Low Energy Electron Diffraction* (Academic, London 1974)
5.15 P. Hohenberg, W. Kohn: Phys. Rev. **136**, 864 (1964)
5.16 W. Kohn, L.J. Sham: Phys. Rev. **140**, 1133 (1965)
5.17 R.D. Cowan: *The Theory of Atomic Structure and Spectra* (University of California Press, Berkeley 1981)
5.18 L.G. Parratt: Rev. Mod. Phys. **31**, 616 (1959)
5.19 R.S. Knox: *Theory of Excitons* (Solid State Phys. Suppl. 5) (Academic, New York 1963)
5.20 J. Zaanen, G.A. Sawatzky, J. Fink, W. Speier, J.C. Fuggle: Phys. Rev. B **32**, 4905 (1985)
5.21 J.J. Rehr, E.A. Stern: Phys. Rev. B **14**, 4413 (1976)
5.22 A. Messiah: *Quantum Mechanics* (North Holland, Amsterdam 1958)
5.23 A. de Shalit, H. Feshbach: *Theoretical Nuclear Physics*, Vol. 1 (Wiley, New York 1974)
5.24 H. Ebert, P. Strange, B.L. Gyorffy: Z. Phys. B **73**, 77 (1988)

5.25 E.A. Stern, J.J. Rehr: Phys. Rev. B **27**, 3351 (1983)
5.26 J.J. Rehr, S.-H. Chou: in *EXAFS and Near Edge Structure*, ed. by A. Bianconi, L. Incoccia, S. Stipcich (Springer, Berlin 1983)
5.27 J.E. Müller, O. Jepsen, O.K. Andersen, J.W. Wilkins: Phys. Rev. Lett. **40**, 720 (1975)
5.28 O.K. Andersen: Phys. Rev. B **12**, 3060 (1975)
5.29 K.H. Johnson: J. Chem. Phys. **45**, 3085 (1966)
5.30 D. Dill, J.L. Dehmer: J. Chem. Phys. **61**, 692 (1974)
5.31 J.L. Dehmer, D. Dill: J. Chem. Phys. **65**, 5327 (1976)
5.32 U. von Barth, G. Grossman: Solid State Commun. **32**, 645 (1979)
5.33 U. von Barth, G. Grossman: Phys. Rev. B **25**, 5150 (1982)
5.34 D.D. Koelling: Rept. Prog. Phys. **44**, 139 (1981)
5.35 T. Loucks: *Augmented Plane-Wave Method* (Benjamin, Reading, MA 1967)
5.36 J. Korringa: Physica **13**, 392 (1947)
5.37 W. Kohn, N. Rostoker: Phys. Rev. **94**, 1111 (1954)
5.38 R.G. Newton: Phys. Rev. Lett. **65**, 2031 (1990)
5.39 H.L. Skriver: *The LMTO Method* (Springer Verlag, Berlin, Heidelberg 1984)
5.40 D.D. Koelling, G. Arbman: J. Phys. F **5**, 2041 (1975)
5.41 B.L. Gyorffy, M.J. Stott: "A One-Electron Theory of Soft X-Ray Emission from Random Alloys," in *Band Structure Spectroscopy of Metals and Alloys*, ed. by D.J. Fabian, L.M. Watson (Academic, London 1973), pp. 385–403
5.42 A. Zangwill: *Physics at Surfaces* (Cambridge University Press, Cambridge 1988) Chap. 2
5.43 M.P. Seah, W.A. Dench: Surface and Interface Analysis **1**, 2 (1979)
5.44 A.L. Fetter, J.D. Walecka: *Quantum Theory of Many-Particle Systems* (McGraw Hill, New York 1971), Chap. 3
5.45 S. Crampin: *Self-Consistent Charge Densities at Isolated Planar Defects in Metals*, Ph.D. thesis, University of London (1989)
5.46 P. Kizler, P. Lamparter, S. Steeb: Z. Naturforsch. A **44**, 189 (1989)
5.47 P.J. Durham, S.S. Hasnain, A. Bianconi, A. Congiu-Castellano, A. Giovanelli, L. Incoccia, S. Morante, J.B. Pendry: The EMBO Journal **2**, 1441 (1983)
5.48 D.D. Vvedensky, D.K. Saldin, J.B. Pendry: Comp. Phys. Commun. **40**, 421 (1986)
5.49 K.H. Johnson: Phys. Lett. A **27**, 138 (1968)
5.50 C.R. Natoli, M. Benefatto: J. Phys. (Paris) C **8**, 11 (1986)
5.51 F.W. Kutzler, C.R. Natoli, D.K. Misemer, S. Doniach, K.O. Hodgson: J. Chem. Phys. **73**, 3274 (1980)
5.52 J.A. Horsley, J. Stöhr, R.J. Koestner: J. Chem. Phys. **83**, 3146 (1985)
5.53 J.J. Rehr, R.C. Albers: Phys. Rev. B **41**, 8139 (1990)
5.54 E.U. Condon, G.H. Shortley: *The Theory of Atomic Spectra* (Cambridge University Press, Cambridge 1951)
5.55 M. Tinkham: *Group Theory and Quantum Mechanics* (McGraw-Hill, New York 1964)
5.56 U. Döbler, K. Baberschke, J. Haase, A. Puschmann: Phys. Rev. Lett. **52**, 1437 (1984)
5.57 K. Baberschke, U. Döbler, L. Wenzel, D. Arvanitis, A. Baratoff, K.H. Rieder: Phys. Rev. **B33**, 5910 (1986)
5.58 R. Zeller: Z. Phys. **72**, 79 (1988)
5.59 W. Speier, J.F. van Acker, R. Zeller: Phys. Rev. B **41**, 2753 (1990)
5.60 P.J.W. Weijs, M.T. Czyżyk, J.F. van Acker, W. Speier, J.B. Goedkoop, H. van Leuken, H.J.M. Hendrix, R.A. de Groot, G. van der Laan, K.H.J. Buschow, G. Weich, J.C. Fuggle: Phys. Rev. B **41**, 11899 (1990)
5.61 P.A. Wolff: Phys. Rev. **124**, 1030 (1961)
5.62 A.M. Clogston: Phys. Rev. **125**, 439 (1962)
5.63 P.S. Bagus, H.F. Schaeffer III: J. Chem. Phys. **56**, 224 (1972)
5.64 G. Loubriel: Phys. Rev. B **20**, 5339 (1979)
5.65 G.D. Mahan: Phys. Rev. **163**, 612 (1967)
5.66 P.W. Anderson: Phys. Rev. Lett. **18**, 1049 (1967)
5.67 P.W. Anderson: Phys. Rev. **164**, 352 (1968)
5.68 P. Nozières, C.T. de Dominicis: Phys. Rev. **178**, 1097 (1969)
5.69 H. Neddermeyer: Phys. Rev. B **13**, 2411 (1976)
5.70 U. von Barth, G. Grossmann: Phys. Scr. **21**, 580 (1980)
5.71 C.A. Swarts, J.D. Dow, C.P. Flynn: Phys. Rev. Lett. **43**, 158 (1980)
5.72 G.D. Mahan: Solid State Physics **29**, 75 (1974)
5.73 J.D. Dow: Comm. Sol. State Phys. **6**, 71 (1975)

5.74 J. Frenkel: Phys. Rev. **137**, 17 (1931)
5.75 J. Frenkel: Phys. Rev. **137**, 1276 (1931)
5.76 G. Wannier: Phys. Rev. **52**, 191 (1937)
5.77 L.A. Grunes, R.D. Leapman, C.N. Wilker, R. Hoffman, A.B. Kunz: Phys. Rev. B **25**, 7157 (1982)
5.78 M. Altarelli, D.L. Dexter: Phys. Rev. Lett. **29**, 1100 (1972)
5.79 R.J. Elliot: Phys. Rev. **108**, 1384 (1957)
5.80 D.K.G. de Boer, C. Haas, G.A. Sawatzky: Phys. Rev. B **29**, 4401 (1984)
5.81 A. Svane, O. Gunnarsson: Phys. Rev. Lett. **65**, 1148 (1990)
5.82 D.D. Vvedensky, J.B. Pendry: Phys. Rev. Lett. **54**, 2725 (1985)
5.83 I. Davoli, A. Marcelli, A. Bianconi, M. Tomellini, M. Fanfoni: Phys. Rev. B **33**, 2979 (1986)
5.84 J.C. Slater: *The Self-Consistent Field for Molecules and Solids*, Vol. 4 of *Quantum Theory of Molecules and Solids* (McGraw-Hill, New York 1974)
5.85 I.I. Sobelman: *Atomic Spectra and Radiative Transitions* (Springer, Berlin, Heidelberg 1979)
5.86 J.D. Jackson: *Classical Electrodynamics* (Wiley, New York 1975)
5.87 J.C. Slater: *Quantum Theory of Atomic Structure*, Vol. 1 (McGraw-Hill, New York 1960)
5.88 B.T. Thole, G. van der Laan, J.C. Fuggle, G.A. Sawatzky, R.C. Karnatak, J.-M. Esteva: Phys. Rev. B **32**, 5107 (1985)
5.89 F.M.F. de Groot, J.C. Fuggle, B.T. Thole, G.A. Sawatzky: Phys. Rev. B **41**, 928 (1990)
5.90 G. van der Laan, B.T. Thole, G.A. Sawatzky, M. Verdageur: Phys. Rev. B **37**, 6587 (1988)
5.91 J.L. Erskine, E.A. Stern: Phys. Rev. B **12**, 5016 (1975)
5.92 B.T. Thole, G. van der Laan, G. Sawatzky: Phys. Rev. Lett. **55**, 2086 (1985)
5.93 G. Schütz, W. Wagner, W. Wilhelm, P. Kienle, R. Zeller, R. Frahm, G. Materlik: Phys. Rev. Lett. **58**, 737 (1987)
5.94 G. Schütz, M. Knülle, R. Wienke, W. Wilhelm, W. Wagner, P. Lienle, R. Frahm: Z. Phys. B **73**, 67 (1988)
5.95 G. Schütz, R. Frahm, P. Mautner, R. Wienke, W. Wagner, W. Wilhelm, P. Kienle: Phys. Rev. Lett. **62**, 2620 (1989)
5.96 E. Fermi: Z. Phys. **59**, 680 (1929)
5.97 U. Fano: Phys. Rev. **178**, 131 (1969)
5.98 J.B. Goedkoop, J.C. Fuggle, B.T. Thole, G. van der Laan, G.A. Sawatzky: J. Appl. Phys. **64**, 5595 (1988)
5.99 G. van der Laan, B.T. Thole, G.A. Sawatzky, J.B. Goedkoop, J.C. Fuggle, J.-M. Esteva, R. Karnatak, J.P. Remeika, H.A. Dabkowska: Phys. Rev. B **34**, 6529 (1986)
5.100 W. Speier, R. Zeller, J.C. Fuggle: Phys. Rev. B **32**, 3597 (1985)
5.101 W. Speier, T.M. Hayes, J.W. Allen, J.B. Boyce, J.C. Fuggle, M. Campagna: Phys. Rev. Lett. **55**, 1693 (1985)
5.102 E. Sobczak, J. Auleytner: Phys. Rev. B **37**, 6251 (1988)
5.103 A. Simunek, J. Vackár, E. Sobczak: Phys. Rev. **38**, 8515 (1988)
5.104 A. Simunek, O. Sipr, J. Vackár: Phys. Rev. Lett. **63**, 2076 (1989)
5.105 P. Aebi, M. Erbudak, F. Vanini, D.D. Vvedensky, G. Kostorz: Phys. Rev. B **41**, 11760 (1990)
5.106 P. Aebi, M. Erbudak, F. Vanini, D.D. Vvedensky, G. Kostorz: Phys. Rev. B **42**, 5369 (1990)

6. X-Ray Absorption Spectroscopy of Small Molecules, Free and Adsorbed on Surfaces

J. Somers

With 16 Figures

Although the earliest X-ray absorption investigations on free atoms and molecules were reported over fifty years ago [6.1], it was not until the late 1960s that the wealth of detail observed in the X-ray absorption spcctra of SF_6 [6.2] caused a revival of interest in this area. A few years later, the first high-resolution inner shell electron-energy loss (ISEELS) measurements of core-level excitation in small molecules began to reveal the large amount of information on the unoccupied molecular states in such spectra [6.3, 4]. Since then, the increased availability of synchrotron radiation sources, the development of grazing incidence monochromators, providing light in the 100–1000 eV energy range, and the extension of the XAS technique to adsorbates has generated a surge in the number of XAS investigations of free and adsorbed molecules.

XAS is now one of the most important techniques from which information on the symmetry and electronic structure of free and adsorbed molecules can be obtained. In this chapter, some basic concepts of the method are introduced. These include the distinction between valence and Rydberg states, symmetry-based selection rules, the Franck-Condon principle, symmetry breaking by core holes, symmetry reduction on adsorption and orientation determination of molecular adsorbates. These basic principles are highlighted by suitable examples.

6.1 Free Molecules

In XAS, the observed transitions correspond to the excitation of a core electron of the particular element under investigation into the unfilled orbitals of the molecule [6.5]. In a linear combination of atomic orbitals (LCAO) model, the unoccupied molecular orbitals of a molecule, containing only second-row elements, are derived from linear combinations of the 2*s* and 2*p* atomic orbitals. These molecular orbitals can be located above and below the photoionisation threshold which, for a free molecule, is defined as the XPS binding energy referenced to the vacuum level. Rydberg-type orbitals are derived from the third and higher principle quantum shells, and form a series which converges on the ionisation threshold with increasing principal quantum number, n. The distinction between the two types of orbital is not always straightforward, however, as mixing of orbitals of the same symmetry to form new states of composite valence-Rydberg character can occur [6.6].

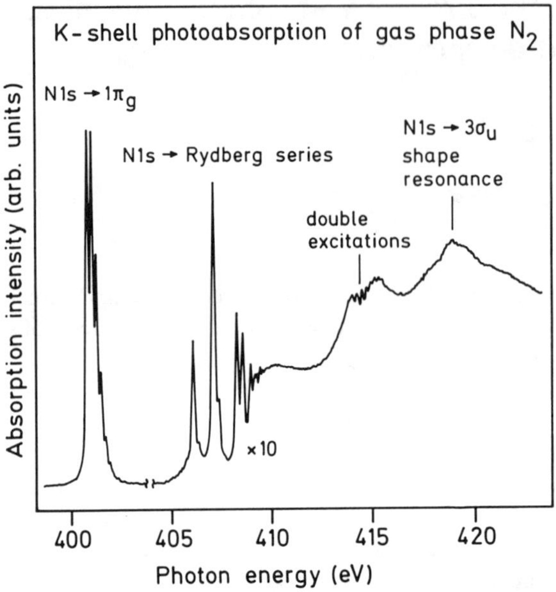

Fig. 6.1. Nitrogen K-edge X-ray absorption spectrum of N_2 in the gas phase. From [6.7]

The X-ray absorption spectrum of the N_2 molecule, recently remeasured by *Chen* et al. [6.7] (see Fig. 6.1), is an exciting example of high-resolution XAS. All the types of transition described above are observed. The features at 401 and 420 eV are due to the excitation of the nitrogen $1s$ electron into the unoccupied $1\pi_g$ and $3\sigma_u$ valence molecular orbitals. The former is doubly degenerate and corresponds to the anti-symmetric combination of the atomic $2p_x$ and $2p_y$ orbitals, while the $3\sigma_u$ is the anti-symmetric combination of the $2p_z$ atomic orbitals. Both orbitals and the relevant coordinate system are shown in the molecular orbital drawings of *Jorgensen* and *Salem* [6.8] in Fig. 6.2. Within the $D_{\infty h}$ point group, appropriate for a homonuclear diatomic molecule, the labels σ and π correspond to the irreducible representations of the point group, defining the symmetry transformation properties of the molecular orbital wave-functions. The subscripts g and u signify that the wave function is either symmetric (gerade) or anti-symmetric (ungerade) with respect to the inversion centre.

The wave functions of molecular Rydberg orbitals also transform according to the irreducible representations of the appropriate molecular point group. The energy of a series member is given by the Rydberg formula

$$\mathrm{TV} = \mathrm{IP} - E_n = \frac{R}{(n-\delta)^2}, \tag{6.1}$$

where TV is known as the term value and corresponds to the difference in energy of the ionisation potential (IP) and the energy (E_n) of the Rydberg orbital with principal quantum number n. R is the Rydberg constant and δ the quantum defect, which is related to the degree of screening and penetration of the Rydberg

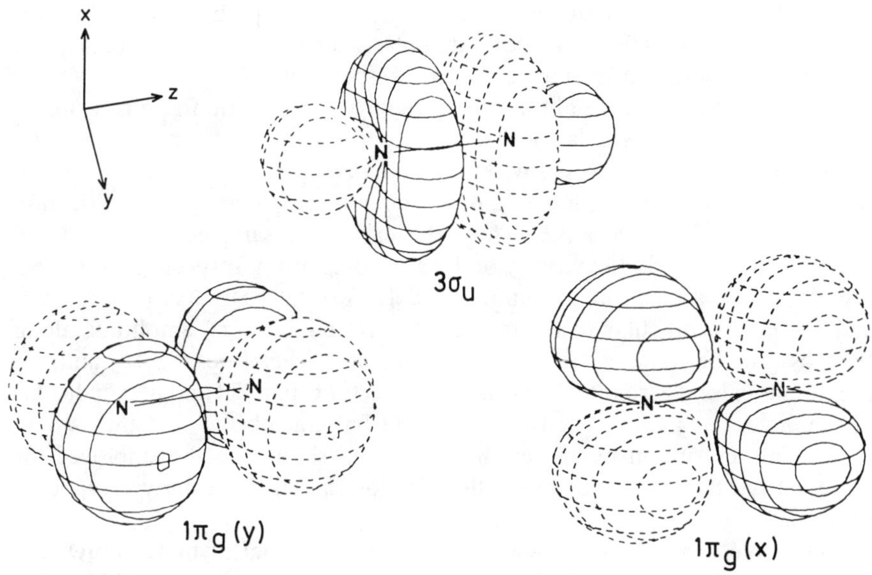

Fig. 6.2. Amplitudes of the N_2 $1\pi_g$ and $3\sigma_u$ molecular orbitals. From [6.8]

orbital with the ion core. This factor is discussed in greater detail in [6.6, 9]. Utilising either empirical or calculated values for δ, this very simple formula greatly assists in the assignment of the large number of Rydberg excitations observed in XAS.

The transitions observed in XAS are not confined to the single-electron excitations of the type discussed above. Multi-electron excitations must be considered also. The absorption feature at 415 eV in Fig. 6.1, for example, is due to a two-electron (shake-up) transition involving the simultaneous excitation of a N 1s electron into the $1\pi_g$ orbital and the excitation of a valence $1\pi_u$ or $3\sigma_g$ electron into one of the unoccupied states. In general, multi-electron excitations exhibit lower absorption cross sections and are most frequently observed in XAS of molecules possessing extended π molecular orbitals.

Having described the types of transition observable in XAS, a discussion of the selection rules in now appropriate. This topic is treated in detail by Vvedensky in Chap 5. The intensity of a given transition is given by

$$I \sim [\boldsymbol{E} \cdot \langle f | \boldsymbol{M}_u | i \rangle]^2, \tag{6.2}$$

where $\boldsymbol{E}$ is the electric field vector of the radiation. $\langle f|$ and $\langle i|$ are the final- and initial-state wave functions and $\boldsymbol{M}_u$ is the dipole moment vector associated with the transition. From an analysis of the symmetry properties of the matrix element $\langle f|\boldsymbol{M}_u|i\rangle$, extremely useful information can be derived from XAS without recourse to detailed calculations. The intensity of a transition will be non-zero only when the direct product of the irreducible representations describing $\langle f|$, $\langle i|$ and one of the cartesian components, $\boldsymbol{u}$, of $\boldsymbol{M}_u$ belongs to, or at least contains, the totally symmetric irreducible representation of the point group. Thus, for N_2, the $1s$ to $1\pi_g$ and $3\sigma_u$ transitions are polarised perpendicular (xy plane) and parallel (z direction) to the molecular axis, respectively.

A more instructive insight into the origin of this selection rule is achieved by reference to the molecular orbital pictures in Fig. 6.2 and the selection rules taken from atomic spectroscopy, where radiative excitation of a $1s$ electron results in a final-state wavefunction of p-symmetry propagating in the direction of the electric field vector, $\boldsymbol{E}$. Thus, the excitation samples the density of unoccupied p-states in the vicinity of the absorbing atom. Inspection of Fig. 6.2 reveals that at a N atom, the unoccupied $1\pi_g$ orbital has maximum p_x or p_y amplitude perpendicular to the molecular axis and zero (p_z) amplitude along the axis. The converse is true for the $3\sigma_u$ orbital. Therefore, the $1\pi_g$ and $3\sigma_u$ excitations exhibit zero intensity when the $\boldsymbol{E}$ vector is parallel and perpendicular to the molecular axis, respectively. It should be remembered that this is only a pictorial description of the selection rule, and should not be employed for more complicated molecules; rather the selection rules must be derived explicitly using (6.2).

For molecules possessing two or more equivalent atoms, symmetry breaking due to localisation of the core hole can occur [6.10, 11]. In the case of N_2, were the core hole localised, the symmetry would be reduced from $D_{\infty h}$ of the ground state to $C_{\infty v}$ in the excited state. As far as the dipole selection rules are concerned, three essentially equivalent treatments of this problem, yielding almost identical results, can be invoked. By allowing the core hole to be delocalised in a molecular orbital derived from a linear combination of the $1s$ atomic orbitals, the analysis can be achieved within the higher symmetry of the ground state. Given the experimental [6.10] and theoretical [6.11] evidence for core-hole localisation, totally ignoring the effect is certainly not realistic. In a second procedure, the core hole can be treated as localised and the selection rules are applied in the lower symmetry configuration of the excited state. Although this method would appear to describe the situation adequately, the localised hole can be treated as non-stationary, and, from a linear combination of all equivalent broken symmetry final-state configurations, a final-state wave function, transforming within the higher symmetry of the ground state, is generated [6.12]. As far as the polarisation dependences of the $1\pi_g$ and $3\sigma_u$ resonances of the N_2 molecule are concerned, each treatment yields identical results, as the rotational axis along the z coordinate is the dominant symmetry element in both the $D_{\infty h}$ and $C_{\infty v}$ point groups. Whether or not symmetry breaking due to core-hole localisation is an important factor in the evaluation of XAS selection rules must be determined on a case-to-case basis. For many highly symmetric molecules, however, it appears to be unimportant as regards the polarisation dependence of the π resonances. In the case of the σ resonances, the different treatments can give rise to subtle differences in the polarisation dependences.

In XAS, the relatively sharp resonances occurring below threshold often exhibit vibrational fine structure, the origin of which is due to the Franck–Condon principle and can be explained by reference to the potential energy

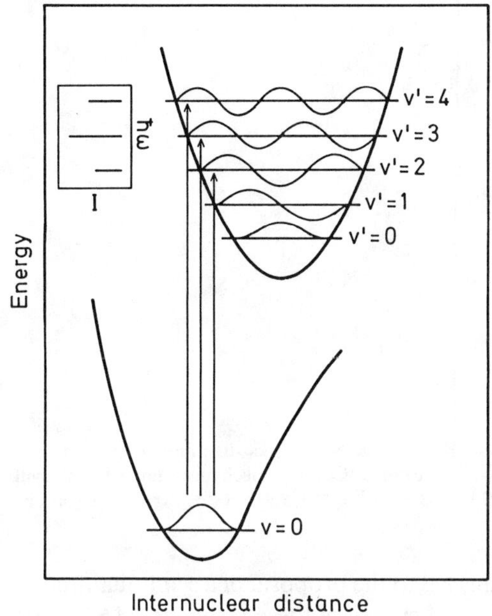

Fig. 6.3. The Franck–Cordon principle, showing transitions from the $v = 0$ level of the electronic ground state to a series of vibrational levels in the electronic excited state. The potential of the electronic excited state is described by a harmonic oscillator, therefore the energy spacing between the various vibrational levels is constant

diagram of the ground and excited states, shown schematically in Fig. 6.3. As an electron is promoted from a non-bonding core orbital into an orbital with anti-bonding character, the minimum in the potential energy of the excited state generally lies at a larger equilibrium internuclear separation than that of the ground state. Given a typical energy spacing of ~ 100–300 meV between the vibrational levels, essentially only the vibrational ground state, $v = 0$, of the electronic ground state is occupied at room temperature. Utilising the Franck–Condon principle, which states that the excitation occurs on a time scale faster than the internuclear motion, vertical transitions are observed from the $v = 0$ level of the ground state to a range of vibrational levels in the excited state. The relative intensities of the vibrational bands observed in XAS will depend sensitively on the change in internuclear distance and on the overlap between the ground-state and excited-state nuclear wavefunctions. Should the energy spacing of the excited-state vibrational levels be small in comparison to the natural width of the resonances, (typically 100 meV), the vibrational splitting will not be resolved, but will give rise to additional broadening.

The broad and asymmetric nature of the resonances observed above threshold is due to a non-Franck–Condon broadening mechanism [6.13, 14]. These quasi-bound states have been described in terms of scattering theory where the excited electron is trapped temporarily by a centrifugal barrier in the molecular potential [6.15, 16]. As this potential is at least partially dependent on the molecular form, the energy of these "shape" resonances is expected to be sensitive to the bond length of the molecule. Indeed, the results of a large

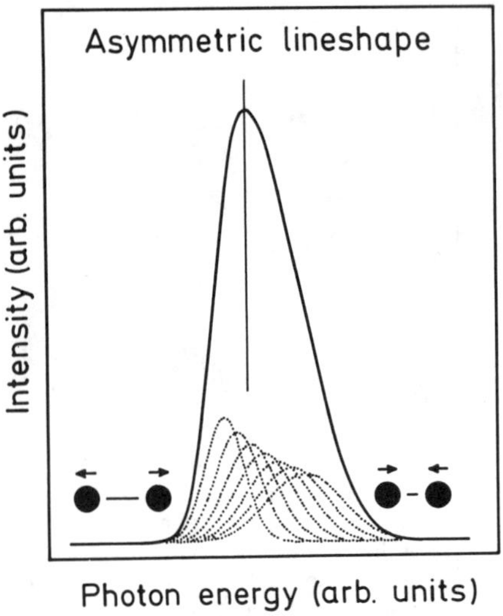

Fig. 6.4. Schematic diagram of how a non-Franck–Condon mechanism imparts an asymmetric line shape to σ resonances. From [6.14]

number of experimental investigations led to the proposal of a semi-quantitative correlation between the bond length and energy of these resonances [6.17–20]. This correlation has subsequently been challenged [6.21–23]. Within the scattering theory, the excitation occurs at various instantaneous bond lengths during the course of a vibrational period. The observed resonance would then correspond to the sum of a series of overlapping symmetric peaks corresponding to the contribution for each instantaneous bond length [6.13], as is illustrated for a diatomic molecule in Fig. 6.4 [6.14]. Generally, the higher the energy above threshold the greater are the asymmetry and width of the sigma resonances.

All the principles of XAS outlined above apply equally well for more complicated molecules. As the molecules under investigation become larger, however, the difficulty in assignment of the spectra increases. In the case of the heteronuclear diatomic molecule CO, only two resonances due to the 2π and 6σ molecular orbitals are expected. For benzene, however, thirty molecular orbitals can be formed from a linear combination of the H $1s$ and C $2s$ and $2p$ atomic orbitals. Fifteen of these are filled by the available valence electrons, leaving fifteen unoccupied states. Even though the actual number of unfilled states is lower, due to the formation of doubly degenerate orbitals within the D_{6h} point group, the problem of assignment remains non-trivial.

Although some assistance in the assignment of complicated spectra can be obtained from polarisation-dependent measurements of partially ordered films, or chemical substitution, there is no substitute for detailed accurate calculations on the excited states. This was recognised almost twenty years ago in the first molecular orbital type calculations on SF_6 [6.5]. In a Hartree–Fock linear combination of atomic orbitals (HF-LCAO) approach, evaluations of the excitation energies are obtained from the difference between the self-consistent ground and excited state energies (ΔSCF). A simplification can be achieved in a Koopmans type approximation where the electrons not involved in the excitation are maintained in identical orbitals of the molecule. The derived energies are poorer than in the ΔSCF method, however, as the valence orbitals are not allowed relax in the presence of the core hole.

The Hartree–Fock approach yields accurate results for the lowest energy state, but is not well suited for the calculation of excited states. Multi-configurational self-consistent field (MC-SCF) calculations give better comparison with experiment. However, a full configuration interaction (CI) calculation should be used to obtain best results. An example is provided by benzene, where, in a molecular orbital scheme, the excitation of a 1s electron into the highest $\pi(b_{2g})$ orbital is very close in energy to the double excitation of a 1s electron, and an electron from the highest occupied π (e_{1g}) orbital into the lowest unoccupied $\pi(e_{2u})$ orbital. In contrast and in good agreement with experiment, MC-SCF calculations find that neither the single- nor the two-particle excitation, of the types mentioned above, exist, rather the intensity is distributed over two states of composite configuration, separated by 5 eV, below and above the ionisation threshold [6.12].

These types of molecular orbital calculations, based on localised orbitals, are not a suitable means to describe states in the continuum. The high-energy continuum states have usually been calculated using the multiple scattering Xα method [6.15]. Since in this approach numerical rather than localised analytical basis sets are used to represent the electron orbitals, continuum and bound states can be described more naturally within the same theoretical framework.

The theoretical methods, employed in the calculation of X-ray absorption spectra, include to some extent the presence of the core hole. In ΔSCF, MC-SCF or CI calculations of the excited state energy, it is included directly. An alternative approximation used in MS-Xα is the so-called "transition state" concept, in which half an electron is removed from the localised core orbital of the absorbing atom [6.24]. Core hole localisation has been the subject of much theoretical interest and is discussed further in [6.11, 12] and [6.25–32]. At present, the most important experimental evidence for localisation is found in the C 1s XPS spectra of benzene [6.10]. The recent availability of high-resolution XAS data, in which vibrational fine structure is resolved [6.33], is an important development, however, as it has already been proposed that symmetry breaking due to core hole localisation is a consequence of vibronic coupling [6.32]. Further theoretical and experimental developments in this area are eagerly awaited.

6.2 Adsorbed Molecules

Much of the interest in XAS of adsorbates stems from the work of *Stöhr* and coworkers [6.19, 20, 34–38], who recognised that, in principle at least, the polarisation dependence of the resonances could be used to determine the orientation of adsorbates with respect to the surface. As an example of the

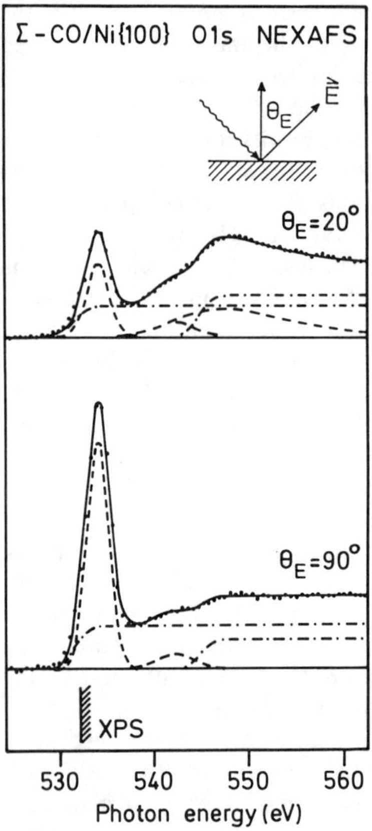

Fig. 6.5. Oxygen K-edge X-ray absorption spectra of CO coadsorbed with H_2 on Ni{100}, recorded at two angles Θ_E, of the $\boldsymbol{E}$ vector of the radiation with respect to the surface normal. In this coadsorption phase the CO molecule is very weakly adsorbed on the surface and is usually referred to as the Σ-CO adsorption state

relative simplicity of this method, the O K-edge X-ray adsorption spectra of CO coadsorbed with H_2 on a Ni{100} surface are shown in Fig. 6.5 [6.39]. The resonances observed at 534 and 548 eV are due to the excitation of the O 1s electron into the 2π and 6σ molecular orbitals. The intensities of the resonances as a function of the angle between the $\boldsymbol{E}$ vector and the surface normal, Θ_E, are then determined by a fitting procedure (described in Sect. 6.3). The results of such a procedure for this adsorption system are shown in Fig. 6.6, along with the expected intensity distributions for perpendicular (solid line) and tilted CO (dashed line) coordination geometries. A perpendicular adsorption geometry of the CO molecule is clearly favoured.

The molecular orbital description of XAS and selection rules presented above apply equally well in the adsorbed phase [6.40, 41]. However, as there is no plane of symmetry parallel to the surface, the number of possible point group symmetries is reduced from thirty two, in the case of the free molecule, to ten [6.42]. In principle, this can cause lifting of degeneracies and modification of selection rules. The calculated curves in Fig. 6.6, for example, were derived on the assumption that the CO molecular axis remains a symmetry element of

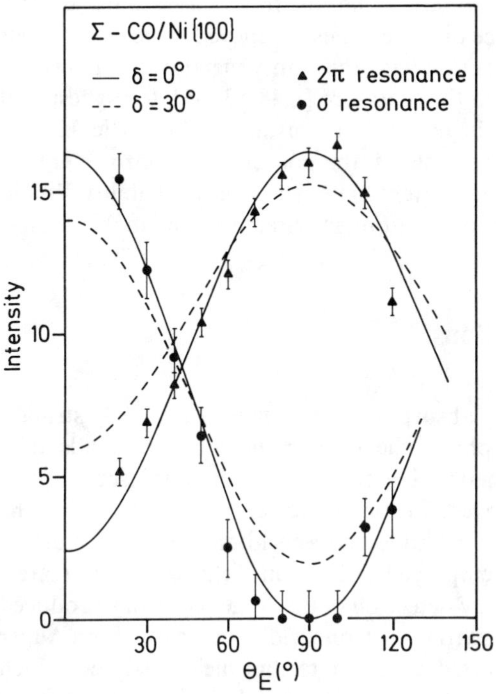

Fig. 6.6. Intensity distributions of the Σ-CO 2π and 6σ resonances as a function of the angle between the ***E*** vector and the surface normal, Θ_E. The solid and dashed curves are the expected distributions for a tilt angle, δ between the CO axis and the surface normal of O and 30° respectively

the effective point group in the adsorbed phase. This assumption poses no problem if the CO molecule is adsorbed in a high-symmetry site with the molecular axis perpendicular to the surface. If the CO were tilted, however, the effective symmetry would be lowered to C_s. A splitting of the 2π resonance into a′ and a″ components would be expected and the selection rules would then be modified accordingly. In view of the weak adsorption bond in the example presented here [6.43, 44], a lowering of the effective symmetry, as far as XAS is concerned, is probably unimportant. Indeed, in many strong chemisorption systems, the effective symmetry in XAS also remains that of the free molecule. An example of this effect, in a strong chemisorption system, includes the σ resonances of the carbonate molecule adsorbed on Ag{110}, which exhibit the expected polarisation dependences for D_{3h} symmetry even though an effective symmetry of C_{3v} or lower might be expected [6.45]. Although symmetry lowering due to the presence of the surface has only been detected in a few cases [6.46, 47], the possibility should not be ignored.

In strong chemisorption systems, new states of mixed metal–adsorbate character must be formed. As such states are expected to be predominantly metal in character, the localised nature of the XAS technique tends to restrict its usefulness as a probe of states of this type. In a small number of cases, however, features in X-ray absorption spectra of adsorbates have been assigned to such unoccupied states [6.48–50].

Transitions into Rydberg type orbitals in the adsorbed phase are also usually not detected [6.51, 52]. The absence of these states is believed to be due to their large radial extent, and their interaction with the empty metal states. However, there are now sufficient examples in the literature [6.48, 53–55], to suggest that Rydberg states can be observed in certain circumstances. In particular, for weakly chemisorbed or physisorbed systems, the poorer coupling to the metal states permits their survival to some extent. The detection probability should also increase if the excitation occurs on an atom not directly bound to the surface.

6.3 Experimental Considerations

As discussed in Chap. 1, the X-ray absorption experiment is relatively straight forward to perform. In the gas phase, the measurement can be made in a conventional transmission geometry. Electron-, ion- or fluorescence-yield measurements may also be employed. In the case of an adsorbed species, the transmission geometry is no longer possible and a yield measurement must be used. By far the most frequently employed is electron yield, whereby greatest surface sensitivity can be achieved by detection of the Auger electrons produced by the decay of the core hole. In a "partial electron yield" experiment, both Auger and secondary electrons are collected with a retarding field analyser, which removes lower kinetic energy electrons originating mainly from the bulk of the sample. The low cross section for fluorescence decay of C, N, O and F core excited states hinders this type of measurement in XAS investigations of these elements. Mass-selected ion-yield measurements, though not widely used, offer high surface sensitivity and the possibility of preferentially probing specific unoccupied orbitals. Chemical fingerprinting of elements of the same type, but in chemically differing environments, could be achieved, in principle, in an Auger yield measurement by electron detection at kinetic energies where the Auger spectrum is dominated by decay channels associated with only one of the inequivalent core holes. Unfortunately, the large number of possible Auger de-excitation channels, even for a diatomic molecule [6.56], sets limits on the success of this experiment.

A detailed account on the quantitative analysis of X-ray absorption spectra has been presented by *Outka* and *Stöhr* [6.14]. The sharp core to bound transitions are best fitted by means of a convolution of Lorentzian and Gaussian line shapes (Voigt profile) which describe the components imparted by the life-time of the excited state and the X-ray monochromator. Until recently, a Gaussian profile has nearly always sufficed as the component imparted by the monochromator resolution (typically 1.0 eV at the C K edge) has usually been substantially larger than the intrinsic line width of the Lorentzian component (typically 0.1 eV). Under such experimental conditions, the ionisation threshold can be described by an error function, which is a convolution of a step and Gaussian function. The asymmetric line shapes due to quasi-bound states above the ionisation threshold may be fitted adequately by an asymmetric Gaussian profile [6.14] or by the lineshape derived from a cylindrical well scattering model [6.57]. However, the advent of better monochromators with better than 0.1 eV resolution even at the O *K* edge is likely to increase the need for more sophisticated lineshape treatments.

Quantitative fitting of XAS data has its widest application for adsorbed species, where orientation information is sought. In the adsorbed phase an additional error function is included to represent the empty states above the metal Fermi level, which become accessible due to the interaction of the molecule with the surface. Such features have been resolved in only two cases [6.14, 49]. As substrate scattering features, which correspond to new states formed on adsorption, would be expected in this energy range, a flat step function is probably a rather poor approximation of this background. Even though the contribution of this background is usually small in comparison to that of the pure molecular resonances, the uncertainty in the evaluation of its intensity could lead to errors in the orientation determination of up to $\pm 15°$ in unfavourable circumstances. The intensity of the molecular resonances, fitted to symmetric or asymmetric line shapes, can be normalised to the absorption edge jump, i.e. the sum of the intensities of the two step functions. An example of such a fitting procedure is shown for CO in Fig. 6.5, where the π and σ resonances have been fitted with symmetric and asymmetric Gaussian functions and the background has been approximated by a single Gaussian and two step functions. In this case, satisfactory fits could only be obtained by inclusion of an extra Gaussian function, without justification, to describe the background. Clearly a better understanding of the backgrounds in XAS of adsorbates is necessary, especially if the technique is to become a useful probe of the adsorbate surface bond.

6.4 Case Studies

6.4.1 Vibrational Splitting of the N_2 $1\pi_g$ Resonance

The origin of vibrational fine structure in XAS has been discussed earlier in Sect. 6.1. As an example, the region of the $1\pi_g$ resonance of the N *K*-edge X-ray absorption spectrum of N_2, recently published by *Chen* et al. [6.7], is presented in Fig. 6.7. From an analysis of this data, the equilibrium internuclear distance of the excited state can be derived. Firstly, a more accurate description of the excited state potential than that of the harmonic oscillator, shown in Fig. 6.3, must be included. This is provided by the Morse potential, shown in Fig. 6.8 which, unlike the harmonic oscillator allows for the dissociation of the molecule at higher energies and larger internuclear separations. The Morse potential energy is given by

$$V = D_e[1 - \exp\{a(R - R_e)\}]^2 \tag{6.3}$$

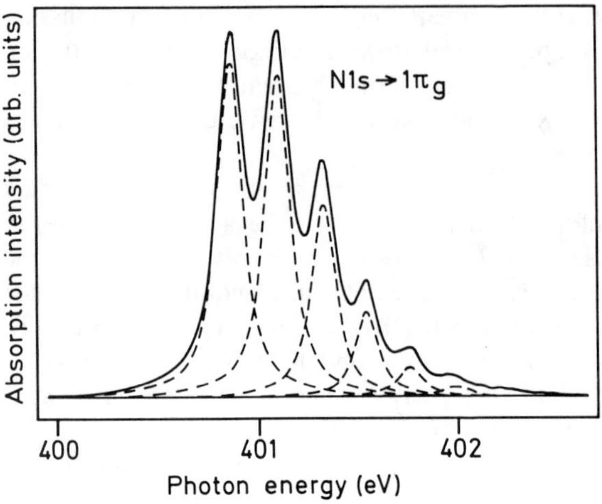

Fig. 6.7. Nitrogen K-edge X-ray absorption spectrum of gas phase N_2 exhibiting the vibrational splitting of the $1\pi_g$ resonance. From [6.7]

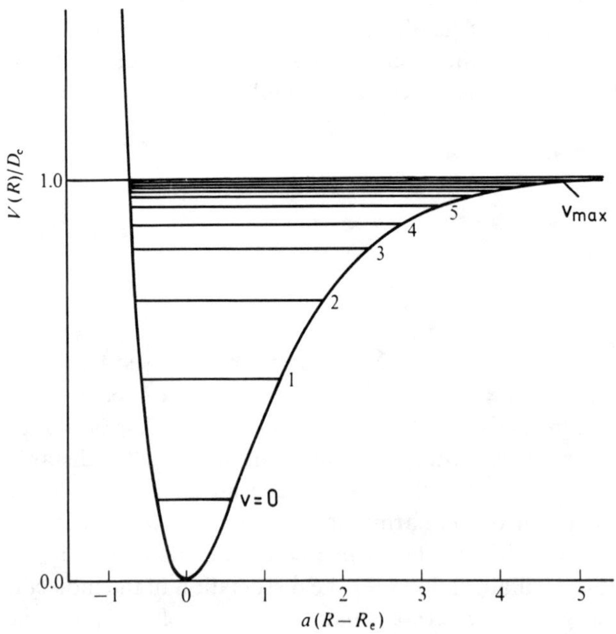

Fig. 6.8. Schematic diagram of the Morse Potential, showing vibrational levels with unequal spacing due to the anharmonicity of the potential

where 'a' is related to both the depth of the potential minimum, D_e, at the equilibrium internuclear separation R_e, and the reduced mass of the molecule. Solution of the Schrödinger equation for this potential then introduces an anharmonicity factor, X_e, such that in XAS the energy, E'_v, of a particular vibrational level, v', is given by

$$E_{v'} = E_{0,0'} + (v' + 1/2)\hbar\omega - (v' + 1/2)^2 X_e \hbar\omega, \tag{6.4}$$

where $E_{0,0'}$ is the lowest possible excitation energy, which corresponds to a transition between the vibrational ground states of the ground and excited electronic states. Using this relationship, the experimental data are analysed to obtain $\hbar\omega$ and X_e, from which the necessary parameters of the excited state Morse potential can be deduced [6.58]. The bond length, R_e, of the excited state is then found by calculating the overlap integrals between the $v = 0$ nuclear wave function of the ground-state and the excited-state nuclear wave functions, corresponding to the observed vibrational levels, for various values of R_e until a best fit to the experimental intensity distribution is obtained. Following such a procedure, an equilibrium internuclear separation of 1.164 Å for the $1s^{-1}$, $1\pi_g^1$ excited state of N_2 has been derived [6.7, 59]. The increase in equilibrium distance by 0.07 Å from that of the ground state is entirely consistent with the occupancy of an antibonding $1\pi_g$ molecular orbital.

6.4.2 Selection Rules in Larger Molecules: Pyrazine

The dipole selection rules, described in Sects. 5.3 and 6.1 above, are easily extended to larger molecules. A particular good example is provided by the X-ray absorption spectra of pyrazine. The first unoccupied π molecular orbitals are of b_{3u} and a_u symmetry. Application of dipole selection rules, either within the D_{2h} point group or within the C_{2v} (N K-edge absorption) or C_s (C K-edge absorption) broken symmetry point groups shows that at the N K edge only excitation into the b_{3u} orbital is allowed, while at the C K edge, both transitions should be observed. The X-ray absorption spectra of condensed pyrazine recently recorded by *Haase* et al. [6.60], are shown in Fig. 6.9. As expected, the resonance due to the a_u unoccupied orbital is absent in the N K-edge spectrum, while both the a_u and b_{3u} resonances are present in the C K-edge data.

6.4.3 Trends Induced by Fluorination of Simple Molecules

Some assistance in the assignment of the features observed in XAS of molecules can be obtained from trends induced by changing the molecule chemically. The substitution of in-plane hydrogen atoms by fluorine in planar molecules gives rise to a relative stabilisation of σ molecular orbitals with respect to those of π character. This is often known as the perfluoro effect, and photoemission studies in the early 1970s confirmed its validity for the occupied electronic states [6.61, 62]. The application of this effect in photoemission is somewhat limited due to the formation of additional occupied states derived from the F $2s$ and

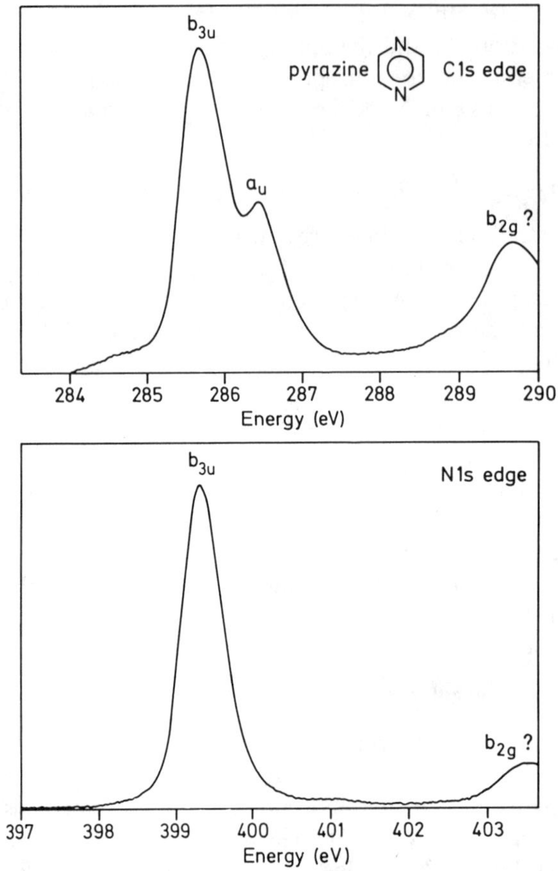

Fig. 6.9. Carbon and nitrogen K-edge X-ray absorption spectra of pyrazine. Only the region of the lowest energy π resonances is shown. After [6.60]

$2p$ orbitals. In contrast, however, no additional unoccupied states are formed on complete fluorination as both H and F are one electron short of a full principal quantum shell. Alterations in the Rydberg series due to fluorination are expected. As yet, very few systematic XAS investigations on the sequential substitution of hydrogen by fluorine in simple molecules have been undertaken, but initial results [6.63, 64] suggest that the effect applies equally well for the unoccupied molecular orbitals.

The perfluoro effect is illustrated in Fig. 6.10, which presents *Hitchcock* and coworker's recent ISEELS investigations of benzene and a series of fluorobenzenes [6.63]. The benzene spectrum itself is complex, and the spectra of the fluorobenzenes are complicated further by the difference in the C 1s ionisation potentials of carbon atoms bonded to H and F, and also by the possible lifting of degeneracy of e-type orbitals on substitution. The most important aspect of this data, however, concerns the π, e_{2u}, excitation (labelled 1 and 2), which is always found at 5.6 ± 0.6 eV below the appropriate K edge as the degree of

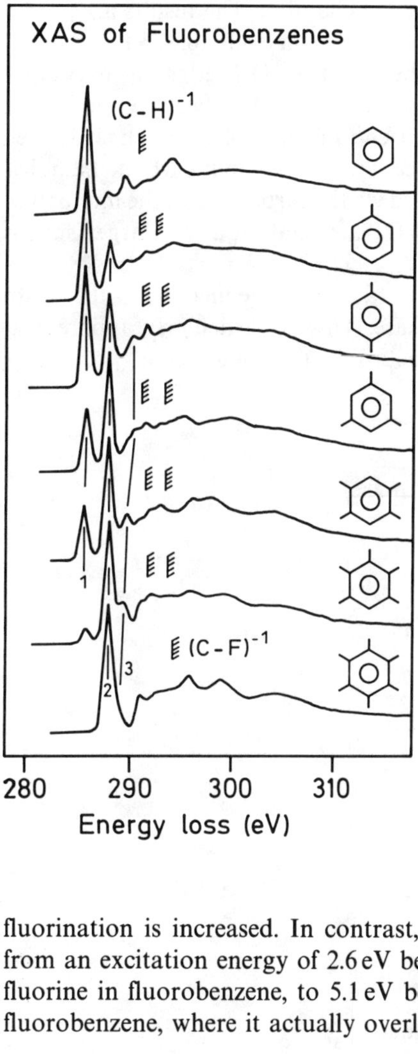

Fig. 6.10. Carbon *K*-edge X-ray absorption spectra of benzene and a series of fluorbenzenes. From [6.63], copyright (1987) American Chemical Society

fluorination is increased. In contrast, a σ unoccupied state (labelled 3) shifts from an excitation energy of 2.6 eV below the threshold for carbon bonded to fluorine in fluorobenzene, to 5.1 eV below threshold in the spectrum of hexafluorobenzene, where it actually overlaps the e_{2u} resonance.

6.4.4 CO in the Gas, Solid, and Adsorbed Phases

So far, investigations of the change in the excitation energy of various transitions upon chemisorption has been largely limited to the high energy σ resonances. This was mainly due to the proposed [6.17–20], and much debated [6.21–23], correlation between internuclear distance and the σ resonance energies. Of more interest as far as the surface-adsorbate bond is concerned, are accurate investigations of the change in energy and width of the core to bound transitions upon chemisorption. One of the first of these investigations was performed by *Himpsel* and *Koch* [6.65], who were in the advantageous position of being able to record high-resolution X-ray absorption spectra of CO in the gas, condensed and adsorbed phase in the same experimental chamber. The results are shown in Fig. 6.11 [6.65]. Upon adsorption, a very small shift (0.3 eV) of the 2π resonance is observed at the C K edge, whereas at the O K edge, the excitation energy decreases by 1.7 eV. The authors attributed this effect to the higher amplitude of the 2π orbital at the carbon end of the molecule, which is then increased and decreased in the presence of oxygen and carbon $1s$ holes, respectively. As the CO molecule is adsorbed via the carbon atom, the interaction with the metal substrate is increased for O $1s$ excitation and a larger shift is expected.

Another noteworthy feature of these data is the large increase in the width of the resonances in the chemisorbed phase. This is particularly apparent at the C K edge where the vibrational structure in the solid phase spectrum is

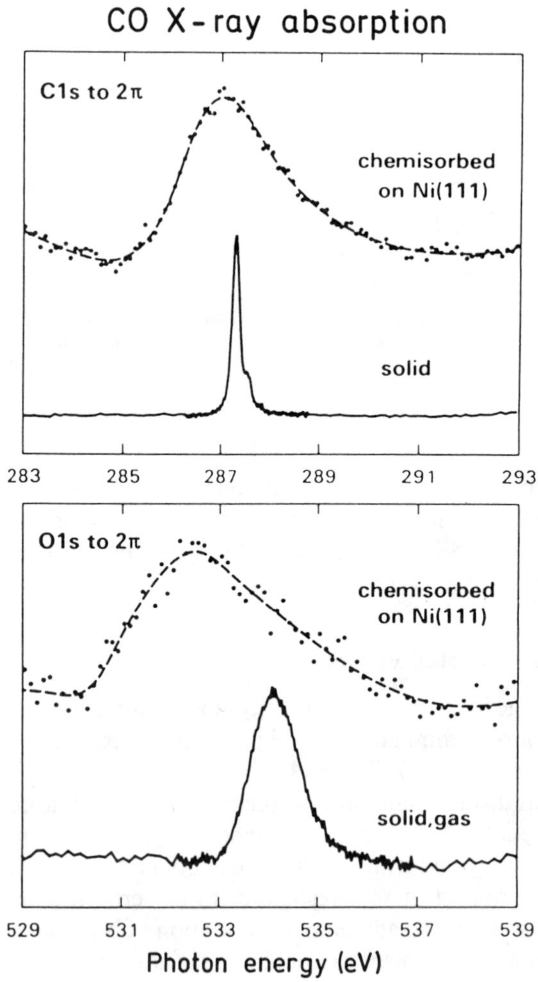

Fig. 6.11. X-ray absorption spectra at the carbon and oxygen K-edges of solid and chemisorbed CO (reproduced from *Jugnet* et al. [6.65] with permission)

Table. 6.1. XPS (C 1*s*, O 1*s*) binding energies and energies for resonant core-to-bound excitation ($1s \rightarrow 2\pi$) for gas-phase CO and various CO adsorbates [6.53]. The XPS energies refer to the peak with the lowest binding energy referenced to the Fermi level except for gas-phase CO

	CO_{gas}	CO/Cu{100}	CO/Ru{001}	CO/Ni{111}	CO + K/Ni{111}
C 1*s*	296.2	286.3	285.7	285.5	285.2
C $1s \rightarrow 2\pi$	287.3	286.4	286.5	287.0	286.7
O 1*s*	542.6	533.3	532.0	531.1	531.0
O $1s \rightarrow 2\pi$	534.1	532.9	532.6	532.6	532.2

clearly resolved. This increased width could be due to hybridisation of the 2π level with the metal *d*-band and concommitant reduction in the excited state lifetime. A further mechanism, which is known to contribute to core level photoemission line widths of adsorbed molecules [6.66, 67], would involve coupling to low-frequency vibrations, corresponding to the frustrated translations and rotations of the free molecule.

A comparison of the $1s \rightarrow 2\pi$ excitation energies with the corresponding XPS binding energies for CO adsorbed on a number of transition metals has recently been reported by *Wurth* et al. [6.53]. The data are reproduced in Table 6.1. Only small differences (< 500 meV) in the $1s \rightarrow 2\pi$ excitation energies are found. In contrast, the C 1*s* and O 1*s* XPS binding energies differ considerably in the various adsorption systems. In the case of strong CO chemisorption, the core-ionised states are lower in energy than the corresponding 2π resonantly excited states, even though the final states in each experiment are usually considered to be very similar. (In XPS the core hole is screened by charge transfer from the metal to the 2π orbital.) For CO on copper surfaces, however, the XPS binding energies are equal to, or even slightly larger than the corresponding $1s \rightarrow 2\pi$ excitation energies. The different behaviour for strong and weakly bound CO is believed to be due to the presence of a less well-screened final state in the weak chemisorption case [6.68]. This has been explained in a static model [6.68], where the lowest energy-screened XPS state is sensitive to the differing contribution of the metal *d*-states to the local density of states at the atom of interest. Alternatively in a dynamic model [6.68], due to the poorer coupling to the metal in the weak chemisorption systems, the lowest energy XPS core-excited state is not reached in the time scale of the photoionisation process.

6.4.5 Formate on Cu{110}

The surface formate species, formed by the de-protonation of formic acid on metal surfaces, has been the subject of numerous investigations as it is a possible surface intermediate in the water-gas shift reaction.

$$CO + H_2O \longrightarrow CO_2 + H_2 .$$

On Cu{110}, the molecule is adsorbed with the molecular plane parallel to the

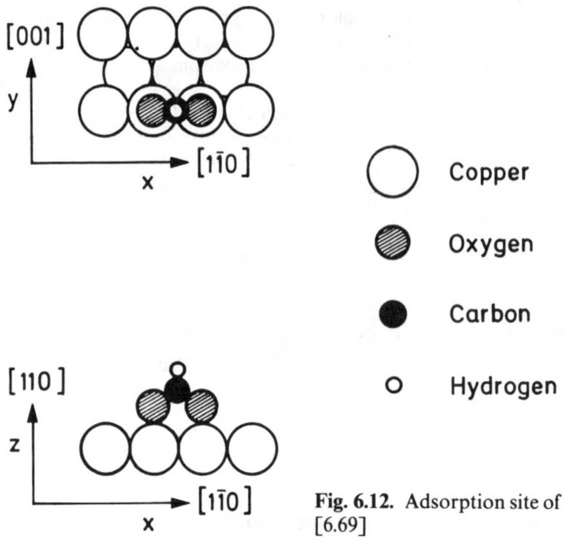

Fig. 6.12. Adsorption site of formate on a Cu{110} surface. After [6.69]

⟨110⟩ azimuth and with the carbon atom above the short bridge site, as shown in Fig. 6.12 [6.69]. Four unoccupied molecular orbitals are expected, of which the $2b_2$ possesses π character and the $7a_1, 8a_1$ and $5b_1$ orbitals are of σ character. The C *K*-edge experimental data, shown in Fig. 6.13 [6.70], exhibit four features at 287.0, 290.4, 297 and 300 eV, which can be assigned to the excitation of a C 1*s* electron ($2a_1$ molecular orbital) into the $2b_2, 7a_1, 8a_1$ and $5b_1$ orbitals. (The polarisation dependence of these resonances is entirely consistent with the selection rule analysis, summarised in Table 6.2.) The 290.4 eV feature lies just below the vacuum level, therefore it could also be assigned to a Rydberg orbital of a_1 symmetry.

In an earlier 'building block' description of XAS [6.51], in which any molecule can be assembled from diatomic or ring subunits, a single σ resonance corresponding to the two equivalent C—O bonds would be expected. The correlation between the internuclear separation and the energy of σ resonances [6.17–20] could then be used to derive structural information on the formate molecule. However, the presence of two high-energy σ resonances of a_1 and b_1 symmetry in the carbon 1*s* X-ray absorption spectrum of formate not only demonstrates that the absorption process must be considered within the framework of the symmetry of the entire molecule, but also indicates that no determination of bond lengths from the σ resonance energies is possible.

We now illustrate the difficulty of isolating effects due to symmetry breaking by a core hole. In the O *K*-edge data presented in Fig. 6.14 [6.70], a broad peak is observed in the energy range of the σ resonances. The small changes in the shape of this feature as a function of the polarisation of the light indicates that a second component is present, but is not resolved as in the C *K*-edge data. Analysis of the selection rules for the O 1*s* excitation shows that both the

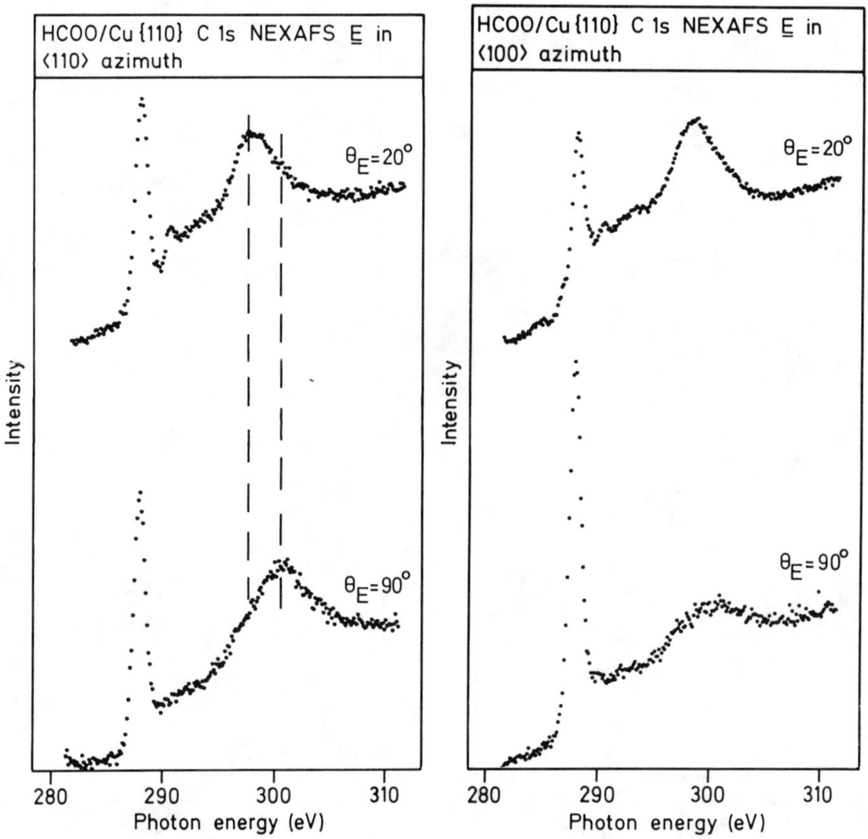

Fig. 6.13. Carbon *K*-edge X-ray absorption spectra of formate on Cu{110}. (**a**) *E* in the ⟨110⟩ azimuth, (**b**) *E* in ⟨100⟩ azimuth. After [6.70]

Table 6.2. Polarisation dependence of the resonances observed in the C *K*-edge XAS of formate on Cu{110}

Energy (eV)	Final-State Symmetry	Polarisation
287.9	B_2	y
290.4	A_1	z
297.6	A_1	z
300.4	B_1	x

expected $8a_1$ and $5b_1$ σ resonances are polarised within the molecular (zx) plane. However, as mentioned in Sect. 6.1, a subtle difference in the polarisation dependences in derived depending on whether the selection rules are applied with the C_s point group, appropriate for a localized core hole, or within the C_{2v} point group, corresponding to a delocalized core hole. In the case of C_s

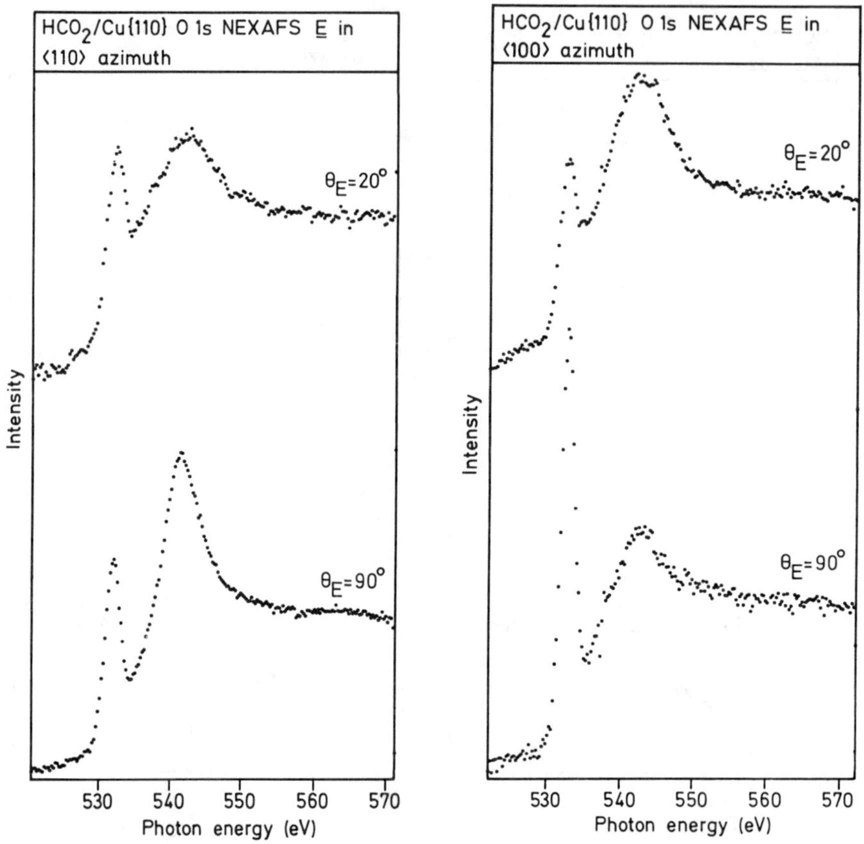

Fig. 6.14. Oxygen K-edge X-ray absorption spectra of formate on Cu{110}. (**a**) E in the ⟨110⟩ azimuth, (**b**) E in the ⟨100⟩ azimuth. After [6.70]

symmetry, the intensity of each σ resonance is given by

$$I_\sigma \sim A[\boldsymbol{E}\cdot\boldsymbol{M}_{zx}]^2.$$

Should the appropriate point group be C_{2v}, the σ resonance intensities are given by

$$I_\sigma \sim B[\boldsymbol{E}\cdot\boldsymbol{M}_x]^2 + C[\boldsymbol{E}\cdot\boldsymbol{M}_z]^2$$

(A, B and C are constants). Therefore, in the latter case, a maximum of the intensity distribution is expected, when the $\boldsymbol{E}$ vector is parallel to one of the z or x directions. Within the C_s point group, however, this condition is relaxed, as the z and x directions are no longer unique as far as the molecule is concerned. They merely define the location of the mirror plane. Unfortunately, due to the presence of two overlapping resonances in the formate O K-edge data, no information on core-hole localization can be extracted in this example.

6.4.6 Bonding Geometry of Pyridine on Ag{111} as a Function of Coverage

Over the past decade, the adsorption of aromatic molecules on metal surfaces has been investigated in order to establish dominant adsorption mechanisms. While benzene (C_6H_6) adsorbs on nearly all transition metals with the molecular plane parallel to the surface, pyridine (C_5H_5N) exhibits a variety of orientations due to the possibility of bonding through the π electrons of the aromatic ring, and/or via the valence lone pair electrons associated with the N atom.

The N *K*-edge X-ray absorption spectrum of pyridine multilayers, measured by *Haase* and co-workers [6.71], is shown in Fig. 6.15. The intense peak at ~400 eV is due to the excitation of an N 1s ($1a_1$) electron into the unoccupied $2b_2(\pi)$ molecular orbital. (A second resonance, due to the unoccupied $2a_2(\pi)$ orbital is forbidden at the N *K* edge). Since the $2b_2$ resonance is excited by the component of the ***E*** vector perpendicular to the ring plane, its intensity can be used to determine the orientation of the molecule in the adsorbed phase. Thus, *Haase* and co-workers have determined the orientation of pyridine on Ag{111} [6.72]. The data as a function of pyridine exposure to the surface are presented in Fig. 6.16 both for the experimental condition where the ***E*** vector is parallel to the surface (I_{90}), and also as the ratio of two angles of the ***E*** vector with respect to the surface (I_{20}/I_{90}). A change in the orientation of the molecule with

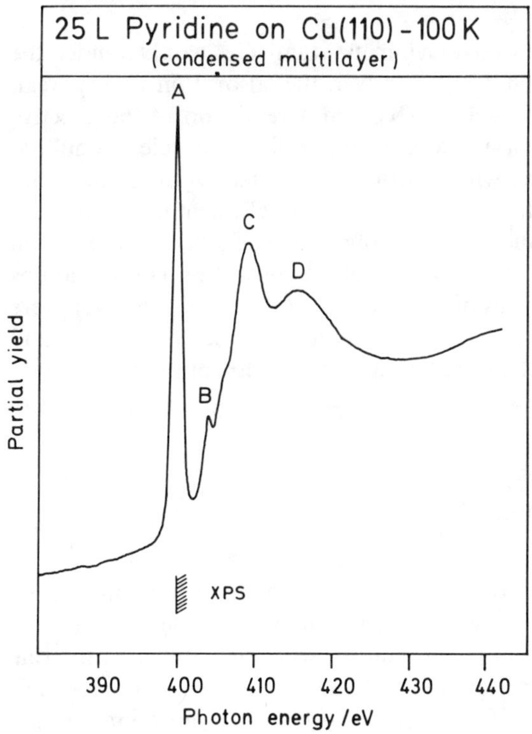

Fig. 6.15. Nitrogen *K*-edge X-ray absorption spectra of multilayers of pyridine adsorbed on Cu{110}. After [6.71]

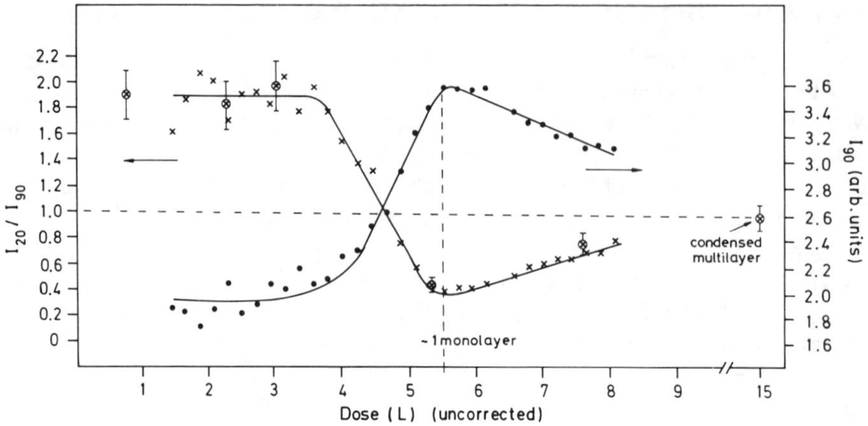

Fig. 6.16. Normalized nitrogen K-edge $2b_2$ resonance (Peak A in Fig. 6.15) intensity for normal incidence ($\theta = 90$), I_{90} (full circles), and the intensity ratio I_{20}/I_{90} (crosses) as a function of dose during permanent exposure of pyridine (~0.1 L/min.) to Ag{111} at 100 K. A phase transition occurs near 4.5 L. Reproduced from [6.72]

respect to the surface occurs at an exposure of about 4 Langmuirs ($1L = 10^{-6}$ torr. s). Further evaluation of the data reveals that at low coverages, the angle between the ring plane and the surface is $45 \pm 5°$ while at monolayer coverages ($5.5L$) this angle increases to $70 \pm 5°$.

In this example, a lower effective symmetry than C_{2v} would render the X-ray absorption analysis invalid. Here, however, the adsorption is very weak (pyridine desorbs from the surface at 200 K), and a reduction of the effective symmetry to C_s, formally appropriate for a tilted pyridine molecule, is unlikely. Measurements at the C K edge, where both the $2b_2$ and $2a_2$ resonances are dipole allowed, could be used to confirm the effective C_{2v} symmetry and also the orientation determination. Should islands of pyridine of different orientation form on the surface, an 'average' orientation of all adsorption configurations would be extracted from an analysis of the resonance intensities. However, were the difference in the resonance excitation energies for the different island orientations sufficiently large, the orientation of each adsorption phase could still be determined individually.

6.5 Conclusions and Outlook

Through the use of XAS a great deal has been learned about the core-excited states of free molecules. High-resolution data, now available due to the development of high-resolution monochromators, will help to unravel the role of vibronic coupling in core-level excitation [6.32]. A particularly important question concerns the influence of coupling to asymmetric vibrations on the polarisation of the transition. As far as adsorbates are concerned, the main role of XAS will remain that of orientation determination. It cannot be stressed too strongly, however, that recognition of the effective symmetry of the molecule on the surface is of the utmost importance if such determinations are to be valid at all.

As a probe of the surface-adsorbate bond, XAS has not, as yet, been particularly successful, even though such information must be present in the spectra. The potential of this aspect of the technique has recently been highlighted by a comparison of XAS data for oxygen chemisorption on copper and nickel surfaces with those of the corresponding bulk oxides [6.73]. The XAS technique is still relatively new, however, and further accurate measurements of bound-state resonance energies of free and adsorbed molecules combined with an improved understanding of the underlying background in adsorbate spectra will undoubtedly lead to significant improvements in this aspect of the technique. Investigations of the core-excited states of organo-metallic compounds, in particular of metal carbonyls, could also assist in the future development of this aspect of XAS.

Acknowledgements. Firstly, I would like to thank Alex Bradshaw for the introduction to XAS and for the many enlightening discussions over the past four years. Thomas Lindner, Andy Robinson and Detlef Ricken, my colleagues on the beamline, also Jochen Haase, Horst Conrad, Klaus Hermann and other members of the Fritz–Haber-Institut are thanked for the many useful discussions on this topic. I would also like to thank Prof. Menzel for providing a copy of [6.68], prior to publication. Finally, many thanks to John Fuggle and John Inglesfield for the invitation to write this chapter.

References

6.1 L.G. Parratt: Phys. Rev. **56**, 295 (1939)
6.2 R.E. LaVilla, R.D. Deslattes: J. Chem. Phys. **44**, 4399 (1966)
6.3 G.R. Wight, C.E. Brion, M. J van der Wiel: J. Elect. Spect. Rel. Phen. **1**, 457 (1972/73)
6.4 G.R. Wight, C.E. Brion: J. Elect. Spect. Rel. Phen. **1**, 191 (1974)
6.5 F.A. Gianturco, C. Guidotti, U. Lamanna: J. Chem. Phys. **57**, 840 (1972)
6.6 M.B. Robin: *Higher excited states in Polyatomic Molecule* (Academic, New York, Vol. 1 (1974), Vol. 2 (1975), Vol. 3 (1985))
6.7 C.T. Chen, Y. Ma, F. Sette: Phys. Rev. A **40**, 6737 (1989)
6.8 J.L. Jorgensen, L. Salem: *The Organic Chemists Book of Orbitals* (Academic, New York 1973)
6.9 see various articles in *Photophysics and Photochemistry in the vaccum Ultraviolet* ed. by S.P. McGlynn, G.L. Findley, R.H. Huebner (Reidel, Dordrecht 1985)
6.10 D. Nordfors, A. Nilsson, N. Martensson, S. Svensson, U. Gelius, S. Lunnell: J. Chem. Phys. **88**, 2630 (1988)
6.11 P.S. Bagus, H.F. Schaefer: J. Chem. Phys. **56**, 224 (1972)
6.12 W.H.E. Schwarz, T.C. Chang, U. Seeger, K.H. Hwang: Chem. Phys. **117**, 73 (1987)
6.13 J.L. Dehmer, D. Dill, S. Wallace: Phys. Rev. Lett. **43**, 1005 (1979)
6.14 D.A. Outka, J. Stöhr: J. Chem. Phys. **88**, 3539 (1988)
6.15 J.L. Dehmer, D. Dill, A.C. Parr: in *Photophysics and Photochemistry in the Vacuum Ultraviolet*, ed. by S.P. McGlynn, G.L. Findley, R.H. Huebner (Reidel, Dordrecht 1985)

6.16 J.L. Dehmer, D. Dill: J. Chem. Phys. **65**, 5327 (1976)
6.17 F. Sette, J. Stöhr, A.P. Hitchcock: J. Chem. Phys. **81**, 4906 (1984)
6.18 F. Sette, J. Stöhr, A.P. Hitchcock: Chem. Phys. Lett. **110**, 517 (1984)
6.19 J. Stöhr, J.L. Gland, W. Eberhardt, D. Outka, R.J. Madix, F. Sette, R.J. Koestner, U. Doebler: Phys. Rev. Lett. **51**, 2414 (1983)
6.20 J. Stöhr, F. Sette, A.L. Johnson: **53**, 1684 (1984)
6.21 M.N. Piancastelli, D.W. Lindle, T.A. Ferrett, D.A. Shirley: J. Chem. Phys. **86**, 2765 (1987)
6.22 A.P. Hitchcock, J. Stöhr: J. Chem. Phys. **87**, 3253 (1987)
6.23 M.N. Piancastelli, D.W. Lindle, T.A. Ferrett, D.A. Shirley: J. Chem. Phys. **87**, 3255 (1987)
6.24 J.C. Slater, K.H. Johnson: Phys. Rev. B **5**, 844 (1972)
6.25 A. Denis, J. Langlet, J.P. Malrieu: Theor. Chem. Acta. **38**, 49 (1975)
6.26 L.S. Cederbaum, W. Domcke: J. Chem. Phys. **66**, 5084 (1977)
6.27 D. Dill, S. Wallace, J. Siegel, J.L. Dehmer: Phys. Rev. Lett. **41**, 1230 (1978)
6.28 R.L. Lozes, O. Goscinski, U.I. Wahlgren: Chem. Phys. Lett. **63**, 77 (1979)
6.29 H. Agren, P.S. Bagus, B.O. Roos: Chem. Phys. Lett. **82**, 515 (1981)
6.30 R. Arneberg, H. Agren, J. Müller, R. Manne: Chem. Phys. Lett. **91**, 362 (1982)
6.31 G.B. Bacskay, G. Bryant, N.S. Hush: Int. J. Quantum Chem. **31**, 471 (1987)
6.32 W. Domcke, L.S. Cederbaum: Chem. Phys. **25**, 189 (1977)
6.33 Y. Ma, F. Sette, G. Meigs, S. Modesti, C.T. Chen: Phys. Rev. Lett. **63**, 2044 (1989)
6.34 D.A. Outka, J. Stöhr: In *Chemistry and Physics of Solid Surfaces* VII, ed. by V.R. Vanselow, R. Howe, Springer Ser. Surf. Sci., Vol. 10 (Springer, Berlin, Heidelberg 1988)
6.35 J. Stöhr, K. Baberschke, R. Jaeger, R. Treichler, S. Brennan: Phys. Rev. Lett. **47**, 381 (1981)
6.36 J. Stöhr, R. Jaeger: Phys. Rev. B **26**, 4111 (1982)
6.37 A.L. Johnson, E.L. Mueterties, J. Stöhr: J. Amer. Chem. Soc. **105**, 7183 (1983)
6.38 A.L. Johnson, E.L. Mueterties, J. Stöhr, F. Sette: J. Phys. Chem. **89**, 4071 (1985)
6.39 J. Somers, S. Kulkarni, Th. Lindner, A.W. Robinson, P. Hollins, G.J. Lapeyre, A.M. Bradshaw: Surface Sci. in press
6.40 A.M. Bradshaw, J. Somers, Th. Lindner: in *Proceedings of Solvay Conference on Surface Science*, Austin, Texas (USA) December 1987, ed. by F.W. de Wette, Springer Ser. Surf. Sci., Vol. 14 (Springer, Berlin, Heidelberg 1988) p. 469
6.41 A.M. Bradshaw, J. Somers: Physica Scripta T **31**, 189 (1990)
6.42 N.V. Richardson, A.M. Bradshaw: in *Electron Spectroscopy: Theory, Techniques and Applications*, Vol. 4, ed. by C.R. Brundle, A.D. Baker (Academic, London 1981) p. 153
6.43 B.E. Koel, D.E. Peebles, J.M. White: Surface Sci. **125**, 709 (1983)
6.44 R. Klauser, M. Surman, Th. Lindner, A.M. Bradshaw: Physica Scripta **35**, 561 (1987)
6.45 M. Bader, B. Hillert, A. Puschmann, J. Haase, A.M. Bradshaw: Europhysics Lett. **5**, 443 (1988)
6.46 J. Somers, M.E. Kordesch, Th. Lindner, H. Conrad, A.M. Bradshaw, G.P. Williams: Surface Sci. **188**, L693 (1987)
6.47 D.W. Moon, S. Cameron, F. Zaera, W. Eberhardt, R. Carr, S.L. Bernasek, J.L. Gland, D.J. Dwyer: Surface Sci. **180**, L123 (1987)
6.48 Th. Lindner, J. Somers, A.M. Bradshaw, A.L.D. Kilcoyne, D.P. Woodruff: Surface Sci. **203**, 333 (1988)
6.49 A.C. Liu, C.M. Friend: J. Chem. Phys. **89**, 4396 (1988)
6.50 D.A. Outka, J. Stöhr, W. Jark, P. Stevens, J. Solomun, R.J. Madix: Phys. Rev. B **35**, 4119 (1987)
6.51 J. Stöhr, D.A. Outka: Phys. Rev. B **36**, 7891 (1987)
6.52 J. Stöhr, D.A. Outka, K. Baberschke, D. Arvanitis, J. Horsley: Phys. Rev. B **36**, 2976 (1987)
6.53 W. Wurth, C. Schneider, R. Treichler, E. Umbach, D. Menzel: Phys. Rev. B **35**, 7741 (1987)
6.54 D. Coulman, A. Puschman, W. Wurth, H.P. Steinruck, D. Menzel: Chem. Phys. Lett. **148**, 371 (1988)
6.55 L. Wenzel, D. Arvanitis, R. Schlögl, M. Muhler, K. Baberschke, G. Ertl: Phys. Rev. B **40**, 6409 (1989)
6.56 E. Umbach, Z. Hussain: Phys. Rev. Lett. **52**, 457 (1984)
6.57 D. Arvanitis, H. Rabus, L. Wenzel, K. Baberschke: Z. Phys. D **11**, 219 (1989)
6.58 D.M. Dennison, K.T. Hecht: In *Quantum Theory*, Vol. 2, ed. by D.R. Bates (Academic Press, New York) 1962 p. 247
6.59 G.C. King, F.H. Read, M. Tronc: Chem. Phys. Lett. **52**, 50 (1977)
6.60 S. Aminpirooz, L. Becker, J. Haase: Surface Sci. **244**, L152 (1991)
6.61 C.R. Brundle, M.B. Robin, N.A. Kueblar, H. Basch: J. Am. Chem. Soc. **94**, 1451 (1972)
6.62 C.R. Brundle, M.B. Robin, N.A. Kueblar: J. Amer. Chem. Soc. **94**, 1466 (1972)

6.63 A.P. Hitchcock, P. Fischer, A. Gedanken, M.B. Robin: J. Phys. Chem. **91**, 531 (1987)
6.64 M.B. Robin, I. Ischii, R. McLaren, A.P. Hitchcock: J. Elect. Spectr. Relat. Phenom **47**, 53 (1988)
6.65 Y. Jugnet, F.J. Himpsel, Ph. Avouris, E.E. Koch: Phys. Rev. Lett. **53**, 198 (1984)
6.66 N. Martenson, A. Nilsson: Surface Sci. **211/212**, 303 (1989)
6.67 J.C. Fuggle, D. Menzel: Surface Sci. **79**, 1 (1979)
6.68 W. Wurth, D. Coulman, A. Puschmann, D. Menzel, E. Umbach: Phys. Rev. B **41**, 12933 (1990)
6.69 D.P. Woodruff, C.F. McConville, A.L.D. Kilcoyne, Th. Lindner, J. Somers, M. Surman, G. Paolucci, A.M. Bradshaw: Surface Sci. **201**, 228 (1988)
6.70 J. Somers, A.W. Robinson, Th. Lindner, D. Ricken, A.M. Bradshaw: Phys. Rev. B **40**, 2053 (1989)
6.71 M. Bader, J. Haase, K.H. Frank, C. Ocal, A. Puschmann: J. de Physique Coll. C **8**, 491 (1986)
6.72 M. Bader, J. Haase, K.H. Frank, A. Puschmann, A. Otto: Phys. Rev. Lett. **56**, 1921 (1986)
6.73 M. Pedio, J.C. Fuggle, J. Somers, E. Umbach, J. Haase, Th. Lindner, U. Höfer, M. Grioni, F. de Groot, B. Hillert, L. Becker, A.W. Robinson: Phys. Rev. B **40**, 7924 (1989)

7. Transmission Electron Energy-Loss Spectroscopy

J. Fink

With 22 Figures

High-energy electron energy-loss spectroscopy (EELS) in transmission has been in almost continuous use for about 50 years to probe the electronic structure of solids. The reason for this is, firstly, that high-energy electrons with a narrow energy distribution can be easily produced and, secondly, that EELS in transmission can be performed under normal vacuum conditions since it is not a surface-sensitive technique such as, e.g. photoemission or inverse photoemission spectroscopy which need ultra-high vacuum conditions. The Franck–Hertz phenomenon in gases which has provided information on electron energy levels in gaseous atoms and molecules was first applied to solids in 1929 by *Rudberg* [7.1] performing EELS measurements in reflection on noble metals. The first high-energy (2–8 keV) EELS measurements in transmission were performed in 1941 by *Ruthemann* [7.2]. In these measurements, discrete excitations in simple metals were recorded which were later explained by *Pines* and *Bohm* [7.3] in terms of collective excitation of conduction electrons, i.e. plasmons. *Ruthemann* reported also in these early days [7.4], excitations of core-level electrons in collodium by EELS in transmission. Both plasmons (for which EELS is the primary source of information) and core-level absorption edges are still the main subjects of investigations by EELS in transmission. Plasmons are caused by a collective excitation of free electrons (intraband plasmons) or a collective excitation of bound electrons (interband plasmons). The latter are related to interband transitions and therefore to the joint density of states between occupied and unoccupied states. Contrary to optical spectroscopy, the momentum between ground state and excited state can be varied systematically in EELS and therefore information on the momentum dependence of the joint density of states is provided. Those results are related to the dispersion of occupied and unoccupied bands. On the other hand, the momentum dependence of the excitation energy of plasmons yields important information on the electron–electron interaction in electron liquids. Finally, excitations of core-level electrons yield information similar to X-ray absorption spectroscopy (XAS, Chaps. 5 and 6) on the unoccupied density of states at the excited atoms, provided that the interaction of the conduction electrons with the core hole is small. If the latter point is not fulfilled, information is at least provided on the total amount of unoccupied states at that site and often on the nature of the ground state. At small momentum transfer, dipole-selection rules apply and therefore only a definite symmetry of unoccupied states or multiplet structure is probed. However, in EELS the momentum transfer can also be increased so that other selection rules apply, which offers the possibility to probe unoccupied states or multiplet structures having different symmetries. Using this technique, the problem is that rather high momentum transfers are needed, leading to low counting rates. This is the reason why this additional possibility (which XAS does not have) has not been used very often until now.

There are numerous review articles on EELS in transmission. Investigations on plasmons, interband transitions and core-level excitations using dedicated spectrometers are reviewed by *Marton* et al. [7.5], *Raether* [7.6–8], *Daniels* et al. [7.9], *Geiger* [7.10], *Schnatterly* [7.11] and *Fink* [7.12]. Electron microscopes with an attached energy analyzer were first used to obtain a quantitative elemental analysis for low-Z elements in thin films measuring absorption edges. During recent years electronic structure studies were also performed using those instruments. Reviews on these measurements were given by *Colliex* [7.13], *Colliex* and *Mory* [7.14], *Colliex* et al. [7.15], *Egerton* [7.16] and *Schattschneider* [7.17].

This chapter describes the principles of EELS measurements and illustrates their use with some examples drawn (for convenience) from our own work. Examples were chosen where unoccupied states (in the sense of a single-particle picture) or excited states of solids have been investigated. In particular, studies on simple metals, synthetic metals and high-T_c superconductors have been selected. Before coming to these examples, the theory of EELS measurements is presented and an introduction to instrumentation is given.

7.1 Theory of Electron Energy-Loss Spectroscopy

7.1.1 The Cross Section for Inelastic Electron Scattering

In EELS in transmission, high-energy ($E \geqq 30$ keV) electrons are transmitted through thin samples having a thickness of about 1000 Å. There are various scattering mechanisms for high-energy electrons in solids. Firstly, there is a certain probability that electrons are scattered elastically, i.e. by Bragg scattering. Furthermore, there is a so-called quasi-elastic scattering where phonons are excited in the scattering process. Since the energy loss due to these excitations is normally much smaller than the energy resolution, these quasi-elastic processes cannot be separated from elastic processes. Finally, we have inelastic electron scattering, in which electrons lose a considerable amount of energy, E. The Coulomb field of the transmitted electrons interacts with charges of the solid and leads to longitudinal excitations. Since this Coulomb field of an electron has Fourier components with frequencies ranging from zero to infinity, electronic excitations in a broad energy range are expected. When the intensity as a function of energy loss $E = \hbar\omega$ is measured, information on the frequency of possible excitations of the solid is obtained. Furthermore, the electrons may be scattered under a certain scattering angle Θ. For finite Θ, a momentum $\hbar \boldsymbol{q}$ is transferred to the solid which is given by $\hbar q = \hbar k_0 - \hbar k_1$, where $\hbar k_0$ and $\hbar k_1$ are the momentum

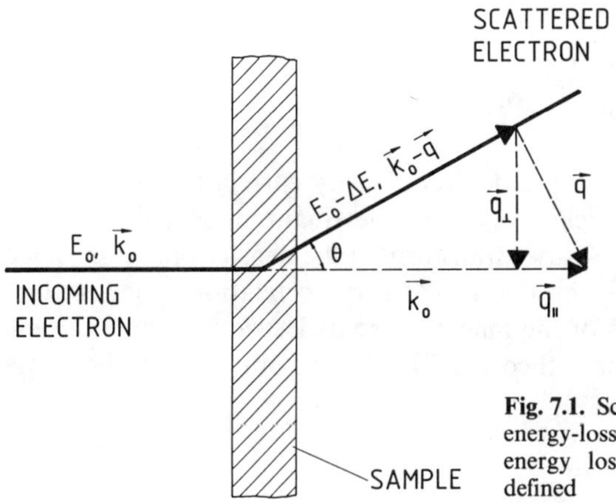

Fig. 7.1. Scattering geomertry of an electron energy-loss experiment in transmission. The energy loss and the momentum transfer are defined

of the incoming and the scattered electron, respectively. Thus, measurements as a function of $\boldsymbol{q}$, which is proportional to the inverse wavelength, provide information on the wavelength of electronic excitations of the solids. Interesting wavelengths in solids range up to 1 Å, corresponding to wave vectors of $\sim 1\,\text{Å}^{-1}$. These wave vectors can be easily achieved by rather small scattering angles. For 200 keV electrons, the wave vector is $\sim 270\,\text{Å}^{-1}$. Therefore, $\boldsymbol{q}$ is much smaller than $\boldsymbol{k}_0$ and $\boldsymbol{q}$ can be decomposed in a component $q_\parallel$ (parallel to $\boldsymbol{k}_0$) and $q_\perp$ (perpendicular to $\boldsymbol{k}_0$) as shown in Fig. 7.1:

$$q_\perp = k_0 \sin \Theta \cong k_0 \Theta \tag{7.1}$$

$$q_\parallel = k_0(\hbar\omega/2E_0) \tag{7.2}$$

$$q^2 = q_\perp^2 + q_\parallel^2 . \tag{7.3}$$

E_0 is the energy of the incoming electron. Using these relations, $q_\perp = 1\,\text{Å}^{-1}$ corresponds to a scattering angle of ~ 5 mrad or $\sim 0.3°$ for $E_0 = 200$ keV. Finally, we remark that energy conservation yields

$$\hbar\omega = E_0 - E_1 = \hbar(k_0^2 - k_1^2)/2m, \tag{7.4}$$

where E_1 is the energy of the scattered electron.

What is actually measured in EELS experiments is the partial differential cross section $d^2\sigma/d\Omega dE$ which gives the fraction of incident electrons with primary energy of E_0 scattered into the solid angle $d\Omega$ with an energy between E_1 and $E_1 + dE$. As in other inelastic scattering theories, e.g. on neutron scattering or X-ray scattering, the cross section can be decomposed into an amplitude factor and a dynamic structure factor [7.18]. The amplitude factor is determined by the interaction of the electron with a free electron, i.e. this factor is given by the Rutherford scattering cross section $(d\sigma/d\Omega)_{\text{Ruth}} = 4/(a_0^2 q^4)$ with a_0 being the Bohr radius. The dynamic structure factor solely depends on the properties of the solid and is given by

$$S(\boldsymbol{q},\omega)=\frac{1}{2\pi\hbar N}\int dt\, e^{-i\omega t}\langle n_{\boldsymbol{q}}(t)n_{-\boldsymbol{q}}(0)\rangle, \tag{7.5}$$

which is the Fourier transform of the density–density correlation function; N is the number of electrons, brackets indicate the quantum mechanical and statistical average and $n_{\boldsymbol{q}}$ is the Fourier transform of the density operator. Since it is not very easy to calculate the density–density correlation function, one often expresses $S(\boldsymbol{q},\omega)$ in terms of the macroscopic dielectric function $\varepsilon_{\mathrm{M}}(\boldsymbol{q},\omega)$. Using the fluctuation-dissipation theorem [7.19] one obtains for the cross section [7.20]

$$\frac{d^2\sigma}{d\Omega\, dE}=\frac{1}{(\pi e a_0)^2}\frac{1}{q^2}\mathrm{Im}\left\{\frac{-1}{\varepsilon_{\mathrm{M}}(\boldsymbol{q},\omega)}\right\}. \tag{7.6}$$

$\mathrm{Im}\{-1/\varepsilon_{\mathrm{M}}(\boldsymbol{q},\omega)\}$ is the so-called loss function. In many cases this function varies only slowly with q. Therefore, the cross section is roughly proportional to the inverse square of the total wave vector. For small energy losses ($E=10\,\mathrm{eV}$), $q_{\parallel}$ is very small ($q_{\parallel}=0.008\,\mathrm{Å}^{-1}$ for $E_0=200\,\mathrm{keV}$) and therefore the momentum dependence of the measured loss function is usually determined by the momentum resolution of the spectrometer ($\Delta q_{1/2}\sim 0.05\,\mathrm{Å}^{-1}$). For losses with higher energies ($E=1000\,\mathrm{eV}$) $q_{\parallel}$ increases (for $E_0=200\,\mathrm{keV}$, $q_{\parallel}=0.79\,\mathrm{Å}^{-1}$) and the momentum dependence of the cross section is a Lorentzian with a full width at half maximum of $2q_{\parallel}$. Therefore, for valence band excitations (low losses) the inelastic scattering events have a strong distribution in the forward direction, while high-energy core-level excitations tend to have wider distributions. For energies less or equal to the plasmon energy ($E\leqq\hbar\omega_{\mathrm{p}}\sim 10\,\mathrm{eV}$), the loss function $\mathrm{Im}\{-1/\varepsilon_{\mathrm{M}}(\boldsymbol{q},\omega)\}$ describes collective excitations of the electron systems. These may be caused by a collective oscillation of free electrons in a metal (intraband plasmon) or by a collective oscillation of bound electrons, e.g. in a semiconductor (interband plasmon). The subscript M on the dielectric function $\varepsilon_{\mathrm{M}}(\boldsymbol{q},\omega)$ indicates that the macroscopic dielectric function should be used which is equal to the microscopic one for homogeneous electron systems but is different for inhomogeneous systems, e.g. electrons in a lattice [7.21]. The difference will be discussed below in more detail. In the following we will treat mainly homogeneous electron systems, and therefore we can omit the subscript M.

The dielectric function $\varepsilon(\boldsymbol{q},\omega)$ is a well-known quantity to describe excitations of an electron system. The real part $\mathrm{Re}\{\varepsilon(\boldsymbol{q},\omega)-1\}\equiv\varepsilon_1(\boldsymbol{q},\omega)-1$ is a measure of the polarizability of the system, while the imaginary part $\mathrm{Im}\{\varepsilon(\boldsymbol{q},\omega)\}\equiv\varepsilon_2(\boldsymbol{q},\omega)$ is a measure for absorptions occurring in the electron system. The loss function $\mathrm{Im}\{-1/\varepsilon\}=\varepsilon_2/(\varepsilon_1^2+\varepsilon_2^2)$ has a peak when $\varepsilon_1=0$ and ε_2 is small. The latter condition is fulfilled when damping of oscillations due to absorption is small. Since the real part ε_1 describes the screening of an external field inside a solid ($E_{\mathrm{int}}=E_{\mathrm{ext}}/\varepsilon_1$) the condition $\varepsilon_1=0$ indicates the instability of the systems against extremely small perturbations E_{ext}. These instabilities are the collective oscillations of the electron system against the ionic background (plasmons) and the plasmon frequency ω_p is given by $\varepsilon_1(\boldsymbol{q}, \omega_p) = 0$. Therefore, the maxima in the loss function at certain frequencies describe collective excitations of the electron system. Far above the plasmon frequency, ε_2 is small and the polarizability goes to zero, i.e. ε_1 is close to 1. Then the loss function $\mathrm{Im}\{-1/\varepsilon(\boldsymbol{q},\omega)\} \sim \varepsilon_2(\boldsymbol{q},\omega)$ is determined only by absorption. For small momentum transfer $\hbar q \to 0$ and $E \gg \hbar\omega_p$, electron energy-loss spectroscopy measures the same as optical-absorption spectroscopy (XAS) namely $\varepsilon_2(0,\omega)$. This will be outlined below in more detail.

Once the loss function $\mathrm{Im}\{-1/\varepsilon(\boldsymbol{q},\omega)\}$ has been determined by EELS, $\mathrm{Re}\{-1/\varepsilon(\boldsymbol{q},\omega)\}$ can be evaluated from the Kramers–Kronig relation

$$\mathrm{Re}\left\{\frac{1}{\varepsilon(\boldsymbol{q},\omega)}\right\} - 1 = \frac{1}{\pi}\int_{-\infty}^{+\infty} \frac{d\omega'}{\omega' - \omega}\,\mathrm{Im}\left\{\frac{1}{\varepsilon(\boldsymbol{q},\omega')}\right\}, \tag{7.7}$$

which follows from the fact that the response of the system is causal. Then, from the real and imaginary part of $1/\varepsilon(\boldsymbol{q},\omega)$, the dielectric functions $\varepsilon_1(\boldsymbol{q},\omega)$ and $\varepsilon_2(\boldsymbol{q},\omega)$ can be calculated.

7.1.2 Models for the Dielectric Function

There are numerous formulations of the dielectric function of solids. Firstly we discuss a phenomenological ansatz called the Drude–Lorentz dielectric function [7.22]. The electrons of the solid are separated into two subgroups, namely the free electrons (Drude part) and the bound electrons (Lorentz part). The equation of motion of the electrons is given by

$$m\left(\frac{d^2x}{dt^2} + \frac{\Upsilon_i dx}{dt} + \omega_i^2 x\right) = -eE_{\mathrm{ext}}(t), \tag{7.8}$$

where Υ_i and ω_i are the damping constants and the eigenfrequencies, respectively. By solving the differential equation, the polarization

$$P = -\sum_i e n_i x \tag{7.9}$$

can be calculated and thus the dielectric functions can be derived to be

$$\varepsilon(\omega) = 1 + 4\pi e^2 \sum_i \frac{n_i}{m(\omega_i^2 - \omega^2 - \mathrm{i}\Upsilon_i\omega)}, \tag{7.10}$$

where n_i are the density of charges having the oscillator energy ω_i. From the condition $\varepsilon_1(\omega) = 0$ we obtain for the plasmon frequency, in the case of only free electrons ($\omega_i = 0$), $\omega_p^2 = 4\pi n e^2/m$. At this energy, the loss function has a peak, the width of which is determined by the damping factor Υ. In the case of a background dielectric constant ε_∞ due to bound electrons with oscillator energies much larger than $\hbar\omega_p$, the plasmon frequency is reduced to $\omega_p^2 = 4\pi n e^2/\varepsilon_\infty m$. When there are only bound electrons with oscillator frequency ω_1 and a charge density n_1 there is a maximum in the loss function at $\omega^2 = \omega_1^2 + 4\pi n_1 e^2/m$. For a small oscillator frequency ω_1 compared to $4\pi n_1 e^2/m$, a maximum is observed close to the free electron plasmon energy $\omega_p = 4\pi n_1 e^2/m$. A typical example is the plasmon of the valence electrons in the semiconductor Si where $\hbar\omega_1 \sim 3$ eV and $\hbar\omega_p \sim 17$ eV. Therefore, the plasmon energy in Si is almost independent of the gap. On the other hand, for core levels the oscillator energy is much larger than $4\pi n_1 e^2/m$ and therefore a maximum appears close to the oscillator energy. In the case of both free and bound electrons there is a mutual interaction of oscillators, i.e. there are hybridized oscillators. The maxima in the loss function are no longer at the plasmon frequency or at the oscillator energy. For example, the free-electron plasmon energy is shifted by adjacent interband transitions. A detailed discussion of the loss function for these cases has been given recently [7.12].

In the simplest microscopic model for non-interacting conduction electrons, i.e. the Sommerfeld model, the dielectric function is determined by intraband transitions from the occupied states inside the Fermi sphere to unoccupied states outside the Fermi sphere. Using time-dependent perturbation theory of the Schrödinger equation, one obtains the relation for the dielectric function

$$\frac{1}{\varepsilon(\boldsymbol{q},\omega)} = 1 + \frac{4\pi}{q^2}\alpha^0(\boldsymbol{q},\omega), \tag{7.11}$$

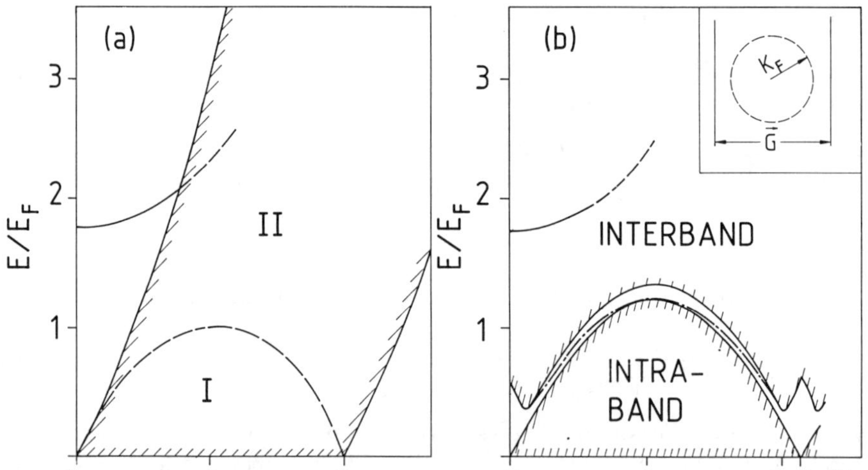

Fig. 7.2. (**a**) Plasmon and range of intraband transitions for an electron liquid in a jellium model. (**b**) Plasmon and range of intraband and interband transitions for a quasi-one-dimensional nearly free electron model with a Fermi surface and a Brillouin zone shown in the insert. Zone boundary collective states in the gap between intraband and interband transitions are indicated by the dash-dotted line

where the susceptibility $\alpha^0(\boldsymbol{q},\omega)$ is given by

$$\alpha^0(\boldsymbol{q},\omega)=\frac{2e^2}{V}\sum_{\boldsymbol{k}}\frac{f(\boldsymbol{k}+\boldsymbol{q})-f(\boldsymbol{k})}{E_{\boldsymbol{k}+\boldsymbol{q}}-E_{\boldsymbol{k}}-\hbar\omega-\mathrm{i}\delta},\tag{7.12}$$

$f(\boldsymbol{k})$ being the Fermi–Dirac distribution function. The loss function

$$\mathrm{Im}\left\{\frac{-1}{\varepsilon(\boldsymbol{q},\omega)}\right\}=\left(\frac{4\pi^2e^2}{q^2}\right)\frac{1}{V}\sum_{\boldsymbol{k}}[f(\boldsymbol{k}+\boldsymbol{q})-f(\boldsymbol{k})]\delta(E_{\boldsymbol{k}+\boldsymbol{q}}-E_{\boldsymbol{k}}-\hbar\omega)\tag{7.13}$$

is different from zero in a band, $2k_\mathrm{F}$ wide, as shown in Fig. 7.2. In range I there is a linear increase of the loss function with increasing energy, while in range II for $q<k_\mathrm{F}$ there is a quadratic decrease [7.23]. Thus for $q<k_\mathrm{F}$, the intraband transition contribution to the loss function at a definite q has the form of a shark fin. The next step in the evolution of a microscopic theory for the dielectric function of a metallic system is to take into consideration the electron–electron interaction in a mean-field approximation. Then we obtain the RPA or Lindhard dielectric function

$$\varepsilon_{\mathbf{RPA}}(\boldsymbol{q},\omega)\equiv\varepsilon_L(\boldsymbol{q},\omega)=1-\frac{4\pi}{q^2}\alpha^0(\boldsymbol{q},\omega).\tag{7.14}$$

Calculating the loss function from $\varepsilon_{\mathbf{RPA}}(\boldsymbol{q},\omega)$ yields, besides the intraband transitions already present in the Sommerfeld model, collective excitations of the electron system, i.e. the intraband plasmons. They appear for $q=0$ at $\omega_\mathrm{p}^2=4\pi ne^2/m$ as in the Drude model. For higher momentum transfer q, the plasmon energy increases and the dispersion relation is given in RPA by

$$E_\mathrm{p}(q)=E_\mathrm{p}(0)+(\hbar^2/m)\alpha q^2+\mathrm{O}(q^4),\tag{7.15}$$

where $E_\mathrm{p}(0)$ is the plasmon energy for zero momentum transfer. The dispersion constant α is given by $\alpha=(3/5)E_\mathrm{F}/E_\mathrm{p}(0)$, E_F being the Fermi energy. There is no plasmon damping in RPA for small-momentum transfer. Only when the plasmon line merges into the range of intraband transitions (see Fig. 7.2) does the plasmon decay into single-particle excitations (Landau damping).

To improve models for the dielectric function of electron liquids, one has to take into account short-range exchange and correlations between electrons. This is achieved by a phenomenological ansatz for the dielectric function yielding

$$\varepsilon(\boldsymbol{q},\omega)=1-\frac{(4\pi/q^2)\alpha^0(\boldsymbol{q},\omega)}{1+G(\boldsymbol{q},\omega)(4\pi/q^2)\alpha^0(\boldsymbol{q},\omega)},\tag{7.16}$$

where $G(\boldsymbol{q},\omega)$ is the local field-correction function. These corrections take into account the local decrease of the electron density around an electron due to correlation effects. There are numerous expressions for $G(\boldsymbol{q},\omega)$ which are reviewed, e.g., by *Ichimaru* [7.24] or *Gorobchenko* et al. [7.25]. RPA is a good approximation at high electron densities, i.e. small r_s values ($r_s=r_0/a_0$ where $2r_0$ is the average distance between electrons and a_0 is the Bohr radius). For smaller densities, i.e. higher r_s values, the ratio of kinetic to potential energy is reduced more and more, and local field corrections become more and more important. Simple metals have r_s values between 2 and 6. A localization of the electrons into a Wigner crystal is expected [7.26] at $r_s = 100 \pm 20$. The local field corrections lead to a reduction of the dispersion coefficient:

$$\alpha/\alpha_{\mathrm{RPA}} = 1 - \frac{5}{12}\xi(r_s)\left(\frac{E_{\mathrm{p}}}{E_{\mathrm{F}}}\right)^2 . \tag{7.17}$$

$\xi(r_s)$ depends on the special form of the local field correction $G(\boldsymbol{q}, \omega)$. Mainly static local field corrections $G(\boldsymbol{q})$ are used for detailed calculations.

In the following, we turn to real crystals, i.e. we take into account the interaction of electrons with the crystal lattice. In this case the general dielectric function $\varepsilon(\boldsymbol{r}, \boldsymbol{r}', t)$ is not just dependent on the difference of $\boldsymbol{r} - \boldsymbol{r}'$ but is lattice periodic, i.e. $\varepsilon(\boldsymbol{r}, \boldsymbol{r}', t) = \varepsilon(\boldsymbol{r} + \boldsymbol{R}, \boldsymbol{r}' + \boldsymbol{R}, t)$. The Fourier transformation of the latter function then gives $\varepsilon(\boldsymbol{k} + \boldsymbol{G}, \boldsymbol{k} + \boldsymbol{G}', \omega)$, where $\boldsymbol{k}$ are wave vectors within the first Brillouin zone and $\boldsymbol{G}$ and $\boldsymbol{G}'$ are reciprocal lattice vectors. This form of the dielectric function implies that an external longitudinal plane-wave perturbation $D(\boldsymbol{k} + \boldsymbol{G}', \omega)$ leads to a response $E(\boldsymbol{k} + \boldsymbol{G}, \omega)$ with the same frequency but with a wavelength $\boldsymbol{k} + \boldsymbol{G}$:

$$E(\boldsymbol{k} + \boldsymbol{G}, \omega) = \sum_{\boldsymbol{G}'} \varepsilon(\boldsymbol{k} + \boldsymbol{G}, \boldsymbol{k} + \boldsymbol{G}', \omega) D(\boldsymbol{k} + \boldsymbol{G}', \omega). \tag{7.18}$$

Besides $\boldsymbol{k} + \boldsymbol{G}'$, Bragg-diffracted components with wave vectors $\boldsymbol{k} + \boldsymbol{G}$ also appear. To a longitudinal perturbation, not only a longitudinal response but also a transverse response appears. The contributions with $\mathbf{G} \neq \mathbf{G}'$, i.e. the off-diagonal elements have been termed local field effects. These should not be confused with local field contributions due to exchange and correlations. The physical origin of the off-diagonal elements is the discrete lattice periodicity and their magnitude depends on how markedly the electronic structure is affected by this periodicity. In simple metals, where the valence electron distribution is very close to a homogeneous electron gas, local field effects are negligible. In transition metals and covalent semiconductors where there is a charge inhomogeneity inside each cell, local field effects are more important. These effects are very important in insulators. Taking into account the lattice local field effects, the macroscopic loss function is given by

$$\mathrm{Im}\left\{\frac{-1}{\varepsilon_{\mathrm{M}}(\boldsymbol{q}, \omega)}\right\} = -\,\mathrm{Im}\left\{\frac{1}{\varepsilon(\boldsymbol{k} + \boldsymbol{G}, \boldsymbol{k} + \boldsymbol{G}', \omega)}\right\}_{\boldsymbol{G} = \boldsymbol{G}'} \tag{7.19}$$

with $\boldsymbol{q} = \boldsymbol{k} + \boldsymbol{G}$. This means that we have to calculate the matrix elements $\boldsymbol{G}, \boldsymbol{G}'$ of $\varepsilon(\boldsymbol{k} + \boldsymbol{G}, \boldsymbol{k} + \boldsymbol{G}', \omega)$, invert the matrix and take the element $\boldsymbol{q} = \boldsymbol{k} + \boldsymbol{G}$. Neglecting off-diagonal elements means neglecting lattice local field effects. In this approximation the inverse macroscopic dielectric function is equal to the microscopic one. The calculation of the total inverse microscopic dielectric function is a major computational task, and the number of such calculations in the literature is still rather small. In this situation approximate methods are very valuable. For momentum transfer $\boldsymbol{q} < \boldsymbol{G}_{\mathrm{Min}}/2$, $\boldsymbol{q}$ is equal to $\boldsymbol{k}$ (band index within the first Brillouin zone) and we can take $\boldsymbol{G}=\boldsymbol{G}'=0$. Then we obtain for the dielectric function a form first derived by *Ehrenreich* and *Cohen* [7.27]

$$\varepsilon(\boldsymbol{q},\boldsymbol{q};\omega)=1-\frac{4\pi e^2}{q^2V}\sum_{\sigma\sigma'}\frac{f(\sigma)-f(\sigma')}{\hbar\omega-E_\sigma+E_{\sigma'}+\mathrm{i}\delta}(\langle\sigma|\mathrm{e}^{-\mathrm{i}\boldsymbol{q}\cdot\boldsymbol{r}}|\sigma'\rangle)^2. \tag{7.20}$$

Here $|\sigma\rangle=|l,\boldsymbol{k}+\boldsymbol{q}\rangle$ and $\sigma'=|l',\boldsymbol{k}\rangle$ are Bloch states and l,l' are band indices. $f(\sigma)$ is again the Fermi–Dirac distribution function. For plane-wave states $|\sigma\rangle=\mathrm{e}^{\mathrm{i}kr}$ with eigenvalues $E=\hbar^2k^2/2m$, one obtains again the RPA dielectric function for free electrons given above. Using the relation $1/(x+\mathrm{i}\delta)=P(1/x)-\mathrm{i}\pi\delta(x)$, the imaginary part of the dielectric function in this approximation is then

$$\varepsilon_2(\boldsymbol{q},\boldsymbol{q};\omega)=\frac{4\pi e^2}{q^2V}\sum_{\sigma\sigma'}(f(\sigma)-f(\sigma'))\delta(\hbar\omega-E_\sigma+E_{\sigma'})|\langle\sigma|\mathrm{e}^{-\mathrm{i}\boldsymbol{q}\cdot\boldsymbol{r}}|\sigma'\rangle|^2. \tag{7.21}$$

As long as the matrix elements do not vary strongly with energy and band index, $\varepsilon_2(\boldsymbol{q},\boldsymbol{q};\omega)$ is determined by the momentum-dependent *joint density of states*

$$\sum_{\sigma\sigma'}\delta(\hbar\omega-E_\sigma+E_{\sigma'}). \tag{7.22}$$

Therefore, if the density of occupied states is known or can be estimated, information on the unoccupied density of states can be derived. For comparison with EELS experiments, either the loss function has to be calculated from the Ehrenreich–Cohen formula (see (7.20)), or a Kramers–Kronig transformation (see (7.7)) of the measured loss function has to be performed giving $\mathrm{Re}\{-1/\varepsilon(\boldsymbol{q},\omega)\}$ and then $\varepsilon_2(\boldsymbol{q},\omega)$. The latter function can be compared with the calculated function $\varepsilon_2(\boldsymbol{q},\omega)$ from (7.21). We emphasize that in optical spectroscopy, only $\varepsilon_2(0,\omega)$ can be determined because the momentum transfer is extremely small compared to the Brillouin zone and cannot be varied. In EELS, one has access to momentum-dependent intraband transitions. Moreover, momentum-dependent interband transitions provide information on dispersion of bands. Excitations such as intraband transitions, interband transitions and plasmons are illustrated in Fig. 7.2b in an ω,q diagram for the case of alkali metals, e.g. Na. In these simple metals the Fermi vector k_F is about $0.88\,G_{110}/2$. For Na, a gap about 0.2 eV wide at the zone boundary between the first and the second band opens at $E\sim1.3\,E_\mathrm{F}$. For $q=0$, interband transitions are expected for $E>0.13\,E_\mathrm{F}$ but no intraband transitions are possible. For $q\neq0$, the diagram shows intraband transitions which remain separated from interband transitions due to the gap at the zone boundary. For $q\sim G_{110}/2-k_\mathrm{F}$, there is a large joint density of states between states at the Fermi level and states at the zone boundary in the second band. This leads to strong interband transitions causing a maximum in the loss function which is termed zone boundary collective state [7.28]. At higher momentum transfer, interband transitions into states close to the zone boundary are reduced and intraband transitions into states close to the zone-boundary are enhanced. Therefore, zone boundary collective states appear for smaller q values at the top of the gap between intra- and interband transitions and then move for $q \sim k_F$ to the bottom of this gap. This is indicated by the dash-dotted line in Fig. 7.2b. Finally, at $E = 1.7\, E_F$ and $q = 0$ the plasmon line starts and shifts to higher energies at higher q. While in the free electron case there was for small q no damping of the plasmon, the plasmon line in the non-free electron case (see Fig. 7.2b) is now in the range of interband transitions. Therefore, the plasmon may decay into interband transitions and a finite width of the plasmon is expected. This is in fact the dominant damping mechanism for plasmons in real metals [7.21].

At the end of this section we mention an important fact for measurements of intra- and interband transitions by EELS. While in the independent-particle model intraband transitions are well pronounced, the interaction of electrons leads below the plasmon frequency to a strong screening of these single-particle excitations. This is visualized by the fact that in the loss function $\mathrm{Im}\{-1/\varepsilon\} = \varepsilon_2/(\varepsilon_1^2 + \varepsilon_2^2)$, ε_2 is divided by $\varepsilon_1^2 + \varepsilon_2^2$, the latter sum being large for energies $0 \leqq E \leqq \hbar\omega_p$. Therefore, in most cases, intraband and interband transitions are not well pronounced in the loss functions and are difficult to measure by EELS. On the other hand, interband transitions (and core-level excitations) for $E > \hbar\omega_p$ are not screened since ε_1 is closed to one.

7.1.3 Core-Level Excitations

In the single-particle approximation, core-level excitations can be immediately described by the Ehrenreich–Cohen formula (7.21). For energies far above the plasmon energy the real part of the dielectric function is close to one and therefore the loss function is

$$\mathrm{Im}\left\{\frac{-1}{\varepsilon(\boldsymbol{q},\omega)}\right\} = \varepsilon_2(\boldsymbol{q},\boldsymbol{q};\omega) = \frac{4\pi e^2}{q^2 V}\sum_\sigma |\langle\sigma|\mathrm{e}^{-\mathrm{i}\boldsymbol{q}\cdot\boldsymbol{r}}|c\rangle|^2\,\delta(\hbar\omega - E_\sigma + E_c). \qquad (7.23)$$

Here $|c\rangle$ and $|\sigma\rangle$ are the core-level wave function and the wave function of unoccupied states, respectively, while E_c and E_σ are their energies. In this approximation the loss function is again determined by the joint density of states between core level and unoccupied states. Since the core level has almost no dispersion in k, the sum over the delta function is related to the density of unoccupied states. We will now consider the matrix element in more detail. For small momentum transfer, the operator $\exp(-\mathrm{i}\boldsymbol{q}\cdot\boldsymbol{r})$ can be expanded into $1 - \mathrm{i}\boldsymbol{q}\cdot\boldsymbol{r} - (\boldsymbol{q}\cdot\boldsymbol{r})^2/2$. Since $|c\rangle$ and $|\sigma\rangle$ are orthogonal, the first term gives no contribution to the matrix element. For $q \ll 1/\langle R_c\rangle$, where $\langle R_c\rangle$ is the mean value of the radius of the core-wave function, only the second term is important. Then the matrix element has the same dipole operator as in XAS but $\boldsymbol{E}$ is replaced by $\boldsymbol{q}$. For the $1s$ level of the light elements (e.g. C, N, O) $\langle R_c\rangle$ is of the order of $0.1\,\text{Å}^{-1}$and therefore $q \ll 1/\langle R_c\rangle$ for q up to several Å^{-1}. The dominance of the dipole operator implies that dipole selection rules should be applied, i.e. the difference in angular momentum between core level and excited state Δl should be ± 1 as in XAS (see Chaps. 5 and 6). Thus, starting from an $1s$ core level only p states can be reached. Starting from a $2p$ core level, s and

d states can be reached but the $\langle 2p|r|s\rangle$ matrix element is an order of magnitude smaller than the $\langle 2p|r|d\rangle$ matrix element. Furthermore, the matrix element is only different from zero when the states $|\sigma\rangle$ overlap with the core-wave function $|c\rangle$. This means that site-selective information is obtained, i.e. the local density of unoccupied states having a distinct angular momentum is probed. Finally, the direction of the momentum transfer relative to the sample can be readily changed. Two possibilities are illustrated in Fig. 7.3. There is a momentum transfer $q_\parallel$ which is determined by the energy loss, i.e. the binding energy of the core level. For zero scattering angle, $q_\perp = 0$ and the total momentum transfer q is parallel to $q_\parallel$. Therefore, in the scattering geometry shown in Fig. 7.3a, the momentum transfer is perpendicular to the sample plane and for a single crystal starting from an 1s level only p_z orbitals could be reached. On the other hand, for measurements at large scattering angles $q_\perp \gg q_\parallel$ and therefore the total momentum transfer is almost in the sample plane. Then for an 1s absorption edge, only $p_{x,y}$ states are probed in a single crystal. A more elegant scattering geometry is shown in Fig. 7.3b where the absolute value of the total momentum transfer remains the same. In this geometry the sample is tilted 45° against the electron beam. Selecting $|q_\perp| = q_\parallel$, the total momentum transfer can be turned into or perpendicular to the sample plane depending on the sign of $q_\perp$. In measurements on single crystals, this again allows differentiation between final states having components in or perpendicular to the sample plane. This illustrates that core-level excitations measured by EELS, just as by XAS, offer

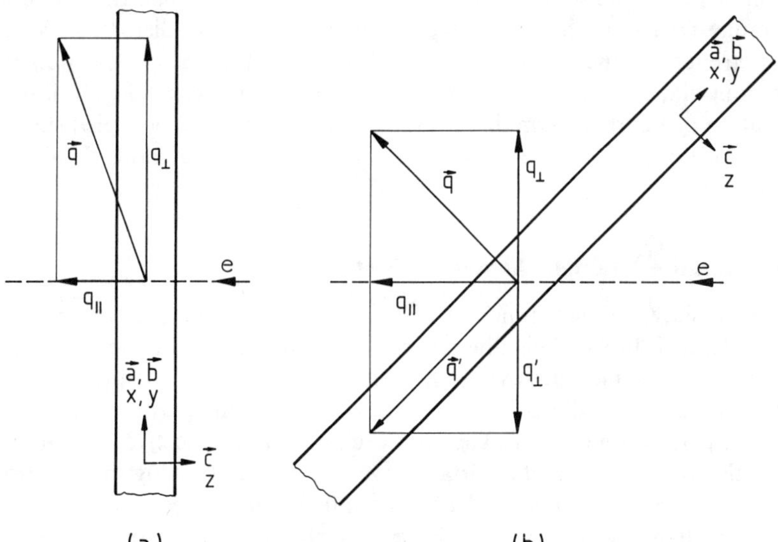

Fig. 7.3. Scattering geometries for doing orientation-dependent electron energy-loss experiments on core levels. (**a**) Sample perpendicular to the electron beam. (**b**) Sample tilted 45° against the electron beam

the possibility of measuring the local density of unoccupied states having a distinct angular momentum and a distinct symmetry. However, there are differences in experimental set-ups which lead to complementarity of EELS and XAS. In particular, we note that EELS measurements in the scattering geometry of Fig. 7.3b can be made on small samples ($\sim 0.5\,\mathrm{mm}^2$) whilst in XAS, using grazing incidence, rather large single crystals must be available to obtain information on the symmetry of unoccupied states.

In principle, EELS offers the interesting possibility to change the selection rules by increasing the momentum transfer. For $q > 1/\langle R_c \rangle$, the $(\boldsymbol{q}\cdot\boldsymbol{r})^2$ term is important leading to monopole and quadrupole transitions. However, very large momentum transfers ($q \geqq 2\,\text{Å}^{-1}$) have to be used to obtain reasonable contributions from these transitions. The high-momentum transfers are related to low counting rates and a high background from multiple scattering, and therefore this possibility has seldom been utilised, although it was used to study the changes of edge singularities in simple metals on variation of selection rules [7.11].

The above-given discussion of core-level excitations is oversimplified, since it takes into account only single-particle excitations. In reality core excitations are strong perturbations to the solid, and many-body effects may appear, as discussed in Chaps. 3–5. The size of these effects depends strongly on the ratio of the Coulomb interaction Q between core hole and conduction band states to the width of the conduction band (W), which is a measure of core-hole screening. Briefly, for $Q/W \ll 1$ the absorption edges are close to the partial, local, unoccupied density of states. For $Q/W \gg 1$ atomic transitions are observed since the influence of the solid (screening of the core hole) is small. In an intermediate range $Q/W \sim 1$, the unoccupied density of states is distorted. With increasing strength of the core-hole potential, more and more spectral weight is shifted from higher energies to the threshold. Further increase of Q/W leads to a separation of a core-exciton line below the continuum of unoccupied states [7.29]. The problem of core excitonic effects is discussed in more detail in Chaps. 4 and 5.

7.1.4 An Illustrative Example: Aluminium Metal

In Fig. 7.4 we show the loss function of an Al single crystal over an energy range from 0.2 to 2000 eV. This illustrates that excitations in an energy range over 4 decades can be measured with one spectrometer. The dominant features in the spectrum are the collective excitations, i.e. the volume plasmon at 15 eV and a surface plasmon near 7 eV. The later excitation is a collective oscillation of the free electrons at the metal surface covered with Al_2O_3. Figure 7.4 also illustrates a problem common to EELS. The interaction of electrons with the solid is larger than one would like. Since it is not possible to prepare free-standing films with a thickness much smaller than the mean free path of the electron, multiple scattering occurs. Even for a 600 Å thick Al film, we clearly realize multiple volume plasmon losses labelled $2P, 3P$ and $4P$ at 30, 45 and

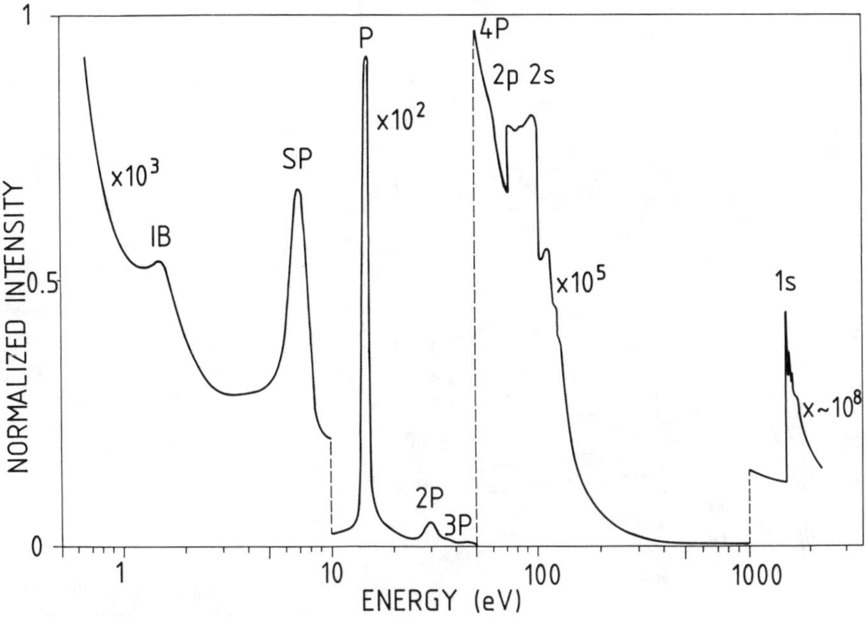

Fig. 7.4. Loss spectrum of a ~600 Å thick Al foil showing an intraband transition (IB), surface plasmon (SP), volume plasmon (P), multiple volume plasmons (2P, 3P, and 4P) and core-level excitations with near-edge structure and extended-edge structure from $2p$ and $2s$, and $1s$ states

60 eV, respectively. At least for smaller momentum transfers, these multiple losses can be corrected and data can be derived which are proportional to the loss function. The essential point for procedures performing double-scattering corrections is that a function which is the spectrum convoluted with itself is subtracted from the original data [7.9, 30]. Recently, procedures have been published [7.16, 31] where multiple losses could be corrected in an approximate way. For core excitations there is also the problem of multiple scattering. This is illustrated in Fig. 7.5, where we show the Al $1s$ absorption edges measured by EELS using samples having different thicknesses [7.32]. For small thickness (200 Å) the measured spectrum is close to a calculation of unoccupied partial density of states of Al metal with p character (not shown). For thicker samples, new thresholds appear in the absorption spectra at an energy which is the threshold energy plus a multiple of the plasmon energy. Therefore, these features can be assigned to a multiple-scattering process involving a $1s$ core excitation plus one or more plasmon excitations. The probability for those multiple-scattering processes is proportional to the solid angle for plasmon losses and the solid angle for core-level losses. The former is given by the momentum resolution, while the latter is given by $q_{\parallel}$ which is proportional to the energy loss. Thus, the probability for multiple scattering in core edges increases with increasing momentum resolution and increasing binding energy of the core

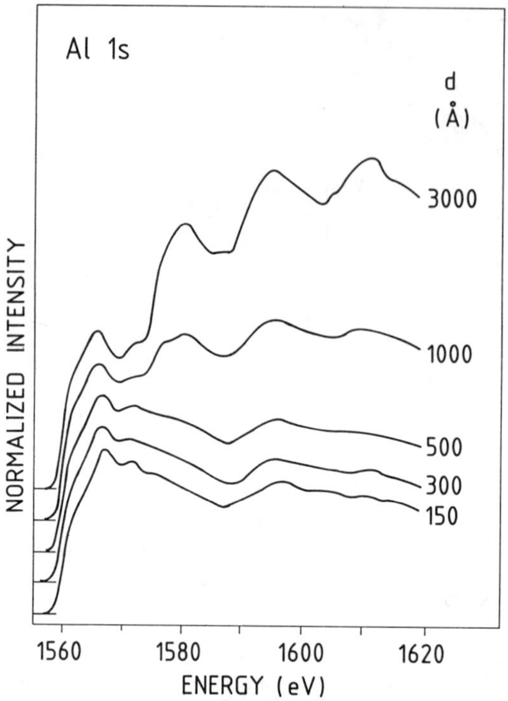

Fig. 7.5. Al 1*s* absorption edges measured on Al films having various thickness. A momentum transfer resolution of $\Delta q_{1/2} = 0.3\,\text{Å}^{-1}$ has been used

level. Finally, it should be noted that in the example given in Fig. 7.5 the multiple-scattering features are particularly pronounced, since Al metal has a narrow plasmon line with a high-scattering cross section. Generally, the plasmon line is much wider. Then multiple-scattering processes yield a broad additional background starting at the plasmon energy ($E_p = 10$–$30\,\text{eV}$). For investigations of unoccupied electronic states close to the Fermi level, multiple-scattering processes are not troublesome.

7.2 Instrumentation

Spectrometers for EELS consist of an electron source, a monochromator (sometimes), an accelerator, a scattering chamber with the sample and deflection plates for momentum transfer analysis, a decelerator, an analyzer and a detector. In Fig. 7.6 we show a sketch of the 170 keV Karlsruhe electron energy-loss spectrometer [7.12]. The most important points are: (i) the sample position is on ground potential and therefore samples can be readily prepared, loaded, cooled or heated, (ii) the energy resolution (and related to it, the electron current) can be varied between 0.08 eV and 0.7 eV using different path energies in the

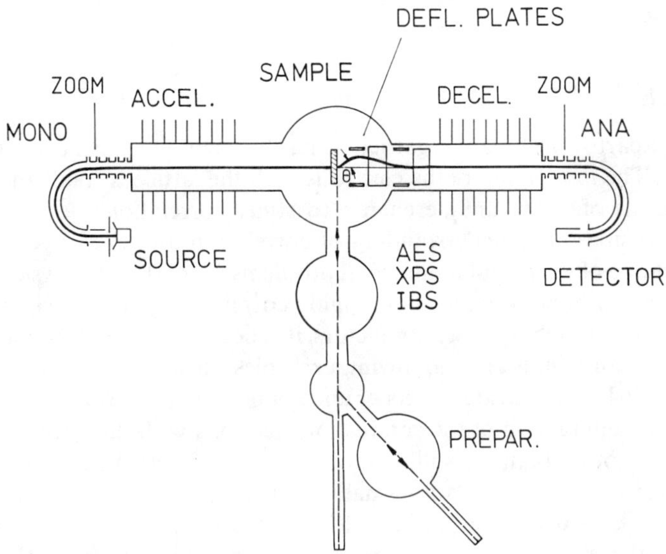

Fig. 7.6. Schematic drawing of the Karlsruhe electron energy-loss spectrometer

electrostatic hemispherical deflection monochromator and analyzer, (iii) the momentum resolution can be varied between 0.03 and 0.3 $Å^{-1}$ by changing the potential on the tube lenses of the zoom lens system. The values for energy and momentum resolution are typical of present generation of purpose-built energy-loss spectrometers although the flexibility of the Karlsruhe design is not reached by other spectrometers.

There is a growing number of energy-loss spectrometers composed of an electron microscope with a magnetic prism energy analyzer plus detector beneath the specimen position. Typically, the energy resolution in these systems is between 1 and 2 eV and there is a high spatial resolution. Those spectrometers are mainly used for the semi-quantitative microanalysis of light elements in electron microscopy specimens by measuring core-level absorption edges. Better resolutions ($\Delta E_{1/2} \sim 0.4$ eV) could be obtained by using field emission tips as electron sources in scanning transmission electron microscopes (STEM). There are also commercial electron microscopes (vacuum generators) now reaching this energy resolution. It should be noted that with these spectrometers a high-momentum resolution (which is crucial for experiments on the electronic structure of solids measuring valence band excitations) could be achieved only in a few microscopes [7.33, 34]. A detailed discussion on EELS spectrometers using electron microscopes is given by *Egerton* [7.16].

7.3 Case Studies

7.3.1 Simple Metals

The alkali metals, apart from Li, are regarded as nature's closest realization of a free-electron gas. The interaction of the electrons with the lattice is small and therefore these metals offer the unique chance to study correlation effects at various electron densities. The understanding of correlation and exchange in metals is at present one of the most fundamental problems in solid-state physics. Probably before we can hope to understand highly correlated systems such as heavy fermion systems or high-T_c superconductors, it is necessary to understand the much simpler alkali metals. The following examples illustrate that recent EELS studies [7.35, 36] have provided a deeper understanding of these metals.

The fact that the interaction of the conduction electrons with the lattice is small in alkali metals, other than Li, is illustrated in Fig. 7.7. There we compare volume plasmons of all alkali metals for small momentum transfer, i.e. large wavelength [7.35–37]. As discussed in Sect. 7.1.2, the plasmon width for $q=0$ is determined by the decay into interband transitions. The stronger the electron–lattice interaction, the stronger is the probability for an excitation of an interband transition given by $\varepsilon_2(\boldsymbol{q}, \omega)$. The plasmon of Li is very broad ($\Delta E_{1/2} \sim 1.5\,\mathrm{eV}$) indicating a large scattering potential of the ions for the

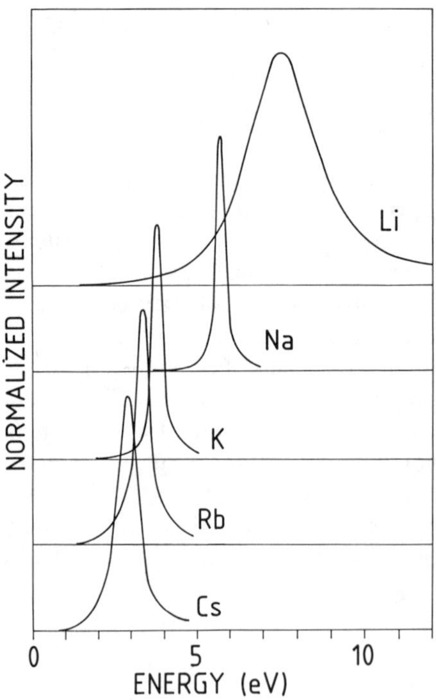

Fig. 7.7. Spectra of volume plasmons of the alkali metals at small momentum transfer $q \sim 0.1\,\text{Å}^{-1}$. Contributions from surface plasmons and double scattering processes have been subtracted

conduction electrons. The plasmons of Na, K, and Rb have an extremely small plasmon width of 0.28, 0.24 and 0.39 eV, respectively. For Cs, the plasmon width again increases to 0.77 eV. The plasmon width of K is, with the exception of Ag, the smallest among all metals indicating a close realization of a free-electron gas where the width of the plasmon should be zero (in the absence of crystal defects and at zero temperature) [7.21].

The plasmon energy for zero momentum transfer is close to the free-electron value. There are small shifts to lower energies (0.2–0.5 eV) which can be almost completely explained by a background dielectric function due to core–electron polarization. Going from Li to Cs, the plasmon energy decreases since the conduction electron density decreases. This is related to the increasing ion size in these monovalent metals having one conduction electron per ion. Among the simple metals, Cs has the lowest electron density, i.e. the highest r_s value. Therefore, among simple metals Cs should exhibit the strongest correlation effects (see Sect. 7.1.2). To investigate the latter, plasmons should be measured at shorter wave lengths, i.e. higher momentum transfers. Those data on the plasmon dispersion are shown in Fig. 7.8. For Na, the positive plasmon dispersion is close to a theoretical curve calculated from the Lindhard–Mermin

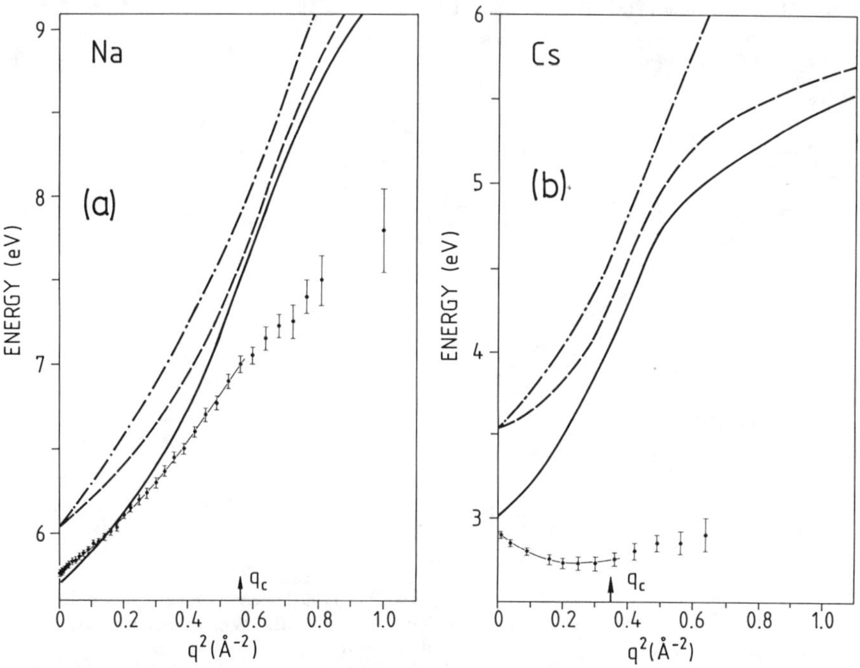

Fig. 7.8. Experimental plasmon dispersion for polycrystalline Na(*a*) and Cs(*b*) compared with theoretical models. Dashed dotted line: Lindhard–Mermin function. Dashed line: local field corrections by *Vashishta* and *Singwi* [7.39] included. Solid line: Core polarizability included. Thin solid line: Least squares fit to the experimental points

dielectric function, i.e. RPA with inclusion of a phenomenologic life time broadening [7.38]. Inclusion of local field effects due to exchange and correlations and core polarization slightly improves the agreement. Going to the heavier alkali metals, the deviations from RPA increase. The plasmon dispersion for small q is more and more reduced. For Cs a negative plasmon dispersion has been observed. As shown in Sect. 7.1.2 the dispersion coefficients α are determined by the slopes of the plasmon curves for $q \to 0$. In Fig. 7.9 we compare the experimental dispersion coefficients α normalized to the RPA value α_{RPA} with theoretical values. Even for Na we have a deviation from RPA but the $\alpha/\alpha_{\mathrm{RPA}}$ is close to a value calculated when local field effects are taken into account. For higher r_s values there are also deviations from the latter theoretical values. When core polarizations are taken into account (open circles) the disagreement even increases. Negative dispersion coefficients for $r_s \geqq 8$ are in principle predicted by current many-body calculations, taking into account local field corrections, while in the alkali metals they occur already for $r_s \geqq 5$. This indicates that the present many-body theories underestimate exchange and correlation effects. We remark that the negative plasmon dispersion may be interpreted by an incipient Wigner crystallization of the electron system. When the electrons are localized to form a lattice, the plasmon is the analogue to an optical phonon of an ion lattice. An optical phonon exhibits a negative dispersion, as observed in the case of Cs.

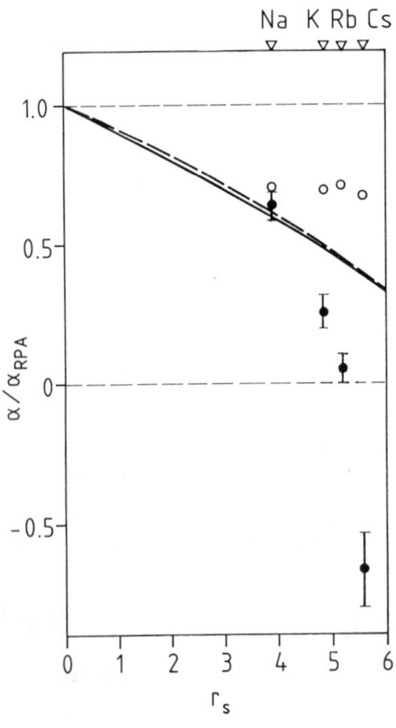

Fig. 7.9. Plasmon dispersion coefficients normalized to RPA values vs. the Wigner–Seitz radius r_s. The solid circles represent experimental values. The solid (dashed) line is the calculated $\alpha/\alpha_{\mathrm{RPA}}$ ratio using static (dynamic) local field corrections by *Vashishta* and *Singwi* [7.39] (by Dabrowski [7.40]). The open circles give the dispersion coefficient when core polarization is accounted for

Besides collective excitations, it was also possible to measure the much weaker intraband and interband transitions in the alkali metals. In Fig. 7.10 we show intraband transition of Na on top of a strong background due to the low-energy tail of the volume plasmon. For small q, the data are affected adversely by the background due to the direct beam. Compared to the expected shark fin (see Sect. 7.1.2) the features are broadened due to a finite relaxation time. The measurements were performed on a Na single crystal with $\boldsymbol{q} \parallel [100]$. The end points of the shark fin, i.e. the intraband transitions with the highest energies, are marked by the solid line in Fig. 7.10a. These transitions are due to excitations from the Fermi level into unoccupied states and therefore their energy is a

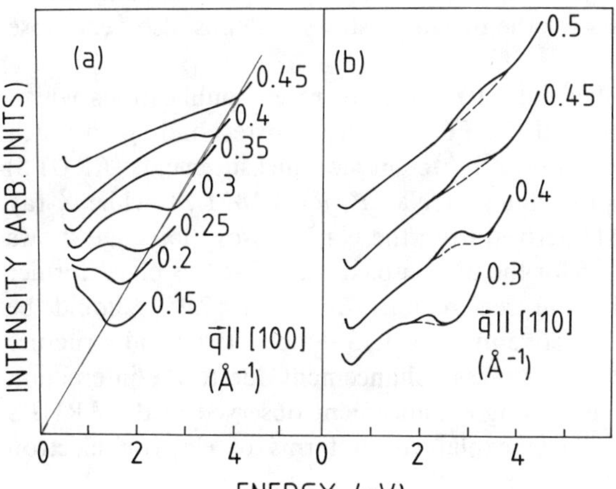

Fig. 7.10. (**a**) Interband transitions of single crystalline Na measured for various momentum transfers parallel to the [100] direction. The solid line indicates the interband transitions with the highest energies. (**b**) Momentum dependence of single-particle excitations for $\boldsymbol{q} \parallel [110]$ (solid line) and $\boldsymbol{q} \parallel [100]$ (dashed line). The difference of the two directions is due to zone boundary collective excitations

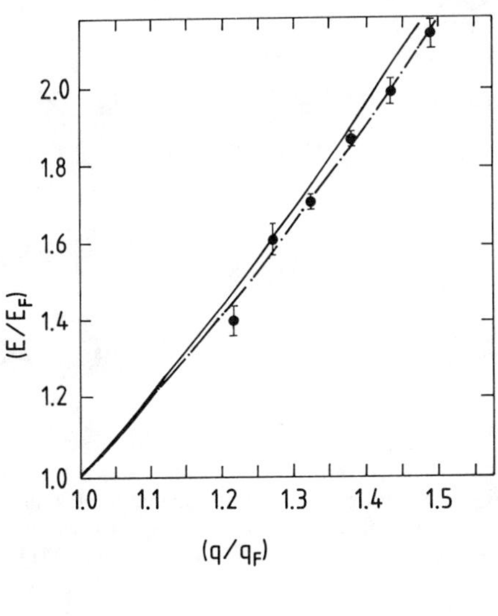

Fig. 7.11. The dispersion of the upper cut-off energy of interband transitions in Na normalized to the free-electron Fermi energy E_F. Symbols with error bars are from fits to the measured spectra (see Fig. 7.10a). Lines are parabolas for free electrons of various band masses being 1.00 (bold line) and 1.05 (dashed–dotted line)

measure of the dispersion of unoccupied states of the conduction band. The data points are given in Fig. 7.11 and they are compared with the free electron parabola. These results indicate that for Na the effective mass for the unoccupied states in the energy range E_F to $E_F + 4\,\mathrm{eV}$ is close to one, namely $m^*/m = 1.05$.

Measurements on single crystalline Na allowed the observation of the zone boundary collective states, i.e. excitations from occupied states into states close to the gap at the (110) zone boundary (see Sect. 7.1.2). The dispersion of these excitations is therefore related to the dispersion of occupied states. In Fig. 7.10b we show loss functions of Na for $\boldsymbol{q} \parallel [110]$. For this direction of the momentum transfer, the zone boundary collective states can be realized on top of the interband transitions. In Fig. 7.12 we compare the dispersion of the peak of these excitations with theoretical calculations [7.41] which have assumed an effective mass $m^*/m = 1$. The perfect agreement between experiment and theory indicate that the effective mass for the occupied states of Na is also very close to one ($m^*/m = 1.0 \pm 0.05$).

These results are in conflict with a number of recent publications which indicate a substantial reduction of the Fermi energy corresponding to much higher band masses. Angular-resolved photoemission measurements (ARUPS) [7.42, 43] yield for the occupied states of Na $m^*/m = 1.18$–1.28, while X-ray absorption spectroscopy [7.44] derived from the Na 1s absorption edge shows an effective mass of $m^*/m = 1.16$ for the unoccupied states. At this place, former optical measurements in the infrared region ($E \leqq 1\,\mathrm{eV}$) [7.45] should be mentioned which derived an effective mass of 1.13. Since LDA band-structure calculations [7.46] indicate that the mass enhancement due to the interaction with the lattice is only 1.03, the strong enhancement observed in the ARUPS and XAS measurements has been explained in terms of electron–electron

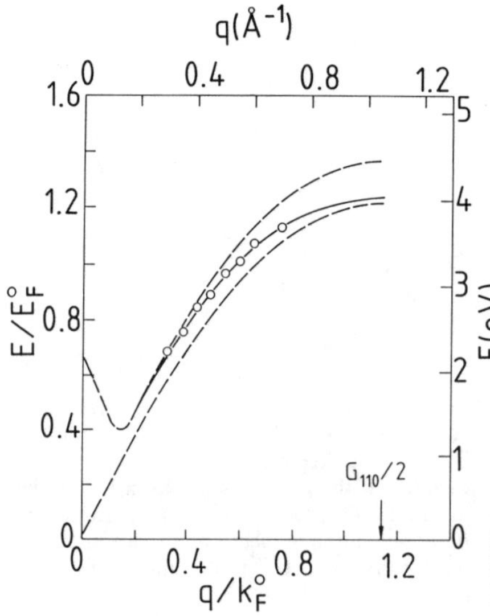

Fig. 7.12. Dispersion of the maximum of zone boundary collective states in Na shown in Fig. 7.10b compared with calculations from [7.41]

interaction, calculating electron self-energy renormalization [7.47, 48]. These calculations predicted effective masses between 1.12 and 1.18 both for occupied and unoccupied states of Na. Thus there is a conflict of the EELS data not only with other experimental data but also with theoretical estimates. On the other hand, XPS data on Na [7.49] yield effective masses close to 1. In addition, there are more refined theoretical calculations on mass changes due to many-body effects yielding even a mass reduction [7.50, 51]. To summarize, the present situation on the effective mass in Na is rather unclear.

7.3.2 Synthetic Metals

The term synthetic metals comprises a new class of materials achieving electrical conductivities which rival those of the best metals. Often the parent compounds are quasi-one-dimensional or quasi-two-dimensional semiconductors. Only after doping these materials, i.e. by removing or adding electrons from or to the parent compounds, are they transformed into highly conducting materials. The subsequent changes in the electronic structures upon doping are the central point to elucidate the mechanism of the high conductivities. In this section we illustrate the use of EELS in this field with two examples, namely electronic structure studies on the conducting polymers and intercalated graphite.

Conducting polymers are conjugated polymers which can be doped by electron acceptors (AsF_6^-, J_3^-, ClO_4^- ⋯ counter ions) or electron donors (Li^+, Na^+, NH_4^+ counter ions). In the doped materials conductivities up to 1.5×10^5 S/cm have been reported [7.52] which are close to that of Cu at room temperature. The prototype of these materials is polyacetylene (PA). It has, as do all conjugated polymers, a σ-bonded planar carbon backbone and a bond-alternated π-electron system originating from C p_z electrons. The band structure shows bonding and antibonding σ and σ^* bands with a gap of about 9 eV, and a bonding π and antibonding π^* band. In a single-particle calculation on a PA chain with equal C—C distances there should be no gap between the π and π^* band. There are at present two explanations for the existing gap: (i) Electron–phonon coupling may open a Peierls gap which leads to a localization of the π-electron density on every second bond, as indicated by a double bond in chemical formula. This model is often called Su–Schrieffer–Heeger (SSH) model [7.53–56]. (ii) Electron–electron interaction may be the dominant factor for dimerization in PA [7.57–59]. At present there is still a lively debate on the size of the two effects [7.60].

Upon $p(n)$-type doping, electrons are removed (added) to the polymer chain. In a rigid band model, electrons would empty (fill) the $\pi(\pi^*)$ band. In reality, the charges are localized, forming defects on the polymer chain leading to defect levels in the gap. For example, for p-type doping there would be no π electron at one C atom, forming a defect in which the double bonds in the zig-zag structure switch from the right-hand slopes to the left-hand slopes. The removed charge leads to a local closing of the gap, forming a mid-gap state which is empty for p-type doping. Then the defect has no spin but a charge. It can be described as a solitary wave [7.53], i.e. it may be mobile. These spinless but charged defects, termed solitons, are supposed to explain the spinless conductivity at low dopant concentration. There are numerous review articles on solitons and other defects (polarons and bipolarons) on these conjugated polymers [7.61–67]. With increasing dopant concentration, there is an overlap of the defects which in reality extend as far as 14 C atoms. This leads to a broadening of the soliton states. A soliton glass is formed and at sufficiently high concentrations the metallic state is reached due to a disorder-induced quenching of the Peierls distortion [7.68]. On the other hand, a first-order transition from a soliton lattice to a polaron lattice has been discussed [7.69]. The detailed electronic structure of the metalic state and the nature of the metal–insulator transition is still under debate.

A prerequisite for the understanding of the heavily doped metallic state is the understanding of the electronic structure of the undoped semiconductor. There is almost no information from the more surface sensitive high-energy spectroscopies such as ARUPS or bremsstrahlung isochromat spectroscopy (BIS) since it is difficult to prepare surfaces representing bulk properties. EELS in transmission is a particularly suitable method for these compounds.

In Fig. 7.13 we show the loss function $\mathrm{Im}\{-1/\varepsilon(\boldsymbol{q},\omega)\}$ and the real and imaginary part $\varepsilon_1(\boldsymbol{q},\omega)$ and $\varepsilon_2(\boldsymbol{q},\omega)$ for highly oriented trans-PA [7.70–72]. The absolute value of the momentum transfer was chosen small ($q = 0.1\,\text{Å}^{-1}$) in comparison with the extension of the Brillouin zone ($q = 1.3\,\text{Å}^{-1}$). Both data for $\boldsymbol{q}$ parallel (solid line) and $\boldsymbol{q}$ perpendicular (dashed line) to the chain orientation $\boldsymbol{c}$ are shown. For $\boldsymbol{q} \parallel \boldsymbol{c}$, two maxima are realized in the loss function, one at 22 eV caused by the collective oscillation of all valence electrons ($\pi + \sigma$) and one at 5 eV related to the π electrons, which is therefore termed the π plasmon. As can be seen from the ε_2 curve for $\boldsymbol{q} \parallel \boldsymbol{c}$ there is a strong maximum at 1.8 eV due to a $\pi - \pi^*$ transition across the fundamental gap at the zone boundary. This oscillator at 1.8 eV causes a zero crossing of ε_1 at 5 eV and because ε_2 is small at that energy, the loss function exhibits a maximum. The π plasmon does not appear at a value calculated in the Drude model ($\hbar\omega_p \sim 5$ eV) taking into account the background dielectric function of the σ electrons ($\varepsilon_\infty \sim 3$). Rather, the π plasmon is a collective excitation of interband transitions, i.e. the π plasmon is an interband plasmon. This will be supported by the momentum dependence of the plasmon which will be discussed below. For $\boldsymbol{q} \perp \boldsymbol{c}$ the π plasmon and the $\pi - \pi^*$ transition is strongly reduced. The anisotropy of the $\pi - \pi^*$ transition is ~ 170 indicating the small polarizability of the π electrons perpendicular to the chain axis. The observed anisotropy is in qualitative agreement with calculations in the framework of the SSH model [7.73]. At higher energies, several $\sigma - \sigma^*$ transitions (for $\boldsymbol{q} \parallel$ and $\perp \boldsymbol{c}$) and a mixed $\pi - \sigma^*$ transition at 9 eV (for $\boldsymbol{q} \perp \boldsymbol{c}$) are observed.

Figure 7.14 shows the transition probabilities $[\propto \varepsilon_2(\boldsymbol{q},\omega) \propto \sigma(\boldsymbol{q},\omega)]$ as a function of the momentum transfer. These data have been obtained via a Kramers–Kronig analysis of the measured loss functions. The absolute size of the loss function has been determined using a sum rule [7.72]. In Fig. 7.15 we

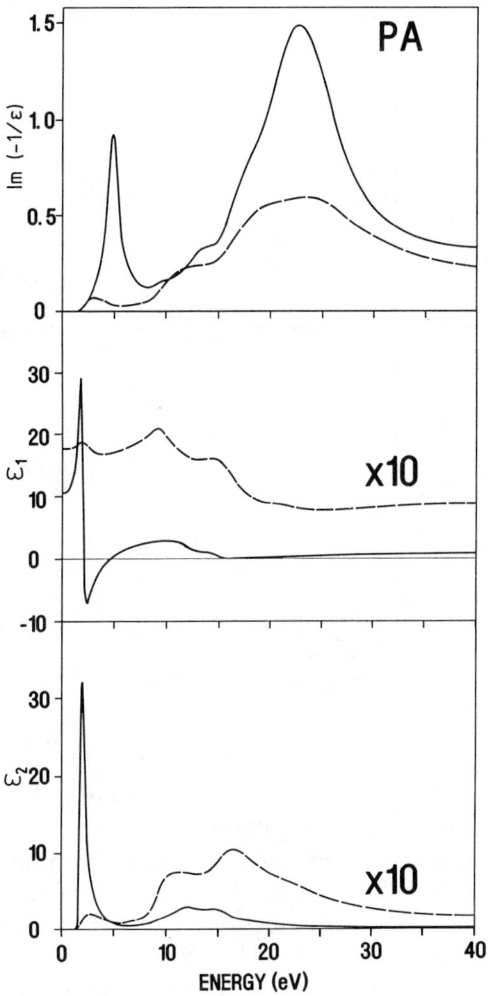

Fig. 7.13. Loss function $\mathrm{Im}\{-1/\varepsilon(\boldsymbol{q},\omega)\}$, $\varepsilon_1(\boldsymbol{q},\omega)$ and $\varepsilon_2(\boldsymbol{q},\omega)$ for undoped trans-polyacetylene for $q = 0.1\,\text{Å}^{-1}$. Solid line: $\boldsymbol{q}$ parallel to the chain axis of the polymer, dashed line: $\mathbf{q}$ perpendicular to the chain axis

compile the excitation energies of plasmons and interband transitions as a function of momentum transfer. The data on the π electrons are compared with theoretical calculations of the $\pi - \pi^*$ transitions (dash-dotted lines) and the π plasmon (dashed line) on the basis of the SSH model including lattice local field corrections [7.74]. A calculation of the momentum-dependent joint density of states of the π and π^* bands [7.70] yields square root singularities on the solid lines shown in Fig. 7.15. The experimental data and the theoretical calculations indicate that inclusion of matrix elements and lattice local field effects strongly suppress the high energy singularity. Furthermore, at higher momentum transfers the lower singularity is broadened and shifted to higher energies. The special shape of the π and π^* bands leads to an almost linear

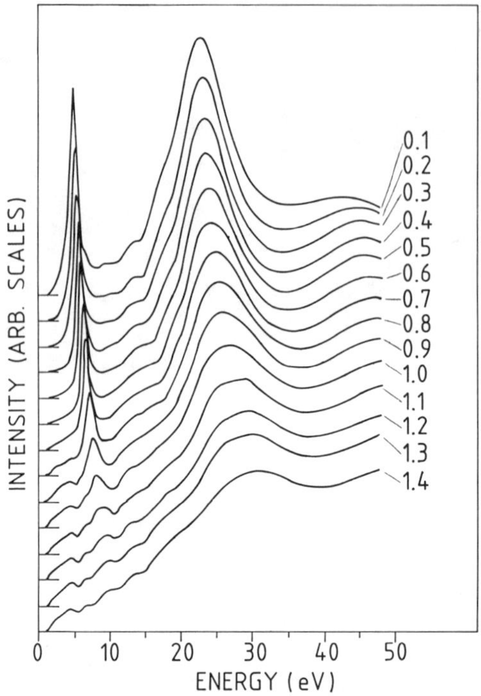

Fig. 7.14. Momentum-dependent transition probabilities of trans-polyacetylene for $\boldsymbol{q}$ parallel to the chain axis

dispersion of the π and π^* transitions for $0 \leqq q \leqq 0.7\,\text{Å}^{-1}$ and a splitting for $q > 0.7\,\text{Å}^{-1}$. Since the π plasmon is related to the $\pi - \pi^*$ transition, it has also an almost linear dispersion. The free parameters for the calculations shown in Fig. 7.15 were the gap energy $E_g = 1.8\,\text{eV}$ and the total band width of the π-electron system $W = 11\,\text{eV}$, the latter value being close to that derived from band-structure calculations [7.75]. The above-mentioned EELS experiments on the π-electron system contribute the first experimental data on the total width of the π-electron bands. Furthermore, they have shown that π-electron bands exist. On the other hand there are also deviations from a simple single-particle model. The gap in the loss function and in the transition probability does not increase with increasing momentum transfer as would be expected, e.g. from the momentum-dependent joint density of states. There is a considerable oscillator strength below the single-particle gap and the onset remains at 1.8 eV. It is remarkable that these tails were not observed in momentum-dependent measurements on graphite (see below) [7.76]. For $q > 0.7\,\text{Å}^{-1}$, the spectral weight below the single-particle gap is mainly due to multiple scattering. For $q < 0.7\,\text{Å}^{-1}$, the tail could be explained in terms of excitonic transitions due to strong correlation effects [7.70]. On the other hand, the spectral weight in the tails may also be caused by a direct generation of charged soliton–antisoliton pairs assisted by quantum fluctuations in the ground state [7.77]. The same

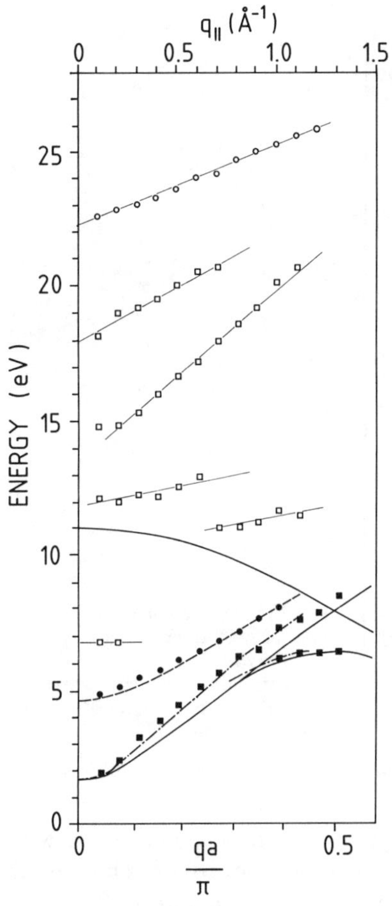

Fig. 7.15. Excitation energies of interband transitions (squares) and plasmons (circles) in trans-polyacetylene as a function of momentum transfer $\boldsymbol{q}$. Full symbols: π-electron system, open symbols: σ-electron system

excitations were observed in measurements of the photoconductivity [7.78] and in measurements of photoinduced absorption [7.79]. Besides excitations of the π-electron system, those from the σ electrons are also shown in Fig. 7.15. Since there are no theoretical calculations available for these excitations, the solid lines are only guides to the eye. There are various $\sigma-\sigma^*$ transitions, the assignment of which is at present speculative. The $\pi+\sigma$ plasmon shows an almost linear dispersion. Since at these high energies all valence electrons are almost free, the dispersion curve represents probably a first example of the predicted linear plasmon dispersion [7.80] of a one-dimensional (quasi)-free-electron plasmon.

An example of EELS studies on the electronic structure of doped conjugated polymers is shown in Fig. 16, which compiles the dispersion of the π plasmon in AsF_5-doped PA for various dopant concentrations. Since it was difficult to determine absolute dopant concentration in 1000 Å thick films, values for the

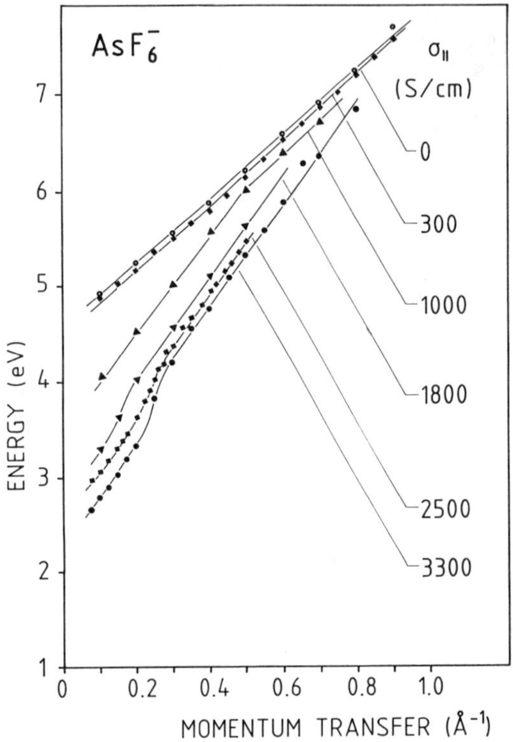

Fig. 7.16. Dispersion of the π plasmon of AsF_6-doped highly oriented polyacetylene for various dopant concentrations (dc conductivities). Uppermost curve: undoped sample, lowest curve: heavily doped sample

conductivity are given. For small q the energy of the π plasmon decreases with increasing dopant concentration, i.e. with increasing conductivity. This can be explained on the basis of calculations performed for a soliton lattice [7.81–84]. In this model, more and more oscillator strength of the $\pi-\pi^*$ transitions is transferred to transitions from the π band into the unoccupied (for p-type doping) soliton band(s). At intermediate dopant concentrations both $\pi-\pi^*$ and π-soliton excitations occur, and are separated by about half the gap energy which broadens the π plasmon. At the highest dopant concentrations the soliton band is strongly broadened and the $\pi-s$ transition is dominant and appears almost at zero energy, again causing a narrow π plasmon. The plasmon energy is reduced by about 2 eV corresponding to the $\pi-\pi^*$ gap. At the highest doping levels, the plasmon dispersion is steeper as expected for a system with a π electron system where the gap is almost closed. An evaluation of the data, although provided with large errors due to ambiguities in the calibration of the loss function, yields a linear dispersion of the $\pi-\pi^*$ transition starting from almost zero energy at zero momentum transfer. This is close to what is expected for an electronic structure of a one-dimensional metal or an electronic structure where the gap is almost completely filled by a wide soliton band. Our data, however, cannot exclude a finite gap between π bands and the soliton band, as derived from optical spectroscopy [7.85]. In fact, the small steps in the dispersion curves of heavily doped PA shown in Fig. 7.16 may be indicative of a ~0.3 eV wide gap between a soliton and a π^* band. Recent theoretical work [7.84] indicates that at small momentum transfer there is a strong oscillator strength for $\pi - s$ transitions. For momentum transfer larger than the size of the Brillouin zone of the soliton lattice, predominantly $\pi - \pi^*$ transitions are observed. With increasing dopant concentration, the Brillouin zone of the soliton lattice is increased. Therefore, the steps between the $\pi - s$ and the $\pi - \pi^*$ transitions should occur at higher momentum transfer with increasing dopant concentration in agreement with the experimental results (see Fig. 7.16). The EELS measurements support a soliton-driven insulator–metal transition for PA. A transition into a metal with **wide** polaron bands [7.69] would also be consistent with the measurements. However, the existence of **narrow** polaron bands in the gap can be excluded for the metallic state of the PA. In the latter case, transitions without dispersion in momentum transfer should be realized.

The described changes of the electronic structure of conjugated polymers upon doping are typical of all polymers, although other defects are created upon doping of other polymers. The doping-induced changes of the band structure are in all cases far away from a rigid band model. Similar studies on conjugated polymers such as polyparaphenylene, polypyrrole, polythiophene, polyaniline and polyphenylenevinylene have been reviewed previously [7.12].

In the following we contrast the results on conducting polymers with investigations on intercalated graphite (compare also with IPS data in Chap. 8). This system exhibits upon doping a change of the electronic structure which is very close to a rigid band model. In pristine graphite, the two-dimensional planes are composed of C atoms in sp^2 configuration. The three trigonally directed σ bonds form the hexagonal lattice structure of the graphite planes. In addition there is weaker π bonding between the carbon atoms in the planes. Between the planes there is only van der Waals bonding. The band structure of a carbon plane is similar to that of PA and shows again bonding and antibonding σ and σ^* bands, and a bonding π and antibonding π^* band. Contrary to PA there is no gap between π and π^* band. Rather, at the corners of the hexagonal Brillouin zone they form electron pockets at the K point and hole pockets at the H point. Thus a semi-metal with about 10^{-4} charge carriers per C atom is formed. At the Fermi level there is a very low density of states which almost linearly increases when going to positive or negative binding energies. As with conjugated polymers, graphite can be doped (intercalated) by acceptors or donors. Different concentrations can be achieved by forming different stages. Stage I means that every plane between the C planes is intercalated. Stage II means that every second plane, and stage III every third is intercalated, and so on. Intercalating leads to a charge removal or additional charges on the graphite planes and therefore to a dramatic increase of charge carriers which causes the high conductivities.

The loss function of pristine graphite shows, as does PA, a $\pi + \sigma$ plasmon of all valence electrons at 27 eV [7.76, 86, 87]. There is a high joint density of states between π and π^* bands at the M point causing a strong $\pi-\pi^*$ interband transition at 4.3 eV. In the loss function the interband transition leads to a pronounced π-interband plasmon at ~ 7 eV (see Fig. 7.17a). There is considerable spectral weight below the π plasmon ranging to zero energy due to $\pi-\pi^*$ transitions around the K point. There are various studies of the momentum dependence of the loss function of graphite [7.76, 86, 87] yielding information on the dispersion of the π bands. In particular, a gap is opened in the loss function which increases with increasing momentum transfer. This is different from PA where the onset of the loss function is independent of q. In graphite, the momentum dependence of the gap provides information on the dispersion of the π and π^* bands at the K point.

In Fig. 7.17b we show the loss function of stage I $FeCl_3$-intercalated graphite [7.88]. Similar but much more extensive studies have been performed by *Ritsko* et al. [7.89, 90]. In the acceptor-intercalated graphite, a lowering of the Fermi level due to the charge transfer to the intercalant is expected in a rigid band model. This is illustrated in the insert of Fig. 7.17b. A gap of ~ 1.7 eV then opens for the $\pi-\pi^*$ intraband transitions, causing the shoulder in $\mathrm{Im}(-1/\varepsilon)$ at 3 eV. The lowering of E_F into the π band allows intraband transitions and an intraband plasmon at 1 eV related to holes in the π band. Since at low q there is a gap of interband transitions which is larger than the plasmon energy, this plasmon is almost undamped and therefore rather narrow. At higher momentum transfer, the plasmon line moves into a range of interband transitions [7.91] and is therefore heavily damped. The 7.0 eV $\pi-\pi^*$ interband plasmon in graphite is shifted to 6 eV in the intercalated compound. This may be explained

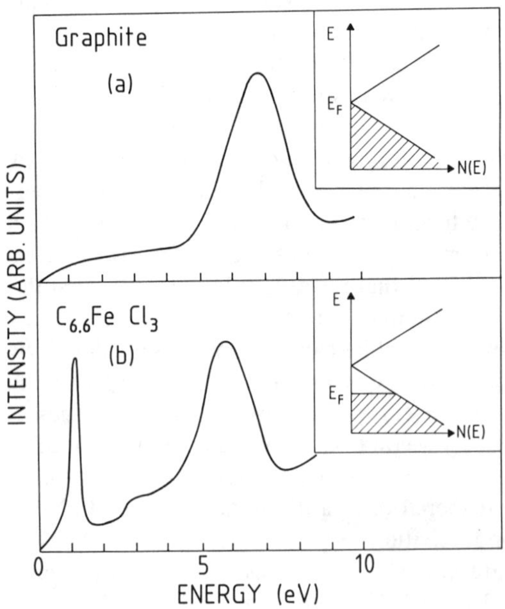

Fig. 7.17. (**a**) loss function $\mathrm{Im}\{-1/\varepsilon(\boldsymbol{q},\omega)\}$ of pristine graphite for $\boldsymbol{q}=0.05\,\mathrm{\AA}^{-1}$ parallel to the graphite planes. (**b**) the same for stage I $FeCl_3$-intercalated graphite ($C_{6.6}FeCl_3$). The inserts show the density of states and the position of the Fermi level.

by a reduced oscillator strength of the $\pi - \pi^*$ transition due to the charge transfer and the reduced density of carbon planes. Performing a Kramers Kronig analysis yields that the energy of the $\pi - \pi^*$ transition at the M point is almost unchanged, indicating that the π-band structure is not significantly altered upon intercalation.

A much more direct observation of the shift of the Fermi level upon doping can be obtained from the C 1s absorption edges shown in Fig. 7.18 [7.88]. Similar data on stages I and II have been published previously [7.92, 93]. Since dipole selection rules apply, the spectra on single crystalline samples oriented perpendicular to the beam axis are for $q_\perp = 0$ related to the unoccupied states having p_z character which make up the graphite π^* bands. For pristine graphite the onset which corresponds to the Fermi level is at 284.2 eV and there is a broad, nearly symmetric peak at 285.4 eV. From the density of unoccupied π^* states, a less steep rise of spectral weight above threshold is expected. A maximum at ~ 2 eV above threshold is expected due to the dispersionless region of the π^* band at the M point causing a logarithmic singularity in the density of states. The observation of a maximum at 1.2 eV above threshold was explained in terms of core excitonic effects, i.e. interactions with the core hole, which enhances the spectral weight at the bottom of the π^* band [7.92]. As shown in the insert of Fig. 7.17b, in the intercalated compounds, the Fermi energy is lowered and new empty states are created in the π band. Exactly these new states are realized

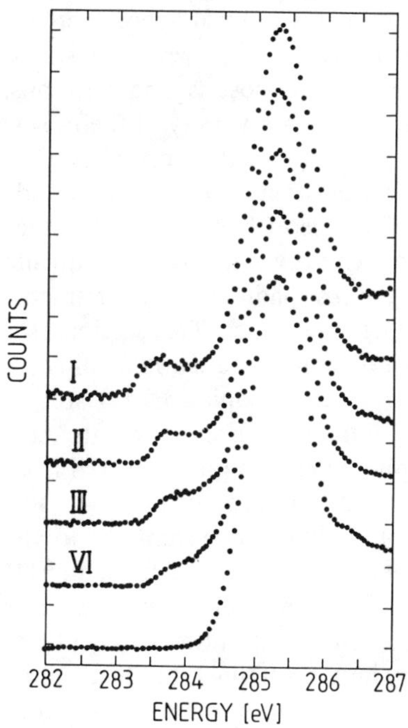

Fig. 7.18. C 1*s* absorption edges of pristine graphite and $FeCl_3$-intercalated graphite, with stage I, stage II, stage III and stage VI

in the C 1s spectra on intercalated graphite compounds. With increasing dopant concentration (lower stages) a more and more pronounced pre-peak appears and the threshold is shifted to lower energy. The pre-peak is a measure of the empty states in the π band. In stage I, the lowering of the Fermi level is about 1 eV. When constant charge transfer per counter ion and a linear density of states is assumed for all stages, the total area under the pre-peak should be proportional to $1/n$, where n is the stage number. The lowering of the Fermi level should then be proportional to $1/\sqrt{n}$ in perfect accordance with the experimental data.

The results of the low-energy loss function and of the C 1s absorption edges indicate that in the intercalated graphite compounds, the π-band structure is not significantly perturbed on intercalation and therefore a rigid band model is a good approximation. More subtle measurements, e.g. optical measurements, indicate small influences of the potential of the counter ions and hybridization of counter ion and graphite orbitals.

7.3.3 High-Temperature Superconductors

Four years after the discovery of cuprate superconductors by *Bednorz* and *Müller* [7.94], the mechanism for high-T_c superconductivity is still completely unclear. Some BCS-like theories attribute the condensation into the superconducting state to electron–phonon interaction, others to excitonic (charge transfer, acoustic plasmons d–d excitation etc.) or magnetic degrees of freedoms. Theories starting from the Hubbard-like Hamiltonians yield superconductivity in the resonating valence bond state, by flux phases, or in a two-dimensional liquid formed by quasiparticles intermediate between bosons and fermions, called anyons. One reason for the development of such a variety of theories is that even the electronic structure of these compounds in the normal state is rather unclear. There is at present a lively debate on whether these compounds behave as Fermi liquids in their normal state (see also Chaps. 4 and 9). Since the understanding of the electronic structure in the normal state is a prerequisite for the fundamental understanding of high-T_c superconductivity, experiments probing the electronic structure are of particular importance. This section uses some examples of EELS investigations of cuprate superconductors as illustrations of the technique. Among all high-energy spectroscopies, EELS studies were rather successful since they probe bulk properties and are less surface sensitive. Other methods such as e.g. photoemission often suffered from the problem of preparing surfaces which represent bulk properties.

At first sight, the cuprate superconductors behave very similarly to the intercalated graphite compounds described in the previous section. In all high-T_c superconductors there are two-dimensional CuO_2 planes. Charge carriers having hole or electron character can be formed on the insulating CuO_2 planes by p- or n-type doping. However, there is a very important difference to graphite compounds. The cuprates belong to the class of transition metal compounds where strong correlation effects are expected. The investigation of the electronic structure of materials with strong electron correlation has been a challenge for solid-state physicists for many decades [7.95].

In a single-particle model, the CuO_2 planes would behave as a paramagnetic metal due to a half-filled $Cu\,3d_{x^2-y^2}$–$O\,2p_{x,y}$ band [7.96]. Correlation effects on Cu sites lead to an antiferromagnetic charge transfer insulator [7.97–99]. The valence band has predominantly O $2p$ character with some admixture of Cu $3d$ states. The conduction band has mainly Cu $3d$ states with some admixture of O $2p$ states. The amount of admixture is a measure of covalency in the CuO_2 layer. Upon p- or n-type doping electrons are removed or added to the CuO_2 layers, respectively. With increasing dopant concentration antiferromagnetism is destroyed and a metallic state exhibiting high-T_c superconductivity is formed. At present, numerous models of the electronic structure in doped compounds are being discussed. In the simplest rigid-band-like model for the charge-transfer insulator, holes in the valence band and electrons in the conduction band will be formed upon p- or n-type doping, respectively. The character of the holes would be mainly O $2p$ and that of the electrons mainly Cu $3d$ like.

As a first example we show in Fig. 7.19 a typical loss function for a superconducting $Bi_2Sr_2CaCu_2O_8$ single crystal [7.100]. The momentum transfer $q = 0.1\,Å^{-1}$ was set parallel to the CuO_2 layers. In addition, the real part of the dielectric function ε_1 and the optical conductivities σ, as derived from a Kramers–Kronig analysis, are shown. The loss spectra are dominated by broad features centered at about 25 eV assigned to a plasmon of all valence electrons. Superimposed on these plasmons are interband transitions and low-energy core-level excitations. There is a marked difference in the loss functions between insulating (undoped) and metallic (doped) compounds at low energies. The insulating parent compounds (not shown) exhibit a clear gap of about 2 eV both in the loss function and the optical conductivity. This gap can be assigned to the charge transfer gap. The superconducting compounds exhibit a further plasmon at about 1 eV which can be ascribed to an in-phase collective motion of the charge carriers in the CuO_2 layers. The situation is similar to that of intercalated graphite (see Sect. 7.3.2) where a plasmon, due to holes created in the two-dimensional graphite layers upon p-type doping, appears in the loss function. A Drude-like decay in the optical conductivity is related to the appearance of the charge carrier plasmon in the loss function. By using the charge carrier density derived from Hall-effect measurements or from chemical arguments and the background dielectric function ε_∞ derived from ε_1 data, the effective mass m^* of the charge carriers can be calculated from the relation $E_p = \hbar(4\pi ne^2/\varepsilon_\infty m^*)^{1/2}$ to be close to the free-electron mass m. This indicates that at energies of ~ 1 eV the charge carriers have to be considered as almost free. However, there may be a considerable interaction of the charge carriers with, e.g., spins on the Cu sites, which may lead to an enhancement of the effective mass at lower energies. The situation is very similar to that of polyacetylene (see Sect. 7.3.2) or Si, where the plasmon of all valence electrons is close to the free-electron value although there is a finite gap at low energies.

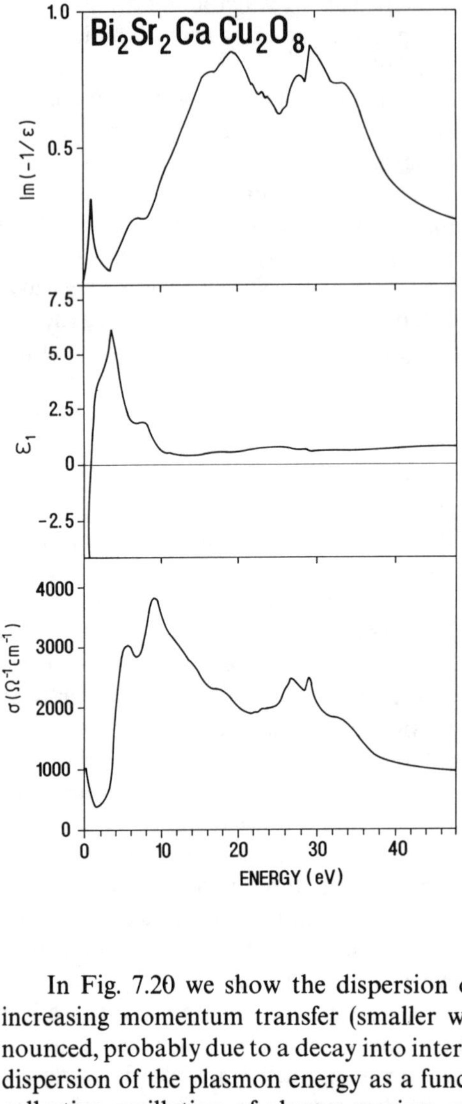

Fig. 7.19. Loss functions $\mathrm{Im}\{-1/\varepsilon(\boldsymbol{q},\omega)\}$, real parts of the dielectric functions $\varepsilon_1(\boldsymbol{q},\omega)$ and optical conductivities $\sigma(\boldsymbol{q},\omega)$ for a single crystal of $Bi_2Sr_2CaCu_2O_8$

In Fig. 7.20 we show the dispersion of the charge-carrier plasmon. With increasing momentum transfer (smaller wavelength) the plasmon is less pronounced, probably due to a decay into interband transitions. There is a quadratic dispersion of the plasmon energy as a function of momentum transfer. For the collective oscillation of charge carriers on two-dimensional sheets, having a distance a, this dispersion relation is expected for $q \ll (2\pi/a)$ and $\mathbf{q}$ parallel to the layers. In RPA the plasmon dispersion is given by $E_p(q) = E_p(0) + Aq^2$ where the dispersion coefficient A is given by $A = 3\cdot[\hbar^2 v_F^2/E_p(0)]/(2d+4)$ where d is the dimension of the system in which the charge carriers move and v_F is the Fermi velocity [7.101]. Using this approximation, the value of the Fermi velocity derived from the experimental plasmon dispersion is in excellent agreement with that derived from LDA band-structure calculations [7.102]. On the other hand, this perfect agreement should not be stressed too much. As discussed above with respect to the plasmon energy, there may be coupling of the charge

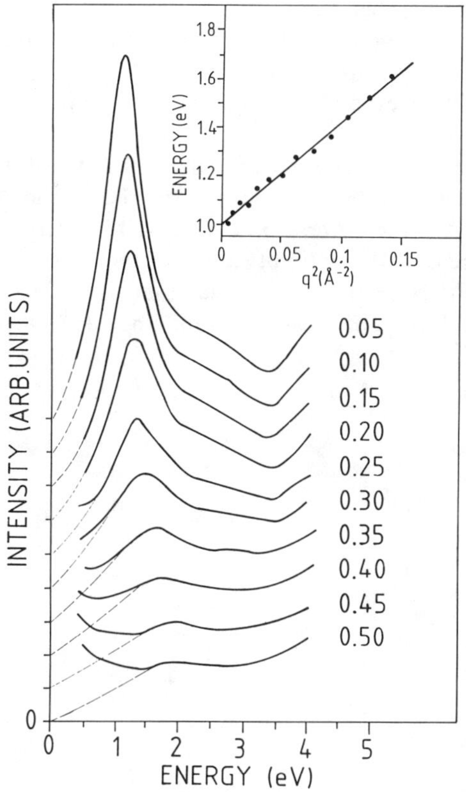

Fig. 7.20. Momentum dependence of the charge carrier plasmon in $Bi_2Sr_2CaCu_2O_8$. The momentum transfer $\boldsymbol{q}$ was chosen parallel to the CuO_2 planes. Inset: Dispersion of the charge carrier plasmon as a function of the squared momentum transfer

carriers at low energy leading to a considerable deviation from LDA band-structure calculations.

Since core-level absorption edges probe unoccupied states, those measurements are particularly well suited to probe the electronic structure of hole states in p-type doped high-T_c superconductors. In Fig. 7.21 we show O 1s absorption edges of $La_{2-x}Sr_xCuO_{4+\delta}$ [7.103–106]. These edges probe the local unoccupied density of states with p symmetry at the O sites provided the interaction of the excited electron with the core hole is small. The latter assumption is supported, at least for metallic cuprates, by the strong similarity of O 1s absorption edges and resonant BIS spectra on $Bi_2Sr_2CaCu_2O_8$ [7.107]. In the O 1s edge of the sample with $x = 0$ and $\delta = 0$, there is a peak **C** at 530.2 eV. This peak is assigned to transitions into unoccupied O 2p states admixed to the conduction band (upper Hubbard band). The wave functions of this band can be written in the form $|\Psi\rangle = \alpha|d^9\rangle + \beta|d^{10}\underline{L}\rangle$ where $|\alpha|^2$ determines the number of 3d holes and $|\beta|^2$ the admixture of empty O 2p states. $\underline{L}$ denotes holes on ligands. The intensity of peak **C** is therefore a measure of covalency of the CuO bond in the CuO_2 layers. For Sr-doped samples ($x > 0$) or samples doped by excess O ($\delta > 0$) a pre-peak V appears at $\sim$528.8 eV, the intensity of which is roughly proportional to the dopant concentration. This feature gives the most direct evidence that unoccupied O 2p states, i.e. holes on O sites, are formed upon p-type doping. From the threshold energy of peak $V(\sim 1.5\,\text{eV}$ below peak **C**) it was concluded that the Fermi level (corresponding to the dashed line in Fig. 7.21) is close to the top of the valence band. Therefore, a picture emerges for the electronic structure of the p-type doped system where there is not just a formation of hole states in the valence band because the spectra clearly show the formation of states in the gap. Furthermore, no indication for narrow "impurity states" is observed, and there is no discontinuity at the metal–insulator transition which occurs at $x = 0.06$. Therefore, it was concluded that for $x \leqq 0.06$ there are localized states at the Fermi level. The localization may be due to the

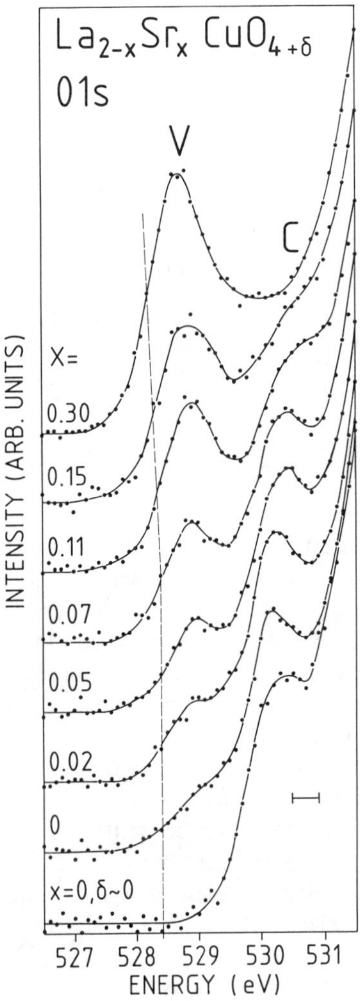

Fig. 7.21. O 1s absorption edges (pre-peak structures) of $La_{2-x}Sr_xCuO_{4+\delta}$ for $0 \leqq x \leqq 0.3$. For $x = 0$, spectra for $\delta \sim 0$ and $\delta \sim 0.015$ are shown

potential of disordered Sr ions. At higher concentrations the metal–insulator transition is driven by a delocalization of these states. It is remarkable that in the superconducting range ($0.11 < x < 0.3$) the peak **C** is still realized indicating that correlation effects are still important. At the highest dopant concentration ($x = 0.3$) peak **C** has almost disappeared. This may indicate that in the high-doping range an electronic structure is formed in which correlation effects are screened and a normal metallic state has developed. The superconducting range is then at the borderline between the highly correlated antiferromagnetic state and the normal metallic state.

The formation of pre-peak structures in O 1*s* absorption edges, indicating the formation of holes on O sites upon *p*-type doping, has also been observed in other systems such as $YBa_2Cu_3O_{7-y}$ [7.104, 108, 109] and $Bi_2Sr_2Ca_{1-x}Y_xCu_2O_8$ [7.110–112]. The observation of a pre-peak in the O 1*s* absorption edge in *n*-type doped systems (e.g. $Nd_{2-x}Ce_xCuO_{4-\delta}$) at first led to confusion about the existence of hole carriers in *n*-type doped high-T_c superconductors [7.113]. In view of the results of the system $La_{2-x}Sr_xCuO_{4+\delta}$ discussed above and systematic measurements of O 1*s* absorption edges in other insulating cuprates, it is now clear [7.114] that the pre-peaks observed in *n*-type doped cuprates and their insulating parent compounds are due to excitations into the O 2*p* states admixed to the conduction band (upper Hubbard band). Compared to

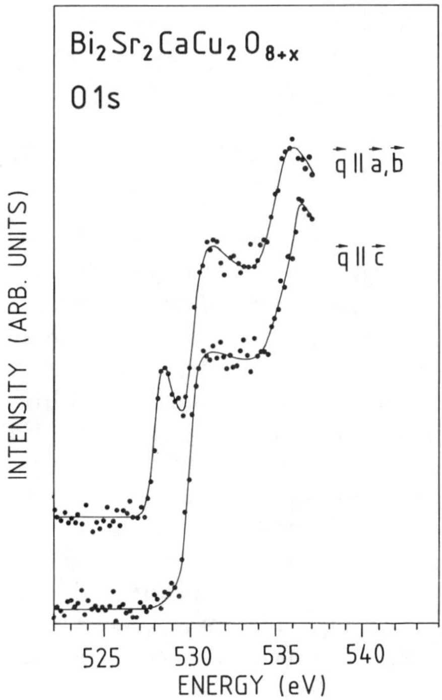

Fig. 7.22. Orientation-dependent O 1*s* absorption edges of a $Bi_2Sr_2CaCu_2O_8$ single crystal

the undoped system there is only a very small enhancement of the pre-peak upon n-type doping and the threshold is almost at the same energy. Since probably all metal states close to the Fermi level are hybridized to O $2p$ states, the O $1s$ absorption edges indicate that upon n-type doping, the Fermi level appears close to the bottom of the conduction band. Further information on the character of charge carriers in n-type doped systems is obtained from Cu $2p$ absorption edges. Although there are considerable excitonic effects in these excitations [7.115, 116] the intensity of the main line is a measure of unoccupied Cu $3d$ states. Upon a 15% n-type doping there is a reduction of empty Cu $3d$ states by ~15%. This clearly indicates that the charge carriers in the n-type doped superconductors have predominantly Cu $3d$ character.

Orientation-dependent measurements of core-level excitations in single crystals provide important information on the symmetry of hole states in high-T_c superconductors [7.117–119]. This is illustrated in Fig. 7.22 where we show O $1s$ absorption edges of a $Bi_2Sr_2CaCu_2O_8$ single crystal for momentum transfer $\boldsymbol{q}$ parallel and perpendicular to the CuO_2 layers in the $\boldsymbol{a}, \boldsymbol{b}$ plane. As outlined in Sect. 7.1.3 different unoccupied orbitals can be reached, e.g., in the O $1s$ spectra, for $\boldsymbol{q} \parallel \boldsymbol{a}, \boldsymbol{b}$ and for $\boldsymbol{q} \parallel \boldsymbol{c}$, O $2p_{x,y}$ and O $2p_z$ orbitals, respectively, are reached. From the spectra shown in Fig. 7.22 it is immediately clear that there are only O $2p_{x,y}$ states at the Fermi level and no O $2p_z$ states. There was considerable discussion on the symmetry of O holes during the last four years. The O $1s$ absorption edges shown in Fig. 7.22 clearly rule out all models for high-T_c superconductivity based on out-of-plane π holes in CuO_2 planes [7.120] and $2p_z$ holes on the apex O atoms [7.121, 122].

7.4 Conclusions

EELS in transmission in the low-energy range ($E \leqq 20$ eV) is in competition with optical spectroscopy and at higher energy ($20 \leqq E \leqq 2500$ eV) with X-ray absorption spectroscopy using synchrotron radiation. Optical spectroscopy has the great advantage of offering much higher energy resolution than EELS. In XAS, energy resolution has only quite recently been achieved in the energy range 250–800 eV, which are superior to that normally available in EELS. However, scattering methods such as EELS or inelastic X-ray scattering offer an additional degree of freedom—momentum transfer—by which not only the frequency but also the wave length of excitations can be studied. Furthermore, the ease with which one can cover the excitation spectrum from the infrared to soft X-ray is still a tremendous advantage even when compared to investigations with powerful synchrotron radiation sources. Therefore, EELS in transmission will also be in the future a powerful tool for investigations of the electronic structure of solids. In particular, new fields can be opened by improving energy and momentum transfer resolution.

Acknowledgements. The author is grateful to M. Alexander, G. Crecelius, A. vom Felde, H. Fritzsche, R. Hott, R. Manzke, S. Nakai, N. Nücker, H. Romberg, B. Scheerer and J. Sprösser-Prou, who have contributed significantly to the work presented in this review. Fruitful collaboration and stimulating discussions with D. Baeriswyl, R. von Baltz, J.C. Fuggle, V. Gorobchenko, G. Leising, H. Rietschel, G.A. Sawatzky, W. Schmatz and W. Weber is acknowledged.

References

7.1 E. Rudberg: K. Svenska Vet. Akad. Handl. **7**, 1 (1929)
7.2 G. Ruthemann: Naturwissenschaften **29**, 648 (1941)
7.3 D. Pines, D. Bohm: Phys. Rev. **85**, 338 (1952)
7.4 G. Ruthemann: Naturwissenschaften **30**, 145 (1942)
7.5 L. Marton, L.B. Lederer, H. Mendlowitz: Adv. Electr. Electron Phys. **7**, 183 (1955)
7.6 H. Raether: In *Springer Tracts in Modern Physics*, Vol. 38 (Springer, Berlin, Heidelberg 1965) p. 85
7.7 H. Raether: In *Physics of Thin Films*, Vol. 9 (Academic, New York 1977) p. 145
7.8 H. Raether: *In Springer Tracts on Modern Physics*, Vol. 88 (Springer, Berlin, Heidelberg 1980) p. 1
7.9 J. Daniels, C. von Festenberg, H. Raether, K. Zeppenfeld: In *Springer Tracts in Modern Physics*, Vol. 54 (Springer, Berlin, Heidelberg 1970) p. 77
7.10 J. Geiger: *Elektronen und Festkörper* (Vieweg, Braunschweig 1968)
7.11 S.E. Schnatterly: Solid State Phys. **34**, 275 (1979)
7.12 J. Fink: Adv. Electr. Electron Phys. **75**, 121 (1989)
7.13 C. Colliex: In *Advances in Optical and Electron Microscopy*, Vol. 9 (Academic, London 1984) p. 65
7.14 C. Colliex, C. Mory: In *Quantitative Electron Microscopy* (SUSSP Publications, Edinburgh 1984) p. 149
7.15 C. Colliex, T. Manoubi, M. Gasguier, L.M. Brown: In *Scanning Electron Microscopy*, Part 2 (SEM Inc., A.M.F. O'Hare, Illinois 1985) p. 489
7.16 R.F. Egerton: *Electron Energy-Loss Spectroscopy in the Electron Microscope* (Plenum, New York 1986)
7.17 P. Schattschneider: Ultramicroscopy **28**, 1 (1989); *Fundamentals of Inelastic Electron Scattering* (Springer, Vienna 1986)
7.18 L. van Hove: Phys. Rev. **95**, 249 (1954)
7.19 R. Kubo: J. Phys. Soc. Jpn. **12**, 570 (1987)
7.20 P.M. Platzman, P.A. Wolff: Solid State Phys. Suppl. **13**, 1 (1973)
7.21 K. Sturm: Adv. Phys. **31**, 1 (1982)
7.22 F. Wooten: *Optical Properties of Solids* (Academic, New York 1972)
7.23 D. Pines: *Elementary Excitations in Solids* (Benjamin, New York 1963)
7.24 S. Ichimaru: Rev. Mod. Phys. **54**, 1017 (1982)
7.25 V.D. Gorobchenko, V.N. Kohn, E.G. Maksimov: In *The Dielectric Function of Condensed systems* ed. by L.V. Keldysh, D.A. Kirzhnitz, A.A. Maradudin (Elsevier, New York 1990) p. 87
7.26 D.M. Ceperley, B.J. Alder: Phys. Rev. Lett. **45**, 566 (1980)
7.27 H. Ehrenreich, M.H. Cohen: Phys. Rev. **115**, 786 (1959)
7.28 E.-Ni Foo, J.J. Hopfield: Phys. Rev. **173**, 635 (1968)
7.29 A. Kotani, Y. Toyozawa: J. Phys. Soc. Jpn. **37**, 912 (1974)
7.30 P. Schattschneider: Phil. Mag. B **47**, 555 (1983)
7.31 P. Livins, T. Aton, S.E. Schnatterly: Phys. Rev. B **38**, 5511 (1988)
7.32 R. Hott: Diplomarbeit, Technical University of Karlsruhe (1988)
7.33 P.E. Batson, J. Silcox: Phys. Rev. B **27**, 5224 (1983)
7.34 P. Schattschneider, F. Födermayer, D.S. Su: Phys. Rev. Lett. **59**, 724 (1987)
7.35 A. vom Felde, J. Fink, Th. Bueche, B. Scheerer, N. Nücker: Europhys. Lett. **4**, 1037 (1987)
7.36 A. vom Felde, J. Sprösser-Prou, J. Fink: Phys. Rev. B **40**, 10181 (1989)

7.37 P.C. Gibbons, S.E. Schnatterly, J.J. Ritsko, J.R. Fields: Phys. Rev. B **13**, 2451 (1976)
7.38 N.D. Mermin: Phys. Rev. B **1**, 2362 (1970)
7.39 P. Vashishta, K.S. Singwi: Phys. Rev. B **6**, 875 (1972)
7.40 B. Dabrowski: Phys. Rev. B **34**, 4989 (1986)
7.41 K. Sturm, L.E. Oliveira: Phys. Rev. B **40**, 3672 (1989)
7.42 E. Jensen, E.W. Plummer: Phys. Rev. Lett. **55**, 1912 (1985)
7.43 In-Whan Lyo, E.W. Plummer: Phys. Rev. Lett. **60**, 1558 (1988)
7.44 P.H. Citrin, G.K. Wertheim, T. Hashizume, F. Sette, A.A. MacDowell, F. Comin: Phys. Rev. Lett. **61**, 1021 (1988)
7.45 N.V. Smith: Phys. Rev. B **183**, 634 (1969)
7.46 W.Y. Ching, J. Callaway: Phys. Rev. B **11**, 1324 (1975)
7.47 J.E. Northrup, M.S. Hybertsen, S.G. Louie: Phys. Rev. Lett. **59**, 819 (1987)
7.48 K.W.-K. Shung, B.E. Sernelius, G.D. Mahan: Phys. Rev. B **36**, 4499 (1987)
7.49 P. Steiner, H. Höchst, S. Hüfner: In *Photoemission in Solids*, ed. by L. Ley, M. Cardona, Topics Appl. Phys., Vol. 27 (Springer, Berlin, Heidelberg 1979) p. 349
7.50 A. Nakano, S. Ichimaru: Phys. Rev. B **39**, 4930 (1989)
7.51 Th. Büche, H. Rietschel: Phys. Rev. B **41**, 8691 (1990)
7.52 H. Naarmann: In *Electronic Properties of Conjugated Polymers* (*Kirchberg II*) ed. by H. Kuzmany, M. Mehring, S. Roth, Springer Ser. Solid-State Sci., Vol. 76 (Springer, Berlin, Heidelberg 1987) p. 12
7.53 W.P. Su, J.R. Schrieffer, A.J. Heeger: Phys. Rev. B **22**, 2099 (1980)
7.54 M.J. Rice: Phys. Lett. **71**, 152 (1979)
7.55 S.A. Brazovskii: JETP Lett. **28**, 606 (1978)
7.56 S.A. Brazovskii: Sov. Phys. JETP **51**, 342 (1980)
7.57 P. Horsch: Phys. Rev. B **24**, 7351 (1981)
7.58 D. Baeriswyl, K. Maki: Phys. Rev. B **31**, 6633 (1985)
7.59 D. Baeriswyl: In *Electronic Properties of Polymers and Related Compounds, Kirchberg II*, ed. by H. Kuzmany, M. Mehring, S. Roth, Springer Ser. Solid-State Sci. Vol. 76 (Springer, Berlin, Heidelberg 1987) p. 198
7.60 S. Kivelson, W.-P. Su, J.R. Schrieffer, A.J. Heeger: Phys. Rev. Lett. **58**, 1899 (1987); **60**, 72 (1988); D. Baeriswyl, P. Horsch, K. Maki: ibid. **60**, 70 (1988); J.T. Gammel, D.K. Campbell: ibid. **60**, 71 (1988)
7.61 J.R. Schrieffer: In *Proceedings of the International School of Physics Enrico Fermi Course LXXXIX*, ed. by B. Bassani, F. Fumi, M.T. Tosi (North-Holland, Amsterdam 1985) p. 300
7.62 D. Baeriswyl: In *Theoretical Aspects of Band Structures and Electronic Properties of Pseudo-One-Dimensional Solids*, ed. by H. Kamimura (Reidel, Dordrecht 1985) p. 1
7.63 S. Etemad, A.J. Heeger, A.G. MacDiarmid: Ann. Rev. Phys. Chem. **33**, 443 (1982)
7.64 H.W. Streitwolf: Phys. Status Solidi B **127**, 11 (1985)
7.65 S. Roth, H. Bleier: Adv. Phys. **36**, 385 (1987)
7.66 A.J. Heeger, S. Kivelson, J.R. Schrieffer, W.-P. Su: Rev. Mod. Phys. **60**, 781 (1988)
7.67 Yu Lu: *Solitons and Polarons in Conducting Polymers* (World Scientific, Singapore 1988)
7.68 E.J. Mele, M.J. Rice: Phys. Rev. B **23**, 5397 (1981)
7.69 S. Kivelson, A.J. Heeger: Phys. Rev. Lett. **55**, 308 (1985)
7.70 J. Fink, G. Leising: Phys. Rev. B **34**, 5320 (1986)
7.71 J. Fink: Synth. Metals **21**, 87 (1987)
7.72 H. Fritzsche, N. Nücker, B. Scheerer, J. Fink, G. Leising: Synth. Metals **28**, D 237 (1989)
7.73 D. Baeriswyl, G. Harbeke, H. Kiess, E. Maier, W. Meyer: Physica B **117**, 617 (1983)
7.74 C.-S. Neumann, R. von Balltz: Phys. Rev. B **35**, 9708 (1987)
7.75 P.M. Grant, I.P. Batra: Synth. Metals **1**, 193 (1979/800)
7.76 U. Büchner: Phys. Stat. Sol. (b) **81**, 227 (1977)
7.77 J. Fink: Synth. Metals **21**, 87 (1987)
7.78 S. Etemad, T. Mitani, O. Ozaki, T.C. Chung, A.J. Heeger, A.G. MacDiarmid: Solid State Commun. **40**, 75 (1981)
7.79 G.B. Blanchet, C.R. Fincher, A.J. Heeger: Phys. Rev. Lett. **51**, 2132 (1983)
7.80 P.F. Williams, A.N. Bloch: Phys. Rev. B **10**, 1097 (1974)
7.81 J.P. Albert, C. Jouanin: Mol. Cryst. and Liq. Cryst. **117**, 283 (1985)
7.82 B. Horovitz: Solid State Commun. **41**, 593 (1982)
7.83 S. Jeyadev, E.M. Conwell: Phys. Rev. B **33**, 2530 (1986)
7.84 H. Puff, H.W. Streitwolf: Phys. Status Sol. B **143**, 765 (1987); and private communication

7.85 X.Q. Yang, D.B. Tanner, M.J. Rice, H.W. Gibson, A. Feldblum, A.J. Epstein: Solid State Commun. **61**, 335 (1987)
7.86 K. Zeppenfeld: Z. Phys. **243**, 229 (1971)
7.87 H. Venghaus: Phys. Stat. Sol. (b) **66**, 145 (1974)
7.88 J. Fink, G. Crecelius, R. Manzke: unpublished results
7.89 J.J. Ritsko, M.J. Rice: Phys. Rev. Lett. **42**, 666 (1979)
7.90 J.J. Ritsko, E.J. Mele: Synth. Met. **3**, 73 (1981)
7.91 E.J. Mele, J.J. Ritsko: Synth. Met. **3**, 89 (1981)
7.92 E.J. Mele, J.J. Ritsko: Phys. Rev. Lett. **43**, 68 (1979)
7.93 J.J. Ritsko, E.J. Mele: Physica B **99**, 425 (1980)
7.94 J.G. Bednorz, K.A. Müller: Z. Phys. B **64**, 189 (1986)
7.95 see e.g. *Narrow Band Phenomena*, ed. by J.C. Fuggle, G.A. Sawatzky, J.W. Allen (Plenum, New York 1988)
7.96 L.F. Mattheiss: Phys. Rev. Lett. **58**, 1028 (1987)
7.97 A. Fujimori, F. Minami: Phys. Rev. B **30**, 957 (1984)
7.98 J. Zaanen, G.A. Sawatzky, J.W. Allen: Phys. Rev. Lett. **55**, 418 (1985)
7.99 S. Hüfner: Z. Phys. B **61**, 135 (1985)
7.100 N. Nücker, H. Romberg, S. Nakai, B. Scheerer, J. Fink, Y.F. Yan, Z.X. Zhao: Phys. Rev. B **39**, 12379 (1989)
7.101 A.L. Fetter: Ann. Phys. **81**, 367 (1973)
7.102 L.F. Mattheiss, D.R. Hamann: Phys. Rev. B **38**, 5012 (1988); L.F. Mattheiss, private communication
7.103 N. Nücker, J. Fink, B. Renker, D. Ewert, C. Politis, P.J.W. Weijs, J.C. Fuggle: Z. Phys. B **67**, 9 (1987)
7.104 N. Nücker, J. Fink, J.C. Fuggle, P.J. Durham, W.M. Temmerman: Phys. Rev. B **37**, 5158 (1988)
7.105 H. Romberg, M. Alexander, N. Nücker, P. Adelmann, J. Fink: In *Electronic Properties of High-T_c Superconductors (Kirchberg IV)* ed. by H. Kuzmany, M. Mehring, J. Fink, Springer Ser. Solid-State Sci. Vol. 99 (Springer, Berlin, Heidelberg 1991)
7.106 H. Romberg, M. Alexander, N. Nücker, P. Adelmann, J. Fink: Phys. Rev. B **42**, 8768 (1990)
7.107 W. Drube, F.J. Himpsel, G.V. Chandrashekar, M.W. Shafer: Phys. Rev. B **39**, 7328 (1989)
7.108 J.A. Yarmoff, D.R. Clarke, W. Drube, U.O. Karlsson, A. Taleb-Ibrahimi, F.J. Himpsel: Phys. Rev. B **36**, 3967 (1987)
7.109 P. Kuiper, G. Kruizinga, J. Ghijsen, M. Grioni, P.J.W. Weijs, F.M.F. deGroot, G.A. Sawatzky, H. Verweij, L.F. Feiner, H. Petersen: Phys. Rev. B **38**, 6483 (1988)
7.110 H. Matsuyama, T. Takahashi, H. Katayama-Yoshida, T. Kashiwakura, Y. Okabe, S. Sato, N. Kosugi, A. Yagishita, K. Tanaka, H. Fujimoto, H. Inokuchi: Physica C **160**, 567 (1989)
7.111 J. Fink, N. Nücker, H. Romberg, M. Alexander, S. Nakai, B. Scheerer, P. Adelmann, D. Ewert: Phys. C **162–164**, 1415 (1989)
7.112 N. Nücker, H. Romberg, M. Alexander, J. Fink: In *Studies of High Temperature Superconductors*, Vol. 6, ed. by A. Narlikar (Nova, Commack, 1991)
7.113 N. Nücker, P. Adelmann, M. Alexander, H. Romberg, S. Nakai, J. Fink, H. Rietschel, G. Roth, H. Schmidt, H. Spille: Z. Phys. B **75**, 421 (1989)
7.114 M. Alexander, H. Romberg, N. Nücker, P. Adelmann, J. Fink, J.T. Markert, B. Maple, S. Uchida, H. Takagi, Y. Tokura, A.C.W.P. James, D.W. Murphy: Phys. Rev. B **43**, 333 (1991)
7.115 J. Fink, Th. Müller-Heinzerling, B. Scheerer, W. Speier, F.U. Hillebrecht, J.C. Fuggle, J. Zaanen, G.A. Sawatzky: Phys. Rev. B **32**, 4899 (1985)
7.116 J. Zaanen, G.A. Sawatzky, J. Fink, W. Speier, J.C. Fuggle: Phys. Rev. B **32**, 4905 (1985)
7.117 N. Nücker, H. Romberg, X.X. Xi, J. Fink, B. Gegenheimer, Z.X. Zhao: Phys. Rev. B **39**, 6619 (1989)
7.118 F.J. Himpsel, G.V. Chandrashekhar, A.B. McLean, M.W. Shafer: Phys. Rev. B **38**, 11946 (1988)
7.119 P. Kuiper, M. Grioni, G.A. Sawatzky, D.B. Mitzi, A. Kapitulnik, A. Santaniello, P. de Padova, P. Thiry: Physica C **157**, 260 (1989)
7.120 K.H. Johnson, M.E. McHenry, C. Counterman, A. Collins, M.M. Donovan, R.C. O'Handley, G. Kalonji: Physica C **153–155**, 1165 (1988)
7.121 H. Kamimura: Jpn. J. Appl. Phys. **26**, 6627 (1987)
7.122 A. Fujimori: Phys. Rev. B **39**, 793 (1989)

8. Inverse Photoemission and How it is Used

P.T. Andrews, I.R. Collins and J.E. Inglesfield

With 21 Figures

Inverse photoemission is widely used to study the wave vector and energy dependence of unoccupied electronic states at the surface and in the bulk of metals and semiconductors [8.1–3], compounds [8.4] and even polymers [8.5]. The importance of measuring the unoccupied states in providing complementary information to photoemission is shown by the inverse photoemission experiments on the high-T_c superconductors [8.6]. Like angle-resolved photoemission (ARPES), $\boldsymbol{k}$-resolved inverse photoemission (or KRIPES) is surface sensitive [8.7], and an important application is the study of the unoccupied surface states [8.8], particularly the image states [8.9, 10]. These are hydrogenic states in which the electron is trapped outside the surface by the image potential, and are interesting because of the information they give about the form of potential outside the surface [8.2, 11]. Electronic states induced by adsorbate atoms are described in Chap. 9, and inverse photoemission has helped to clarify the bonding of molecules such as CO to surfaces [8.12]. In this chapter we concentrate on the general principles of inverse photoemission, illustrating these with applications to graphite, a classic two-dimensional system, alkali metals which are nearly free-electron like, W in which d-electrons dominate the physical properties, and finally the image states.

The use of bremsstrahlung isochromat spectroscopy with low energy photons was largely initiated by the experimental work of *Dose* [8.13, 14], and the theoretical work of *Pendry* [8.7, 15] who drew attention to the close analogies with ultraviolet photoemission spectroscopy—hence inverse photoemission. The predicted count rates were low (Sect. 8.1.1), and at first there was some concern among experimentalists that other processes yielding low-energy photons would swamp the inverse photoemission signal. All such processes have cross sections which are comparable to that of inverse photoemission—the most serious source of background is from bremsstrahlung transitions following inelastic scattering of the incoming electrons [8.16]. Nevertheless, many research groups subsequently built instruments for KRIPES, and the method became established as an invaluable tool for mapping unoccupied states in the bulk and at the surface.

As discussed in Chap. 1, KRIPES spectra are recorded in one of two ways, or possibly a combination of these. In isochromat mode the yield of photons of constant energy is measured as a function of incident electron energy, and in fluorescence mode the surface is bombarded with electrons at fixed energy and the resulting photon spectrum is measured. As *Pendry* pointed out [8.15], KRIPES is time-reversed ARPES and the considerable experience of analysing

ARPES spectra can be brought to bear on KRIPES analysis, for determining the energy and wave vector of the states involved in the transition and for identifying which transitions involve bulk and which involve surface states. The isochromat mode is the time-reversed equivalent of the usual way of recording ARPES spectra by measuring an electron spectrum with fixed photon energy, and the fluorescence method corresponds to constant final state spectroscopy in photoemission.

The aim of inverse photoemission—in fact, the theme of this book—is to characterise the final state, initially unoccupied, into which the incident electron can drop in the transition. KRIPES is particularly important because the final state can be characterised quite completely, both its energy and lifetime and its $\boldsymbol{k}$ dispersion. Measurements in which a spin-polarized electron beam is used enable the spin dependence of the final state to be determined [8.17]. The energy of this final state, measured relative to the Fermi energy, can be found because for a particular experimental set-up the potential difference between the electron gun and the Fermi energy of the specimen is fixed. So the Fermi edge occurs at the same incident electron energy, for a given photon energy, for all specimens, and this energy can be found by using specimens such as polycrystalline gold with a readily identified Fermi edge. The component of the Bloch wave vector $\boldsymbol{k}$ parallel to the surface, $\boldsymbol{K}$, is conserved in the transition, because the crystal potential parallel to the surface is periodic (Sect. 8.1.2), and this can be found from the energy and angle of the incident electrons. This enables us to characterise completely the electronic states in essentially two-dimensional systems, such as layer compounds (Sect. 8.2) or surface states localised at the surface (Sects. 8.4.2, 8.5). In fact, inverse photoemission has provided real insight into the various categories of surface state (Shockley, Tamm and image states [8.2]). As with ARPES, there is a serious problem in identifying the wave vector component perpendicular to the surface, k_z, for a full characterisation of bulk states. The problem is that k_z in the final state is different from k_z of the initial-state electrons outside the solid—which is what we know. There are several ways round this; for example, it is often assumed that the incident electron is in a free-electron state inside the solid (Sect. 8.1.2).

The analogies between KRIPES and ARPES mean that the computational methods which have been developed for calculating photoemission spectra [8.18] can be applied without modification to inverse photoemission [8.15]. By comparison of the calculated spectra with experiment it is possible to identify transitions as bulk direct transitions, or surface transitions, and in cases where the band structure is complicated, the computations can indicate which transitions are likely to be intense and which ones of neglible intensity. This helps with the problem of band mapping, though in cases of high symmetry some transitions can be eliminated simply on symmetry grounds (Chap. 9). In Sect. 8.1.3 we shall describe the computational methods, and throughout the chapter we shall indicate their application. These methods are still undergoing development largely as a result of increasing computer power [8.19], and the analysis of KRIPES spectra is a particularly good test of any program because it is dealing with states at higher energy than the occupied states which are involved in ARPES (Chap. 2). Of course the computer programs are based on band-structure ideas, in which the wavefunctions of the electron in its initial and final state are calculated as accurately as possible, and in the end it is the differences between the theoretical and observed bands which matter—the self-energy.

8.1 The Physics of Transitions in Inverse Photoemission

8.1.1 Matrix Elements and the Golden Rule

For practical and computational purposes, inverse photoemission and photoemission really are inverse processes [8.7, 15]. By time-reversal symmetry, the probability of an electron incident on a solid in state $|j\rangle$ undergoing a transition to an unoccupied state $|i\rangle$ and emitting a particular photon is just the same as the reverse photoemission process, in which this photon excites the electron from $|i\rangle$ to $|j\rangle$ (complex-conjugated). Of course we measure currents rather than individual states, and this leads to a phase-space factor between the expressions for the photon flux in inverse photoemission and the electron flux in photoemission.

Let us start with the golden rule expression for photoemission [8.20, 21]. The photocurrent of states reaching the detector with wave vector $\boldsymbol{k}$, per unit solid angle and per unit energy range is[1]

$$J_{\mathrm{el}}(\boldsymbol{k}) = \frac{k}{4\pi^2} \sum_i |\langle \psi_j | \delta H | \psi_i \rangle|^2 \delta(E_j - E_i - \omega), \tag{8.1}$$

where the sum is over initial states i, and δH is the perturbation due to the photon field

$$\delta H = \frac{1}{2c} (\boldsymbol{A} \cdot \boldsymbol{p} + \boldsymbol{p} \cdot \boldsymbol{A}). \tag{8.2}$$

$\boldsymbol{p}$ is the momentum operator and $\boldsymbol{A}$ is the vector potential of the light, with frequency ω; the delta-function in (8.1) ensures conservation of energy, so that the energy E_j of the photoelectrons equals the initial state energy E_i plus the photon energy. For our purposes, it is the prefector of (8.1) which is most interesting: the energy density of final states is given by $k/8\pi^3$ per unit solid angle, which when multiplied by the standard factor of 2π in the golden rule gives us (8.1). Now the intensity of the photon field perturbing the electrons in (8.1), A^2, is proportional to ω/c times the photon flux. So the photocurrent per

[1] We use atomic units, with $e = \hbar = m = 1$. The atomic unit of energy is the Hartree (1 Hartree = 27.2 eV)

unit photon flux is given by

$$J_{\mathrm{el}}(\boldsymbol{k}) = \frac{k\omega}{4\pi^2 c} \sum_i |\langle \psi_j | \delta h | \psi_i \rangle|^2 \delta(E_j - E_i - \omega), \tag{8.3}$$

where δh is the normalized perturbation.

In the case of inverse photoemission, the energy density of photon states is $\omega^2/8\pi^3 c^3$ per unit solid angle. This gives a prefactor of $\omega^2/4\pi^2 c^3$ in the golden rule expression, and we get another factor of ω from the intensity of the electromagnetic field in the perturbation. So the photon flux reaching the detector (with wave vector $\boldsymbol{q}$) is given by

$$J_{\mathrm{ph}}(\boldsymbol{q}) = \frac{\omega^3}{4\pi^2 c^3} \sum_i |\langle \psi_j | \delta h | \psi_i \rangle|^2 \delta(E_j - E_i - \omega) \tag{8.4}$$

per unit solid angle and energy range. The incident electron flux is proportional to k, so the photon flux per unit electron flux is

$$J_{\mathrm{ph}}(\boldsymbol{q}) = \frac{\omega^3}{4\pi^2 k c^3} \sum_i |\langle \psi_j | \delta h | \psi_i \rangle|^2 \delta(E_j - E_i - \omega). \tag{8.5}$$

Hence the photon flux per unit electron flux in inverse photoemission, and the photoelectron flux per unit photon flux in photoemission, are related by [8.7]

$$J_{\mathrm{ph}} = \frac{\omega^2}{2E_j c^2} J_{\mathrm{el}}, \tag{8.6}$$

or equivalently

$$J_{\mathrm{ph}} = \left(\frac{q}{k}\right)^2 J_{\mathrm{el}}. \tag{8.7}$$

This factor is very small, typically 10^{-5} in a UV experiment [8.2], and this is the reason that inverse photoemission is the more difficult experiment. It also results in an inherently poorer resolution in KRIPES than in ARPES, because to obtain a useful count rate a large incident current is needed, in which the electron–electron interaction broadens the electron energy distribution.

These expressions for photoemission and inverse photoemission are sometimes described as one-step, because the matrix element contains the whole of the process. This is in comparison with the three-step picture of, let us say, photoemission in which the process is notionally separated into the absorption of a photon, the propagation of the photoelectron through the crystal, and its passage through the surface to the detector [8.22]. The application of the three-step model to inverse photoemission [8.23] is discussed in Chap. 9. It is, in fact, not strictly valid to separate the process in this way, because the propagation of the electron through the crystal and its passage through the surface are linked together in its wavefunction. In Sect. 8.1.3 we shall see how the one-step expression can be calculated accurately in a one-electron framework.

8.1.2 Conservation Rules and Symmetry

As in photoemission, a great deal can be learnt from inverse photoemission spectra without detailed calculation, by applying conservation rules. The two quantities which are fully conserved are firstly energy

$$E_j - E_i = \omega, \tag{8.8}$$

which is built into the golden rule (8.5), and secondly the component of the electron wave vector parallel to the surface. If the surface is periodic, the wavefunctions ψ_i, ψ_j have two-dimensional Bloch periodicity, such that when they are displaced through a surface lattice vector $\boldsymbol{R}_l$ they are multiplied by phase factors $\exp(\mathrm{i}\boldsymbol{K}_i \cdot \boldsymbol{R}_l), \exp(\mathrm{i}\boldsymbol{K}_j \cdot \boldsymbol{R}_l)$ [8.24]. As the momentum of the light is very small, this means that the matrix element in (8.5) vanishes unless

$$\boldsymbol{K}_j = \boldsymbol{K}_i + \boldsymbol{G}, \tag{8.9}$$

where $\boldsymbol{G}$ is a surface reciprocal lattice vector. These rules by themselves enable two-dimensional bands to be mapped, either the bands of two-dimensional systems such as graphite (Sect. 8.2) or surface state bands (Sects. 8.4.2, 8.5).

Transitions from the incident electronic state to unoccupied *bulk* states also occur, partially conserving the perpendicular component of the wave vector. The incident electron is in a LEED state, corresponding to shooting the plane wave $\exp(\mathrm{i}\boldsymbol{k}\cdot\boldsymbol{r})$ at the surface, with the resulting reflected waves in the vacuum, and transmitted waves inside the solid. The transmitted waves are Bloch solutions travelling away from the surface with the same wave vector component parallel to the surface as the incident wave, possibly scattered through a surface reciprocal lattice vector, together with evanescent waves decaying into the crystal from the surface. We would then expect the matrix element in (8.5) to be large for transitions to bulk states $|i\rangle$ which have the same *perpendicular* component of wave vector as one of the travelling waves in the incident state and the same parallel component (8.9). This corresponds to a direct transition, and if the perpendicular component of the wave vector is known, the final-state bands can then be mapped out (Sects. 8.2, 8.4). Simplifying approximations are commonly made for the incident-electron state inside the solid: for instance it may be assumed to be in a free-electron state, accelerated by the inner potential as it crosses the surface into the bulk [8.1, 2]. While this approach has had some success in particular cases, the assumptions made are not correct in general.

The presence of the surface means that the perpendicular component of the wave vector is not strictly a good quantum number—which is why the wavefunctions, both initial and final state, have evanescent components near the surface. Moreover, the electrons in the initial incident state have a finite mean-free path, typically 10 Å [8.25], so the travelling wave components also decay into the solid. The finite mean-free path means that the incident electron, which has a definite energy, has a smeared-out wave vector component perpendicular to the surface [8.26]. The electron in the final state has a finite lifetime, corresponding to a smeared-out energy. Smearing out the wave vector in the initial incident state and the energy in the final state both contribute to fuzzy energy bands, and broaden the energy at which the direct transition occurs [8.27]. The evanescent initial state also increases the surface sensitivity of inverse photoemission, and the spectrum reflects the surface density of final states, the continua of bulk states at the surface and surface states localised at the surface, with fixed wave vector component $\boldsymbol{K}$ parallel to the surface (Sect. 8.4).

Symmetry considerations provide extra information about the final-state wavefunction [8.28]. The electromagnetic perturbation $\boldsymbol{A}\cdot\boldsymbol{p}+\boldsymbol{p}\cdot\boldsymbol{A}$ in (8.2), which couples the incident electron to the final state, has the symmetry transformation properties of a dipole aligned along the electric field vector ("dipole-allowed" transitions). Hence electrons incident normally, which are essentially uniform over the surface (the $x-y$ plane, say) are not coupled to final states made up of d_{xy} or $d_{x^2-y^2}$ orbitals in a four-fold symmetric environment for example (Sect. 8.4.1): for all directions of light polarization the matrix element vanishes. As another example, electrons incident in a mirror plane of the sample have a wavefunction which is even with respect to reflection in the mirror plane. This means that transitions to even final states are only possible if the polarization vector of the light also lies in the mirror plane; conversely transitions to odd final states only occur with a polarization vector perpendicular to the mirror plane. This consideration is widely used in photoemission to help with the identification of states [8.29], because synchrotron light sources are polarized, but unfortunately, in inverse photoemission the polarization of the emitted photons is not normally measured. Varying the angle of detection of the light changes the ratio of surface to perpendicular polarization, and this can sometimes be used to provide information about the orbitals in the final state [8.1, 30].

8.1.3 Calculating Inverse Photoemission

A good calculation of the inverse photoemission spectrum can help with the identification of features as either surface or bulk, and by "tuning" the potential felt by the electrons to obtain optimal agreement with experiment, information can be obtained about the self-energy—the energy shift and lifetime broadening of the energy bands caused by electron–electron interactions (Chap. 3) [8.26]. The necessary ingredients of this calculation are an accurate way of finding the electronic states both in the bulk and at the surface, a proper evaluation of the matrix elements, and a way of putting in the effects of finite mean-free path for the incident states and finite lifetime for the final states. This exists in a widely used computer package developed by *Pendry* and co-workers [8.18].

The starting point is to rewrite (8.5) in terms of the Green's function for the electrons [8.31], which is given by

$$G(\boldsymbol{r},\boldsymbol{r}';E)=\sum_i\frac{\psi_i^*(\boldsymbol{r})\psi_i(\boldsymbol{r}')}{E_i-E}. \tag{8.10}$$

(The Green's function is usually much easier to deal with than individual wavefunctions in infinite systems like solids, with continua of energy levels.) Then the sum over states i in (8.5) is given by

$$\sum_i \psi_i^*(\boldsymbol{r})\psi(\boldsymbol{r}')\delta(E-E_i)=\frac{1}{\pi}\,\mathrm{Im}\,\{G(\boldsymbol{r},\boldsymbol{r}';E+\mathrm{i}\varepsilon)\}, \tag{8.11}$$

so we can write (8.5) as

$$J_{\mathrm{ph}}\propto \mathrm{Im}\,\{\textstyle\int d^3\boldsymbol{r}\int d^3\boldsymbol{r}'\psi_j^*(\boldsymbol{r})\delta h G(\boldsymbol{r},\boldsymbol{r}';E_j-\omega+\mathrm{i}\varepsilon)\delta h\psi_j(\boldsymbol{r}')\}. \tag{8.12}$$

This Green's function expression has the advantage of making contact with many-body formalism. Equation (8.12) can still be used for inverse photoemission when electron–electron interactions are included (within well-defined approximations) [8.32], if G is replaced by the Green's function for the interacting electrons. This describes the propagation of the extra electron in the lower energy state, in the presence of the screening cloud due to all the other electrons. $\mathrm{Im}\,\{G\}$—the spectral function—has peaks corresponding to the quasiparticle energy bands with the form

$$\mathrm{Im}\,\{G(E)\}\propto\frac{\mathrm{Im}\,\{\Sigma\}}{(E-E_0(\boldsymbol{k})-\mathrm{Re}\,\{\Sigma\})^2+\mathrm{Im}\,\{\Sigma\}^2}. \tag{8.13}$$

Here $E_0(\boldsymbol{k})$ is the density-functional or Hartree–Fock band structure, and Σ is the self-energy whose real part gives an energy shift and whose imaginary part gives a lifetime broadening to the quasiparticle energy bands [8.26] (Chap. 3). The self-energy behaves like a complex, energy-dependent potential which can be added to V in the Schrödinger equation in the program. The incident electron, in state ψ_j in (8.12), also feels a self-energy, but because this electron is at a definite energy the effect of $\mathrm{Im}\,\{\Sigma\}$ is to introduce a mean-free path, equivalent to momentum broadening [8.26].

In evaluating (8.12) on the computer, several approximations have to be made. Firstly, it is assumed that the vector potential in δh is constant spatially, neglecting the screening of the photofield. This means that δh can be transformed to grad V form [8.18], where V is the potential felt by the electrons. Next it is assumed that V has muffin-tin form, a spherically symmetric atomic-like potential at each atomic site, and a flat potential in the interstitial region between the atomic muffin tins. The muffin-tin form of potential gives the electronic states very well for reasonably close-packed systems, but the atomic muffin tins need to be augmented with empty muffin tins, not containing an atomic nucleus in open structures, such as those of the diamond-type semiconductors. At the surface a simplified (one-dimensional) form of the surface barrier is used—usually a step barrier. This sometimes leads to errors in the description of the surface density of states, and in recent work a more realistic form of barrier potential taken from a self-consistent surface electronic structure calculation has been used [8.19].

With these simplifications, the way that the program works is as follows [8.18]. The incident electron state ψ_j, a LEED state, is calculated using a layer-scattering approach in which the solid is divided up into layers of atoms. ψ_j is expanded in plane waves between the layers, and the reflection and transmission properties of each layer are calculated, giving the probability amplitudes for a plane wave with parallel wave vector component $\boldsymbol{K}$ to be reflected and transmitted into plane waves $\boldsymbol{K}+\boldsymbol{G}$. By repeated reflection and transmission operations, the full wavefunction of the electron incident on the whole semi-infinite crystal can be determined. Knowing ψ_j, $\int d^3\boldsymbol{r}' G(\boldsymbol{r},\boldsymbol{r}';E_i)\delta h\psi_j(\boldsymbol{r}')$ can be found—in this expression, $\delta h\psi_j$ acts as a source of electrons for which G describes the propagation in the lower energy state (c.f. the three-step model). An immediate simplification is that δh (i.e. grad V) is only non-zero inside the muffin tins, and at the surface potential step, and it is then straightforward to use the layer-adapted multiple-scattering technique to find the whole wavefield of $\int d^3\boldsymbol{r}' G\delta h\psi$.

This approach is very successful in describing the inverse photoemission and photoemission spectra of many close-packed metals, especially for the fairly close-packed surfaces [8.33] (Sect. 8.4.1). Figure 8.1 shows inverse photoemission spectra from Ag(001) for different angles of electron incidence, calculated with the self-consistent surface barrier [8.19], compared with experiment [8.34]: agreement is remarkably good. Problems arise on more open surfaces because the lattice planes parallel to an open surface are very close together, and the

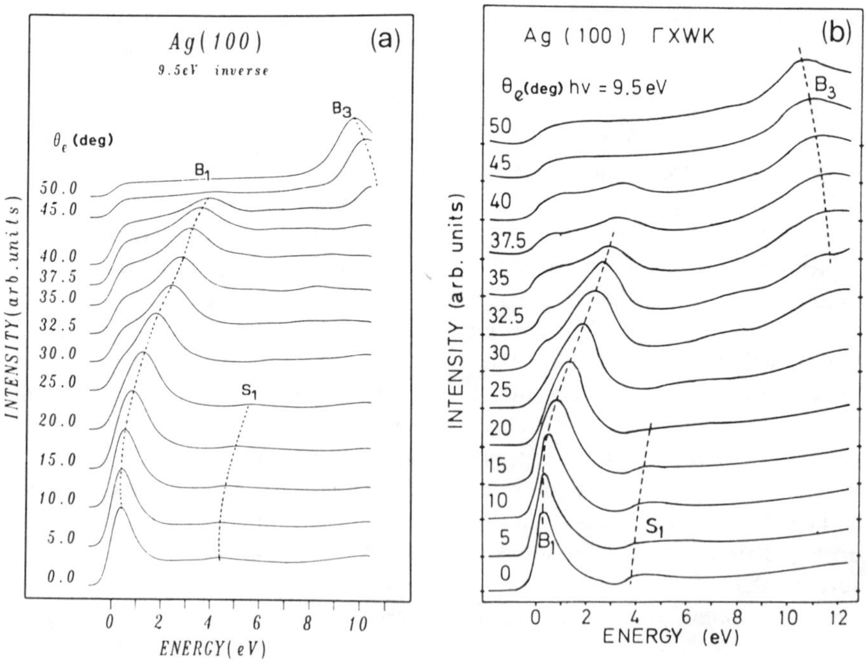

Fig. 8.1. (**a**) Calculated inverse photoemission spectra from Ag(001) at a photon energy of 9.5 eV, for varying angles of electron incidence [8.19], (**b**) experimental spectra [8.34]

expansions for the wavefunctions in the interstitial region between the planes may not converge. The simplification which is most difficult to improve on is the use of a constant photon field, with no refraction or screening and the same field inside as outside the crystal. The problem is that the surface dielectric response of metals and semiconductors is only known very approximately, because it depends in a complicated way on the wavefunctions [8.35]. This is likely to be an area of major interest in the future.

8.2 Two-Dimensional Systems: Graphite

Two-dimensional systems are easier to study than three-dimensional systems because the electronic states are characterised by a two-dimensional Bloch wave vector, and the conservation rules given in Sect. 8.1.2 enable the bands to be mapped out. This is particularly important at surfaces (Sect. 8.4.2, Chap. 9). To illustrate the principles, we shall concentrate here on graphite, which like other layer compounds consists of sheets of atoms within which the bonding is strong, separated from each other by relatively large distances with weak interlayer bonding.

In graphite, hexagonal layers of C atoms are stacked in an *ABAB*.... stacking sequence, with half the atoms lying directly above those in the adjacent plane, and half lying above the centres of the hexagons. The strong bonding within the layers comes mainly from sp^2 hybridized $2s, 2p_x$ and $2p_y$ atomic orbitals; these σ bands disperse with wave vector component $\boldsymbol{K}$ parallel to the layers, but because there is little overlap in the perpendicular direction they show hardly any k_z dispersion [8.36, 37]. The π states, resulting from the overlap between $2p_z$ orbitals within the layers, also contribute to the intralayer bonding. As the $2p_z$ orbitals stick out of the layers, the π states show some dispersion with k_z [8.36, 37]. The resulting band structure consists of occupied bonding and unoccupied antibonding σ states, and π states which cross the Fermi energy and provide the conductivity. But in addition, the stacking of the graphite layers introduces a new type of state, an *interlayer* state above the Fermi energy in which electrons move in a free-electron-like way between the layers [8.37, 38]. The interlayer band disperses strongly with k_z.

Here we describe experiments on good quality single crystals of graphite [8.39], motivated by the substantial differences between graphite band structures calculated using different methods [8.36, 37, 40]. Like other layer compounds, graphite cleaves easily parallel to the layers, and this is the only surface available for measurement. The electron energy corresponding to E_F was determined by measuring spectra from a clean polycrystalline gold specimen electrically attached to the graphite, and E_F is subsequently used as the zero of energy. To identify the intralayer and interlayer bands, normal incidence electrons with $\boldsymbol{K}=0$ were used and the spectra recorded at different photon energies. For a particular final-state energy, the energy and wave vector k_z of the incident electron vary with photon energy, so any final state peak which is independent of photon energy corresponds to a state which does not disperse with k_z. This is either an intralayer state or a surface state. On the other hand, a peak which varies in energy with photon energy must involve a direct transition to a state dispersing with k_z—an interlayer state.

The normal incidence spectra are shown in Fig. 8.2, for photon energies between 14 and 26 eV. They show a background rising from E_F to a strong peak at 9.5 eV, with other small peaks at lower and higher energies. The only peak which changes energy as the photon energy is altered moves down from about 8 eV at $\hbar\omega = 14$ eV to 5.5 eV in the $\hbar\omega = 20$ eV spectrum—this feature is in an energy region where the unoccupied *interlayer* band is expected (Fig. 8.3).

The peak at an energy of 3.9 eV, independent of $\hbar\omega$ (Figs. 8.2, 8.3), is an example of an image-potential induced surface state (Sect. 8.5) [8.41]. It can be identified as such for the following reasons: the fact that its energy is independent of $\hbar\omega$ means that it has two-dimensional character; it lies in a bulk band gap, so it must be a surface state; its energy is tied to the vacuum level, and its

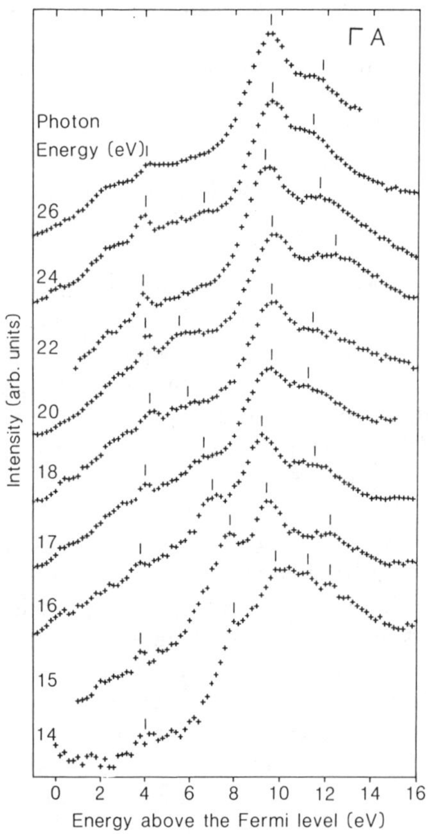

Fig. 8.2. Normal incidence KRIPES spectra from graphite, for different photon energies [8.39]

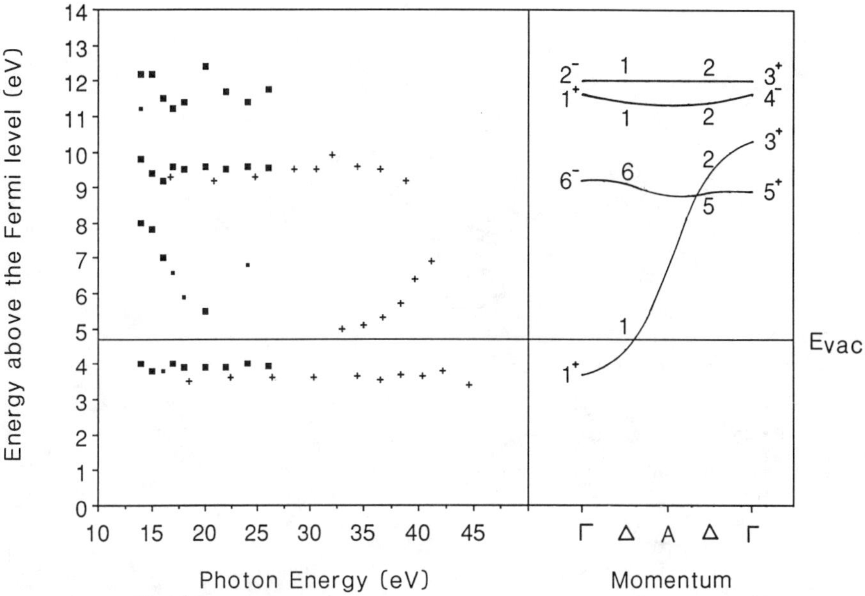

Fig. 8.3. Binding energies of the peaks in the normal incidence spectra (indicated by tick marks in Fig. 8.2) plotted against photon energy [8.39]. Large (small) squares represent strong (weak) features; + indicates the results of *Fauster* et al. [8.42]. The right-hand panel gives the unfolded bands along ΓA calculated by *Holzwarth* et al. [8.37]

dispersion with the wave vector component parallel to the surface is free-electron-like—characteristic of an image state. The other constant energy peaks correspond to *intralayer* bands, whose dispersion with $\boldsymbol{K}$ in the ΓK and ΓM directions has been measured by using initial-state electrons incident in the ΓAHK and ΓALM planes respectively (Fig. 8.4). The energies of the peaks as a function of $\boldsymbol{K}$ are shown in Fig. 8.5, together with the band structure calculated by *Holzwarth* et al. [8.37]. The discrepancies between the measured and calculated intralayer bands are only about 1 eV up to 10 eV above E_F. It remains to be seen whether these discrepancies are related to self-energy effects, because the self-energy has not yet been calculated in anything as complicated or as asymmetric as graphite.

Let us turn now to the interlayer state and its dispersion with k_z. This has also been studied by *Fauster* et al. [8.42] in a series of normal incidence KRIPES measurements on polycrystalline, highly oriented pyrolitic graphite. They observed a state dispersing between 5 and 7 eV above E_F for photon energies between 33 and 41 eV. To find k_z in these transitions, Fauster assumed that the incident electron is in a free-electron state, putting the bottom of the band at the valence-band minimum. Extrapolating the resulting band structure of the interlayer band, they concluded that it has a minimum at 4.0 ± 0.5 eV above E_F. To determine the minimum of the interlayer band in a more straightforward

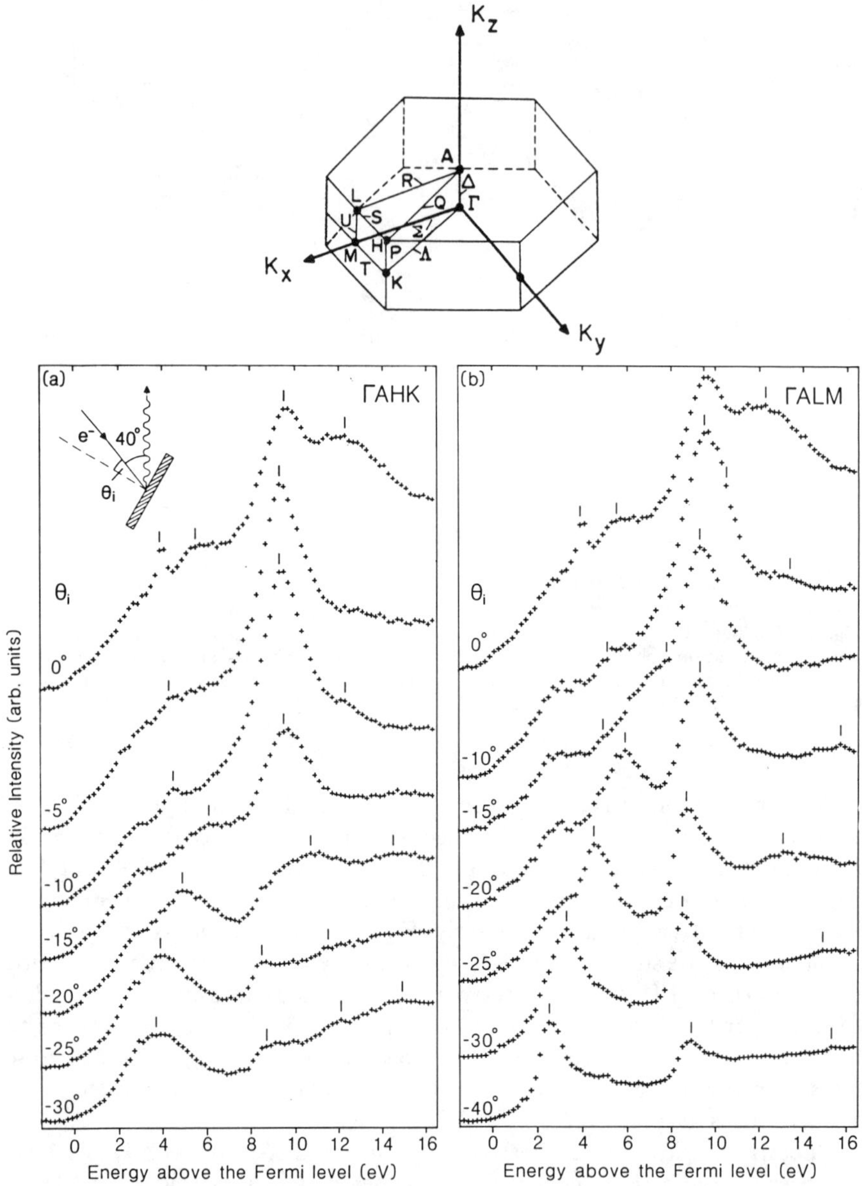

Fig. 8.4. KRIPES spectra from graphite at a photon energy of 20 eV, for varying angles of electron incidence in (**a**) the *ΓAHK* plane and (**b**) the *ΓALM* plane [8.39]. The Brillouin zone for graphite is also shown

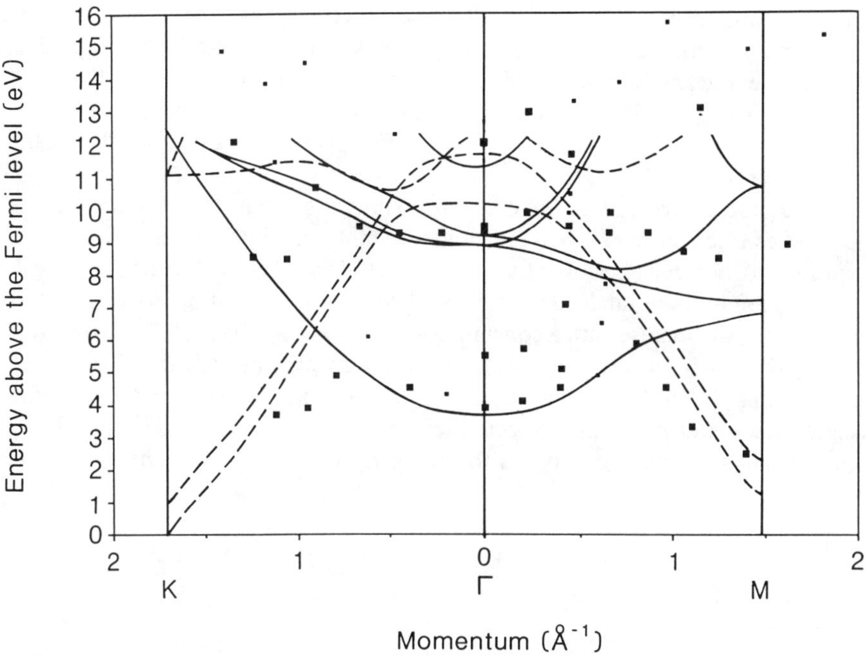

Fig. 8.5. Comparison of the experimentally determined band structure with the calculations of *Holzwarth* et al. [8.37]; dashed (solid) lines denote $\pi(\sigma)$ bands, and large (small) squares denote strong (weak) features [8.39]

way, *Collins* et al. [8.39] plotted the final state energies observed in their spectra and those of Fauster against photon energy, and found that both sets of results correspond to the same band (Fig. 8.3), with a minimum in the dispersion at about 5 eV. The most recent band-structure calculation puts this minimum at 3.8 eV [8.40], but there is considerable variation between calculations [8.39] so again it is premature to conclude that the difference is due to self-energy effects.

8.3 Nearly-Free-Electron Metals: Alkalis

The electrons in alkali metals feel only a weak pseudopotential [8.43], and they should behave like a nearly free-electron gas. However, photoemission experiments from Na(110) show that self-energy effects reduce the bandwidth by 22%, and also show an unexpected peak at the Fermi energy [8.44]; this has been controversially interpreted in terms of charge density waves [8.45]. All these aspects make it worthwhile to study alkalis with inverse photoemission: *Collins* et al. [8.46] have studied Na(110) grown on a Ni(001) surface, and *Woodruff* and *Smith* [8.47] have investigated polycrystalline Na, K, Rb and Cs grown on

Cu. The most striking feature of all these spectra is that they are extremely weak and structureless, with no strong peaks due to direct transitions. The extreme weakness of alkali spectra is illustrated in Fig. 8.6, where the KRIPES spectrum from Na(110) is compared with other materials recorded at 20 eV photon energy, in the same apparatus, all the spectra being normalized to the same electron beam charge.

The states involved in inverse photoemission from Na(110) are in the nearly-free-electron-like *s–p* band, and direct transitions to these states are expected to be weak because the atomic pseudopotential is weak. The weak atomic pseudopotential leads to a small atomic contribution to the matrix element in (8.5), because with a constant $\boldsymbol{A}$-field, $\boldsymbol{A}\cdot\boldsymbol{p} \propto \boldsymbol{A}\cdot\nabla V/\omega$ [8.18]. Moreover the rapid dispersion of the *s–p* band leads to a low joint density of states. In fact, Collins did observe some excess intensity from Na(110) at the energies where direct transitions are expected from band-structure calculations [8.48], but the peaks are small compared to the background (Fig. 8.7). The count rate

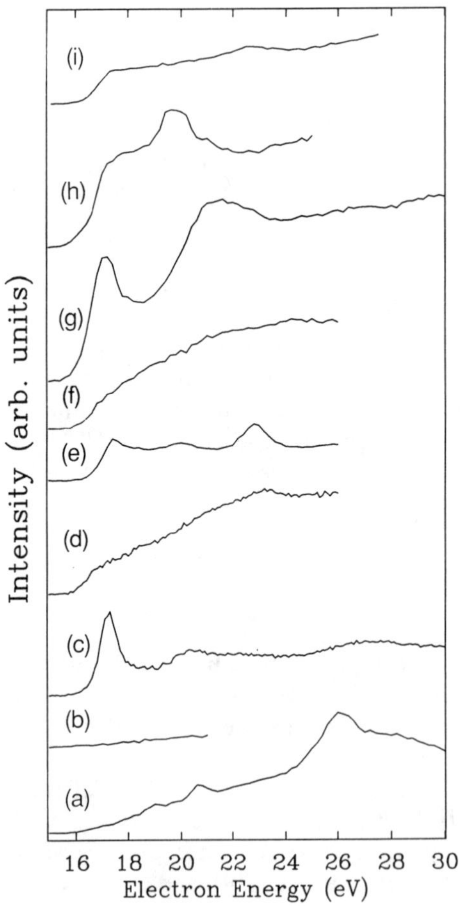

Fig. 8.6. Normal incidence inverse photoemission spectra from various systems, all (except Y(0001)) recorded at 20 eV photon energy; Y(0001) was recorded at 19 eV, as the 20 eV spectrum was dominated by fluorescence photons. (a) single-crystal graphite, (b) Na(110), (c) Ni(110), (d) Y(0001), (e) Pd(111), (f) Ho(0001), (g) Ta(001), (h) W(001), (i) polycrystalline Au

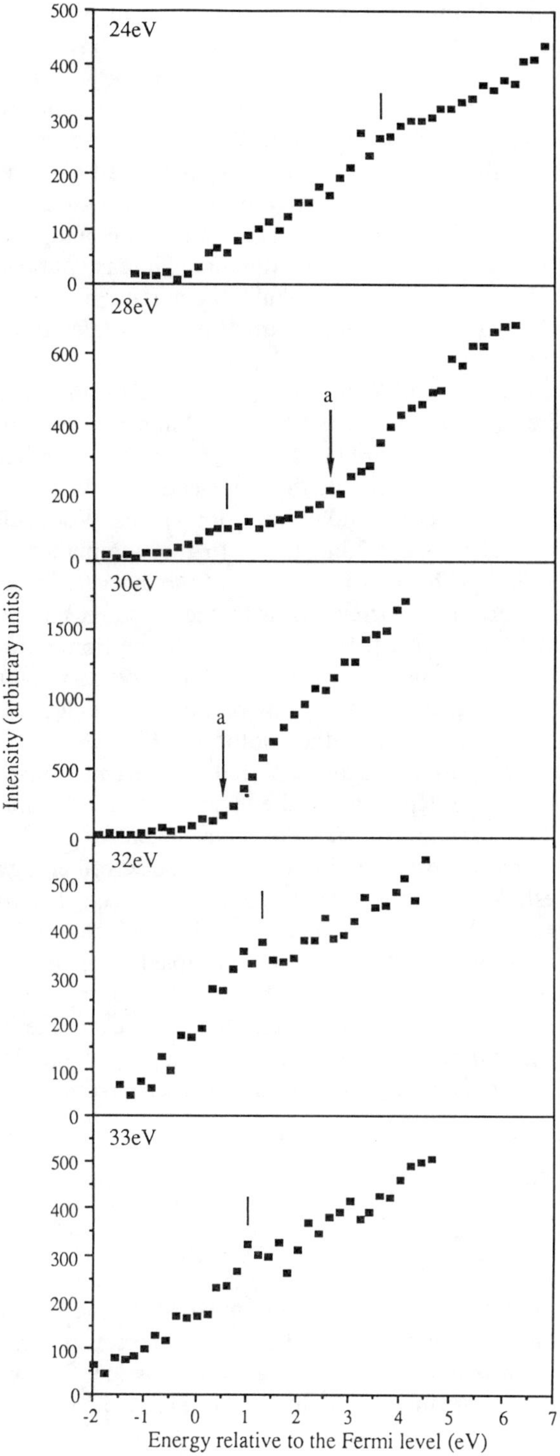

Fig. 8.7. Normal incidence isochromat inverse photoemission spectra from Na(110), for different photon energies [8.46]. Direct transitions are indicated by bars; the *a* arrows mark the onset of enhanced emission resulting from excitation of the 2*p* core electrons

from Na(110) is much lower than from noble metals such as Cu and Ag which also have an *s–p* band dispersing through the Fermi energy—in these cases, good spectra with a clear step at E_F are readily observed [8.34, 49]. This is presumably because the atomic pseudopotential for the *s–p* electrons in the noble metals is stronger than in the alkalis [8.50]. In nearly-free-electron systems in which the atomic contribution to the matrix element is small, grad V at the surface barrier is expected to dominate the matrix element [8.51], so the spectrum should reflect the density of states at the surface. In the case of inverse photoemission from Al(111) for example, there is a continuum of intensity above a very clear Fermi edge [8.52]. However, in the alkalis the surface barrier is itself very small [8.53], and the Fermi edge is not apparent in Fig. 8.7, let alone surface density of states features.

Somewhat surprisingly, *Woodruff* and *Smith* [8.47] found little sign of the *d*-bands which band-structure calculations predict to lie within a few volts of E_F in K, Rb and Cs [8.54, 55]. Because of their high density of states, and the fact that they feel a stronger effective potential than the *s–p* electrons, we would expect the *d*-states to give strong peaks as they do in transition metals. Woodruff and Smith ascribe the absence of strong *d*-band peaks to the dominance of inelastic scattering processes [8.47]. These authors find that the inelastic background rises rapidly above a threshold correlating with the plasmon energy (which is low in alkalis), and they suggest that plasmon scattering dominates the inelastic loss processes in these materials. In most systems the dominant inelastic process leading to the background is scattering of the incident electron by electron–hole pairs prior to the emission of the photon [8.16].

Processes other than inverse photoemission can lead to the emission of photons, which are detected in these experiments: the arrows *a* in the 28 and 30 eV spectra from Na(110) (Fig. 8.7) show the onset of enhanced emission from a fluorescence process. These arrows correspond to an incident electron energy of 30.5 eV, which is the threshold energy for exciting a 2*p* core electron into the unoccupied states above E_F. This leaves behind a core hole which can be filled by the radiative transition of an electron from the occupied states below E_F. As the occupied band width is about 2.5 eV, the radiative decay gives photons in the energy range 28.0–30.5 eV, which are detected in the experiment. This is the same process as in *appearance potential spectroscopy* [8.56]. The possibility of APS-type effects should always be considered when planning (or interpreting) inverse photoemission experiments.

8.4 Transition Metals: Tungsten

The physical properties of the transition metals are dominated by the rather tightly bound *d*-electrons. In W the *d*-band is half filled, so there is a large density of unoccupied states to be studied using inverse photoemission. Surface effects are particularly interesting, because the high surface density of states at the Fermi energy drives a reconstruction of W(001) [8.57–60]. Unoccupied states in W had already been studied before the use of inverse photoemission—band edges lying above the vacuum zero were identified by *Willis* and *Christensen* [8.61, 62] and by *Schäfer* et al. [8.63] using secondary electron emission, and were in good agreement with relativistic band-structure calculations [8.64]. Here we shall describe inverse photoemission experiments from W(001) and (110), using the analysis to illustrate general points about the relationship between inverse photoemission and surface and bulk electronic structure.

8.4.1 Analysis of the W Spectra

One of the reasons for using KRIPES spectra to study the bulk electronic structure is to test the accuracy of band-structure calculations, and to see to what extent self-energy corrections—expected to increase with energy above E_F [8.65]—are important. As we saw in Sect. 8.2, the difficulty in finding the bulk band structure experimentally is the determination of k_z, the perpendicular component of the wave vector. In a direct transition, which is generally characterised by a photon-energy dependence of the peak position, k_z in the final state is the same as in the initial incident state; the problem is that we generally need the initial-state band structure inside the solid in order to determine the final state. When the same transition can be measured from different surfaces, the method of energy coincidence (triangulation) may be used to pinpoint the wave vector of the transition without prior assumptions about the initial-state band structure (Chap. 9) [8.2]. This has proved effective in some cases [8.66], but when using this method, there is the possibility of confusion if a peak in the spectrum is the result of several transitions lying close together in energy, whose relative intensity may vary with the surface from which the measurements are made.

In order to determine k_z in direct transitions it is often a good approximation to use a free-electron-like initial state [8.2], fixing the bottom of the band by some suitable choice of inner potential [8.67], or taking the lowest energy of the valence band as the origin of the free-electron dispersion [8.42]. This can be a good approximation, because the finite mean-free path of the incident state, due to inelastic scattering, smears out band gaps [8.51]. An alternative assumption is to assume that the momentum broadening of the initial state is so substantial that the spectrum represents the one-dimensional density of states of the final states, in other words the energy distribution of the final states with wave vector component $\boldsymbol{K}$ parallel to the surface.

In the discussion which follows, we use full photoemission-type calculations (Sect. 8.1.3) to interpret experimental data from W [8.68]. These have been applied rather successfully to systems such as Ni and Cu [8.33], but the complicated nature of the initial incident states in W makes this a sensitive test of the methods and programs. Isochromat spectra for W(001) and W(110) surfaces were calculated using a relativistic version of the photoemission program [8.69], for a range of photon energies between 15 and 30 eV. It is necessary to use a

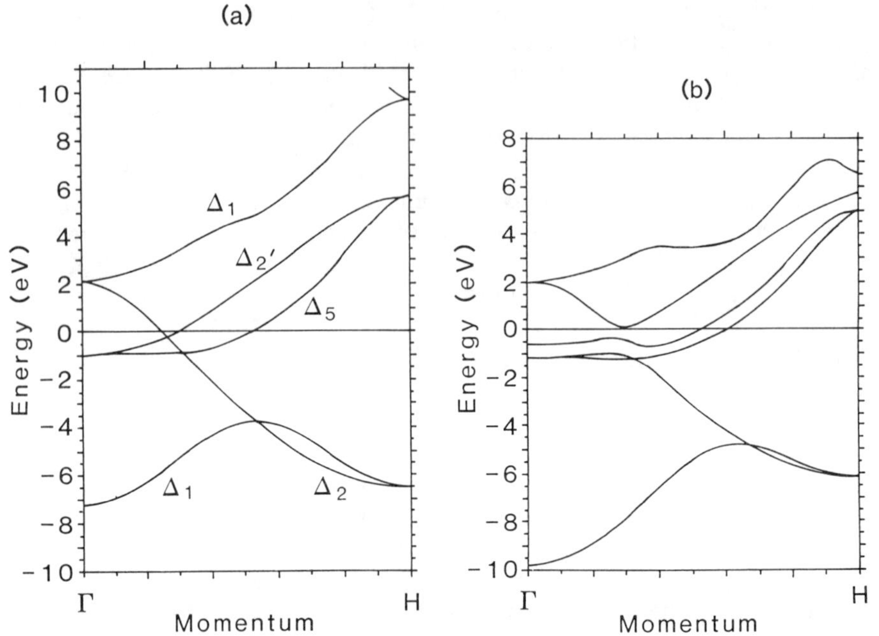

Fig. 8.8. Band structure for W in the ΓH [001] direction. (**a**) Non-relativistic calculation, (**b**) relativistic calculation, showing spin-orbit gaps opened up

relativistic program, because W is quite a heavy element, and spin–orbit coupling affects the band structure (Fig. 8.8). (Spin–orbit gaps are of the order of 0.5 eV in W, compared with 0.1 eV in Ni, for example.) The potential for W was taken from a separate self-consistent band-structure calculation, for which the energy bands were in close agreement with those of *Christensen* and *Willis* [8.62] for energies as high as 30 eV above E_F. The position of the surface potential step was adjusted to move the prominent surface state on the (001) surface just below E_F at normal incidence. This adjustment is necessary because in the Pendry photoemission program the surface barrier is represented simply as a potential step, though in subsequent work this approximation has been removed [8.19]. Deficiencies in the treatment of the surface barrier also lead to errors in the relative intensities of transitions to bulk bands.

The predicted W(001) spectra for all energies show a broad distribution of intensity from the Fermi edge to where the plots end at 8 eV above E_F (Fig. 8.9). To decide which transitions are involved in this intensity, we can use symmetry arguments. Transitions to the Δ_2 and $\Delta_{2'}$ bands, in the non-relativistic limit (Fig. 8.8), are forbidden for normal-incidence electrons; normal-incidence electrons travelling in the z-direction are uniform in the $x - y$ plane, whereas the Δ_2 and $\Delta_{2'}$ bands are made up from $x^2 - y^2$ and xy orbitals respectively, and the dipole matrix element entering (8.5) cannot couple them. This means that

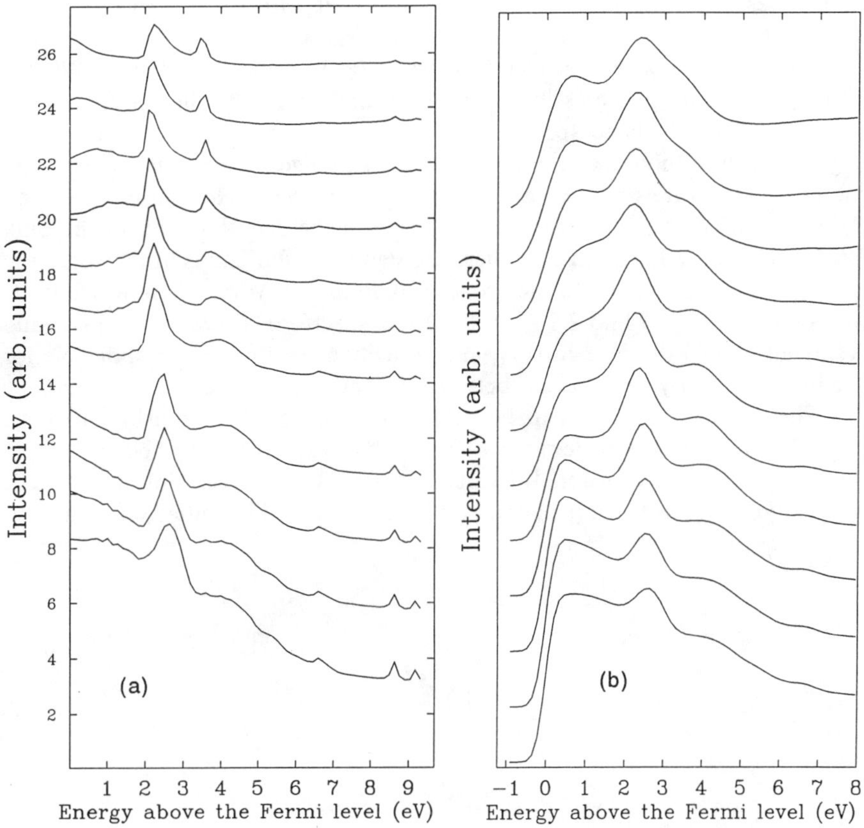

Fig. 8.9. (**a**) Calculated normal incidence isochromat inverse photoemission spectra from W(001), for photon energies of 15, 16, 17, 18, 20, 21, 22, 24, 26, 28 and 30 eV (lowest energy at the bottom), (**b**) calculated spectra truncated at E_F and convoluted with instrumental broadening

the intensity between E_F and 2 eV, where the unoccupied Δ_1 band begins, must be coming from transitions over a wide range of k_z to the Δ_5 band. The main features in the calculated spectra are the peaks varying between 2.6 eV and 2.5 eV, and between 4.3 eV and 3.6 eV as the photon energy varies between 15 and 30 eV (Fig. 8.9b). Both peaks come from transitions to the Δ_1 band, and inspection of the calculated spectra before convolution with the instrument function (Fig. 8.9a) shows that the lower energy peak approaches the Δ_1 band edge at 2 eV in the high-photon-energy spectra. The other feature comes, at high photon energies, from transitions to a flat region of the Δ_1 band at 3.5 eV, broadening and shifting from this energy as the photon energy is reduced. The final appearance of the calculated spectra is considerably modified by the effect of finite energy resolution (compare Figs. 8.9a and 8.9b). This conceals the changing shapes of the peaks in the spectra, and their energy shifts as the photon energy is changed, which is apparent in Fig. 8.9a. Simpler behaviour is shown by the weak peak at 6.8 eV, which does not shift with photon energy; this is just due to a high density of states in the Δ_1 band.

The situation with the W(110) calculation is similar to that of W(001), although here all the accessible final states are symmetry allowed at normal incidence. The calculated spectra (Fig. 8.10) all show intensity from the Fermi edge upwards, although as the photon energy changes, the relative intensity in the broad step varies relative to the structure near 2 eV. Normal incidence on the (110) surface corresponds to states lying in the ΓN direction in the bulk Brillouin zone, and around 2 eV there are two fairly flat bands in this direction (Fig. 8.11). These give rise to the broad peak in the 30 eV spectrum, which splits into two peaks at 1.7 and 2.6 eV in the 15 eV spectrum. There are also (Σ_1 and Σ_3) bands which disperse rapidly, and transitions to these are responsible for the broad intensity distribution between E_F and the 2 eV peak.

The predicted spectra can be summarised as showing transitions to the bands above E_F over a wide range of k_z, but they cannot be described as simply reflecting the one-dimensional density of states. There is certainly too wide a range of transitions to be explained by a free-electron-like initial state, but too

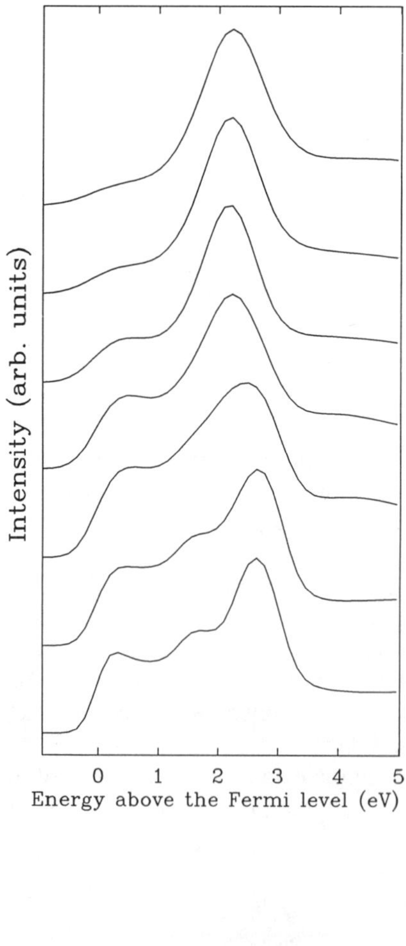

Fig. 8.10. Calculated normal incidence isochromat inverse photoemission spectra from W(110), for photon energies of 15, 17, 20, 22, 25, 28 and 30 eV

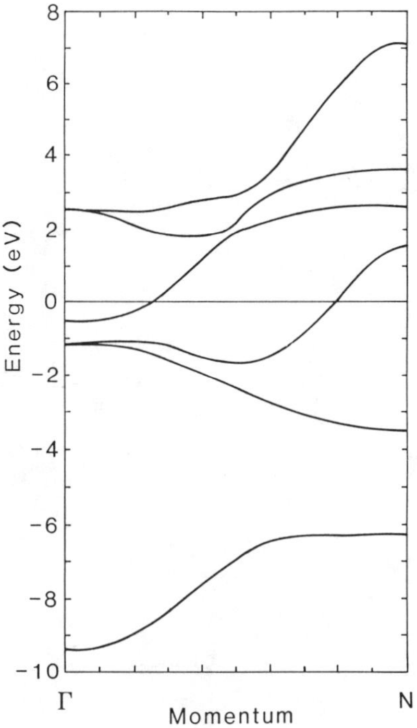

Fig. 8.11. Band structure for W in the ΓN [110] direction

few to treat the spectra as the densities of states. In some ways this is bad news for the experimentalist, as it means that simple interpretations of the spectra are not very useful. Rather the calculations confirm that in the initial state, several bands are populated, and it is necessary to perform a detailed calculation in order to determine the different transition probabilities and to compare the band structure with the experimental data.

Let us now see how well the calculated spectra compare with experiment. The calculated spectra for W(110) (Fig. 8.10) appear to correspond well with the experimental results (Fig. 8.12), both in predicting the shape of the spectra, and the changes with photon energy. The relative intensity of the low energy regions near E_F and the peaks around 2–3 eV are accurately predicted, as are the changes in energies of the peaks with photon energy. The relatively small discrepancies in peak energies are a combination of self-energy shifts and deficiencies in the calculation—it is hard to disentangle the two—but the results give us confidence in the quality of the band structure up to about 30 eV above E_F. The calculated spectra for W(001) (Fig. 8.9b) appear to be displaced from the experimental results (Fig. 8.13) by about 0.5 eV to lower energies, but in other respects reproduce quite well the intensity distribution in the spectra. The shifts may be due to small errors in the momentum of the initial state, which lead to errors in the energy of direct transitions especially when $\partial E/\partial k_z$ is large, as it

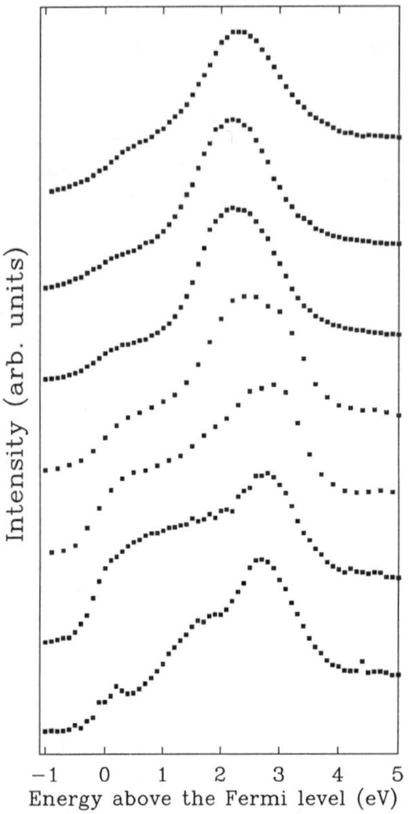

Fig. 8.12. Experimental normal incidence isochromat inverse photoemission spectra from W(110), for photon energies of 15, 17, 20, 22, 25, 28 and 30 eV

is for the Δ_1 band contributing to the features in the (001) spectra. These shifts may also be caused by inaccuracies in the treatment of the potential at the surface [8.19], because transitions to bulk electronic states also have a large contribution from the surface region. The use of the bulk potential up to the surface is likely to be better at the (110) surface, which is fairly close packed, than at the more open (001) surface.

The analysis of any inverse photoemission spectrum involving bulk states relies on having a band structure as its starting point. However, we have shown that when the system is complicated with many bands at the energy of the incident electron, there is no reliable set of assumptions that one can safely make to simplify the comparison between the experimental spectra and the band structure. In favourable cases a band crossing or a band maximum or minimum may be evident in the data, but this is not often the case with data taken over a limited range of photon energies. A full multiple-scattering calculation, such as we have described here, can give a good prediction for the measured spectrum, and this is probably the best available confirmation of the quality of the potentials used in the band structure calculation.

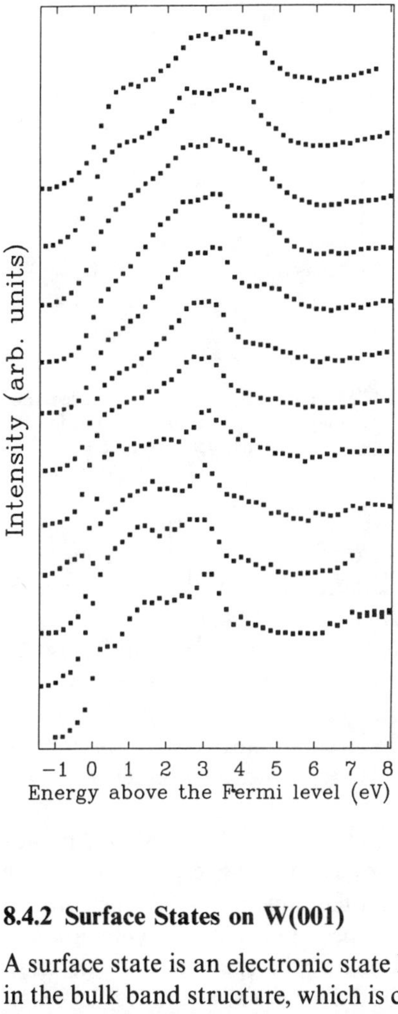

Fig. 8.13. Experimental normal incidence isochromat inverse photoemission spectra from W(001), for photon energies of 15, 16, 17, 18, 20, 21, 22, 24, 26, 28 and 30 eV

8.4.2 Surface States on W(001)

A surface state is an electronic state lying in an energy gap (or a symmetry gap) in the bulk band structure, which is consequently localized at the surface [8.24]. Only the wave vector component parallel to the surface is well defined, so their dispersion can be entirely determined from photoemission and inverse photoemission experiments using the conservation rules given in Sect. 8.1.2. The high surface density of states of the surface states means that they can play a large part in determining surface properties, and in the case of W(001) it has been suggested that they drive the room temperature surface reconstruction (Fig. 8.14a) [8.57, 58, 60, 70]. The reconstruction produces new surface Brillouin zone boundaries halfway along the $\bar{\Gamma}\bar{M}$ lines (Fig. 8.14b), at which energy gaps open up. If the surface-state band crosses E_F at this point, the reconstruction is stabilized by the reduction in energy of all the occupied part of the surface-state band. Photoemission experiments cast doubt on this Peierls (or charge-density wave) mechanism for the reconstruction: *Holmes* and *Gustafsson* [8.71] found that a band of surface states reaches E_F at about $0.43 \times \bar{\Gamma}\bar{M}$, and more recently

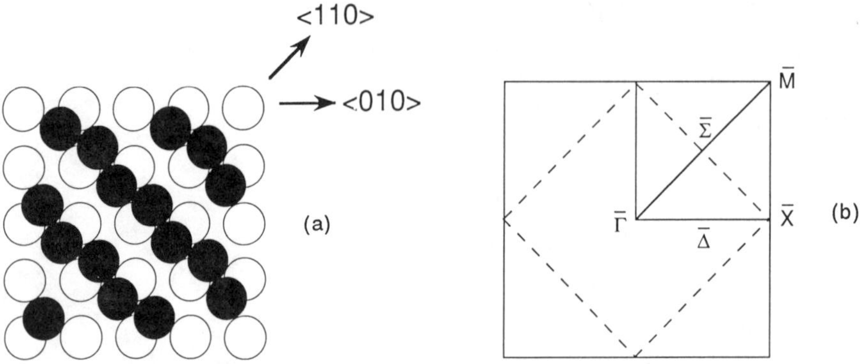

Fig. 8.14. (**a**) Reconstruction of the W(001) surface, the filled circles representing the surface atoms and the open circles the sub-surface atoms [8.59], (**b**) solid line—the Brillouin zone for the unreconstructed surface, dashed line—the Brillouin zone for the reconstructed surface

Smith et al. [8.72] found that a surface state dispersed towards E_F at about 1/3 $\bar{\Gamma}\bar{M}$ but were unable to conclude whether or not it crossed the Fermi level. In order to study the continuation of this surface state above E_F, inverse photoemission measurements have been carried out by *Collins* et al. [8.73]. The KRIPES measurements were made in isochromat mode at a photon energy of 20 eV, close to the 18, 20 and 22 eV used for the ARPES experiments so that the initial states in inverse photoemission are the same as the final states in photoemission. (It is possible that initial-state effects are responsible for the non-observation of the surface state in inverse photoemission experiments carried out by *Drude* et al. [8.74].)

There are three established criteria for identifying surface states in photoemission and inverse photoemission. These are the photon-energy independence of peak position (because k_z is not a good quantum number), sensitivity to surface contamination, and the location of the peak in a suitable symmetry gap in the bulk band structure. In these experiments, the spectra from the clean surface were compared with those from a H-dosed surface at saturation coverage; H has the advantage over other gases such as O in removing surface states, in that it introduces relatively few new states itself to complicate the spectrum. The symmetry aspects of the surface states can be emphasised (along symmetry directions in the surface Brillouin zone) by using photons emitted in different directions, which changes $\boldsymbol{A}$ in the matrix element (8.2). Although the apparatus used by Collins did not have facilities for varying the angle between the electron beam and the light emission direction, it was possible because of the symmetry of the surface to rotate the crystal between two positions with the same polar angle for the electrons, but different polar angles for the light emission (Fig. 8.15).

Figure 8.16 shows the resulting spectra for varying angles of electron incidence for both the clean and H-dosed W(001) surface. The difference spectra reveal the presence of a surface state which first appears above E_F at 17.5° angle of incidence for the initial-state electrons, and disperses upwards as the angle

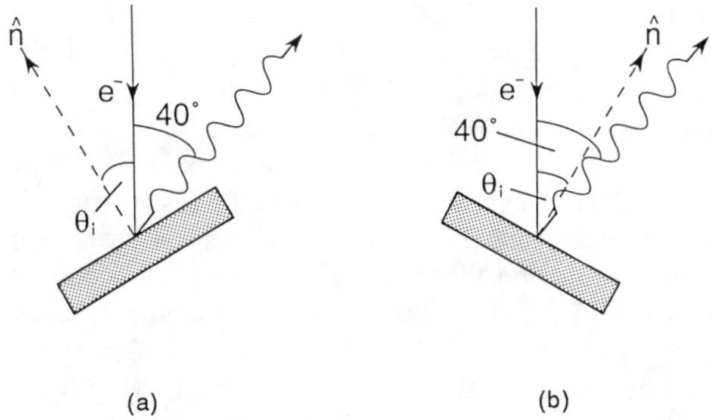

Fig. 8.15. (**a**) Experimental geometry for negative angle data, (**b**) geometry for positive angle data

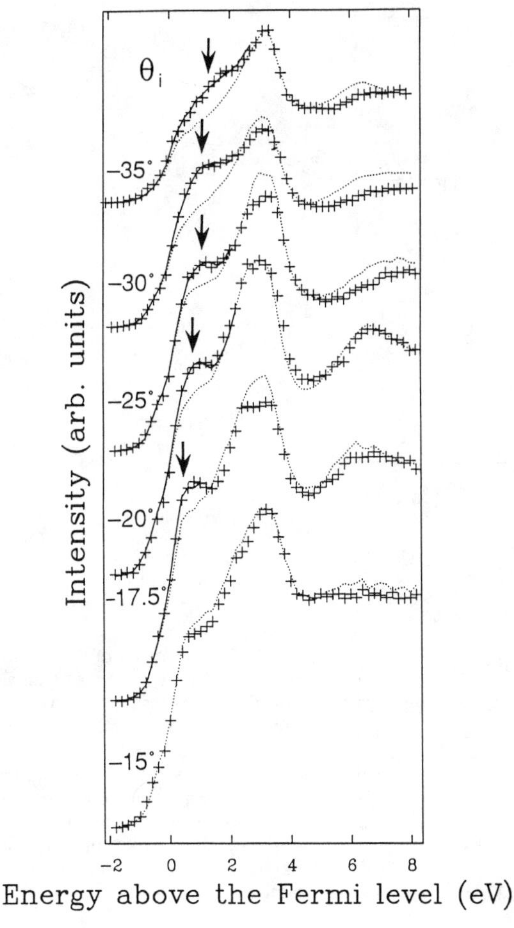

Fig. 8.16. Isochromat inverse photoemission spectra from W(001), for a photon energy of 20 eV, at different angles of electron incidence. Crosses (dashed lines) denote clean (H-dosed) surfaces. Full lines show fits to the difference spectra. The arrows show the energies of surface states deduced from the computer fits

of incidence (hence $\boldsymbol{K}$) increases. The energy of the state is plotted as a function of $\boldsymbol{K}$ in Fig. 8.17, together with the results for the occupied part of the surface-state band measured by *Smith* et al. [8.72]. The measured dispersion of the unoccupied surface state is consistent with it being the continuation of the occupied surface state measured in photoemission, crossing E_F before the $\bar{\Gamma}\bar{M}$ midpoint. The actual point at which the surface state crosses E_F is very difficult to pinpoint with certainty, because when the experimental peak is near E_F it corresponds to a partially filled state and the maximum in this peak occurs at a higher energy than the energy of the state itself. ARPES experiments show another surface state crossing E_F at $0.84 \times \bar{\Gamma}\bar{M}$ [8.71], and this may correspond to the weak shoulder at 0.8 eV above E_F in the $+35°$ and $-35°$ spectra (Fig. 8.18).

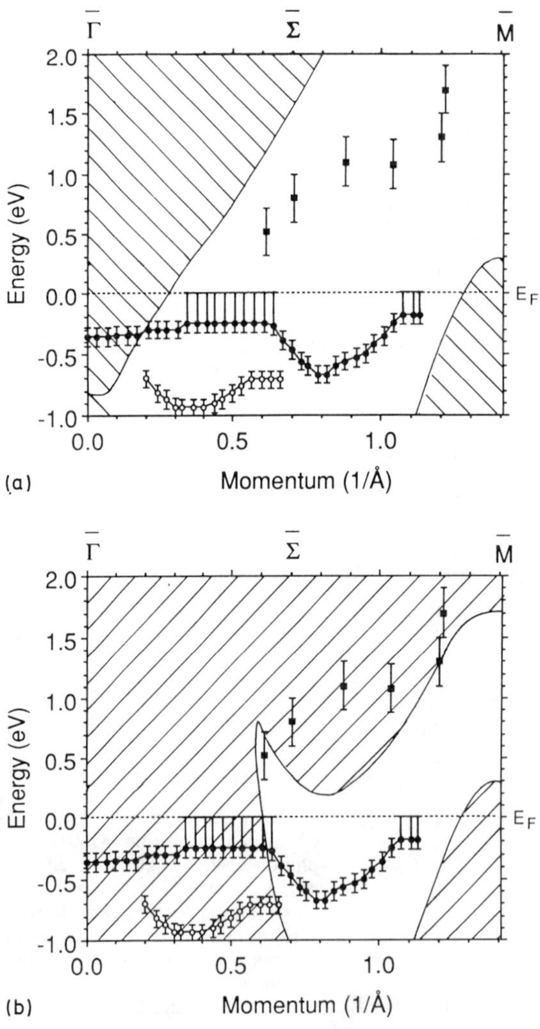

Fig. 8.17. Dispersion of the surface states on W(001) in the $\bar{\Gamma}\bar{M}$ direction. Circles are results from ARPES experiments [8.72], and squares are results from the KRIPES experiments described here. Projections onto the surface Brillouin zone of W(001) bulk bands are shown, (**a**) showing the projection of odd bands and (**b**) the projection of even bands

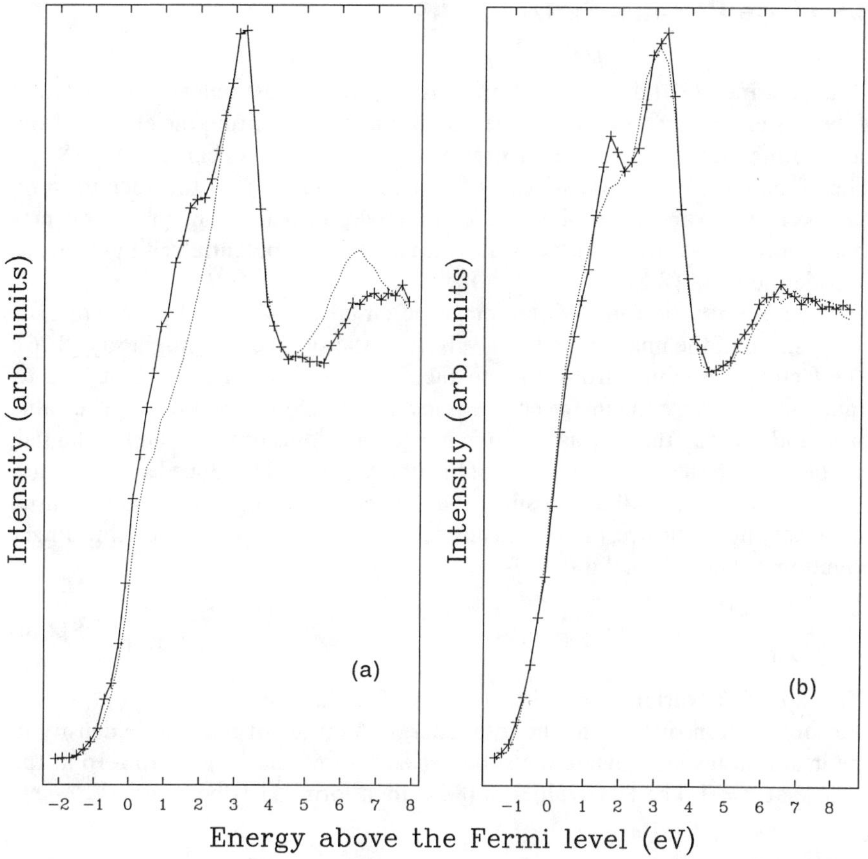

Fig. 8.18. Isochromat inverse photoemission spectra from W(001), for a photon energy of 20 eV, at angles of incidence of (**a**) $-35°$ and (**b**) $+35°$. The line (dots) denote clean (H-dosed) surfaces

These two states crossing E_F have different symmetry [8.71], and this may be reflected in the different relative intensities of the features in the $+35°$ and $-35°$ spectra (Sect. 8.1.2).

The conclusion from this work, that the surface states cross E_F away from the $\bar{\Gamma}\bar{M}$ mid-point, means that the Peierls mechanism for driving the surface reconstruction seems less likely. From the point of view of this volume its interest is in the determination of surface-state dispersion, and in the continuation of a surface-state band from the occupied region into the unoccupied region.

8.5 Image Potential Surface States

Outside a metal surface an electron feels the Coulombic image potential, and if there is an energy gap around the vacuum level, preventing the electron from penetrating into the solid, a Rydberg series of surface states can develop [8.75]. The measurement of these states, with an energy just below the vacuum zero, has been a rewarding application of inverse photoemission, providing new insight into the nature of surface states and the form of potential felt by electrons outside the solid [8.2].

The asymptotic form of the image potential is $-1/4|z-z_0|$, where z_0 is the position of the image plane—the electrostatic surface, one might say [8.76]. The factor of 4 comes from 2 (from the distance between the charge and its image) × 2 (because the image charge is *induced* by the electron itself). The value of z_0, and the way that the asymptotic image form joins onto the bulk potential, can be fitted to the precise energies of the image-potential-induced surface states [8.77]. If we put $z_0=0$ and assume that the electron feels the image potential all the way up to surface, then its wavefunction satisfies a hydrogenic Schrödinger equation outside the solid ($z>0$)

$$-\frac{1}{2}\frac{d^2\psi}{dz^2}-\frac{1}{4z}\psi=\left(E-\frac{K^2}{2}\right)\psi. \tag{8.14}$$

The $\exp(\mathrm{i}\boldsymbol{K}\cdot\boldsymbol{R})$ variation parallel to the surface has been factored out, giving the contribution of $K^2/2$ to the total energy. As most of the charge density in the image states lies outside the solid, let us assume that ψ goes to zero at the surface at $z=0$. The eigenvalues of (8.14) then form a Rydberg series

$$E_n=-\frac{1}{32n^2}+\frac{K^2}{2}. \tag{8.15}$$

The nth solution has n nodes, including the node at the origin (Fig. 8.19). Corrections to (8.15) arise because of deviations of the potential from the Coulombic form near the surface, and the fact that the semi-infinite solid does not behave like an infinite barrier in reflecting the electrons at $z=0$ [8.75].

There are several noble metal and transition metal surfaces for which conditions are right for the occurrence of image states, with a bandgap in the s–p band around the vacuum zero of energy [8.2]. One important consideration which makes it possible to observe the image states is the fact that many-electron effects do not wipe them out [8.75]. Image states typically lie several volts above E_F, and in the bulk metal a typical lifetime broadening for states at this energy may be of the order of a volt [8.26]. However the broadening of image states is much less than this, as is illustrated by the observation of several states in the Rydberg series [8.77]; the intrinsic width of the first two image states on Ag(001) has recently been measured (using two-photon photoemission) as 21 and 5.5 meV respectively [8.78]. The reason for the long lifetimes is that the image state wavefunctions lie well outside the surface, with very little overlap

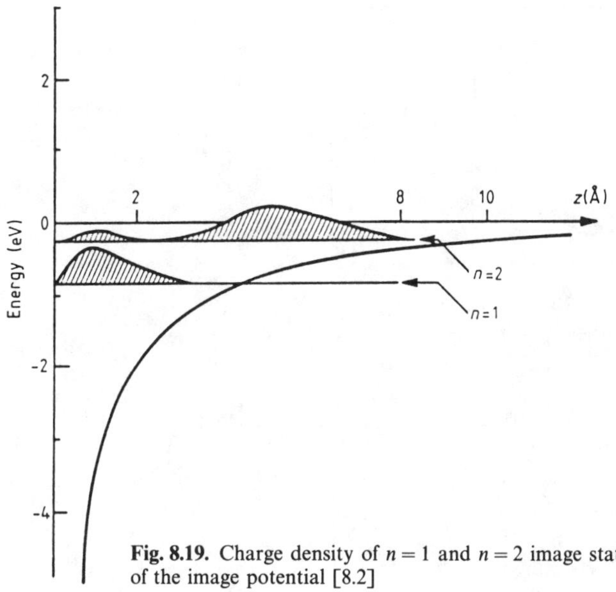

Fig. 8.19. Charge density of $n=1$ and $n=2$ image state wavefunctions, in the field of the image potential [8.2]

with the occupied electron states from which electron–hole excitations can cause loss of energy [8.75, 79]. The squared overlap of the image states with the occupied states, hence lifetime broadening, actually decreases as $1/n^3$—the same as the separation of the states [8.75]. So the whole series should in principle be observable.

8.5.1 Observation of Image States

The KRIPES spectrum for states with $\boldsymbol{K}=0$ on Cu(001) is shown in Fig. 8.20 [8.10]. The large peak just above the Fermi energy is due to interband transitions, but structure close to the vacuum zero, due to the image states, is clearly visible: the $n=1$ state is measured at $E=-0.64\,\mathrm{eV}$ relative to the vacuum zero, and the higher states are apparent as weak structure just under the vacuum level. The dispersion of the $n=1$ state with $\boldsymbol{K}$ has been measured [8.80] (Fig. 8.21), and corresponds closely to $E=K^2/2m^*$, with an effective mass close to 1—not surprising for states which lie mostly outside the crystal potential. These surface states can also be measured using two-photon photoemission [8.81], with greater resolution than in KRIPES. In this experiment, electrons are excited from below E_F into the surface state by absorbing the first photon, and are then ionized by the second photon and measured. These extraordinarily precise measurements give energies of $-0.57\,\mathrm{eV}$ for the $n=1$ state and $-0.18\,\mathrm{eV}$ for the $n=2$ image state at $\boldsymbol{K}=0$ on Cu(001) [8.81], and an effective mass of $0.9\pm0.1\,m_e$ [8.82].

As well as the image states, the s–p band gap can also support conventional Shockley surface states [8.83]. Unlike the image states, which are a direct

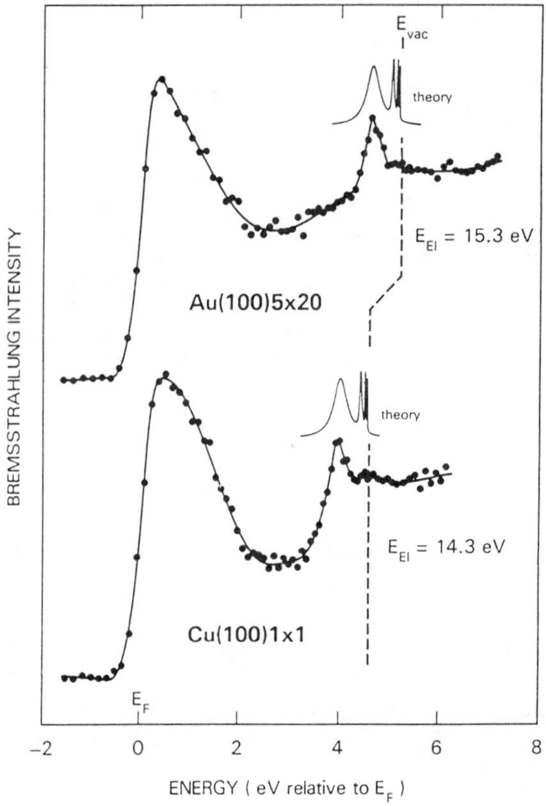

Fig. 8.20. Inverse photoemission spectra at normal incidence from Cu(001) and Au(001) [8.10]. The vertical dashed lines give the vacuum zero

consequence of the image potential, the Shockley states are a consequence of the crystal potential itself, and they extend much less into the vacuum. Nevertheless, they can be classified as $n = 0$ states—they are states with *no* nodes in the surface region, actually peaking in amplitude at around the surface itself (the $n = 0$ classification is put on a firmer footing in a phase shift analysis [8.2]). On Cu(001) the Shockley state actually becomes a surface resonance just below the edge of the band gap, but it is still measurable in KRIPES experiments as a weak shoulder on the bulk transition [8.84]. As the Shockley states are closely tied to the crystal potential, their dispersion with $\boldsymbol{K}$ can be very different from free-electron like. On Cu(111) the Shockley state, near the bottom of the band gap, has an effective mass of $m^* \simeq 0.5\, m_e$, for example [8.85].

Even image states can show an effective mass differing significantly from 1 if they lie close to the band edge, where they tend to track the band edge dispersion. On Ag(111), for example, it is found that $m^* = 1.3\, m_e$ for the $n = 1$ state [8.82], which lies just below a rather flat band edge. A few years ago there was considerable discussion in the literature about the origin of deviations of m^* from unity—in terms of many-electron corrections to the energy (i.e. self-energy effects) [8.1], or the role of corrugations in the surface potential

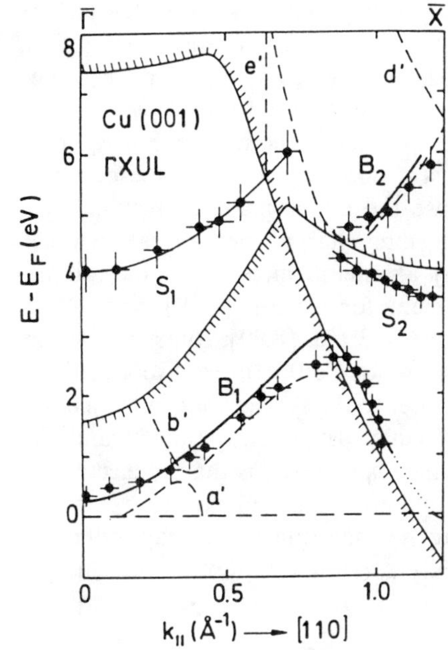

Fig. 8.21. Dispersion of spectral features with $\boldsymbol{K}$ in isochromat inverse photoemission spectra from Cu(001) [8.80]. S_1 is the $n = 1$ image state, the hatched lines indicate bulk band edges

[8.86]. In fact both of these effects can only change the image state dispersion by a few per cent [8.87]; the major effect is the variation of the reflectivity of the semi-infinite crystal through the band gap, and the dispersion of the band gap with $\boldsymbol{K}$.

8.5.2 Determining the Potential in the Vacuum

From the precise energies of the image states (the *quantum defect* correcting the Rydberg series of energies (8.15)), information can be found about the form of potential felt by electrons outside the solid [8.77]. This is conveniently parametrized by a functional form such as

$$
\begin{aligned}
V(z) &= -\frac{1}{4(z - z_0)}(1 - \mathrm{e}^{-\lambda(z - z_0)}), \quad && z > z_0 \\
&= -\frac{U_0}{1 + A\mathrm{e}^{\beta(z - z_0)}}, && z < z_0,
\end{aligned}
\tag{8.16}
$$

in which z_0, the bulk inner potential U_0 and the cross-over length λ are adjustable parameters, and A and β are chosen to ensure continuity of V and dV/dz at z_0 [8.88]. The solution of the Schrödinger equation with this potential in the surface region can be matched onto solutions of the bulk Schrödinger equation in the energy gap, which decay exponentially into the solid, and in this way the energy of the surface states can be determined.

By varying the parameters in (8.16) to find the image and Shockley surface state energies, *Smith* et al. [8.77] have determined z_0 for several systems. For Cu(001) they find z_0 at $+0.47$ a.u., measured from the geometrical surface (sometimes called the jellium edge—the plane at which the solid is chopped in two), the plus sign indicating that z_0 is on the *vacuum* side of the surface. For Ag(001), on the other hand, z_0 is at -0.18 a.u., just inside the surface. These numbers can be compared with calculation, because the image plane corresponds (at least classically) to the centroid of the screening charge when a weak external electric field is applied to the surface [8.76], a system which is tractable computationally. Computation gives $z_0 = +0.97$ a.u. for Ag(001) [8.89], and $+1.35$ a.u. for jellium with an electron density of $r_s = 3$ a.u. [8.90], sometimes taken as a good model for Ag. Equivalent calculations for Cu(001) are not yet available. Experiment certainly suggests that z_0 lies closer to the surface than in the jellium model, but there also seems a real discrepancy between experiment and the detailed computation for Ag(001). A possible explanation is that an electron is a dynamical, quantum-mechanical object, and not a classical fixed charge which the computation is assuming in determining the image plane: the image potential felt by the electron outside the solid is really a continuation of the self-energy, and this remains to be calculated accurately.

References

8.1 V. Dose: Surface Science Reports **5**, 337 (1985)
8.2 N.V. Smith: Rep. Prog. Physics **51**, 1227 (1988)
8.3 F.J. Himpsel: Surface Sci. Reports **12**, 3 (1990)
8.4 B. Reihl, J.G. Bednorz, K.A. Müller, Y. Jugnet, G. Landgren, J.F. Morar: Phys. Rev. B **30**, 803 (1984)
8.5 H.M. Meyer III, T.J. Wagener, J.H. Weaver, M.W. Feyereisen, J. Almlöf: Chem. Phys. Lett. **164**, 527 (1989)
8.6 A.J. Viescas, J.M. Tranquada, A.R. Moodenbaugh, P.D. Johnson: Phys. Rev. B **37**, 3738 (1988)
8.7 J.B. Pendry: Phys. Rev. Lett. **45**, 1356 (1980)
8.8 B. Reihl, R.R. Schlittler, H. Neff: Phys. Rev. Lett. **52**, 1826 (1984)
8.9 V. Dose, W. Altmann, A. Goldmann, U. Kolac, J. Rogozik: Phys. Rev. Lett. **52**, 1919 (1984)
8.10 D. Straub, F.J. Himpsel: Phys. Rev. Lett. **52**, 1922 (1984)
8.11 P.M. Echenique, J.B. Pendry: Prog. Surf. Sci. **32**, 111 (1990)
8.12 F.J. Himpsel: J. Phys. Chem. Solids **49**, 3 (1988)
8.13 V. Dose: Appl. Phys. **14**, 117 (1977)
8.14 G. Denninger, V. Dose, H. Scheidt: Appl. Phys. **18**, 375 (1979)
8.15 J.B. Pendry: J. Phys. C **14**, 1381 (1981)
8.16 V. Dose, G. Reusing: Appl. Phys. **23**, 131 (1980)
8.17 M. Donath: Appl. Phys. A **49**, 351 (1989)
8.18 J.F.L. Hopkinson, J.B. Pendry, D.J. Titterington: Comp. Phys. Commun. **19**, 69 (1980)
8.19 U. König, P. Weinberger, J. Redinger, H. Erschbaumer, A.J. Freeman: Phys. Rev. B **39**, 7492 (1989)
8.20 G.D. Mahan: Phys. Rev. B **2**, 4334 (1970)
8.21 P.J. Feibelman, D.E. Eastman: Phys. Rev. B **10**, 4932 (1974)
8.22 C.N. Berglund, W.E. Spicer: Phys. Rev. **136**, A1030 (1964)
8.23 D.P. Woodruff, N.V. Smith, P.D. Johnson, W.A. Royer: Phys. Rev. B **26**, 2943 (1982)
8.24 A.J. Zangwill: *Physics at Surfaces* (Cambridge University Press, Cambridge 1988)
8.25 C.J. Powell: J. Electron Spectrosc. **47**, 197 (1988)

8.26 L. Hedin, S. Lundqvist: In *Solid State Physics*, Vol. 23, ed. by F. Seitz, D. Turnbull, H. Ehrenreich (Academic, New York 1969) p. 1
8.27 J.A. Knapp, F.J. Himpsel, D.E. Eastman: Phys. Rev. B **19**, 4952 (1979)
8.28 J. Hermanson: Solid State Commun. **22**, 9 (1977)
8.29 A. Goldmann, E. Bartels: Surface Sci. **122**, L629 (1982)
8.30 W. Altmann, K. Desinger, M. Donath, V. Dose, A. Goldmann, H. Scheidt: Surface Sci. **151**, L185 (1985)
8.31 E.N. Economou: *Green's Functions in Quantum Physics*, Springer Ser. Solid-State Sci. Vol. 7 (Springer, Berlin, Heidelberg 1983)
8.32 G. Borstel: Appl. Phys. A **38**, 193 (1985)
8.33 G. Borstel, G. Thörner: Surface Sci. Reports **8**, 1 (1987)
8.34 W. Altmann, V. Dose, A. Goldmann: Z. Phys. B **65**, 171 (1980)
8.35 P.J. Feibelman: Phys. Rev. B **12**, 1319 (1975).
8.36 R.C. Tatar, S. Rabii: Phys. Rev. B **25**, 4126 (1982)
8.37 N.A.W. Holzwarth, S.G. Louie, S. Rabii: Phys. Rev. B **26**, 5382 (1982)
8.38 M. Posternak, A. Baldereschi, A.J. Freeman, E. Wimmer, M. Weinert: Phys. Rev. Lett. **50**, 761 (1983)
8.39 I.R. Collins, P.T. Andrews, A.R. Law: Phys. Rev. B **38**, 13348 (1988)
8.40 H.J.F. Jansen, A.J. Freeman: Phys. Rev. B **35**, 8027 (1987)
8.41 I. Schäfer, M. Schlüter, M. Skibowski: Phys. Rev. B **35**, 7663 (1987)
8.42 Th. Fauster, F.J. Himpsel, J.E. Fischer, E.W. Plummer: Phys. Rev. Lett. **51**, 430 (1983)
8.43 V. Heine: In *Solid State Physics*, Vol. 24, ed. by H. Ehrenreich, F. Seitz, D. Turnbull (Academic, New York 1970) p. 1
8.44 E. Jensen, E.W. Plummer: Phys. Rev. Letters **55**, 1912 (1985)
8.45 A.W. Overhauser: Phys. Rev. Lett. **55**, 1916 (1985)
8.46 I.R. Collins, A.R. Law, P.T. Andrews: J. Phys. C **21**, L655 (1988)
8.47 D.P. Woodruff, N.V. Smith: Phys. Rev. B **41**, 8150 (1990)
8.48 W.Y. Ching, J. Callaway: Phys. Rev. B **11**, 1324 (1975)
8.49 W. Altmann, V. Dose, A. Goldmann, U. Kolac, J. Rogozik: Phys. Rev. B **29**, 3015 (1984)
8.50 M.L. Cohen, V. Heine: In *Solid State Physics*, Vol. 24, ed. by H. Ehrenreich, F. Seitz, D. Turnbull (Academic, New York 1970) p. 37
8.51 E.W. Plummer: Surf. Sci. **152/153**, 162 (1985)
8.52 D. Heskett, K.-H. Frank, E.E. Koch, H.-J. Freund: Phys. Rev. B **36**, 1276 (1987)
8.53 N.D. Lang, W. Kohn: Phys. Rev. B **3**, 1215 (1971)
8.54 D.A. Papaconstantopoulos: *Handbook of the Band Structure of Elemental Solids* (Plenum, New York 1986)
8.55 R. de Groot: Private communication
8.56 D.P. Woodruff, T.A. Delchar: *Modern Techniques of Surface Science* (Cambridge University Press, Cambridge 1986)
8.57 T.E. Felter, R.A. Barker, P.J. Estrup: Phys. Rev. Lett. **38**, 1138 (1977)
8.58 M.K. Debe, D.A. King: J. Phys. C **10**, L303 (1977)
8.59 M.K. Debe, D.A. King: Surf. Sci. **81**, 193 (1979)
8.60 C.L. Fu, A.J. Freeman, E. Wimmer, M. Weinert: Phys. Rev. Lett. **54**, 2261 (1985)
8.61 R.F. Willis, N.E. Christensen: Phys. Rev. B **18**, 5140 (1978)
8.62 N.E. Christensen, R.F. Willis: J. Phys. C **12**, 167 (1979)
8.63 J. Schäfer, R. Schoppe, J. Hölzl, R. Feder: Surf. Sci. **107**, 290 (1981)
8.64 N.E. Christensen, B. Feuerbacher: Phys. Rev. B **10**, 2349 (1974)
8.65 W. Speier, R. Zeller, J.C. Fuggle: Phys. Rev. B **32**, 3597 (1985)
8.66 U. Kolac, Th. Fauster, V. Dose, W. Altmann: Solid State Commun. **54**, 791 (1985)
8.67 L. Ilver, A. Kovacs, J. Kanski, P.O. Nilsson, E. Sobczak: Physica Scripta **35**, 726 (1987)
8.68 I.R. Collins, A.D. Laine, P.T. Andrews, P.J. Durham: J. Phys.: Condensed Matter **3**, 5307 (1991)
8.69 B. Ginatempo, P.J. Durham, B.L. Gyorffy: J. Phys.: Condensed Matter **1**, 6483 (1989)
8.70 J.E. Inglesfield: Prog. Surf. Sci. **20**, 105 (1985)
8.71 M.I. Holmes, T. Gustafsson: Phys. Rev. Lett. **47**, 443 (1981)
8.72 K.E. Smith, G.S. Elliott, S.D. Kevan: Phys. Rev. B **42**, 5385 (1990)
8.73 I.R. Collins, A.D. Laine, P.T. Andrews: To be published
8.74 W. Drube, D. Straub, F.J. Himpsel, P. Soukiassian, C.L. Fu, A.J. Freeman: Phys. Rev. B **34**, 8989 (1986)
8.75 P.M. Echenique, J.B. Pendry: J. Phys. C **11**, 2065 (1978)
8.76 N.D. Lang, W. Kohn: Phys. Rev. B **7**, 3541 (1973)

8.77 N.V. Smith, C.T. Chen, M. Weinert: Phys. Rev. B **40**, 7565 (1990)
8.78 S. Schuppler, N. Fischer, Th. Fauster, W. Steinmann: To be published
8.79 P.M. Echenique, F. Flores, F. Sols: Phys. Rev. Lett. **55**, 2348 (1985)
8.80 M. Donath, M. Glöbl, B. Senftinger, V. Dose: Solid State Commun. **60**, 237 (1986)
8.81 K. Giesen, F. Hage, F.J. Himpsel, H.J. Riess, W. Steinmann: Phys. Rev. B **35**, 971 (1987)
8.82 K. Giesen, F. Hage, F.J. Himpsel, H.J. Riess, W. Steinmann, N.V. Smith: Phys. Rev. B **35**, 975 (1987)
8.83 J.E. Inglesfield: Rep. Prog. Phys. **45**, 223 (1982)
8.84 D.P. Woodruff, S.L. Hulbert, P.D. Johnson, N.V. Smith: Phys. Rev. B **31**, 4046 (1985)
8.85 S.L. Hulbert, P.D. Johnson, N.G. Stoffel, W.A. Royer, N.V. Smith: Phys. Rev. B **31**, 6815 (1985)
8.86 N. Garcia, B. Reihl, K.H. Frank, A.R. Williams: Phys. Rev. Lett. **54**, 591 (1985)
8.87 J.B. Pendry, C.G. Larsson, P.M. Echenique: Surf. Sci. **166**, 57 (1986)
8.88 R.O. Jones, P.J. Jennings, O. Jepsen: Phys. Rev. B **29**, 6474 (1984)
8.89 G.C. Aers, J.E. Inglesfield: Surf. Sci. **217**, 367 (1989)
8.90 M. Weber, A. Liebsch: Phys. Rev. B **35**, 7411 (1987)

9. Further Topics in Low-Energy Inverse Photoemission

R. Schneider and V. Dose

With 22 Figures

More than a decade of work on inverse photoemission (IPE) has considerably advanced the field of research and several surveys at the respective level of maturity have appeared [9.1–4]. Initially the development was oriented along known results from ordinary photoemission (PES), which were simply reproduced for the case of unoccupied levels. Gradually, problems arose and were tackled which had no precedent in ordinary PES [9.5]. The present situation is characterized by a full equivalency of PES and IPE. Consequently, the physical problem dominates over the choice of technique. In this sense we extend Chap. 8 by raising some recent topics in band-structure spectroscopy, which in the present context draw exclusively on IPE data.

9.1 Synopsis

9.1.1 The Basic Concept of IPE

IPE uses the effect that electrons impinging on a solid surface may emit radiation (some experimental details are given in Chap. 1). This phenomenon has been known for ninety years as X-ray emission. The X-ray emission may be divided into bremsstrahlung emission, associated with relaxation of the high-energy electrons to lower energies, and fluorescence, associated with radiative decay of deep holes created by the electrons. This chapter describes features associated with the bremsstrahlung below about 30 eV and because of the energy-dependent small elastic mean-free path of 10 to 20 Å for low-energy electrons with kinetic energies below 30 eV [9.6], IPE is very surface sensitive.

In IPE electrons with well-defined kinetic energies E_{Kin} and well-defined angle of incidence θ impinge on the sample and couple to bulk states of the solid, which are lying above the vacuum level E_{Vac} of the sample. From the initial state with energy E_{i} and wave vector $\boldsymbol{k}_{\mathrm{i}}$ the electrons can decay via radiative transitions to other lower lying unoccupied electronic final states with energy E_{f} and wave vector $\boldsymbol{k}_{\mathrm{f}}$. In the IPE experiment the emitted photons with a quantum energy $\hbar\omega$ are detected for a photon take-off angle α. For this process the conservation of energy reads

$$E_{\mathrm{i}} = E_{\mathrm{f}} + \hbar\omega. \qquad (9.1)$$

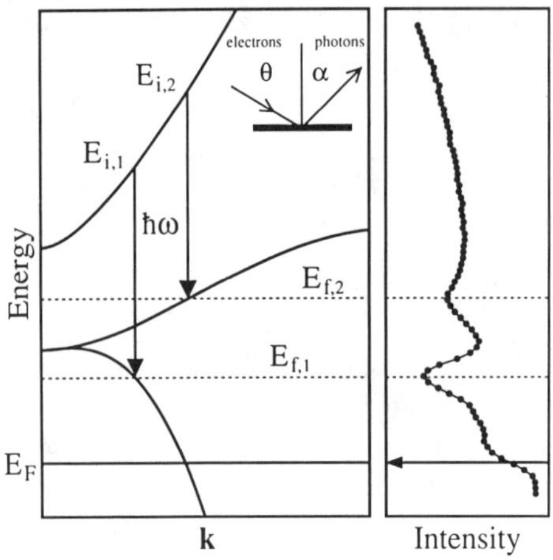

Fig. 9.1. Principle of inverse photoemission. A peak appears in the measured spectra, when the energy difference between initial and final state is equal to the quantum energy

Using the momentum conservation for the radiative transition, we obtain

$$\boldsymbol{k}_i = \boldsymbol{k}_f + \boldsymbol{G} + \boldsymbol{q}, \tag{9.2}$$

where $\boldsymbol{G}$ is a reciprocal lattice vector. For photon energies below about 100 eV the wave vector of the photon $\boldsymbol{q}$ remains small compared to the size of the Brillouin zone, e.g. $\boldsymbol{G}$ and $\boldsymbol{q}$ can be neglected in the momentum balance. Therefore, in a reduced zone scheme (left part of Fig. 9.1), the radiative transition occurs vertically ($\boldsymbol{k}_i = \boldsymbol{k}_f$) as a so-called direct transition. In an IPE experiment with fixed photon energy $\hbar\omega$ the intensity of the radiation (count rate of the emitted photons) is measured as a function of the final-state energy E_f (right part of Fig. 9.1) for an angle of incidence θ of the electrons and a photon take-off angle α. In such a spectrum a peak should appear at energies E_f, where a direct transition between two bands separated by $\hbar\omega$ is possible. The width of the peaks is determined by the finite lifetimes of the final states. The measured spectrum has its onset at the Fermi level E_F and shows an energy-dependent, nearly structureless background, which stems mainly from radiative transitions after electron–hole pair production. Superimposed on this background, Fig. 9.1 shows two direct transitions.

9.1.2 The Aim of this Chapter

The main aim of this chapter is to show that the complete system (bulk with surface and vacuum) must be regarded to fully understand the IPE measurements.

In Sect. 9.2 we will discuss the band mapping in mirror planes for bulk and surface states. The existence of the surface results in the fact that only the wave vector parallel to the surface $\boldsymbol{k}_\parallel$ is a good quantum number, whereas the wave vector perpendicular to the surface $\boldsymbol{k}_\perp$ of the incoming electrons changes. The observed bulk transitions will be compared with theoretical predictions for the optical curves in projected bulk band structure of a combined-interpolation-scheme calculation. If an identical transition can be observed on different surfaces this transition can be triangulated and $E(\boldsymbol{k})$ can be obtained.

Measurements of the angular distribution of the radiation in IPE require the variation of photon take-off angle α. As will be shown in Sect. 9.3, the radiation angular distribution follows quite well the dipole model, resulting in a squared sine function dependence of the intensity of the transition from the photon take-off angle relative to the orientation of the transition dipole. In addition the data are modified by the transmission probability of photons through the solid–vacuum interface, qualitatively following the classical Fresnel formulas.

Electronic states created by adsorbates are discussed in Sect. 9.4 as generalized "surface" states. Our definition includes surface umklapp processes. These occur because of the change of the surface geometry and symmetry by an adsorbate leading to different diffraction possibilities prior to the radiative transition. Bulk transitions can become visible through surface umklapps. For the identification of this effect, measurements with variable photon energy are necessary. The influence of the bulk on adsorbate states is also discussed. This discussion shows clearly that the separation into bulk, surface and adsorbate states is purely artificial.

Temperature-dependent IPE transitions will be presented in Sect. 9.5. The temperature dependence of intensities will be shown to be mainly due to surface vibrations. This is proved by comparing the mean square displacements extracted from the temperature dependence of IPE bulk transitions on different surfaces and from scattering experiments, which are sensitive to vibrations in the surface.

We conclude our chapter with a discussion of recent results from IPE experiments with spin-polarized electrons impinging on ferromagnetic samples.

9.2 Band Mapping in Mirror Planes

We recall from Sect. 9.1 that IPE in the ultraviolet spectral range is suitable for determining the final state energies E_f as a function of the angle of incidence θ of the electrons. The wave vectors $\boldsymbol{k}_\mathrm{i}$ and $\boldsymbol{k}_\mathrm{f}$ entering momentum conservation (9.2) refer to the bulk solid whereas the initial conditions of the experiment specify only the wave vector $\boldsymbol{K}$ in vacuum. Upon penetration of the vacuum–solid boundary, the incident electron experiences an attractive force of unknown magnitude normal to the surface. Therefore, the component of the wave vector in the solid perpendicular to the surface $\boldsymbol{k}_\perp$ is greater than $\boldsymbol{K}_\perp$ in the vacuum

$$\boldsymbol{k}_\perp > \boldsymbol{K}_\perp . \tag{9.3}$$

The forces parallel to the surface are periodic and lead to diffraction such that

$$\boldsymbol{k}_{\mathrm{i},\parallel} = \boldsymbol{K}_{\mathrm{i},\parallel} + \boldsymbol{G}_\parallel , \tag{9.4}$$

where $\boldsymbol{G}_{\|}$ is a surface reciprocal lattice vector. At sufficiently low quantum energy $\hbar\omega$ and hence initial energy, the requirement that $|\boldsymbol{k}_{i,\|} - \boldsymbol{G}_{\|}| \leqq |\boldsymbol{K}_i|$ leads to a preference for $\boldsymbol{G}_{\|} = 0$. $K_{i,\|}$ is given by

$$K_{i,\|} = (2m/\hbar^2(E_f + \hbar\omega - \phi_P))^{1/2} \sin\theta, \tag{9.5}$$

where E_f is the final-state energy with respect to the sample Fermi level and ϕ_P is the sample work function. Therefore, one can extract from the experimental data the final-state energies E_f as a function of the wave vector parallel to the surface, $E_f(k_{\|})$. Comparing the measured $E_f(k_{\|})$ data with predicted transitions from band-structure calculations is a very important test for these calculations. Since the band structure of even simple crystals is quite complicated at general points in reciprocal space, the experiments are usually done in mirror planes. Figure 9.2 shows the bulk Brillouin zone of an fcc lattice. The shaded planes are the ΓXUL and ΓXWK mirror planes of the crystal. In these mirror planes different bands may become degenerate. This reduces the complexity of the observed spectra. Also only transitions from even initial states can occur, because the wavefunction of the incoming electron is a plane wave and has even parity. It can therefore couple only to those bulk electronic wavefunctions which have also even parity [9.7]. This leads again to a reduction of the number of possible transitions and a comcomitant simplification of the observed spectra.

The volume Brillouin zone is projected onto the surface Brillouin zone as shown in Fig. 9.2 for a (100) surface of an fcc crystal. The surface reciprocal mesh is the projection of the bulk reciprocal lattice on to the surface plane. The two mirror planes ΓXUL and ΓXWK correspond to the $\overline{\Gamma X}$ and $\overline{\Gamma M}$ directions of the surface Brillouin zone in Fig. 9.2. Figure 9.3 shows experimental

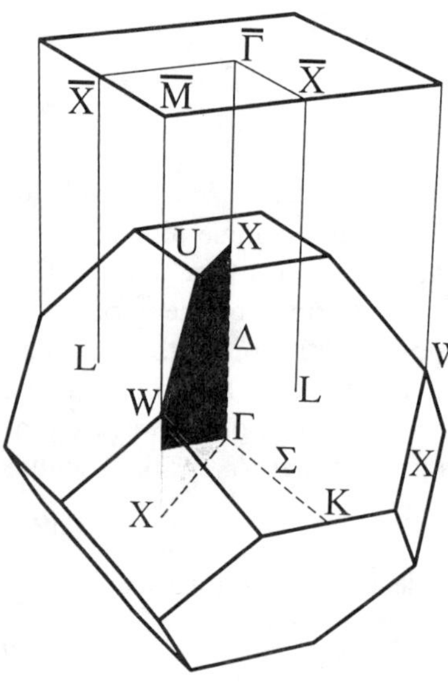

Fig. 9.2. Volume and surface Brillouin zone of a face centered cubic (100) crystal

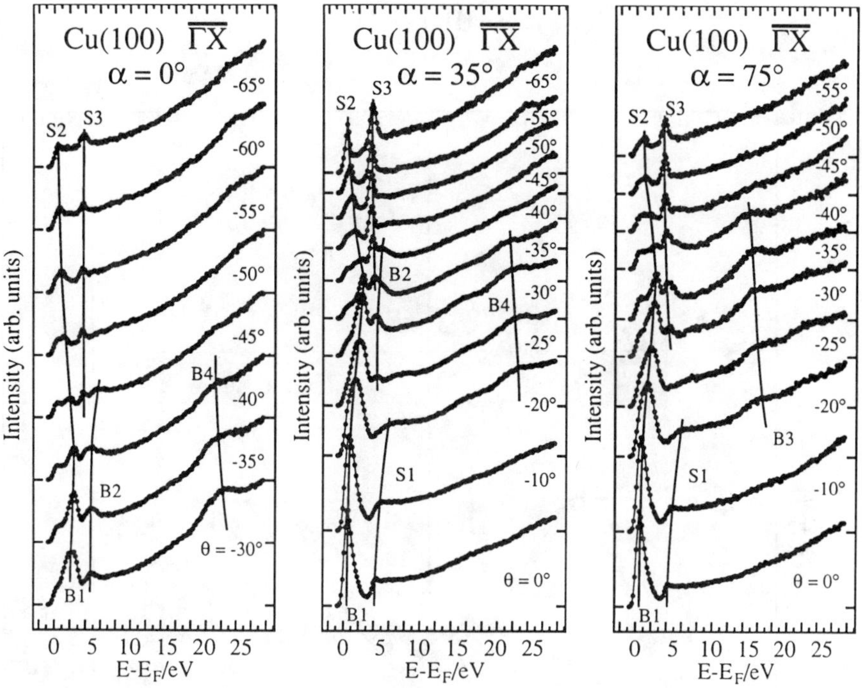

Fig. 9.3. IPE spectra from Cu(100) at various polar angles θ of the electrons in the $\overline{\Gamma X}$ direction for three photon take-off angles $\alpha = 0°$, 35° and 75°

IPE spectra for Cu (100). The quantum energy is 9.6 eV [9.8]. Three sets of experimental spectra are shown for three photon take-off angles $\alpha = 0°, 35°, 75°$ and a variation of the polar angle of incidence θ in the $\overline{\Gamma X}$ direction. Measurements with constant photon take-off angle α are possible by a rotatable electron gun and an independently rotatable sample [9.9]. The intensity of the observed radiation is displayed as a function of final-state energy E_f. Seven different emission features are present. The peaks B1–B4 are transitions into bulk states, whereas S1–S3 are transitions into surface states. This identification will be discussed later. The measurement at different photon take-off angle α allows a separation of B2 and S3 [9.10], which have strongly different radiation angular distributions (see Sect. 9.3). From the spectra of Fig. 9.3 one can extract the experimental $E_f(k_\parallel)$ data. These are shown in Fig. 9.4 for energies up to 8 eV as filled and open symbols with error bars for transitions into bulk and surface states, respectively. Data for the ΓXWK mirror plane are also included. Open symbols without error bars are transitions predicted by a combined-interpolation-scheme calculation [9.9, 11]. The size of the symbols indicates the intensity of the transition. We shall now attempt to interpret the measured bulk transitions in the frame of the well-known three-step model [9.12]. This model was developed

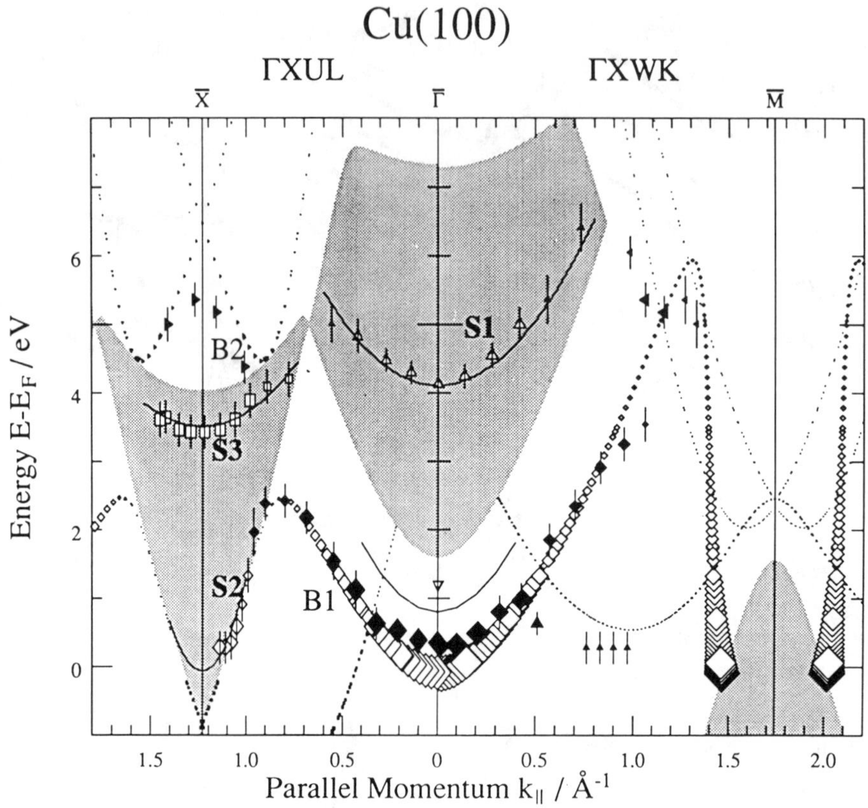

Fig. 9.4. $E(k_{\|})$ diagram for Cu(100). Transitions predicted from the combined-interpolation-scheme calculation are shown by the open symbols without error bars. The different symbols belong to transitions between different bands, characterized by the dominant plane wave component of the wavefunction of the initial and final state $\boldsymbol{G}_i$ and $\boldsymbol{G}_f$. The experimental data for bulk transitions are shown as filled symbols with error bars. The size of the symbols indicates the intensity of the transition. The symbols of the measured bulk transitions were chosen to correspond to the calculated transitions. The band gaps of the projected bulk band structure are shown as grey shaded areas. The data for transitions into surface states are marked as open symbols with error bars

for PES [9.13] and then transferred to IPE [9.14]. It gives a good framework for initial discussion, although the three steps are not strictly separable.

1. The incoming electrons couple to a bulk band of the crystal.
2. From the initial state $E_i(\boldsymbol{k}_i)$ electrons are lost by non-radiative decay in the crystal.
3. A direct optical transition occurs.

The intensity I for photons with energy $\hbar\omega$ and for incoming electrons with energy E can be expressed as

$$I(\hbar\omega, E) \sim \sum_{i,f} \int_{\Omega} d^3\boldsymbol{k} |M_{fi}|^2 c_i \delta(E_i(\boldsymbol{k}) - E_f(\boldsymbol{k}) - \hbar\omega) \delta(E - E_i(\boldsymbol{k})). \quad (9.6)$$

The first δ-function in (9.6) generates the conservation of energy (9.1). The second δ-function defines the energy of the initial state as the energy of the incoming electron. The coupling probability to bulk bands is expressed by the coefficients c_i. These coefficients are the overlap integrals between the initial state wavefunction and free-electron states in the crystal. Therefore, they are large if a bulk band is free-electron like [9.11]. The transition matrix elements for PES and IPE M_{fi} are determined by the operator $\boldsymbol{A}\cdot\boldsymbol{p}$ [9.4] (see Sect. 9.3), the scalar product of the vector potential $\boldsymbol{A}$ and the momentum operator $\boldsymbol{p}$. For radiation with wavelength λ such that $\lambda/a \ll 1$ where a is the lattice constant of the crystal, the variation of the vector potential across the unit cell can be neglected and the dipole approximation is obtained [9.2]. Its practical validity extends to well beyond 100 eV. The transition matrix element is then determined by the momentum matrix element $\boldsymbol{p}_{fi}$. Calculating $\boldsymbol{p}_{fi}$ gives the intensity of the transition and the orientation of the transition dipole. All subsequent calculations of the momentum matrix elements $\boldsymbol{p}_{fi}$ are based on the combined interpolation scheme [9.15], which was fitted at high-symmetry points to a self-consistent band-structure calculation of *Bross* and *Schiekel* [9.16].

Figure 9.5 displays the k-space location of the possible transitions with a photon energy of 9.6 eV from even initial states into final states in the energy range -1 to 8 eV in the two mirror planes. The transitions from odd initial states have been omitted since they cannot be observed in the IPE experiments (see previous discussion). The arrows in Fig. 9.5 indicate the direction of the surface normals for the three low-index surfaces (100), (110), and (111). The different symbols belong to transitions between different bands, characterized by the respective dominant plane wave component of the wavefunction of the initial and final state. Most of the calculated transitions have too small matrix elements to be visible in the experiment. The main transition is the bulk transition marked with a diamond (B1) [$\boldsymbol{G}_i = (0\,0\,\bar{2})$, $\boldsymbol{G}_f = (0\,0\,0)$]. The second experimentally observed bulk transition B2 has $\boldsymbol{G}_i = (\bar{1}\,\bar{1}\,1)$ and $\boldsymbol{G}_f = (\bar{1}\,\bar{1}\,\bar{1})$. Within a

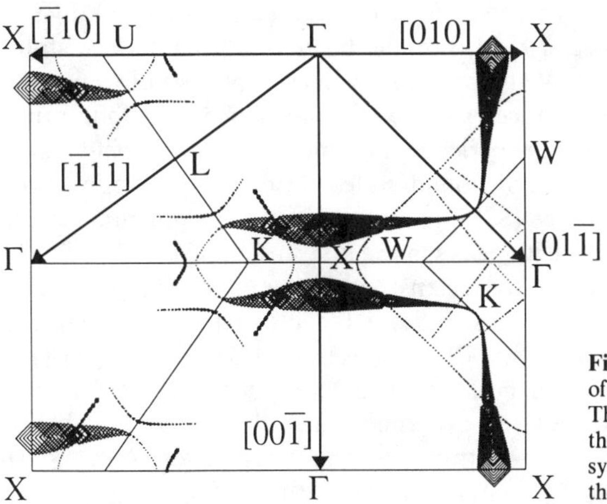

Fig. 9.5. Predicted k-space location of the transitions displayed in Fig. 9.4. The arrows indicate the directions of the surface normals. The size of the symbols is scaled with the square of the calculated matrix elements

simple two-band model [9.14] the momentum matrix element $\boldsymbol{p}$ is determined by the exchanged reciprocal lattice vector $\boldsymbol{G}$, which is the difference between the initial-state reciprocal lattice vector $\boldsymbol{G}_\mathrm{i}$ and the final-state reciprocal lattice vector $\boldsymbol{G}_\mathrm{f}$

$$\boldsymbol{p} = \boldsymbol{G} V_G / \omega . \tag{9.7}$$

V_G is the coefficient of the plane wave $\exp(\mathrm{i}\boldsymbol{G}\cdot\boldsymbol{r})$ in the Fourier expansion of the crystal potential. $V_G \exp(\mathrm{i}\boldsymbol{G}\cdot\boldsymbol{r})$ mediates a transition between the two free-electron-like *sp*-bands. The size of the symbols in Figs. 9.4 and 9.5 is scaled with the square of the calculated dipole matrix element. For transitions characterized by identical exchanged reciprocal lattice vector, intensity changes indicate the inadequacy of the two-band model, where one would expect from (9.7) a constant intensity for these transitions.

The predicted dipole orientation for the transitions can be checked experimentally as will be discussed in the next section. Together with a very good overall agreement in the $E(k_\parallel)$-curves for the bulk transitions, a highly reliable determination of the reciprocal lattice vectors $\boldsymbol{G}$, exchanged in the transition, is obtained. The results of this analysis provides the basis for the discussion of Sect. 9.5.

The existence of the surface complicates the discussion. The coupling of the incoming electron to the initial state can change the observed intensity compared to the predicted intensity. This effect can be quite large, such that the order of intensities as obtained from the combined interpolation scheme calculation can even be reversed. An example of the importance of initial-state coupling effects is contained in Fig. 9.4. Choosing a normalization such that observed and calculated intensities for bulk transition B1 match approximately we find a considerable discrepancy for B2. Obviously, the initial-state coupling probability is much higher for B2 than for B1.

Band gaps of the projected bulk band structure are shown in Fig. 9.4 as grey shaded areas. The transitions into surface states are marked as open symbols with error bars. The surface states, lying in or close to the band gaps, can be divided into two classes [9.17]. For the barrier-induced surface states (S1) at normal incidence at 4.1 eV the probability density peaks in front of the surface in the vacuum. Barrier-induced surface states result from the long-range image potential which an electron experiences in front of a surface. Briefly, these image states form Rydberg series of bound states close to the vacuum level. The crystal-induced surface states (S2, S3) are sensitive to the potential at the surface and their probability density peaks in the first surface layers. Both kinds of surface states have been discussed extensively in the literature [9.2, 4]. In Fig. 9.4 the continuous-line parabolas indicate the dispersion of the surface states. For these parabolas, if necessary, PES results have also been used [9.18].

If a particular transition can be seen on different surfaces the full $E(\boldsymbol{k})$ dispersion can be compared between experiment and theory [9.19–21]. For copper we find at higher energy a bulk transition which can be seen on Cu(110) and (111). On Cu(111) it shows up only in the $\overline{\Gamma M}$ (ΓKL) direction due to the initial state effect discussed before, because the coupling of the initial state wavefunction to the bulk states is different for $\overline{\Gamma M}$ and $\overline{\Gamma M'}$ (ΓXUL). This results in the $\overline{\Gamma M'}$ direction in a total invisibility of this transition, although the matrix element is the same as in the $\overline{\Gamma M}$ direction. In Fig. 9.5 the $E(k_{\|})$ dispersion of this bulk state on Cu (110) (squares) and on Cu (111) in the $\overline{\Gamma M}$ direction (diamonds) is shown.

The triangulation is performed in the following way: From Fig. 9.6 one extracts pairs of $k_{\|}$ values for the (110) and (111) surface for a given final state energy (hence the alternative appellation 'energy coincidence method'). The respective $\boldsymbol{k}_{\|}$ defines a (dashed) line in Fig. 9.7 parallel to the (110) and (111) direction indicated by the light arrows. The fat arrows indicate the $k_{\|}$ directions for the (110) and (111) surface in the mirror plane. The filled diamonds at the intersections of these lines are the desired absolute $\boldsymbol{k}$-space locations of the transitions. The size of the diamonds indicates the experimental error in the triangulation, resulting from the uncertainty in the energy and $\boldsymbol{k}_{\|}$. The uncertainty in the energy is shown in Fig. 9.6, whereas the error in $\boldsymbol{k}_{\|}$ is 0.1 Å^{-1} due to the limited angular resolution of the electron gun of $\pm 2°$. The open symbols in Figs. 9.6 and 9.7 represent calculated results. From these calculations the initial- and final-state characters have also been identified to be predominantly $(1\,1\,\bar{1})$ and $(\bar{1}\,1\,\bar{1})$ respectively. In contrast to the experimental observation two branches are predicted for the (111) surface. Failure to observe the predicted second

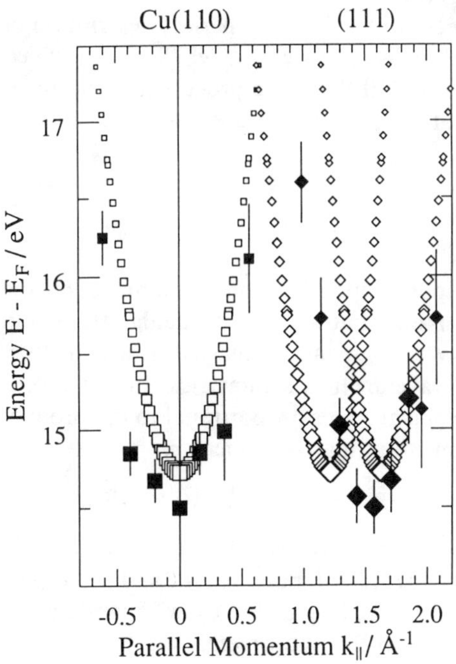

Fig. 9.6. $E(k_{\|})$ diagram for the higher energy bulk transition, which can be observed on Cu(110) (squares) and Cu(111) (diamonds). The open symbols are the results of the combined-interpolation-scheme calculation, the filled symbols with error bars show the experimental data

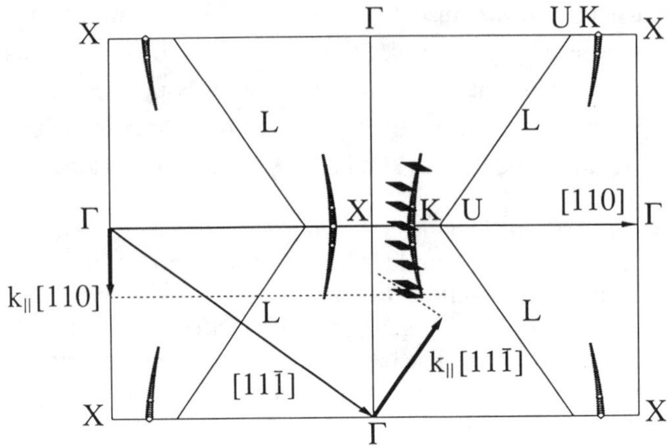

Fig. 9.7. Triangulation of the higher energy bulk transition of Fig. 9.6 in the ΓXUL mirror plane. The fat arrows indicate the $k_{\|}$-directions for the (110) and (111) surface in the mirror plane, which are perpendicular to the surface normals marked as arrows. The results of the combined-interpolation-scheme calculation are shown as open diamonds

branch is attributed to unfavourable initial-state coupling. The experimental triangulation results show a good agreement with the combined-interpolation-scheme calculation (open diamonds). This very same transition will be used in Sect. 9.5 to discuss the strong influence of surface effects on the temperature dependence of intensities of bulk transitions.

Finally, we want to note in this section that a nearly perfect description of the measured spectra can be obtained in the one-step model [9.22, 23] which treats the measurement process of IPE (or PES) as one process in a multiple-scattering formalism, including the scattering at the surface barrier.

9.3 Radiation Angular Distribution

An experimental validation of the dipole approximation is obtained by the measurement of the radiation angular distribution of a particular transition [9.24]. Such measurements also allow a polarization analysis which in turn provides a further stringent test on wavefunction symmetries. In the dipole approximation the transition matrix element M_{fi} is proportional to the product of the vector potential $\boldsymbol{A}$ and the momentum matrix element $\boldsymbol{p}_{\mathrm{fi}}$

$$M_{\mathrm{fi}} \sim \boldsymbol{A} \cdot \boldsymbol{p}_{\mathrm{fi}} = A_x \langle \psi_{\mathrm{f}} | \partial/\partial x | \psi_{\mathrm{i}} \rangle + A_y \langle \psi_{\mathrm{f}} | \partial/\partial y | \psi_{\mathrm{fi}} \rangle + A_z \langle \psi_{\mathrm{f}} | \partial/\partial z | \psi_{\mathrm{fi}} \rangle . \tag{9.8}$$

In Fig. 9.8 the expected angular dependence for pure A_x, A_y and A_z dipole transitions is shown. They should be described by a $\sin^2 \delta$ law. δ is the angle between the dipole axis and the direction of the photon emission.

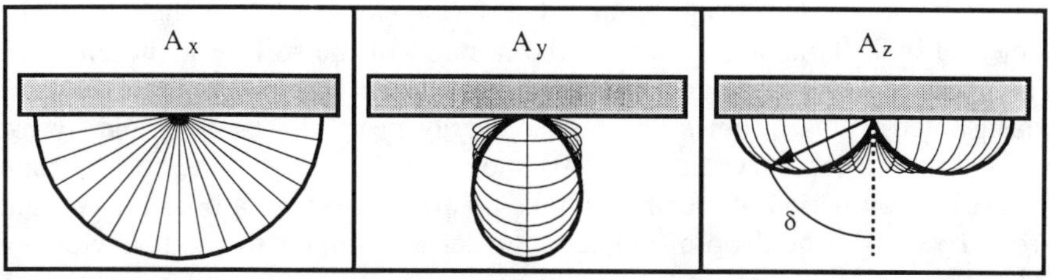

Fig. 9.8. Radiation angular distribution for pure A_x, A_y and A_z dipole transitions. δ is the angle between the dipole axis and the direction of the photon emission

Figure 9.9 shows a set of inverse photoemission spectra for various photon take-off angles α normalized to equal intensity at a final-state energy E_f of 5.7 eV. For the measurement of the angular distribution of the emitted radiation a rotatable electron gun was used. The electrons impinged at an angle $\theta = -60°$ relative to the surface normal of the Cu(100) crystal. Electrons and photons propagated in the ΓXUL mirror plane. We can identify two peaks in the spectra at 0.6 eV and 3.5 eV above the Fermi level. These transitions have been presented

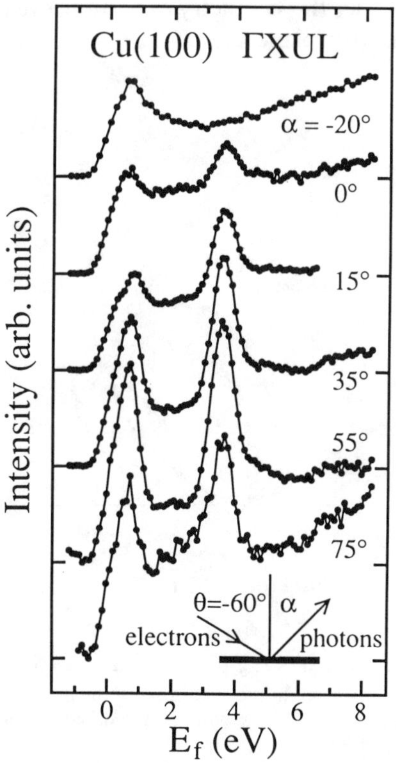

Fig. 9.9. Set of IPE spectra for various photon take-off angles α normalized to equal intensity at a final-state energy of $E_f = 5.7$ eV. The electrons impinged at an angle $\theta = -60°$ relative to the surface normal of the Cu(100) crystal. Electrons and photons propagated in the ΓXUL mirror plane

in Sect. 9.2 as the two crystal-induced surface states (S2, S3) in the gap of the projected bulk band structure near the $\bar{X}$ point of the surface Brillouin zone. The peaks remain stationary with changing photon take-off angle α but show strong intensity variations. The higher energy peak vanishes completely for $\alpha = -20°$. The top curve in Fig. 9.10a shows the intensity of the surface state at 3.5 eV as a function of the photon take-off angle α. We see a broad maximum around $\alpha = 30°$. The drop-off for large angles α is similar to the behaviour of the background measured at 5.7 eV. We now assume that the angular dependence of the background represents the transmission probability for the photons through the solid–vacuum interface. If we then normalize the intensity of the surface state emission to the background, we obtain the dotted curve in the bottom of Fig. 9.10a. This curve should represent the emission probability for this transition as a function of photon take-off angle. A least squares fit to $f(\delta) = A_x + A_{y,x} \sin^2 \delta$ reproduces the measured data excellently. In Fig. 9.10b this data correction is shown as a polar plot. In this particular case we obtain further $A_x = 0$ as could be anticipated from Fig. 9.9, since the intensity of the transition disappears completely for $\alpha = -20°$. The fit result $A_x = 0$ implies that the transition dipole has no component perpendicular to the ΓXUL mirror plane. Symmetry-selection rules for dipole transitions [9.25] in a mirror plane require that A_x or $A_{y,z}$ is equal zero. The initial-state wavefunction in the vacuum must always be even under reflection with respect to the mirror plane [9.26]. Since the matrix element must be invariant under the symmetry operation, even final states have a dipole axis lying in the mirror plane ($A_x = 0; A_{y,z} \neq 0$). Odd

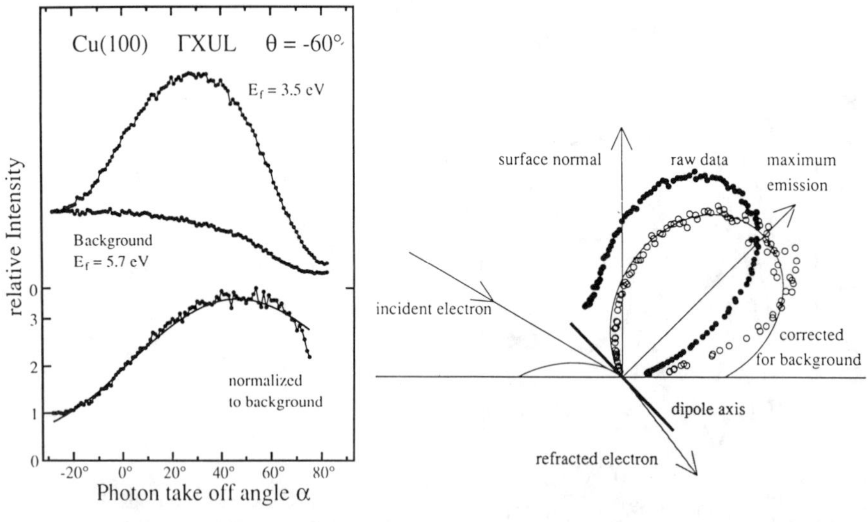

Fig. 9.10. (**a**) Intensity of the surface state at $E_f = 3.5$ eV and of the background at $E_f = 5.7$ eV of Fig. 9.9 as a function of the photon take-off angle α (upper part). In the lower part of the intensity of the surface state normalized to the background as a function of the photon take-off angle α is shown. (**b**) Polar plot of the data for the surface of Fig. 9.10a

final states have a dipole axis oriented normal to the mirror plane ($A_x \neq 0$; $A_{y,z} = 0$). The data show conclusively, therefore, that this particular transition can be described as a dipole transition into an even final state and that there is no odd state at the same energy. The dipole axis encloses an angle of $\sim 45°$ with the surface normal. At the first glance this is somewhat surprising for a crystal-induced surface state which owes its existence mainly to the potential step at the surface and should, therefore, have a dipole axis normal to the surface. However, away from normal incidence surface corrugation [9.27] introduces also x, y components in the transition dipole, resulting in a dipole axis tilted away from the surface normal. For the particular case for the transition in Fig. 9.10, the dipole axis is oriented approximately parallel to the incident (or refracted) electron beam, reminiscent of the classical picture of bremsstrahlung emission by a decelerated electron (see Fig. 9.10b).

In Fig. 9.11 the experimental data on Cu(100) in the ΓXUL mirror plane for the dispersion of transitions with 9.6 eV photon energy (lower part) and the orientation of the dipole axis of these transitions (upper part) are compared with the results of the combined-interpolation scheme. The shaded area indicates the projection of the bulk energy bands onto the Cu(100) surface. The combined-interpolation scheme was used to calculate all bulk transitions at 9.6 eV photon

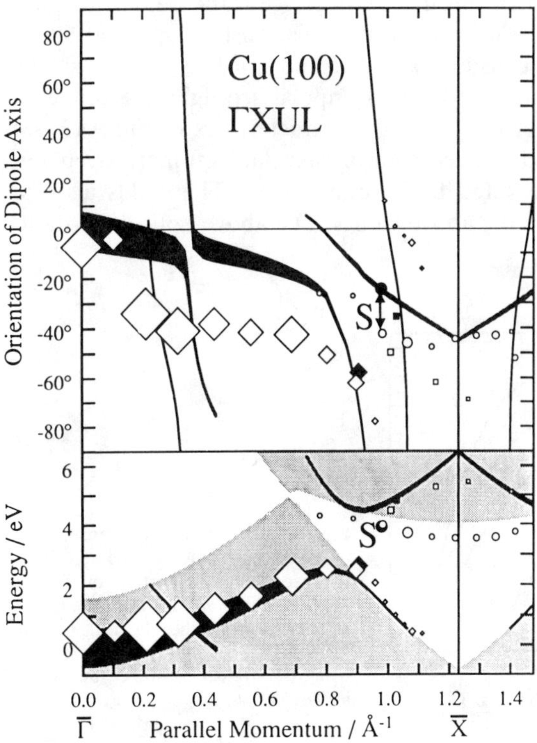

Fig. 9.11. Experimental data on Cu(100) in the ΓXUL mirror plane for the dispersion of transitions with 9.6 eV photon energy (lower part) and the orientation of the dipole axis of these transitions (upper part) are compared to the results of the combined-interpolation scheme

energy with even initial state. These are shown by the dark bands in Fig. 9.11. The dispersion agrees well with the measured bulk states. The width of the bands corresponds to the calculated intensity of the transition. All observed transitions are into even final states which means all the dipole axes lie in the ΓXUL mirror plane. For the experimental data the dipole axis is determined by fits such as the one shown in the bottom part of Fig. 9.10a. These fits yield also the maximum intensity of the dipole emission, which is represented by the size of the symbols. The intensities agree qualitatively with the calculations. For the bulk transitions the measured orientation of the dipole axis follows approximately the calculated lines, but is shifted $\sim 20°$ away from the surface normal. This discrepancy for bulk states is not a shortcoming of the combined-interpolation scheme but persists also with theoretical results from the IPE one-step model [9.23]. Such data are shown by filled symbols in Fig. 9.11. For the surface states (data points outside the shaded regions in the bottom part of Fig. 9.11) only one calculated data point (S) is available showing also a $\sim 20°$ deviation from the experimentally determined dipole orientation. Since this discrepancy is observed for almost all data points, we have to discuss the possibility of a systematic error or omission in the data analysis. The quality of the fits allows the determination of the angle to 5° or better (see Fig. 9.10a). The most critical assumption is the normalization of the data to the background, as shown in Fig. 9.10a. In order to access the background angular distribution, the simplest model for the transmission of light through an interface is the classical Fresnel formulae [9.28]. In Fig. 9.12 the calculated transmission for copper at 9.6 eV photon energy with the optical constants $\varepsilon_r = 0.35$ and $\varepsilon_i = 1.75$ [9.29] is shown for s ($A_x = 0, A_{y,z} \neq 0$), p ($A_x \neq 0, A_{y,z} = 0$) and unpolarised light. Assuming an unpolarised background, we get the radiation characteristics of Fig. 9.13 (see Fig. 9.8 for comparison). The calculated transmission falls off more steeply at angles above 60° than the measured background curve. This holds also for optical constants far away from the above values. The above optical constants

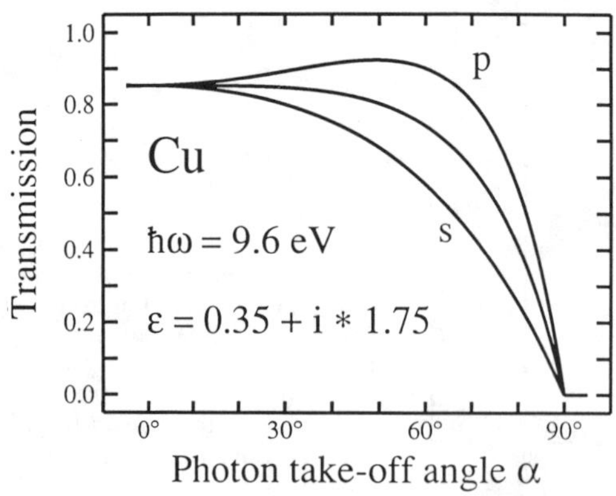

Fig. 9.12. Probability for transmission of a photon created in the solid into the vacuum as a function of photon take-off angle α. The transmission is calculated from the classical Fresnel formulas for s, p and unpolarised light using the optical constants for copper at 9.6 eV

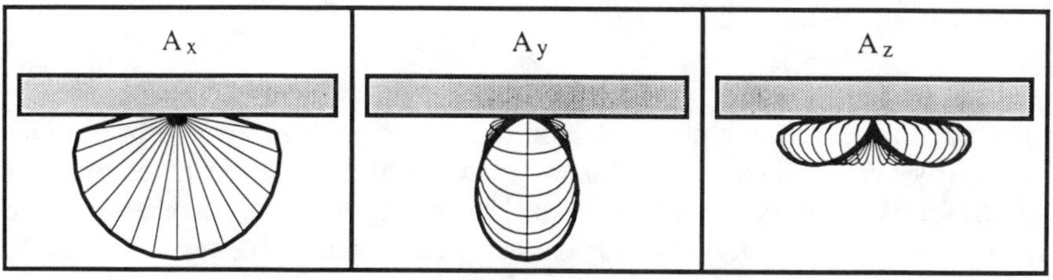

Fig. 9.13. Radiation angular distribution for pure A_x, A_y and A_z dipole transitions corrected with the transmission for unpolarised background

give total reflection for light impinging at an angle larger than 37° onto the solid–vacuum interface. We would observe, therefore, mainly transitions with a dipole axis oriented far away from the surface normal. This is in contrast to the observation which gives strong intensity for $k_{\parallel} = 0$. Symmetry considerations alone dictate a dipole axis parallel to the surface normal for $k_{\parallel} = 0$. If we include refraction, as implied by Fresnel formulae in the data analysis procedure, then we obtain considerably worse fits and an even larger discrepancy to the calculations. Fresnel corrections are obviously not suitable for describing our experimental results for inverse photoemission originating from the first few atomic layers of the surface. Local-field theories of the electromagnetic field near a surface [9.30, 31] give only small corrections to the Fresnel formulae which would not resolve the discrepancies.

For a possible explanation, we have to remember how the electromagnetic field is introduced into the Hamiltonian of the system. The momentum operator $\boldsymbol{p}$ of the system without electromagnetic field is replaced by the generalised momentum $\boldsymbol{p} - e/c\boldsymbol{A}$. Neglecting the quadratic term in $\boldsymbol{A}$ gives the transition matrix element M_{fi}

$$M_{\mathrm{fi}} = \langle \psi_{\mathrm{f}} | \boldsymbol{A} \cdot \boldsymbol{p} + \boldsymbol{p} \cdot \boldsymbol{A} | \psi_{\mathrm{i}} \rangle , \tag{9.9}$$

where ψ_{f} and ψ_{i} are the exact many-body wavefunctions of the system. Equation (9.9) can be rewritten to

$$M_{\mathrm{fi}} = 2 \langle \psi_{\mathrm{f}} | \boldsymbol{A} \cdot \boldsymbol{p} | \psi_{\mathrm{i}} \rangle - \mathrm{i}\hbar \langle \psi_{\mathrm{f}} | \nabla \boldsymbol{A} | \psi_{\mathrm{i}} \rangle . \tag{9.10}$$

In practical calculations the exact wavefunctions are replaced by one-particle wavefunctions and the dipole approximation is used. In the normally used Coulomb gauge, the second term of (9.10) is zero and the transition matrix element depends on $\boldsymbol{A} \cdot \boldsymbol{p}$ as discussed before. Many-body effects can be included in the vector potential $\boldsymbol{A}$, which is the dielectric response of the system to the radiation field [9.32]. But the second term is not equal to zero if longitudinal electromagnetic waves (plasmons in solids) or rapidly changing transversal fields (due to the surface barrier) exist. Therefore the second term can contribute to the transition matrix element and perhaps this gives the correction for the theoretical calculated data.

9.4 Adsorbate-Induced States

The study of the bonding of an adsorbate to a surface is motivated by the fact that nearly every chemical reaction with technical application takes place at the surface of catalysts. The first step of these reactions is the adsorption of the reaction partners. This very first step is a subject of considerable current research. IPE adds the possibility of studying the adsorbate-induced unoccupied electronic states at surfaces. For a reduction of the complexity of the problem mostly well-ordered systems are studied. For the purpose of this chapter we define adsorbate-induced unoccupied electronic states as all states which emerge in the spectra after adsorption in addition to features of the clean surface. The adsorbate-induced states are a consequence of the additional three-dimensional potential of the adsorbate layer. In principle these states are equivalent to surface states which are produced by the approximately one-dimensional surface barrier. Several 'types' of adsorbate-induced states can occur, although this classification is artificial because they are all produced by the total three-dimensional system adsorbate plus substrate.

1. The new geometry of the surface can produce surface umklapp processes. By the exchange of a reciprocal lattice vector of the adsorbate-covered surface, bulk states can appear in the spectra which are absent in the clean surface spectra.
2. The adsorbate's intrinsic electronic states are influenced by direct adsorbate–adsorbate interaction at the surface or by indirect adsorbate–adsorbate interaction mediated by the substrate. This latter case has frequently been slighted by surface scientists. In the case of substrate-mediated adsorbate–adsorbate interaction, electronic states may exhibit three-dimensional dispersion. The experimental results are then not only determined by the adsorbate layer symmetry, but also by the symmetry of the underlying substrate.
3. Due to the changed surface geometry the coupling of the incoming electron to the initial states changes. Normally this results in a strong intensity decrease of bulk direct transitions observed on the clean crystal. But sometimes bulk transitions with high transition matrix elements become visible upon formation of the adsorbate layer. Such states have escaped detection on the clean surface by the low coupling probability. This coupling probability is increased by the adsorbate.
4. The adsorbate–substrate system has a new surface potential compared to the clean surface. Therefore, surface states can shift, disappear or appear (a very recent example constitutes the hydrogen adsorption on Ni(110) [9.33]).

The remainder of this section will deal with examples for points one and two given above.

9.4.1 Surface Umklapp

As a first example illustrating the three dimensionality of the problem, we consider bulk direct transitions on Ni(110) both on the clean and adsorbate-covered surface. A flexible apparatus is required allowing measurements at arbitrary k-points of the bulk Brillouin zone. In particular k-points differing only by the exchange of a surface reciprocal lattice vector must be accessible. In IPE (as in photoelectron spectroscopy) this requires variable quantum energy by using a monochromator [9.34, 35]. In Fig. 9.14 IPE spectra of a Ni(110) surface with a sulphur [$c(2 \times 2) - \mathrm{S}$], chlorine [$c(2 \times 2) - \mathrm{Cl}$] and oxygen [$(2 \times 1) - \mathrm{O}$] adsorbate structure are shown. The spectra are measured with a monochromator with variable quantum energy for constant initial-state energies from 16.8 eV up to 23.8 eV [9.36]. The spectra of the clean surface are shown as full lines for comparison. In the spectra of the clean surface with 23.8 eV initial-state energy, a direct bulk transition can be seen as a weak structure at 2.9 eV above the Fermi energy [9.37]. The adsorbtion of sulphur, chlorine and oxygen suppresses this bulk transition. In the Ni(110) $- c(2 \times 2)$S spectra, another peak at 4.8 eV shows no variation of its peak position with the energy of the initial state. It is an adsorbate state of sulphur and will not be discussed here. We will focus on the additional emission at about 2 eV final-state energy. This feature shows dispersion with the initial-state energy E_i and therefore, because of the normal incidence condition, with the wave vector perpendicular to the surface. This is a typical behaviour of a bulk electronic state. However, for normal incidence at $\bar{\Gamma}$ ($\Gamma(\Sigma)K$—line of the bulk) no bulk transitions in this final-state energy range exist. Analogous spectra with final-state energies up to 3.5 eV for the $c(2 \times 2)$Cl and the $c(2 \times 2)$S structure, are identical to those observed on $c(2 \times 2)$S. For the (2×1)O structure the additional emission at about 2 eV does not show up. We conclude that this emission is only determined by the periodicity of the surface. Due to the $c(2 \times 2)$ structure a surface umklapp can occur from $\bar{\Gamma}$ to $\bar{S}$ ($L(Q)W$—line of the volume). This is again a $\bar{\Gamma}$ point of the SBZ of the adsorbate-covered surface, hence equivalent to normal incidence for the adsorbate. The observed feature could therefore originate from $\bar{\Gamma}$ or $\bar{S}$, or both. For the substrate, however, $\bar{S}$ is only accessible for normally incident electrons via diffraction exchanging a surface reciprocal lattice vector $\boldsymbol{G}$. A series of spectra of the clean surface for the $\bar{S}$ point measured by suitable choice of the angle of incidence of the electrons for each initial-state energy is shown in the lower right panel of Fig. 9.14. The arrows mark the transitions which are expected from combined-interpolation-scheme calculations. They agree well with observations. However, more important is the fact that the spectra at the $\bar{S}$ point of the clean surface are identical to the spectra of the $c(2 \times 2)$S and $c(2 \times 2)$Cl structure. From this the adsorbate-induced emission at the $\bar{S}$ point can be identified as a surface umklapp process into a bulk band, inaccessible on the clean surface with normal incidence [9.36].

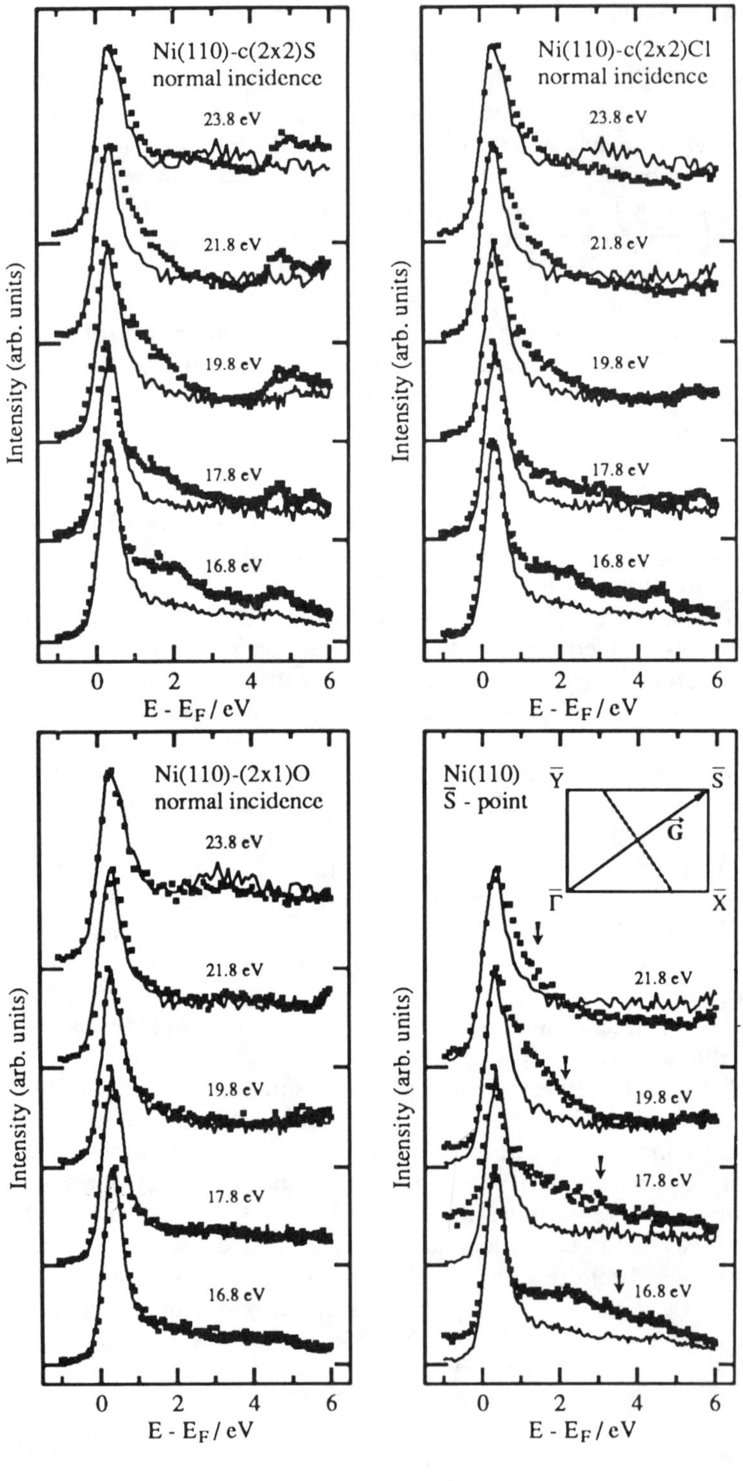
Ni(110)-c(2x2)S
normal incidence
23.8 eV
21.8 eV
19.8 eV
17.8 eV
16.8 eV
Intensity (arb. units)
E - E_F / eV
Ni(110)-c(2x2)Cl
normal incidence
23.8 eV
21.8 eV
19.8 eV
17.8 eV
16.8 eV
Intensity (arb. units)
E - E_F / eV
Ni(110)-(2x1)O
normal incidence
23.8 eV
21.8 eV
19.8 eV
17.8 eV
16.8 eV
Intensity (arb. units)
E - E_F / eV
Ni(110)
S̄ - point
Ȳ
S̄
G
Γ̄
X̄
21.8 eV
19.8 eV
17.8 eV
16.8 eV
Intensity (arb. units)
E - E_F / eV

9.4.2 Ni(111) $p(2 \times 2)$S

The adsorbate-induced states are not only determined by the symmetry of the adsorbate structure, but also by the bulk symmetry due to the interaction of the adsorbate with the substrate [9.38]. In Fig. 9.15 IPE spectra for the Ni(111) $p(2 \times 2)$S structure measured with fixed photon energy of 9.6 eV are shown [9.38]. The transition close to the Fermi level is a transition into a d band of nickel [9.39]. The structure at 1.7 eV at normal incidence is an adsorbate-induced state. This state disperses differently in the $\overline{\Gamma M}$ and $\overline{\Gamma M'}$ direction. This could be explained by two transitions at different energies, which have different radiation characteristics (different dipole orientations). In order to rule out this possibility, spectra for an angle of incidence of the electrons of $\theta = -20°$ in the $\overline{\Gamma M}$ and $\overline{\Gamma M'}$ direction were taken for the identical photon take-off angle $\alpha = 25°$. In the experimental set-up for these measurements it was not possible to rotate the electron gun. Fortunately, the sample could be rotated azimuthally about 180° from the $\overline{\Gamma M}$ to the $\overline{\Gamma M'}$ direction enabling measurements for identical α. The spectra clearly show (Fig. 9.15) that the asymmetric dispersion of the adsorbate-induced state is coupled to the coordinate system of the crystal and is not the result of different dipole orientations. The adsorbate layer has

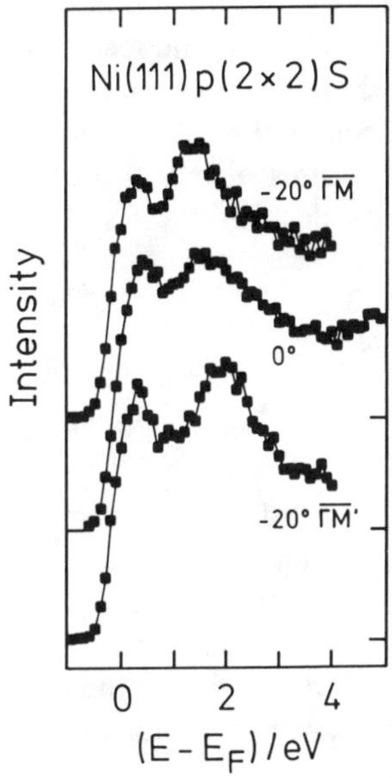

Fig. 9.15. IPE spectra from a $p(2 \times 2)$ sulphur-covered Ni(111) surface

Fig. 9.14. Normal incidence IPE spectra, at constant initial-state energies, for Ni(110). The upper left panel shows spectra for $c(2 \times 2)$S overlayer, the upper right panel spectra for $c(2 \times 2)$Cl overlayer and the lower left panel spectra for (2×1)O overlayer. The spectra of the clean surface are shown as full lines. The spectra of the clean surface at the $\bar{S}$ point can be seen in the lower right panel

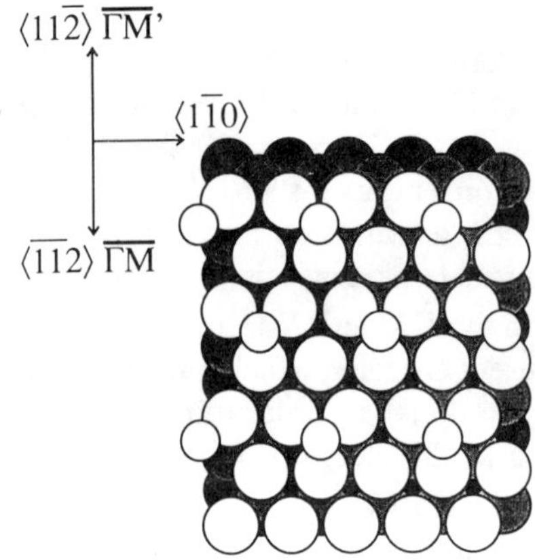

Fig. 9.16. Face-centered cubic (111) surface with a $p(2 \times 2)$ adsorbate structure

six-fold symmetry, whereas the [111] axis of the crystal has only three-fold symmetry (Fig. 9.16). The asymmetric dispersion cannot be explained by the potential of the two-dimensional adsorbate layer (direct adsorbate–adsorbate interaction). The time-reversal symmetry gives symmetric dispersion for a two-dimensional system, although the potential may be asymmetric, because a mirror plane is missing due to the three-fold coordinated adsorption site of the sulphur. The conclusion is that the asymmetric dispersion of the adsorbate-induced state can only be explained by an interaction of the adsorbate with the bulk. This shows clearly the fact that adsorbtion problems are normally three dimensional. Proper understanding of adsorbate-induced states requires also a thorough consideration of substrate bulk states since IPE probes electronic states of the total system.

9.5 Temperature Dependence of IPE

Especially in spin-polarised IPE, temperature-dependent studies are important for the study of magnetic phase transitions [9.40–42]. An understanding of the non-magnetic temperature effects is necessary for a discussion of the magnetic effects. In this section we will discuss the temperature dependence of IPE from the copper surfaces Cu(100), (110) and (111). An extensive discussion of the temperature dependence of IPE is given in [9.18]. We will focus here on the importance of the surface.

In Fig. 9.17 a typical set of temperature-dependent IPE measurements is shown. For the temperature-dependent measurements, the photon take-off angle α was optimized to obtain maximum count rates. The main temperature effect is the reduction of the intensities of the transitions with increasing temperature. The background, indicated by the dashed line in Fig. 9.17, is found to be inde-

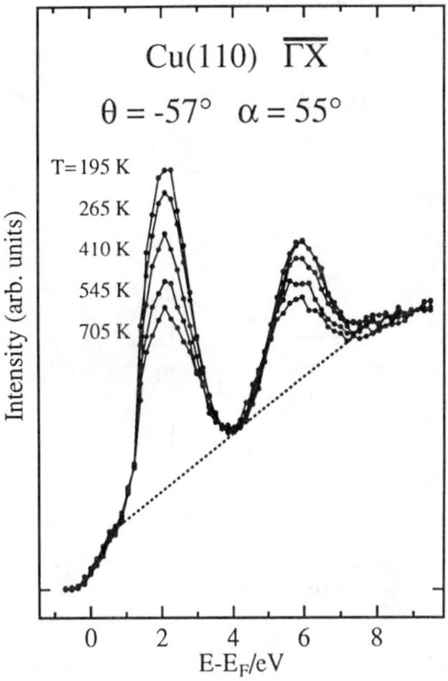

Fig. 9.17. Set of IPE spectra for Cu(110) for several temperatures. The estimated background is indicated by the dashed line

pendent of temperature within the statistical error limits. The peak width and shape also do not show a temperature dependence, although the experimental set-up was not able to detect small changes because of its limited resolution (full width at half maximum of the apparatus function is 0.7 eV [9.8]).

In Fig. 9.18 the temperature dependence of the direct transition into a bulk state, which has been triangulated from Cu(111) and (110) in Sect. 9.2, is shown. We would expect the same temperature dependence for both surfaces if bulk effects are dominating the temperature dependence, because it is the identical bulk transition observed on different surfaces. The intensity on Cu(110) decreases much more strongly than on Cu(111). In Fig. 9.18 the peak intensities of the triangulated transition and a bulk transition on Cu(100) were measured as a function of temperature during cooling down after heating the sample to 900 K. Several independent runs were accumulated to reduce the statistical error. The intensities were corrected for the background, which was found to be independent of temperature within the statistical error limits. Possible bulk impurity segregation and contamination during experiment could be excluded by the comparison of IPE spectra before and after the temperature-dependent measurements. The data points are plotted semilogarithmically versus temperature and can be fitted well by straight lines for the whole temperature range for Cu(111) and (100) and for temperatures below about 400 K for Cu(110). The statistical uncertainty is indicated by error bars, which are in most cases smaller than the size of the symbols. The normalization of the data points comes from the extrapolation of the straight line fits to $T=0$ K.

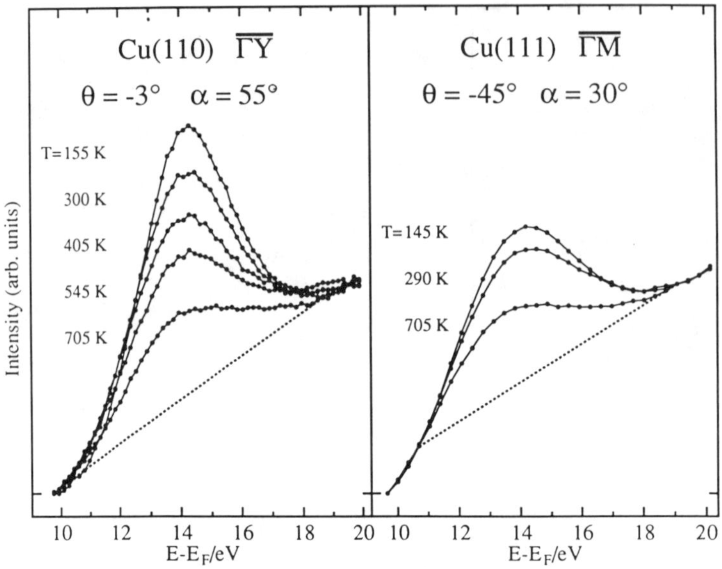

Fig. 9.18. Inverse photoemission spectra for a selected bulk transition observed on two different surfaces for various sample temperatures

As will be explained below, the interpretation of the results can be attempted within a simple Debye–Waller ansatz [9.43]

$$I = I_0 \exp(-\langle u^2 \rangle \Delta \boldsymbol{k}^2), \tag{9.11}$$

where $\langle u^2 \rangle$ is the mean square displacement of the atoms and $\Delta \boldsymbol{k}$ is the momentum exchanged in the transition. The mean square displacement can be expressed in the Debye model [9.44] as

$$\langle u^2 \rangle = 3\hbar^2 T/(M k_B \Theta_D^2). \tag{9.12}$$

Here, M is the mass of the atom, Θ_D is the Debye temperature and k_B is Boltzmann's constant. The combination of (9.11) and (9.12) gives immediately the exponential decreases of the intensity with temperature as shown by the straight line fits in Fig. 9.18. The momentum $\Delta \boldsymbol{k}$ is obtained from the calculations with the combined interpolation scheme of Sect. 9.2 and by comparison of the experimental data with the calculated bulk transitions. Using (9.11) one can calculate the temperature dependence of the mean square displacements on the three copper surfaces from the IPE data of Figs. 9.18 and 9.19. The data for Cu(111) follow a linear dependence up to 800 K whereas the data for Cu(110) deviate from that linear dependence above 400 K. Inspection of Fig. 9.20 shows a surprisingly good agreement between IPE and results from helium atom [9.45, 46] and ion scattering [9.47]. The helium atom scattering data are corrected for correlated vibrations following [9.45]. For Cu(100) a deviation from the linear behavior above 800 K

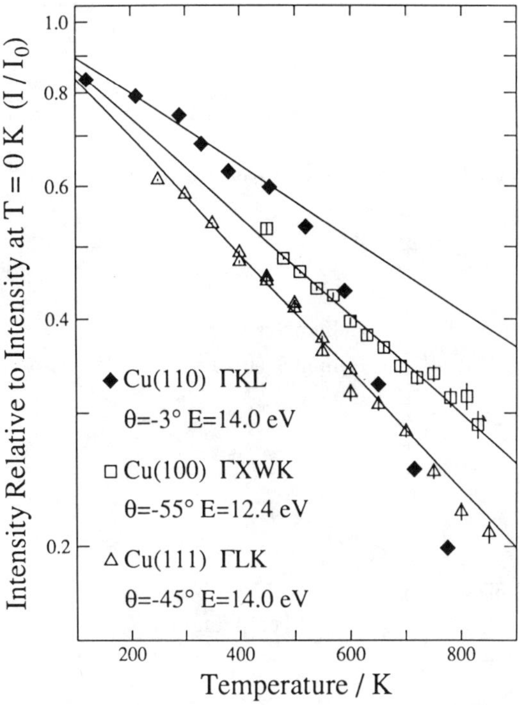

Fig. 9.19. Semilogarithmic plot of the temperature dependence of the intensity of the transitions of Fig. 9.18 and of a bulk transition on Cu(100). The straight lines are linear fits for the temperature range from 200 K to 900 K for Cu(111) and Cu(100). The fitting range is restricted to temperatures less than 400 K for Cu(110)

can be found. Unfortunately one is not able to obtain IPE data in this temperature range because of evaporation of copper onto the entrance window of the detector. The IPE data on Cu(111) and (110) are multiplied by a factor of 0.35 and on Cu(001), which is a different transition, by a factor of 0.75 in order to correct for multiple-scattering effects. The multiple-scattering effects in IPE are similar to those known from LEED. They are not included in the simple model of (9.11). Multiple scattering enhances the mean square displacements as extracted from the IPE data, because if the incoming electron is multiply scattering it 'sums up' vibrations of many atoms and not only that of a single atom. The IPE data on Cu(111), where the scaling factor for this transition was determined by comparison with the results of the scattering experiments on Cu(110), also agree quite well with the mean square displacements of surface atoms, which are the results of phonon slab calculations shown as full lines in Fig. 9.20 [9.48]. The deviation from the linear dependence may be due to the anharmonicity of the potential at the surface. This results in increased mean square displacements at the surfaces of Cu(001) and (110) above 800 K and 400 K, respectively.

Since the scattering experiments probe only the top surface layer, the good agreement shows conclusively that the temperature dependence of IPE is influenced strongly by vibrations of the atoms in the surface layer. We would like to note here that the mean free path of electrons in our energy range is

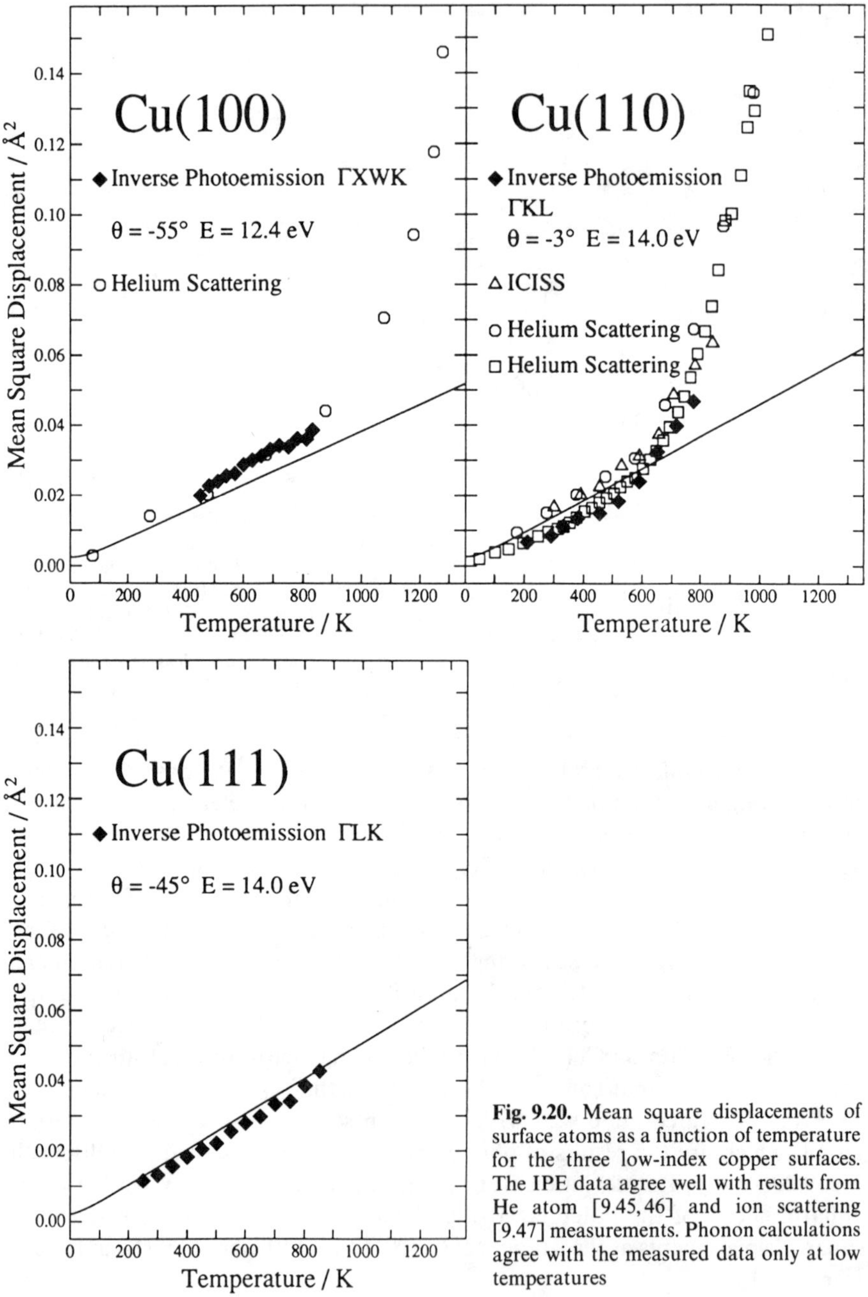

Fig. 9.20. Mean square displacements of surface atoms as a function of temperature for the three low-index copper surfaces. The IPE data agree well with results from He atom [9.45,46] and ion scattering [9.47] measurements. Phonon calculations agree with the measured data only at low temperatures

about 10 Å corresponding to up to 5 layers. Surface-specific effects should be dominating only in the first two or three layers with deeper layers showing essentially bulk behaviour.

The temperature dependence of IPE can be interpreted in the following way: In an observed direct transition in IPE the incoming electron couples to the initial bulk state and decays to the final state by producing a photon with the photon energy of the detector. The temperature introduces vibrations of the lattice atoms. Therefore, the increasing scattering of electrons out of the initial state by the vibrating atoms with increasing temperature is the dominant temperature-dependent process and results in a decrease of the intensity of the transition. This scattering is a multiple-scattering process. The scattered electrons can decay to lower energies via radiative transitions with different photon energies or via Fermi surface excitations. Fermi surface excitations preceeding radiative transitions are believed to be responsible for the background in IPE spectra. The decreasing intensity of the direct transition need not result in an observable increase in the background, because in the decay of the scattered electron via a direct transition of different photon energy, the electron is not detected and in the decay via processes producing the background of the IPE spectra the gain of intensity for the background is distributed over the whole solid angle 4π.

9.6 Spin-Resolved IPE

For ferromagnetic materials such as iron or nickel, the spin degeneracy is removed from the electronic states and each band of the hypothetically non-magnetic material appears as a twin in the magnetic state. A comprehensive description of electronic states in ferromagnetic crystalline solids requires therefore spin resolution of energy versus momentum band dispersions. Spin-resolved IPE requires a spin-polarised electron beam, which can be produced by photoemission from GaAS [9.49]. In this section we will present spin-resolved IPE data on Ni(110) for an *sp*-bulk transition and for a transition into a crystal-induced surface state. Experimental details and a more extensive discussion is given in [9.40].

The magnetic moment per atom in bulk nickel is approximately $0.55\,\mu_B$, where μ_B is the Bohr magneton [9.50]. The non-integer value of the magnetic moment proves that Ni is a band ferromagnet. At zero temperature the majority 3*d*-band states are fully occupied while the associated minority states with average occupation number 4.45 extend to energies slightly above the Fermi energy. Theoretical calculations predict a spin splitting of bulk transitions not only for *d*- but also for *sp*-like final bands due to hybridization [9.51]. In the case of a flat final band the measured spin splitting reflects directly the exchange splitting of the band. For steep final bands the measured spin splitting depends not only on the exchange splitting of the final band and its slope, but also on a possible though generally much smaller splitting of the initial state [9.52]. Furthermore,

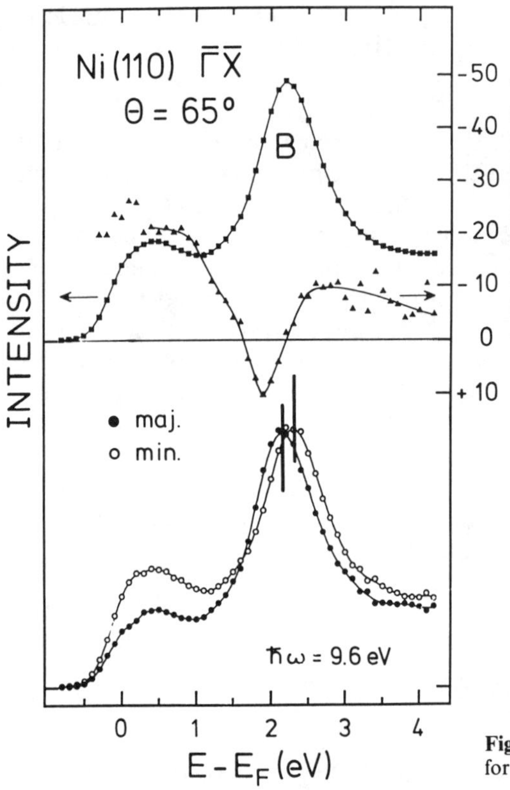

Fig. 9.21. Spin splitting of a bulk transition for an *sp*-like final band

off normal-incidence spin-split features belong to slightly different values of $k_{\parallel}$ due to the different final-state energies. As an example of the spin splitting of an *sp*-band, Fig. 9.21 shows a bulk transition B on Ni(110) for an angle of incidence θ of 65° [9.52]. The spin-resolved data of Fig. 9.21 refer to a hypothetical polarization of the incident electron beam of 100%. Let $n_{\uparrow,\downarrow}$ represent the count rates for a beam with polarization P, and $N_{\uparrow,\downarrow}$ the count rates expected for a hypothetical 100% polarized beam. The asymmetry A is then defined as

$$A = (N_{\uparrow} - N_{\downarrow})/(N_{\downarrow} + N_{\downarrow}) = (n_{\uparrow} - n_{\downarrow})/(n_{\uparrow} + n_{\downarrow})P \cos \phi); \qquad (9.13)$$

using $N = N_{\uparrow} + N_{\downarrow} = n_{\uparrow} + n_{\downarrow}$ we obtain

$$N_{\uparrow,\downarrow} = 0.5\, N(1 \pm A). \qquad (9.14)$$

ϕ in the above formula is the angle between the domain magnetization and the spin polarization vector. We want to emphasize here that the magnetization structure of the surface must be properly accounted for in all spin-resolved electron spectroscopies. This point has not received proper attention in the past. The magnetic domain orientation on Ni(110) was verified by ex-situ magneto-optic Kerr measurements to point into the ⟨111⟩ direction [9.52]. Let γ be the angle

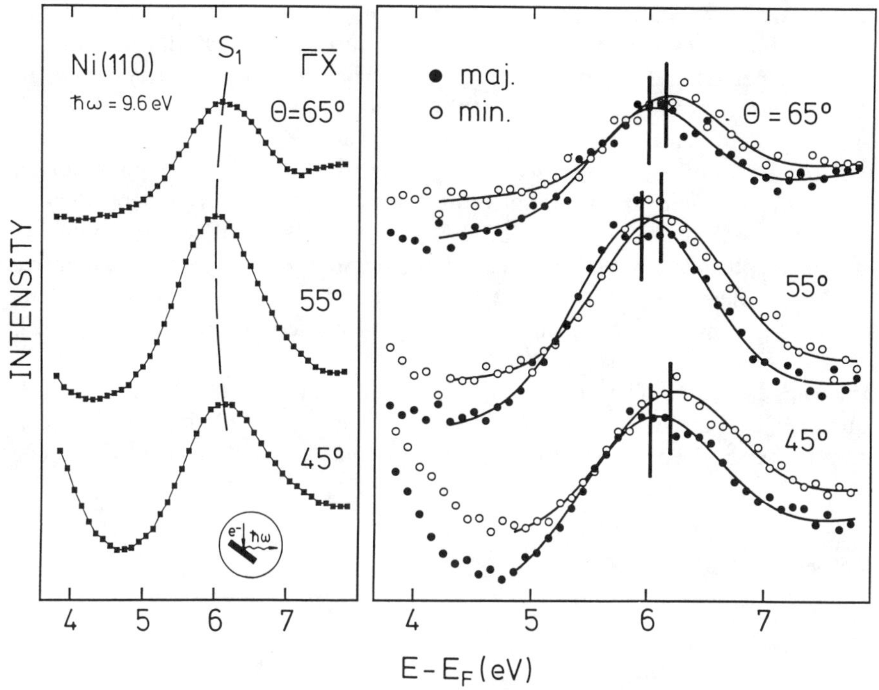

Fig. 9.22. Spin splitting of a crystal-induced surface state

between ⟨111⟩ and ⟨110⟩ direction, $\gamma \approx 35°$, then by using the addition theorem for spherical harmonics $\cos\phi$ may be expressed as

$$\cos\phi = \cos\gamma\cos\theta, \tag{9.15}$$

where θ is the angle of incidence for the electrons. The spin polarization of the GaAs photoemitter used for the IPE measurements as the spin-polarised electron source was $P = 33\% = 0.27/\cos\gamma$. From theory one would expect for the observed bulk transition at this value of $k_{\|}$ a spectral splitting of 250 meV. The measurement is, however, carried out at constant θ, not at constant $k_{\|}$. This reduces the expected splitting to about 150 meV in reasonable with the observed(140 ± 20) meV Closer inspection of Fig. 9.21 shows that the line width of the spin "down" (↓ denotes majority) line is slightly larger than the spin "up" (↑ denotes minority). The shorter lifetime of the spin-down state follows from the high density of empty spin-down states near to the Fermi energy [9.53].

In Fig. 9.21 the asymmetry A is also shown as a function of the final-state energy. It is important to note A changes sign in the region of the direct transition. This rules out an apparent spin splitting simulated by an unpolarized line superimposed on a polarized background.

Finally, we will discuss spin-resolved IPE data for a transition into a crystal-induced surface state on Ni(110). As discussed before, crystal-induced surface states arise from a modification of the bulk electronic structure near the surface. Consequently they depend on the band structure of the material in question and are expected to show spin splitting [9.23]. The probability density of crystal-induced surface states peaks usually in the first atomic layer. The state S_1 in Fig. 9.22 has been identified as such a crystal-induced surface state. Figure 9.22 shows spectra of S_1, both spin averaged and spin-resolved near $\bar{X}$. The spin-resolved data show a definite splitting. Since the energy of S_1 is nearly independent of $k_{\parallel}$ at $\bar{X}$, the observed splitting of (170 ± 30) meV is equal to the final-state exchange splitting. This is considerable magnetic effect for a state 6 eV above the Fermi energy. Moreover, since S_1 is localised in the first atomic layer, its observed spin splitting refutes magnetically dead surface layers on Ni(110) [9.52].

Acknowledgements. The authors want to thank K. Desinger, M. Donath, H. Dürr and Th. Fauster for their contributions to this chapter. Continuous financial support by the Deutsche Forschungsgemeinschaft is gratefully acknowledged.

References

9.1 F.J. Himpsel: Sulf. Sci. Rep. **12**, 3 (1990)
9.2 V.Dose: Surf. Sci.Rep. **5**, 337 (1985)
9.3 F.J. Himpsel: Comm. Cond. Mat. Phys.**12**, 199 (1986)
9.4 N.V. Smith: Rep. Prog. Phys. **51**, 1227 (1988)
9.5 V. Dose: J. Vac. Sci. Technol. A **5**, 2032 (1987)
9.6 M.P. Seah, W. A. Dench: Surf. Interface Analysis **1**, 2 (1979)
9.7 J. Hermanson: Solid State Commun. **22**, 9 (1977)
9.8 V. Dose, Th. Fauster, R. Schneider: Appl. Phys. A **40**, 203 (1986)
9.9 Th. Fauster, R. Schneider, H. Dürr: Phys. Rev. B **40**, 7981 (1989)
9.10 M. Donath, M. Glöbl, B. Senftinger, V. Dose: Solid State Commun. **60**, 237 (1986)
9.11 N. V. Smith: Phys. Rev. B **19**, 5019 (1979)
9.12 V. Dose: Prog. Surf. Sci. **13**, 225 (1983)
9.13 C.N. Berglund, W.E. Spicer: Phys. Rev. **136**, A 1030 (1964)
9.14 D.P. Woodruff, N.V. Smith, P.D. Johnson, W.A. Royer: Phys. Rev. B **26**, 2943 (1982)
9.15 N.V. Smith, L.F. Matheis: Phys. Rev. B **9**, 1341 (1974)
9.16 H. Bross, B. Schiekel: Private communication
9.17 P.M. Echenique, J.B. Pendry: J. Phys. C **11**, 2065 (1978)
9.18 R.Schneider, H. Dürr, Th. Fauster, V. Dose: Phys. Rev. B **42**, 1638(1990)
9.19 E.O. Kane: Phys. Rev. Lett. **12**, 97 (1964)
9.20 R.R. Turtle, T.A. Callcott: Phys. Rev. Lett. **34**, 86 (1975)
9.21 A. Goldmann, M. Donath, W. Altmann, V. Dose: Phys. Rev. B **32**, 837 (1985)
9.22 J.B. Pendry: J. Phys. C **14**, 1381 (1981)
9.23 G. Borstel, G. Thörner: Surf. Sci. Rep. **8**, 1 (1987)
9.24 Th. Fauster, R. Schneider, H. Dürr: Phys. Rev. B **40**, 7981 (1989)
9.25 W. Eberhardt, F.J. Himpsel: Phys. Rev. B **21**, 5572 (1980) and Phys. Rev. B **23**, 5650 (1981)
9.26 J. Hermanson: Solid State Commun. **22**, 9 (1977)
9.27 N.V. Smith: Phys. Rev. B **14**, 3549 (1985)
9.28 J.D. Jackson: *Classical Electrodynamics* (Wiley, New York 1975)
9.29 J.H. Weaver, C. Krafka, D. W. Lynch, E.E. Koch: DESY Report **41**, HASYLAB 81/05 (1981)
9.30 F. Forstmann, R.R. Gerhardts: *Metal Optics Near the Plasma Frequency, Springer Tracts Mod. Phys.*, Vol. 109 (Springer, Berlin, Heidelberg (1986)
9.31 P.J. Feibelman: Prog. Surf. Sci. **12**, 287 (1987)
9.32 N. Barberan, J.E. Inglesfield: J. Phys. C **14**, 3114 (1981)

9.33 G. Rangelov, N. Memmel, E. Bertel, V. Dose: Submitted to Surf. Sci.
9.34 Th. Fauster, D. Straub, J.J. Donelon, D. Grimm, A. Marx, F.J. Himpsel: Rev. Sci. Instrum. **56**, 1212 (1985)
9.35 V. Dose: "Instrumentation in Inverse Photoemission", in *Lectures on Surface Science*, ed. by G.R. Castro, M. Cardona (Springer, Berlin, Heidelberg 1987)
9.36 K. Desinger, W. Altmann, V. Dose: Surf. Sci. **201**, L491 (1988)
9.37 W. Altmann, K. Desinger, V. Dose, A. Goldmann: Solid State Commun. **65**, 1411 (1988)
9.38 K. Desinger: IPP-Report **9/72**, (1989)
9.39 A. Goldmann, M. Donath, W. Altmann, V. Dose: Phys. Rev. B **32**, 837 (1985)
9.40 M. Donath: Appl. Phys. A **49**, 351 (1989)
9.41 M. Donath, V. Dose: Europhys. Lett. **9**, 821 (1989)
9.42 W. Nolting, W. Borgiel, V. Dose, Th. Fauster: Phys. Rev. B **40**, 5015 (1989)
9.43 N. J. Shevchik: Phys. Rev. B **16**, 3428 (1977), J. Phys. C, L555 (1977), Phys. Rev. B **20**, 3020 (1979)
9.44 N.W. Ashcroft, N.D. Mermin: *Solid State Physics* (Holt Saunders, London 1976)
9.45 D. Gorse, J. Lapujoulade: Surf. Sci. **162**, 847 (1985)
9.46 P. Zeppenfeld, K. Kern, R. David, G. Comsa: Phys. Rev. Lett. **62**, 63 (1989)
9.47 H. Dürr, R. Schneider, Th. Fauster: Vacuum **41**, 376 (1990)
9.48 R. Schneider: IPP-Report **9/75**, (1989)
9.49 U. Kolac, M. Donath, K. Ertl, H. Liebl, V. Dose: Rev. Sci. Instrum. **59**, 1933 (1988)
9.50 N. F. Mott: Adv. Phys. **13**, 325 (1964), and references therein
9.51 J. Noffke: Private Communication
9.52 M. Donath, V. Dose, K. Ertl, U. Kolac: in course of publication in phys. Rev. B **41**, (1990)
9.53 R. Feder, A. Rodriguez: Solid State Commun. **50**, 1033 (1984)

10. Bremsstrahlung Isochromat Spectroscopy (BIS or High-Energy Inverse Photoemission)

John C. Fuggle

With 17 Figures

The unoccupied states of a solid are characterized by a wavefunction, an energy and a self-energy ε with its real (energy) and imaginary (lifetime broadening) parts (Chap. 1). For delocalized states a $\boldsymbol{k}$, or quasi-momentum quantum number is also relevant. Bremsstrahlung isochromat spectroscopy (BIS) can say little about $\boldsymbol{k}$, but it is one of the most direct probes of the density of unoccupied states (UDOS), and of the energies of those states. By comparison of BIS spectra with suitable UDOS calculations, one may obtain data on the real and imaginary parts of the self-energies. In addition, for solids with highly localized electrons, the comparison of PS and BIS gives the most direct measure of the effective Coulomb correlation energy between electrons (Hubbard's U_{eff}). Our goal here is to explain the general characteristics of BIS and its use. We do not give details of the experimental methods here, as these were discussed in Chap. 1.

10.1 Two Illustrations of Typical BIS Spectra: Ag and Gd

In Fig. 10.1 two BIS spectra, of Ag [10.1] and Gd [10.2], are shown to illustrate two extremes of behaviour seen in BIS. The Ag spectrum shows a sharp rise at E_F and then a series of shoulders and peaks which may be correlated with the Ag band structure [10.1, 3, 4]. The relationship between the Ag spectrum and the broadened density of states is quite close and may be taken as typical for systems in which the electron–electron correlation is not very strong. However there are differences in the energies of the features which are significant and are due to the variations in the self-energy [10.1, 5, 6], which is dominated by the energy dependence of the exchange-correlation term in the potential [10.7]. These effects are described further in Chaps. 1 and 3.

The structure in the BIS spectrum of Ag is analogous to the XANES structure in XAS. Close to E_F the structure represents the critical points of the band structure [10.3, 8], whilst at high energies the electron mean free path is short and the structure would be quite well reproduced by cluster calculations. At still higher energies one even sees weak oscillations in the intensity which are analogous to EXAFS [10.9, 10]. There are, however several differences with respect to XAS. Firstly, one does not have to consider the influence of a core-hole potential in BIS because no core hole is created. Secondly, BIS is not subject

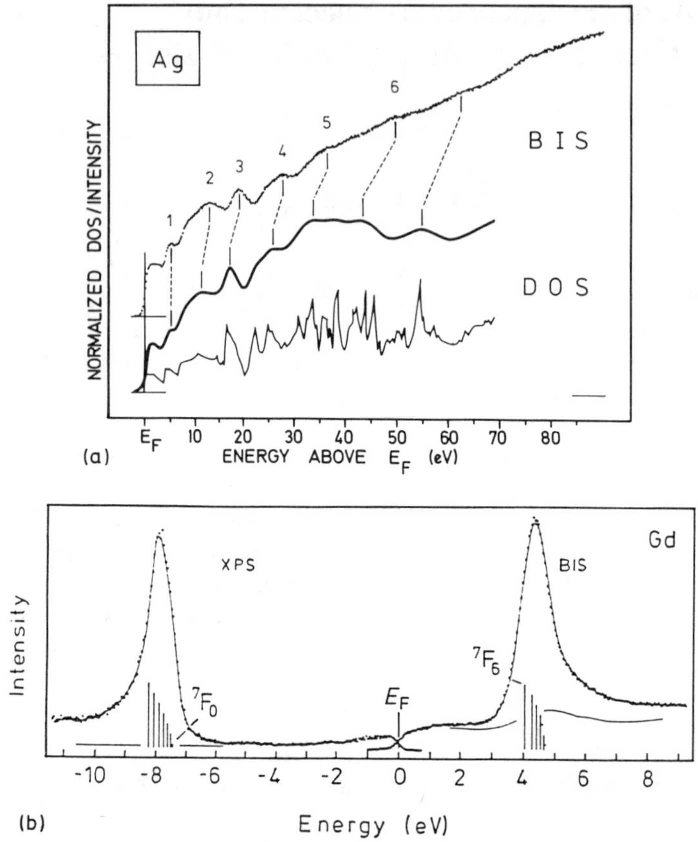

Fig. 10.1. (**a**) BIS spectrum and calculated DOS for Ag. The correspondence of the structures in the measured spectrum (dots) with features in the broadened DOS (solid line) is indicated. The DOS was broadened with a Lorentzian of FWHM $0.7 + 0.12(E - E_f)$ (from [10.1]), (**b**) XPS/BIS spectra of Gd (Adapted from [10.2])

to the same selection rules and samples all the unoccupied states, regardless of the angular momentum quantum number, l, (although the matrix elements are generally different for different l). Finally BIS is not site selective. As a consequence, the extended fine structure in BIS is unlikely to ever be as useful as EXAFS for studies of geometric structure. It may, however, be useful for clarification of some questions concerning the phase shifts associated with core holes in EXAFS (see also 10.1, 10–13).

The spectrum of Gd falls into a different class and is typical of the spectrum from highly correlated materials. A band structure calculation would show a narrow band of $4f$ states pinned at E_F. However the XPS/BIS spectra of the valence bands show no $4f$ states at E_F. Instead the $4f$ states behave like localized, atomic states with a very large difference between the first ionization potential and the first electron affinity. This difference reflects very closely the effective

Coulomb correlation energy between the $4f$ states. Only the $5d$ and $6s$ states are found near E_F. This behaviour is typical of the $4f$ states in all the rare earths, although more complex behaviour is found in some early rare earths and mixed valence materials, as will be discussed in Sect. 10.6.

10.2 Matrix Element Effects

Inverse photoemission is normally considered as the time-reversed version of photoemission [10.14–16], which is acceptable for most purposes [10.16, 17]. As described in Chap. 8, a full treatment of the cross sections for PS and IPS or BIS leads to the result that their ratio is the square of the ratio of the electron and photon wavelengths, i.e.

$$r(\mathrm{PS/IPS}) = (\lambda_e/\lambda_p)^2. \tag{10.1}$$

This is also known as the Milne relation [10.18, 19]. It means that for 10 eV electrons, $\lambda_e \sim 2$–4 Å, depending on the potential in the solid, and $\lambda_p \sim 1200$ Å, so that r is $\sim 10^{-5}$. Even at 1500 eV r is $\sim 10^{-3}$, so that inverse photoemission and BIS signals are very weak.

10.2.1 The Atomic Approximation to BIS Matrix Elements and Partial Cross Sections

BIS is not a local probe in the sense of X-ray absorption, but probes all the unoccupied states of a material. Thus there is no inherent site and symmetry selectivity as a result of selection rules. However the weight of any given state in a spectrum is given by a transition matrix element M_{fi} between the initial and final states

$$M_{fi} = \langle \phi_f | t | \phi_i \rangle. \tag{10.2}$$

The simplest approximation to (10.2) is to use the relative values of the photoemission matrix elements for the valence electrons from photoelectron spectroscopy. To illustrate this approach we have listed in Table 10.1 some partial cross sections (per electron) for photoionization by 1487 eV radiation. Also listed is the product of the number of unoccupied states per atom (N) of a given character in the element and the partial cross section according to [10.20]. This last column allows us to see that one may not simply use the number of unoccupied states per atom as a guide to the contribution of that atom to the spectrum. For instance, in the case of alloys such as ScNi or ScPd, it is clear that there are many more states of metal d character on the Sc sites than on the Ni or Pd sites. Nevertheless, it is the Ni or Pd character which will dominate the BIS spectra close to the Fermi level, E_F.

If one wishes to go beyond the approach to partial cross sections illustrated with Table 10.1, it is necessary to take into account the whole process and also

Table 10.1. Some partial ionization cross sections σ for valence electrons [10.20] in units of 13,600 barns. Also given are the number of unoccupied states per atom (N) and $\sigma \times N$

Element	Level	σ	N	$\sigma \times N$
O	$2p$	0.0048	2	0.0096
Al	$3p$	0.0033	5	0.0165
Sc	$4s$	0.0157	0.4	0.0063
Sc	$3d$	0.0042	8.6	0.036
Ni	$4s$	0.028	0.8	0.022
Ni	$3d$	0.048	0.6*	0.064
Pd	$4d$	0.123	0.4*	0.049

* This only includes the unoccupied states in the metal d band. There is also some unoccupied metal d character hybridized with the metal sp bands

the wavefunctions of the electrons in the solid. We illustrate this with the approach of *Speier* et al. [10.21] who argued that *Winter's* XPS formalism [10.22] was appropriate to BIS because inverse photoemission is the time inverse of photoemission [10.14–16]. The underlying principle is the dipole approximation of the golden rule for radiative transitions

$$W_\mu = 2\pi \sum_{\mathrm{f}} |\langle \phi_{\mathrm{f}} | \boldsymbol{A} \cdot \boldsymbol{p} | \phi_{\mathrm{i}} \rangle|^2 \delta(E_{\mathrm{f}} - E_{\mathrm{i}} \pm h\nu) \tag{10.3}$$

between the initial and final states ϕ_{i} and ϕ_{f}. In the single-electron (or independent-electron) appproximation the BIS final state is described by Bloch states of the periodic solid-state potential. The incident electron in BIS is described by a plane wave modified by the crystal potential. Multiple scattering is neglected in view of the high energies (typically 1.5 keV) used in BIS and this leads to a localised description of the transition (the so-called single-site approximation [10.22]). Such a description allows separation of the transition rate into an atomic cross section, σ, and band structure effects

$$I_E = \sum_N \sum_l \sigma_l^N(E) n_l^N(E), \tag{10.4}$$

where l is the angular momentum quantum number, N is the type of atom, n_l^N is the partial DOS and σ_l^N is the transition matrix element given in its simplest form as

$$\sigma_l^N(E) = \left| \sum_{l'} \delta_{l',l\pm 1} \frac{1}{2(E+\omega)} \int_0^{R_{\mathrm{MT}}} R_{l'}^N(E+\omega, r) \mathrm{e}^{\mathrm{i}\delta_l(E+\omega)} \right.$$
$$\left. \cdot \sin[\delta_l(E+\omega)] \frac{dV^N}{dr} \tilde{R}_l^N(E,r) r^2 dr \right|^2, \tag{10.5}$$

where $\tilde{R}_l(E,r)$ is the low-energy electron radial wavefunction, normalized inside

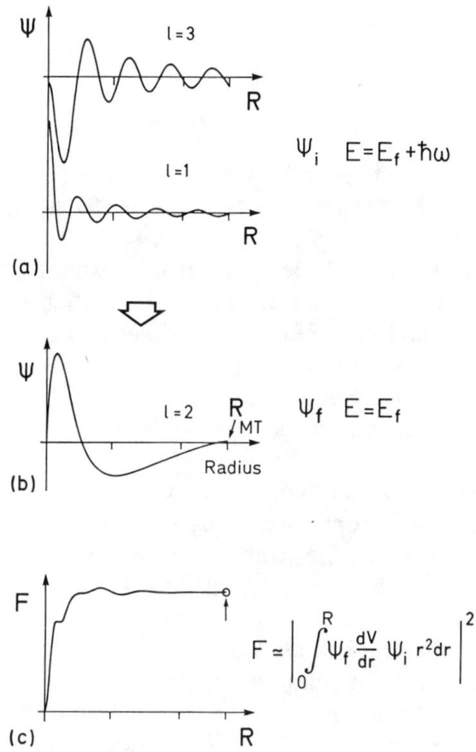

Fig. 10.2. Illustration of matrix element calculations for element 41 (Nb). For further explanation, see text

the Wigner-Seitz sphere, $R_{l'}(E+\omega, r)$ is the high-energy radial wavefunction, $V^N(r)$ is the muffin-tin potential, R_{MT} is the muffin-tin radius, $\delta_l(E+\omega)$ is the phase shift of the high-energy state and $\delta_{l',l\pm 1}$ is the dipole selection rule. The radial wavefunctions of the initial and final states are the exact solutions of the Schrödinger equation inside the muffin-tin potential.

Figure 10.2 illustrates the consequences of (10.5) for a 1486.7 eV incident electron in BIS of Nb. The dipole selection rules dictate that, for the transition to a Nb 4*d* final state, it is necessary to consider the $l=1$ and $l=3$ incident electron wavefunctions in the Wigner-Seitz sphere. However the contribution of the incident wave with the lower l value is smaller because the large number of radial holes suppresses the amplitude of the wavefunction in the core region where grad V is highest.

The matrix element variation as a function of the distance from the centre of the atom is then given in Fig. 10.2. The transition is possible from the 1486.7 eV $l=1$ and $l=3$ initial states ψ_i is shown in Fig. 10.2, curve (a). The BIS final state (ψ_f) in (b) is an anti-bonding *d* state from the top of the *d* band for which the real part of the wavefunction is shown in the middle. The development of the matrix element with outward integration is shown in curve (c). It is the sum of contributions for ψ_i with $l=1$ and $l=3$

$$F=\left|\int_0^R \psi_f \frac{dV}{dr}\psi_i r^2 dr^2\right|. \tag{10.6}$$

The value of the matrix element at the Wigner-Seitz boundary is the one relevant for cross sections and is shown by the arrow at the right of the curve in Fig. 10.2. However, to rationalize solid-state effects it will be desirable to consider the *shape* of curves such as (c) in Fig. 10.2. Here one must be extremely cautious, as we now explain. In theory the matrix element can be written in three forms: the field gradient or dipole acceleration form, as used here, a length form and a momentum form [10.23]. These forms should integrate up to the same value at the Wigner-Seitz radius, but the *shape* of the curve will be different in different forms and the intensity may appear to have its origin in different regions of space [10.24].

In the dipole acceleration form the contribution of the outer regions of the atom is small because dV/dr is weaker away from the nucleus and because the oscillations in the wavefunctions of the incoming electron have a shorter period than the final-state wavefunctions so that the contributions cancel. This is a general effect and as a consequence the major contribution to the matrix element comes from the core of the atom in the dipole acceleration form. This also illustrates the semi-classical concept that the bremsstrahlung transition arises through the retardation near the core of the atom where the electric field (grad V) is strong enough to give the large momentum transfer. Below we will actually use the fact that the contribution of the core region dominates the matrix elements, although we stress the need to make similar calculations in the length and momentum form of (10.5–6).

We consider now, in connection with Fig. 10.3, the variation of the XPS/BIS matrix elements for the $3d$ and $4d$ states in transition metals as calculated using (10.5). The effective cross sections per electron are given in arbitrary units for different final-state symmetries and for different energies above the muffin-tin zero. We note that the lower energy part of the curves is appropriate to the XPS process and the higher part, above E_F, is appropriate to BIS. It is immediately obvious that the partial cross sections vary very strongly with element; for instance the d-state matrix element varies by about a factor of 20 between Sc and Pd and this is a reflection of the variation of the atomic partial cross sections, discussed in connection with Table 10.1.

Figure 10.3 illustrates two solid-state effects not found in the atomic cross-section approach. Firstly we note that the calculated intensities are the products of matrix elements, which may be evaluated for any energy, and state densities, which are zero for regions in which there are no solid-state eigenstates. Thus we see that, for instance, in Fig. 10.3 the matrix elements actually peak outside the d bands.

The second solid-state effect in connection with Fig. 10.3 is the variation of matrix element with position in the band. This effect is clearly important for

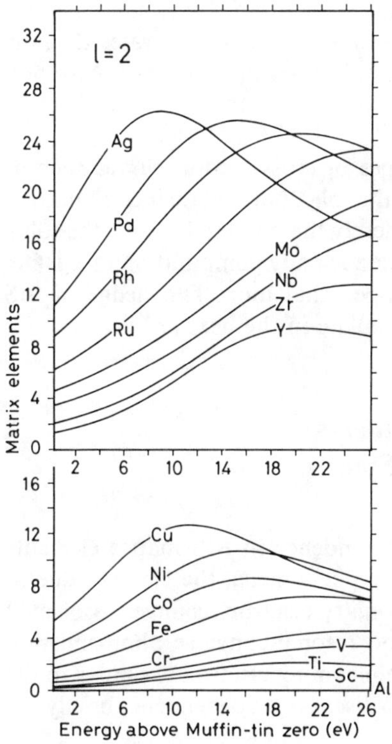

Fig. 10.3. XPS/BIS matrix elements for $3d$ and $4d$ states in Sc to Cu and Y to Ag as a function of energy above the muffin-tin zero: XPS is below E_F and BIS above E_F. The approximate positions of the d-band onset, the Fermi level and the top of the d bands can be derived from data in Table 10.2

evaluation of BIS spectra from broad bands. Within the dipole acceleration form of (10.5) the energy dependence can be rationalized using Fig. 10.4. The figure gives the normalized $4d$ wavefunctions for bonding and anti-bonding states of Ag at the bottom and top of the $4d$ band, respectively. In the anti-bonding state the wavefunction is zero at the edge of the muffin tin (this is the criterion for the top of the band). At the bottom of the band the value of the wavefunction at the muffin-tin edge is maximised. At first sight this change is not expected to make any difference because in the dipole acceleration

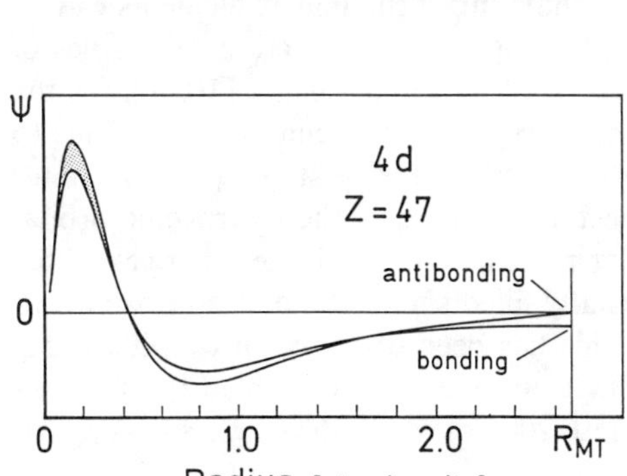

Fig. 10.4. Illustration of the origin of the differences between XPS/BIS matrix elements for bonding and anti-bonding $4d$ ($Z = 47$) states. The changes in wavefunction (real part) at the edges of the atom lead to renormalization in the core of the atom

form of the matrix element the outer regions of the atom contribute relatively little to the matrix element (Fig. 10.2). However, Fig. 10.4 illustrates that the changes in the external region of the atom result in significant renormalization of the wavefunction in the core region of the atom, and this does lead to significant variations in matrix element.

As a final point we note that the shapes of cross-section curves such as those in Fig. 10.3 do depend on the nature of the solid, but that their computation does not require major effort. The input is merely the potentials inside the solid, which are precisely the same potentials needed for any computation of a DOS. It is not a major task. Thus it is to be hoped that more knowledge of BIS matrix elements in the solid state will accumulate in the next few years.

10.3 Dependence of BIS Matrix Elements on Primary Beam Energy and Spin

We will describe three sources of energy dependence in BIS matrix elements here. These are, in order of sharpness of the phenomena, the changes due to the normal changes in wavelength of the primary electron, changes associated with the phenomena of Cooper minima and resonant inverse photoemission which occurs when the kinetic energy of the incoming electron is similar to the binding energy of a core electron. We will also discuss spin dependence briefly.

10.3.1 Broad Trends in Energy Dependence of the Matrix Elements

It has long been recognized that BIS spectra measured at different isochromat energies show different distributions of the overall intensity [10.25–28]. Some of these effects are probably related to the inelastic scattering effects discussed in Sect. 10.4, but some are simply the result of a change in the single-particle matrix elements due to a change in the wavelength of the incoming primary electron. This can be seen by reference to Fig. 10.2 and (10.5), where it is clear that if the wavelength of the primary electron changes, then so too does the matrix element. General trends in the behaviour of the matrix elements can be predicted using calculations based on hydrogenic wavefunctions, or tables of photoelectric cross sections, such as those of *Yeh* and *Lindau* [10.29], for the partially filled shells. Intuitively one must expect hydrogenic wavefunctions to be less appropriate, in the solid state, for high-energy states (say for states 10–100 eV above E_F because the expectation radius of the hydrogenic orbital is usually larger than the Wigner-Seitz radius. Thus for larger energies of the final-state electrons, and even for partially filled shells, the best approach is to take into account solid-state effects. This has been done by *Speier* [10.21, 28] for high-energy BIS, but to our knowledge nobody has attempted this for low- and intermediate-energy isochromats (50–200 eV) where the single-site approximation breaks down.

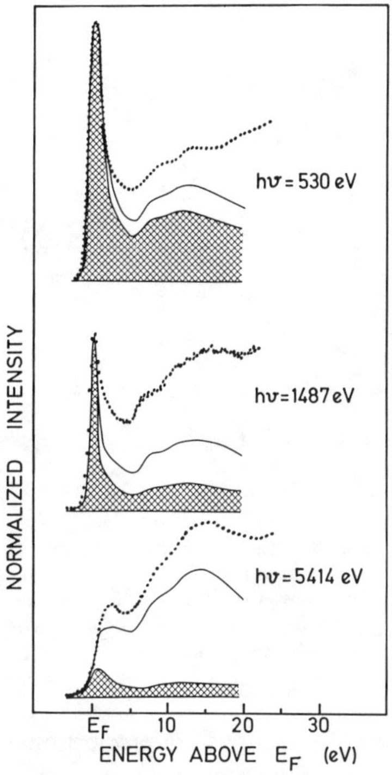

Fig. 10.5. BIS spectra of Ni at different isochromat energies compared with theoretical results incorporating the matrix elements and solid-state effects. The *d* contribution to the sum of the *l*-symmetries has been shaded. The theoretical curves have been adjusted for the different experimental resolutions, so that no conclusions should be drawn from the *d* peak-to-step intensity ratios. Figure taken from [10.28]

We give just one illustration of the influence of this form of energy dependence in the matrix elements, for Ni from [10.28]. The peak at the Fermi level must clearly be attributed to the Ni 3*d* band and it clearly has greater prominence at 530 eV than at 5414 eV. This is reflected also in the theoretical curves. The BIS matrix elements decrease, with isochromat energy in this energy region, for all states. However the decrease is clearly larger for the *d* states than for the *sp* states. This effect can, in principle, be used to separate the contributions from different *l* states, but unfortunately few laboratories have the capability to measure BIS spectra at different isochromat energies.

10.3.2 Cooper Minima

The Cooper minimum is a well-known phenomenon in photoemission which should also occur for some unoccupied states in inverse photoemission or BIS. The Cooper minimum occurs only for electronic states with a node in the radial part of the wavefunction (e.g. 4*d*, 5*d*, 6*d* states, but not 3*d*) and causes the partial cross section to decrease by a factor of typically 10–100 across a broad energy region [10.30]. It is thus very useful for extraction of information on the partial density of states in a solid [10.31, 32].

The origin of the Cooper minimum is illustrated in Fig. 10.6 where the radial part of a typical atomic wavefunction with a node is compared with a

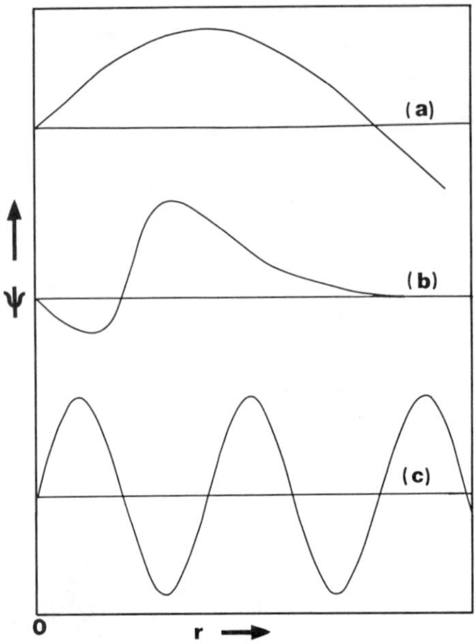

Fig. 10.6. Illustration of the origin of the Cooper minimum in PS and BIS cross section: radial wavefunctions for (**a**) low-energy electron, (**b**) atomic orbital with node, (**c**) high-energy electron

low- and high-energy electron wavefunction. The cross sections are proportional to the square of the overlap between the radial parts of the atomic wavefunction and the incoming (BIS) or outgoing (PS) wave (10.5). It is easy to see that for the low-energy wave the overlap with the positive part of the atomic wave dominates and the total overlap is positive. For the high-energy wave the wavelength of the travelling electron has been chosen so that the negative part of the atomic function overlaps the positive part of the travelling wave, and vice versa. The total overlap integral is thus negative and there has been a change in sign with respect to the low-energy wave. This would indicate that the overlap integral, and hence cross section would pass through zero at a given energy. This is not quite the case because there are always two outgoing waves with $\delta l = \pm 1$ and with differing radial wavefunctions. However as the $\delta l = +1$ channel dominates, the result of the change in sign of the $\delta l = +1$ overlap integral is a broad, deep minimum in the cross section. The Cooper minimum has been hardly exploited in BIS, if at all, but a description has been included here because several laboratories are building (or have built) spectrometers capable of exploiting the effect.

10.3.3 Resonant Inverse Photoemission

The last effect leading to energy dependence in the BIS or IPS matrix elements is known as "resonance inverse photoemission" and can lead to very strong variations in matrix element for very small changes in isochromat energy. The first observations of this sort were probably those for La in the early 1970s

[10.33–37] and although there has been some confusion concerning the origin of the effect, the basic physics is now quite clear. It is related to the Fano effect in PS [10.38, 39], as stated in the original papers.

The origin of these resonances can be explained by reference to BIS of La, which has a $4f^0$ configuration in the ground state. The BIS spectrum has a strong peak about 4 eV above E_F due to transitions to the $4f^1$ final state, i.e.

$$\mathrm{La}\,(4f^0) + e \rightarrow \mathrm{La}\,(4f^1) + h\nu. \tag{10.7}$$

This channel can occur regardless of the energy of the incoming electron. However for E_P of about 850 eV another channel can operate. This involves a "first" step of simultaneous promotion of a $3d$ electron to the $4f$ shell and the absorption of the primary electron into the $4f$ shell. This step is then "followed" by fluorescence decay in which one of the $4f$ electrons falls into the $3d$ hole. We warn that the naive step-wise representation is over simplistic, but for our purposes we can represent this second channel as

$$\mathrm{La}\,(4f^0) + e \rightarrow \mathrm{La}\,(\underline{3d}, 4f^2) \rightarrow \mathrm{La}\,(4f^1) + h\nu. \tag{10.8}$$

These two channels can interfere when E_P is close to, or equal to, the energy required to promote an electron from the $3d$ to $4f$ shell, because at this energy the initial and final states of the two channels are indistinguishable. The result

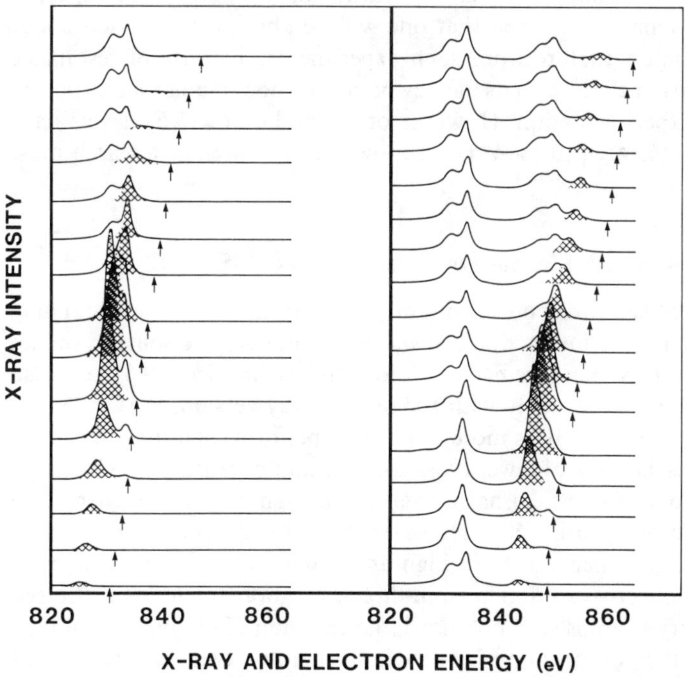

Fig. 10.7. X-ray emission spectra from La as a function of incident electron energy from ~820 to ~860 eV. The figure was redrawn from data in [10.33, 34] and is explained in the text

can be a strong enhancement of the isochromat X-ray intensity, as illustrated for the case of La in Fig. 10.7 [10.40].

Figure 10.7 gives the X-ray spectra from La measured with electron energies as indicated by the arrows under each curve. Starting at $E_P = 830.5\,\mathrm{eV}$ the spectrum is very weak and corresponds to BIS with an electron falling into the $4f^1$ level $\sim 4\,\mathrm{eV}$ above E_F. As E_P is increased the BIS emission tracks the electron energy, as indicated schematically by the shaded area in the figure. However the thresholds for creation of $3d_{3/2}$ and $3d_{5/2}$ core holes also lie in this energy region. They are centred about 833 and 849 eV respectively, but are actually two manifolds of many thresholds as a result of atomic multiplet effects and shake-up satellites [10.41, 42]. These new thresholds give rise to two effects: new X-ray lines appear in the spectra at constant energy as a result of fluorescence decay of the core hole, and the BIS emission is enhanced as the electron energy matches the threshold. We have represented the characteristic emission as shaded area in Fig. 10.7, but, of course, it is inherent to the nature of these processes that a complete separation of BIS and characteristic emission is not always possible.

One should note that a fully quantitative explanation of resonant BIS for La must take account of the Coulomb and multiplet interaction between $4f$ electrons. As noted in the last paragraph, these interactions can shift the energy of the resonance with respect to the XPS binding energies, cause multiple threshold effects, and complicate the appearance of the fluorescence spectra. Nevertheless we remain convinced that one will be able to learn much about solid-state electronic structure from such experiments. The use of resonance inverse photoemission effects is in its infancy because it does require sophisticated and specialized experimentation. However other studies, of U $5f$ levels have begun to appear recently [10.43–45] and growth in this area of research must be expected.

10.3.4 BIS Combined with Electrons of Polarized Spin (BISCEPS)

Spin-polarized IPS measurements are desirable for a microscopic understanding of the origin of magnetic interactions. *Schneider* and *Dose* give some examples of spin-polarized IPS measurements at low-energy in Sect. 9.6 (see also [10.46–48]). Here we deal briefly with the high-energy version, BISCEPS.

The BISCEPS experiment is more difficult to perform than the low-energy (SPIPES) analogue because of lower cross sections and because the high-energy monochromator/detector system has a smaller throughput. Thus higher beam currents are required [10.49, 50]. However, in other respects the basis of spin-polarized measurements at both high and low energies is the same, with both needing the low cross section for spin-flip, or magnon, scattering processes. Thus if the primary beam is spin polarized, the emission of the bremsstrahlung radiation reflects the polarization of the available empty states. Electron beams with about 30–40% polarization may be obtained from GaAs or AlGaAs photocathodes with circularly polarized light [10.51, 52].

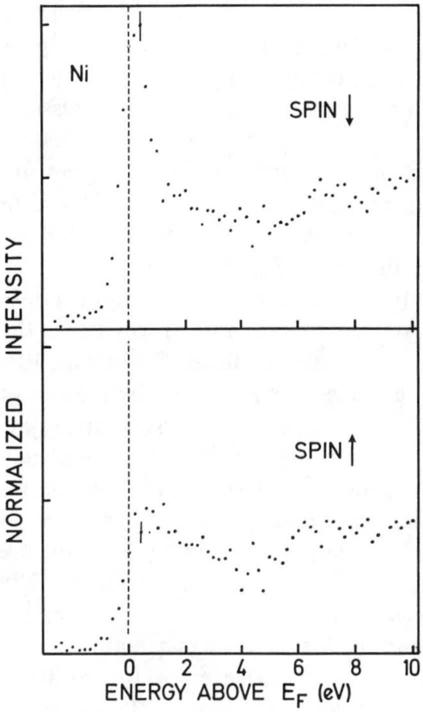

Fig. 10.8. BIS spectra of Ni resolved into spin-up and spin-down components by the use of a spin-polarized primary beam. Figure taken from [10.50]

A BISCEPS result for Ni is shown in Fig. 10.8. The minority (spin-up) band shows a peak near E_F which is due to the Ni 3*d* band. This peak is absent in the majority (spin-down) band because Ni is a strong ferromagnet with no majority states in the *d* band above E_F. The statistics in this spectrum are not sufficiently good to unambiguously detect the small exchange splitting in the edge due to the critical point about 6 eV above E_F (see [10.53] and Sect. 9.6). The BISCEPS method is, in principle, complimentary to low-energy SPIPES measurements in the same way that XPS and UPS are complimentary for the occupied states: the high-energy methods are not $\boldsymbol{k}$ resolved, have different cross-section ratios, and may have larger probing depths. However the yield and stability of the polarized electron source must clearly be improved before the full potential of the technique can be exploited.

10.4 Inelastic Losses and BIS of Nearly Free Electron Metals

Here we discuss the role of inelastic scattering of the primary beam in BIS, before the radiative decay with its consequent emission of bremsstrahlung. We will also discuss the role of intrinsic energy losses. The idea was first put forward in 1917 by *Webster* [10.54] that the observed bremsstrahlung from a solid material was the sum of a thin target bremsspectrum, in which the primary electron beam had suffered no inelastic collisions, and a series of overlapping replicas of this spectrum shifted to lower photon energies by inelastic scattering of the primary beam. This describes the physics in that, just as in photoemission, the transitions carrying the relevant information about the state densities do occur predominantly near the surface (i.e. most weight in the elastic spectrum is from a slab whose thickness is comparable to the mean free path length of the electrons). Those transitions with one or more inelastic scattering have a higher probability of occurring deeper in the solid.

The concept of replica spectra, shifted by the loss energy, is most appropriate to free-electron-like solids where the predominant elastic loss process is via the creation of well-defined plasmons. Such a spectrum is illustrated in Fig. 10.9 for Al. The spectrum exhibits a step at the Fermi level, followed by a series of steps which could be seen as replicas of the step at E_F after creation of one, two, three or more plasmons. However the first impression is not completely true, and there is other structure hidden under the plasmon structure, as we will see below. We note first a significant difference between the problems created by inelastic losses in BIS and XPS studies of valence bands: in XPS the valence bands have limited width and most of the loss features in the spectrum actually occur outside the band. This is not the case in BIS, where states characteristic of the solid exist up to very high energy and the loss structures can always hide some information. Thus their treatment is of greater relative importance for BIS.

In order to strip out the influence of inelastic scattering in BIS spectra, it is necessary to know the loss function, that is, the modification of the electron energy distribution before radiative decay due to inelastic scattering of the monoenergetic beam. In the case of free-electron metals this is not too difficult and will be illustrated as an example for what must be done in more complicated systems. The appropriate loss function for BIS is taken from photoemission for

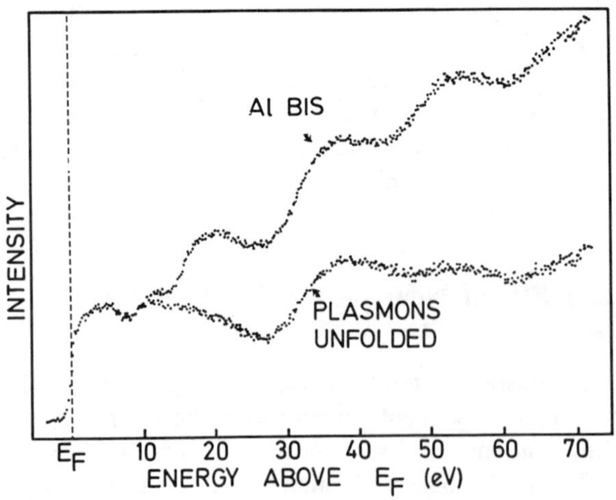

Fig. 10.9. Illustration of the BIS spectrum of Al and the plasmon subtraction procedure. The weight of the plasmons subtracted was calculated according to (10.11). For the lower curve, $\alpha = 0.62$, $\beta = 0.1$

similar electron kinetic energies [10.55, 56], because BIS is the inverse of the photoemission process. The weight of the extrinsic losses is thus the same [10.57]. The weight of losses intrinsic to the photoemission and BIS transition can, in principle, be different but they do not dominate the observed losses in photoemission (see e.g. [10.58]) and we assume the same is true for BIS.

The loss part of the BIS spectrum can be subtracted from the measured spectrum, $B(E)$, by deconvolution, if the loss function, $L(E)$, is known. Briefly, the spectrum corrected for losses, $BC(E)$, is given by

$$B(E) = BC(E) + BC(E)*L(E). \tag{10.9}$$

The loss function in this case is given by

$$L(E) = \sum_{n=1}^{\infty} D_n(E) + D_S(E) + \sum_{n=1}^{\infty} D_n(E)*D_S(E), \tag{10.10}$$

where $D_n(E)$ and $D_S(E)$ are the contributions from the nth bulk and surface plasmon respectively and the last terms are due to the creation of both a surface and a bulk plasmon by the same electron. The intensity of $D_n(E)$ is given as [10.59] and references therein)

$$D_n = \alpha^n \sum_{m=0}^{n} (\beta/\alpha)m/m!. \tag{10.11}$$

Here, $\alpha = (\lambda + 1/L) - 1$ with λ being the mean free path of the electrons for extrinsic plasmon losses and L is the mean free path for other inelastic losses. The parameter β is a measure of the excitation of intrinsic plasmon losses.

In Fig. 10.9 an attempt was made to subtract the plasmon losses using (10.9–11) with β held constant at the value of 0.1 found for valence band photoemission [10.58, 59]. α, representing the strength of intrinsic losses, is set at 0.62, in agreement with the value from XPS, in the second lowest curve. This leads to suppression of the shoulder at ~16 eV in the raw spectrum, but a second shoulder at ~33 eV cannot be suppressed, even if the extrinsic loss weight is strongly varied. It must thus be concluded that this feature is related to a density-of-states feature and not to intrinsic losses. This is confirmed by comparison of the BIS spectra to DOS calculations in Fig. 10.10.

The features in the corrected Al BIS spectrum of Fig. 10.10 correspond quite closely to the DOS features; namely the small peaks at 3.4 eV (theory 3.4 eV), at 11.5 eV (theory 10.3), the minimum at 27.7 eV (theory 26.1) a maximum at around 40 eV and another broad feature at about 55.1 eV (theory 50.1 eV). The most marked feature in the spectrum is the minimum at ~28 eV and the subsequent rise in intensity to about 40 eV. This is a strong deviation from a free-electron-like E dependence of the DOS and is attributed to the d- and f-states [10.60, 61]. It is interesting to note that it was not possible to identify the strong minimum at ~28 eV in inverse photoemission studies [10.62] and that BIS was necessary for that purpose.

There are four possible reasons why there is not an exact 1:1 correspondence between the (corrected) BIS and DOS curves of Al: a) Ambiguities over the correction for the plasmon losses, b) Technical problems in the DOS calculation, c) Variations in the matrix elements for the states of different symmetry and energy and d) Variations in the self-energy. The first of these is almost unavoidable, but it is our judgement that it is not important here. Considering the second point, one should not assume that all DOS calculations are suitable for detailed comparison with high-energy BIS, but with appropriate caution the DOS should be sufficiently reliable. The third point is relatively trivial (see Sect. 10.2) but has not been studied for Al. The fourth point, the self-energy, is by far the most interesting at present. As repeatedly stated, the calculated eigenenergies from a normal band structure have, strictly speaking, no physical meaning as individuals and it is an approximation to compare them with measured ionization potentials (PS) or electron affinities (BIS). They should really be corrected by an energy-dependent self-energy which has both real and imaginary parts. In broadening the calculated DOS for Fig. 10.10, *Hoekstra* et al. [10.59] were, in fact, making an approximation to account for the imaginary part of the self-energy, which is equivalent to the lifetime broadening. The discrepancies in energy between the calculated DOS and measured BIS features in Fig. 10.10 must be at least partly attributed to the real part of the self-energy,

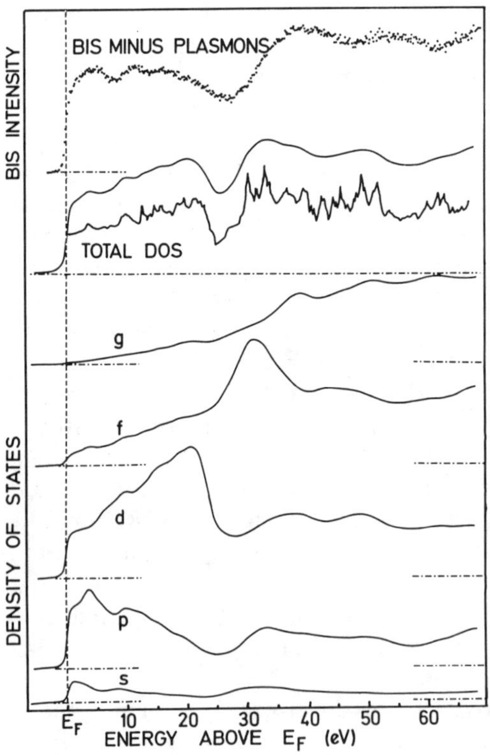

Fig. 10.10. The partial and total density-of-states curve for Al compared to the measured BIS spectrum after correction for plasmon losses. With the exception of one total DOS curve, the DOS have been broadened to account for instrumental and lifetime broadening effects

but because of the ambiguities associated with points a) and b), we should not be too dogmatic about the sizes of the real part of the self-energy.

The case of electron energy losses in Al BIS was one of the more easy to handle. In general, non-free-electron metals have broader and more complex inelastic energy losses [10.63]. In addition, more complex materials will have more complex intrinsic losses, or shake-up processes, and it will not always be appropriate to assume that the intrinsic losses of BIS and XPS are the same. As an example of how serious this problem can be, we cite the examples of Ni and La or Ce, where respectively approximately 40–45% and up to 90% of the weight is in the high-energy XPS peaks normally known as "shake up". In some cases, such as Ce, it is certainly not appropriate to subtract out the intrinsic losses from the BIS spectrum and one should regard the whole XPS/BIS spectrum with peaks due to many-body excitations, as representing information of the one-particle Green's function of the valence electrons [10.64].

10.5 BIS of Transition Metals

The transition metals (TM) have strong, well-defined, *d* bands which dominate XPS and BIS spectra within about 10 eV of E_F. These bands have calculated widths of between ~4 and ~10 eV [10.65, 66] Table 10.2, with the smallest widths being found at the top right of the TM block in the periodic system and the largest widths at the bottom left. XPS/BIS studies must be expected to show the filling of the *d*-states from left to right in the periodic system, and the change of the shape of the bands with changes in crystal structure of the TM. We must also expect increased many-body effects in the more strongly correlated bands of the elements at the top right of the TM block.

BIS studies exist for most of the transition metal elements [10.55, 66], but the ideas of the paragraph above, and the influence of the principles described in Sects. 10.1–3 can be illustrated by reference to just a few examples: we choose Mo, Ru and Ni. In Fig. 10.11 the DOS of Mo shows a pronounced minimum near E_F. This minimum is typical of bcc TM and separates the mainly bonding and mainly anti-bonding states. The Ru DOS shows four "groups" of peaks, separated by minima at ~4.5, ~3 and ~0 eV below E_F. Again this structure is typical of hcp TM but the structure may be weaker in some elements, like Y [10.66], as a result of hybridization with *sp* bands. Of course the spectra of Ru and Pd do not show all the structure of the DOS because of lifetime broadening of the final states of XPS and BIS, as discussed in Chap. 3. However the main features of the DOS can be recognized in the spectra, such as the *d*-band width and the filling of the *d* band from Mo to Ru, with consequent increase in occupied band width and decrease in unoccupied band width. We can also observe some features in the DOS associated with the crystal structure, such as the position of the Fermi level of Mo in the pronounced minimum between bonding and anti-bonding states, or shoulders at 0.5 and 5 eV below E_F in the XPS of Ru.

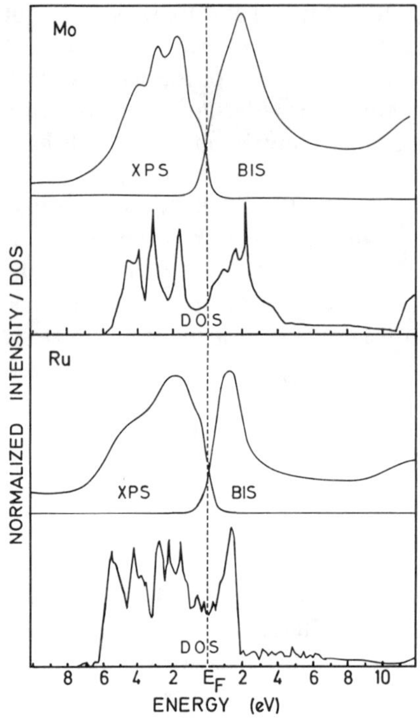

Fig. 10.11. XPS/BIS spectra of bcc Mo and hcp Ru from data given in [10.66–68]. The spectra have been normalized so that the areas of the XPS and BIS peaks are similar to the ratio of areas of the occupied and unoccupied DOS. Also shown are single-particle DOS from [10.66]

The spectra in Fig. 10.11 show no significant satellites and many-body effects in the excitation process seem to be weak. For all the transition-metal BIS spectra there is relatively good agreement between the calculated and observed unoccupied band widths. However there do seem to be some effects due to the dependence of the single-particle matrix elements (ME) on the energy of the electron or hole added. In particular, for the XPS spectra it is probably safe to attribute the relative weakness of the spectral weight from high BE states to ME effects.

It has been noted that, in the unoccupied states, the intensity above the *d* bands is much higher than in the DOS [10.66]. This is probably not entirely due to inelastic losses, as evidenced by the low weight to the high BE side of the *d* bands in XPS. The region above the *d* band has a high contribution of 5*sp* states whose ME are, however, not larger than those of the 5*d* states at this primary beam energy (at $\sim 5\,\mathrm{kV}$ this is no longer the case [10.28]). In my opinion this problem deserves more systematic study.

The next example we choose is that of Ni, which has been a test case for many photoemission studies because of its many-body satellites (see e.g. [10.69–71]). Ni has the smallest band width and the largest effective, on-site, Coulomb correlation energy (U_{eff}) of all the transition metals [10.72]. This, combined with its nearly filled band, results in strong distortion of the observed XPS spectrum in comparison to the independent-particle DOS. A satellite, corresponding quite

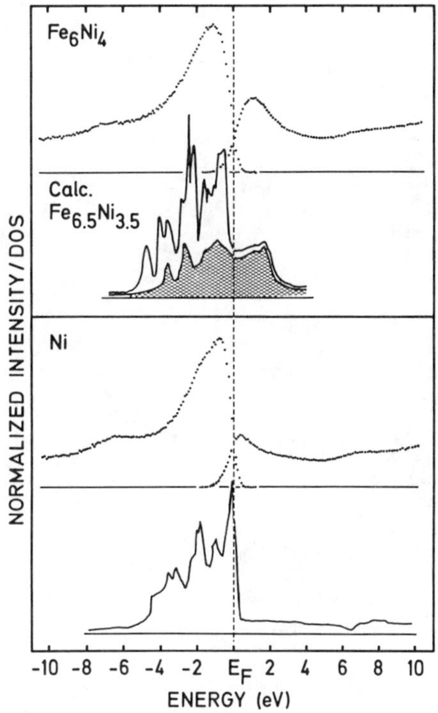

Fig. 10.12. Comparison of XPS/BIS spectra of Ni and $Fe_{6.5}Ni_{3.5}$ with the calculated density of states for Ni. The spectra have been normalized so that the areas of the XPS and BIS peaks are similar to the ratio of the areas of the occupied and unoccupied DOS. Also shown are single-particle DOS from [10.66]. Figure from [10.73]

closely to local d^8 final-state configurations, is pushed out of the band to about 6 eV below E_F by the strong d–d interaction (see Fig. 10.12). In addition the Ni d band in photoemission is far narrower than in the single-particle DOS [10.69–71]. Similar effects are found for $Fe_{6.5}Ni_{3.5}$.

As for all the other transition metals, no strong satellites have been found in BIS of Ni, [10.66] as illustrated in Fig. 10.12. The reason for this apparent discrepancy is too complex to explain in detail here, but we can explain the effects. Strong satellites are typically found in XPS or BIS when U_{eff} is between about one and ten times the band width, and for pure TM only Ni falls in this category, although satellites are common in TM compounds. However, the satellite intensity is also influenced by the degree of band filling. If the band is nearly full, satellites will tend to be larger in XPS. If the band is nearly empty, satellites will tend to be larger in BIS. More details of this effect can be found in, for instance, [10.64, 74–76] under the keyword "degeneracy".

Figure 10.12 also confronts us with the problem of spin-resolved densities of states, which are important for understanding magnetic properties. $Fe_{6.5}Ni_{3.5}$ is a disordered alloy whose unoccupied band width is larger than that of Ni, as shown by both calculations [10.77] and BIS experiments [10.73]. In Fig. 10.12 its minority states are shaded in the DOS and the unoccupied states are seen to be strongly spin polarized. These effects can only be distinguished by the spin-polarized experiments, which were described in Sect. 10.3.4.

Table 10.2. Theoretical (theor.) values for occupied (W^-) and unoccupied* (W^+) d band widths as compared to experimental (exp.) BIS values for W^+ [10.66]. All values in eV. Dual values indicate the experimental uncertainty. Also given are the values, $E_F - M$, of the Fermi level with respect to the muffin-tin zeros from [10.21]

Element	W^- (theor.)	W^+ (theor.)	W^+ (exp.)	$E_F - M$
Sc	1.7	4.0	4.4–4.7	6.6
Ti	3.1	3.7	3.2–3.8	8.7
V	3.2	4.0	3.4–3.8	10.4
Cr	4.5	2.5	1.8–2.3	11.6
Mn			2.9–3.5	
Fe	5.3	2.6	2.9–3.5	11.3
Co	5.5	1.2	1.4	9.2
Ni	5.0	0.4	0.4	9.2
Y	2.0	5.3	6.1	6.3
Zr	3.1	5.3	5.2–5.7	8.2
Nb	3.5	5.6	5.2–5.9	9.2
Mo	5.4	3.9	3.9–4.5	10.9
Ru	6.2	2.0	2.0	9.7
Rh	5.8	1.2	1.2	8.7
Pd	4.8	0.4	0.4	6.8

* This only includes the unoccupied states in the metal d band. There is also some unoccupied metal d character hybridized with the metal sp bands

10.6 BIS of Rare Earths and Actinides

The largest contribution to the XPS/BIS spectra of rare earths and actinides is usually that of the $4f$ and $5f$ states. In the rare earths, U_{eff} for the $4f$ states was shown experimentally to range from 5 to 11.5 eV in the classic work of *Lang* et al. [10.2]. In no case do the rare earths have $4f$ band widths greater than ~1 eV so that atomic correlations always dominate the spectra. However the small hopping integrals of the $4f$ electrons, and mixed valence effects which give rise to non-integral f-counts, lead to complicated effects in the spectra [10.40, 78–85]. These effects are not only interesting in themselves, but also for the insight they give into the fascinating ground-state properties of rare earths. Actinides are intermediate between rare earths and TM in their behaviour and also show very interesting effects, but of course only a small number of the series are available for study.

We start our discussion of the effects which can occur in rare earths and actinides, with a consideration of the BIS spectra of several Ce intermetallic compounds with Ni. At room temperature γ Ce has about $0.98f$ electrons per atom so that the BIS spectrum is dominated by the peak associated with transitions from the ground state, with mostly f^1 character to the final state with mostly f^2 character, 4–6 eV above E_F. The BIS peaks are broadened by unresolved multiplet structure [10.2], as shown. As more and more Ni is added,

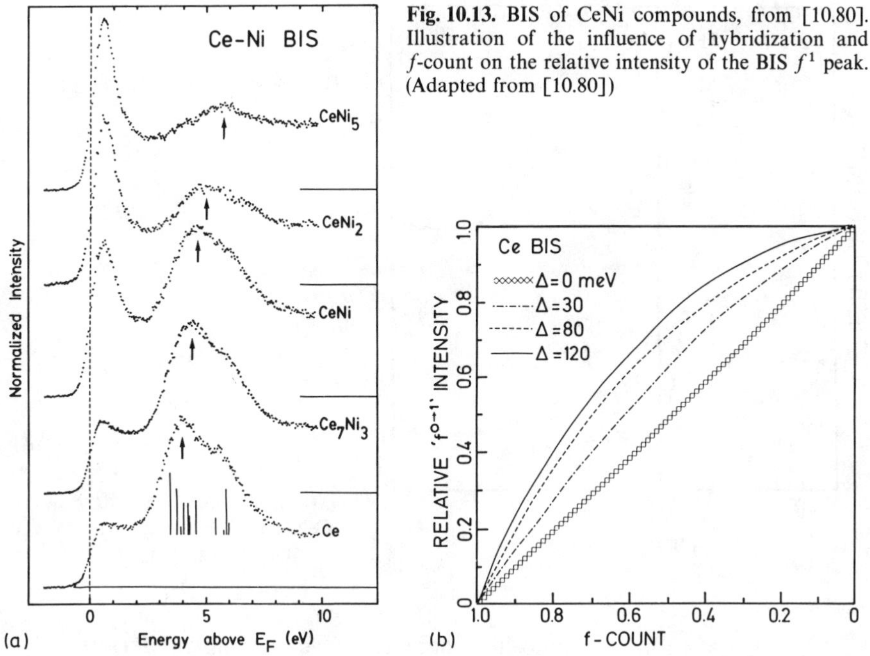

Fig. 10.13. BIS of CeNi compounds, from [10.80]. Illustration of the influence of hybridization and f-count on the relative intensity of the BIS f^1 peak. (Adapted from [10.80])

a new peak is seen to grow at E_F. Comparison with the corresponding La–Ni compounds shows that this is not due to Ni states, which have a lower cross section, but to transitions which may be essentially described as $\sim f^0$ to f^1. These transitions arise because the f-count in the ground state of the materials is now less than one. However, in the presence of hybridization one may not simply take the intensity ratio of the f^1 to f^2 peaks as directly proportional to the weight of the f^0 configuration in the ground state. This is illustrated in the right-hand side of Fig. 10.13, where it is seen that for a given f-count a quite modest level of hybridization leads to strong variation of the peak ratios.

One should also be aware that the peaks which arise in XPS/BIS spectra of highly correlated materials as a result of hybridization do not always have simple descriptions. To illustrate such effects we choose two other examples from work on Ce in Fig. 10.14. The XPS spectra of CeN and $CeSi_2$ are shown. Ce in $CeSi_2$ has 0.97 $4f$ electrons per atom and the strength of the hybridization is small. The intensity scale of the XPS and BIS spectra can only be compared when one realizes that the integrated XPS intensity should be ~ 13 times less than that in BIS because the $4f$ shell occupancy is only 1/14. For $CeSi_2$ the XPS spectrum is dominated by the transition from f^1 to f^0 at about 2 eV below E_F, whilst the BIS spectrum is dominated by the peaks due to transitions from f^1 to f^2. Note however the small peaks at E_F which are due to hybridization with the valence states. Their relative intensities are larger in CeN because the hybridization is larger there (~ 15 meV as opposed to 3 meV for $CeSi_2$). These

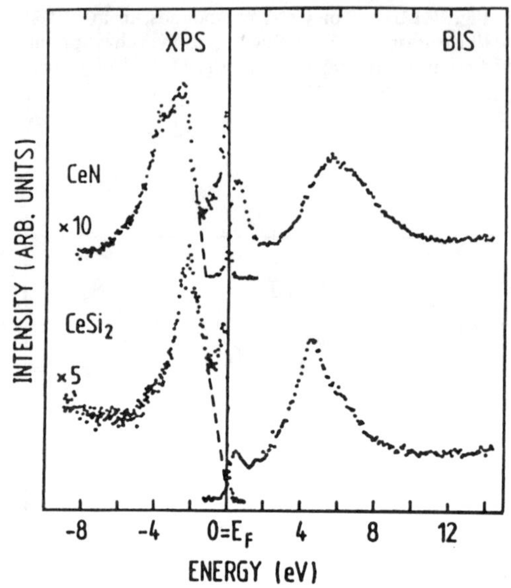

Fig. 10.14. Combined XPS/BIS spectra of CeN and $CeSi_2$. Figure taken, with permission, from [10.84]

peaks, which often have Kondo-like character, have no simple description, but their matrix elements behave like those of the $4f$ states [10.76]. As the hybridization of the $4f$ levels increases, eventually the description of the other peaks in terms of simplistic labels also becomes problematic.

Studies of spectra such as those of the Ce compounds illustrated here have given much insight into the electronic structure of the so-called mixed valence compounds of the rare earths and actinides in recent years. For work on actinides, for which the principles are very similar to those for Ce studies but with larger hybridization, we refer to [10.86–90]. We note, however, that more work is still to be done. There are still questions concerning the width of the peaks [10.89, 91, 92] and many important materials with anomalous ground-state properties have still to be studied. The interpretation of the data is complex, and involves solutions of Anderson impurity and lattice Hamiltonians which are unfamiliar to many spectroscopists [10.93]. However it is now accepted that these high-energy spectroscopies do give us unique information relevant to the enormous numbers of anomalies in the ground-state electronic and cohesive properties. The same may also be said of $3d$ transition metal compounds where the $3d$ states are also often highly correlated. As examples here we refer to work on NiO in [10.94] and Sect. 4.2, Cu-oxides, [10.95] and high-T_c superconductors, [10.96, 97].

10.7 BIS of Compounds

We have, in fact, dealt with some compounds in Sect. 10.6, but those were examples where the main contributions to the XPS/BIS spectra arose from narrow bands of highly correlated states. Here we wish to present a case study of some silicides which illustrate the properties of compounds with broad bands and in which correlation effects are less strong. As stressed in Sect. 10.2, in compounds the XPS/BIS valence band spectra contain contributions from all the elements present, and the contributions from the different *l* partial densities of states will be, in general, different. In addition there will be some variations of the partial cross sections as a function of energy within the band, and the possibility of contributions to the shape from self-energies, many-body satellites and inelastic losses.

The data we wish to work with, adapted from [10.98, 99] is reproduced in Fig. 10.15. The lowest section shows the theoretical DOS and partial DOS of $MoSi_2$ and these are compared with the theoretical BIS spectra, in which the matrix elements are taken into account. It is striking that the Si *d* states, which are very strong in the unoccupied DOS, contribute negligible weight to the BIS. Also the influence of the Mo 4*d* states is enhanced by the matrix elements. This latter is not the case in the corresponding 3*d* silicides, where the metal 3*d* and Si *sp* cross sections are comparable [10.100].

Moving up Fig. 10.15, we see that the theoretical spectra reproduce the measured spectra rather well. Any shifts due to self-energy effects are hardly noticeable on this scale and there are no significant many-body satellites. Note, however, that the spectra have already been constructed from broadened DOS, so that the lifetime broadening of the final state (i.e. the imaginary part of the self-energy) is important. We may also note a small discrepancy between the theoretical and experimental intensity at the high-energy end of the BIS spectrum. This is probably due to the influence of inelastic scattering.

The XPS/BIS spectra of transition metal silicides are basically hardly influenced by many-body effects, which allows us to study the influence of chemical bonding. As an example we have compared the XPS/BIS spectra of $MoSi_2$ and WSi_2 in Fig. 10.15. There is seen to be a clear increase in separation of the two main peaks, which is especially interesting because the two materials are isostructural with almost identical lattice constants. Because of the large weight of the metal *d*-state contribution to the spectra, we can trace the origin of the shifts to these *d* states. It is thus clear that the increase in orbital radius from 4*d* to 5*d* states results in increasing overlap and hence increased splitting between the bonding and anti-bonding states.

Transition-metal silicides are ideal for XPS/BIS studies because they are insensitive to the large electron beam doses needed for BIS. It should, however, be noted that large numbers of materials show similar insensitivity and would be suitable for studies similar to those illustrated here for silicides.

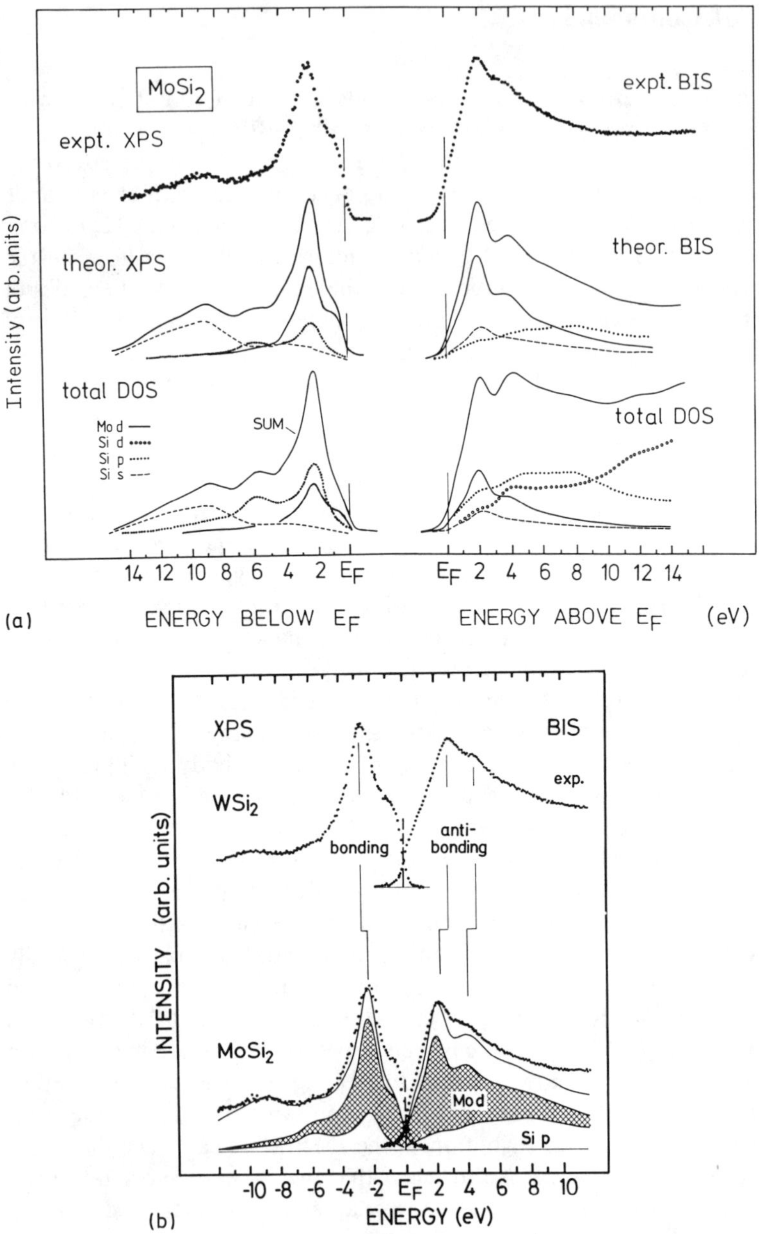

Fig. 10.15. **(a)** XPS and BIS spectra for $MoSi_2$ compared with the broadened DOS and calculated spectra including the matrix elements. **(b)** The XPS/BIS spectra of $MoSi_2$ and WSi_2 are shown to illustrate the greater splitting of bonding and anti-bonding levels in the 5*d* silicide. Figure from [10.98]

10.8 BIS and Optical Properties

It is clear that the resolution attained in studies of optical properties in the IR and visible region, far exceeds that feasible in present-day BIS experiments. However, the optical response is influenced by *both* the occupied and unoccupied DOS (i.e. the joint DOS or JDOS) *and* the complicated optical transition matrix elements. In XPS/BIS the matrix elements are simple by comparison with those influencing the optical response and the JDOS problems do not exist. Thus XPS/BIS studies can be used to significantly enhance the utility of optical studies. As an example we will consider the cases of crystalline and amorphous Si [10.101].

The problems of deducing properties from optical spectra are illustrated in Fig. 10.16. On the left-hand side of Fig. 10.16, the XPS/BIS spectra of Si are shown. One may easily see that the optical transitions correspond to excitation of electrons from the occupied (XPS) to the unoccupied (BIS) states. For excitation from a given valence state in crystalline materials very few transitions are allowed because of ***k*** conservation, as illustrated in the right-hand side of the figure. However, as the dashed lines indicate, even in this case many transitions may contribute to the optical response at any given energy, so that the matrix element

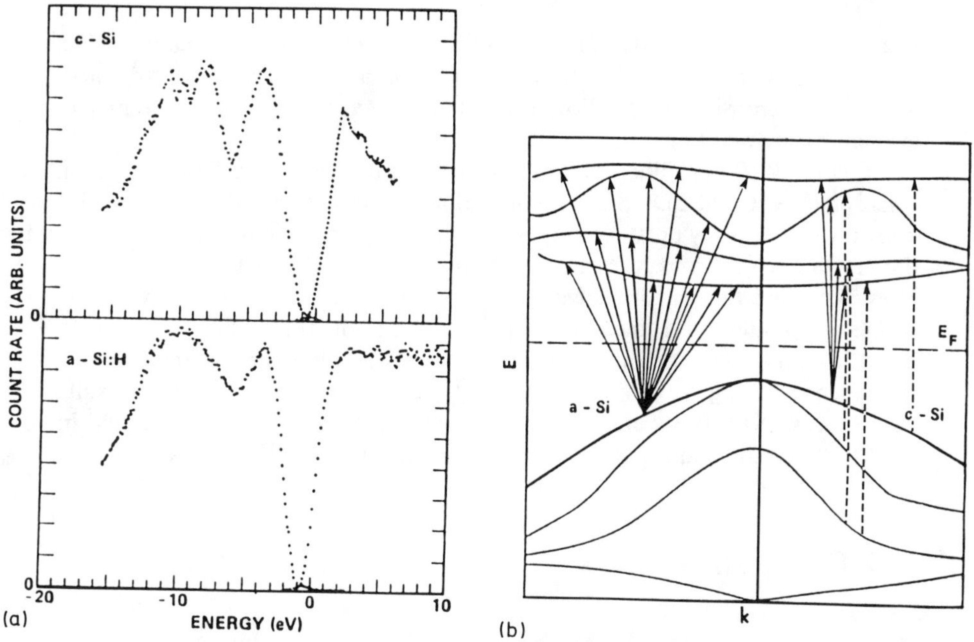

Fig. 10.16. (**a**) XPS and BIS spectra for crystalline and amorphous Si. (**b**) non-zero optical matrix elements for amorphous Si:H and crystalline Si from a particular valence state. For further explanation, see text. Both figures are adapted from [10.101]

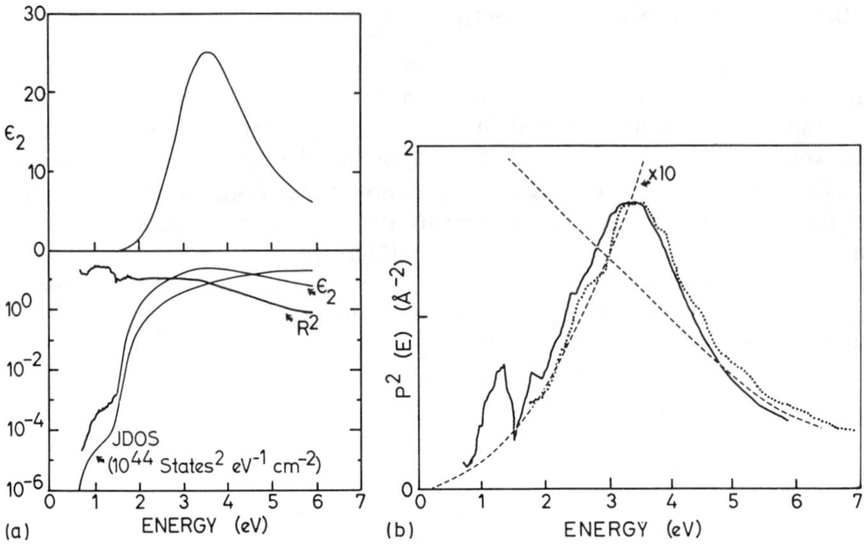

Fig. 10.17. (**a**) Linear and logarithmic plots of the dielectric function, ε_2, and a logarithmic plot of the JDOS (here $J(E)$). (**b**) Comparison of the energy dependence of the optical matrix elements, $P^2(E)$ [10.103], drawn as a full line, with previous results (dashed and dotted lines). For further details see text. Figure adapted from [10.101]

is a very complex quantity. This is even more true for amorphous materials such as *a*Si:H, where there is no $\boldsymbol{k}$ conservation and indeed $\boldsymbol{k}$ is not defined. Thus all the transitions are allowed but the energy dependence of the matrix element is not well known.

Figure 10.17 (left) illustrates the dielectric function, ε_2, of amorphous Si. This is related to the JDOS by the optical matrix elements and these were not known with certainty until recently. In fact, on a logarithmic scale, ε_2 resembles the JDOS derived from XPS measurements quite closely, as Fig. 10.17 shows. However when the matrix elements, derived from comparison of ε_2 with the JDOS are plotted on a linear scale, (Fig. 10.17, right) the variation is seen to be rather large with a roughly linear increase at low energies and an E^{-5} dependence for $E \geqq 4\,\text{eV}$. Whilst one may argue that the experimental results derived from XPS/BIS were in agreement with the dotted result, calculated in 1973 [10.102], they are quite different from two other earlier results.

10.9 Concluding Remarks

In one form or another, BIS has been around for 70 years but compared, for example, to photoemission it is still not a widely used technique. If it has not achieved this status after 70 years, it probably never will. The reasons for this

Table 10.3. Comparison of XAS, IPS, and BIS with UPS and XPS

Property	XAS	IPS	BIS	UPS	XPS
Probe of DOS/UDOS	UDOS	UDOS	UDOS	DOS	DOS
Atom selective	yes	no	no	no	no
Symmetry selective	yes	no	no	no	no
k selective	no	yes	no	yes	no
Self-energy effects	yes	yes	yes	yes	yes
Core-hole effects	yes	no	no	no	no
Extrinsic inelastic losses	no	yes	yes	yes	yes

are complex, but include the facts that the technique is not suitable for chemical analysis in the normal sense and it does not have the element specificity of techniques like XAS. However the fact that BIS is not widely used should not be taken as an indication of its lack of importance in the overall picture. This can be illustrated by reference to Table 10.3. For instance, XAS, IPS and BIS are the most direct probes of unoccupied states but XAS is strongly influenced by core-hole effects, whilst BIS is not and IPS is strongly ***k*** conserving, whilst BIS is not. Thus for maximum understanding of the unoccupied (anti-bonding) states one should combine *all three* techniques. Certainly the techniques should *not* be seen as being in competition with each other. Rather one should recognize that they are complementary and the information from any one technique can be enhanced by that from others if one makes good use of their individual strengths. We will finish this chapter by recalling two aspects of the use of the complementarity to illustrate the role BIS can play.

One obvious example of the complementarity is the combination of XAS and BIS to try to separate the influence of the energy dependence of the exchange-correlation term, ExC, from core-hole effects in XAS. The ExC term can be probed by BIS, as indicated in Sect. 10.1. In XAS one must know about the behaviour of ExC before one can start to probe the core-hole effects directly. Now, in 1990, we have begun to get a quantitative understanding of self-energies and ExC for simple systems (see also Chap. 3), but the understanding of core-hole effects in XAS is clearly a much larger task and will require considerable input from BIS and XPS/BIS studies.

The second example we choose is the combination of XPS and BIS to study electron–electron correlation in narrow-band systems. Here, as discussed in Chap. 4 and Sects. 10.1 and 10.6, the dominant term in any electronic property of such materials is the effective Coulomb correlation energy, U_{eff}, which is usually expressed as the difference of the ionization energy and the electron affinity. The first of these is given by photoemission and the second by BIS [10.2]. The only real competition to the XPS/BIS combination for direct determination of U_{eff} is Auger spectroscopy (AES) using the Cini–Sawatzky methodology [10.104–106]. However even here we find complementarity because AES and XPS/BIS are generally not applicable to the same systems. Thus the XPS/BIS combination is really essential for studies of many highly correlated systems.

In both these examples, the BIS study is a small component of the total research, but the small BIS component has greatly enhanced the utility of other work. We expect this will continue to be the case. In this chapter we have stressed the principles of BIS and the complications in its interpretation. However is must be stressed that the problems are less dramatic than for many other techniques and the methods used to overcome them are quite simple and reliable to a high degree.

Acknowledgements. It is a pleasure to thank all the people who have helped in the preparation of this chapter by discussing problems, sending me reprints and commenting on various sections of the text, as well as the people who have worked with me on BIS over the years. These people include M. Abbate, J.F. van Acker, J.W. Allen, Y. Baer, H. Borsje, A. Burr, M. Campagna, V. Dose, O. Gunnarsson, F.U. Hillebrecht, J.E. Inglesfield, J. Keppels, C. Laubschat, I. Lindeyer, B. Reihl, G.A. Sawatzky, W.-D. Schneider, W. Speier, J.H. Weaver, P.J.W. Weijs, D. Whitehead and VSW, E. Umbach, J. Zaanen, W. Szweryn and all of my group in Nijmegen.

References

10.1 W. Speier, R. Zeller, J.C. Fuggle: Phys. Rev. B **32**, 3597 (1985)
10.2 J.K. Lang, Y. Baer, P.A. Cox: J. Phys. F **11**, 121 (1981)
10.3 D. van de Marel, G.A. Sawatzky, R. Zeller, F.U. Hillebrecht, J.C. Fuggle: Sol. St. Commun. **50**, 47 (1984)
10.4 *Narrow Band Phenomena*, ed. by J.C. Fuggle, G.A. Sawatzky, J.W. Allen (Plenum, New York 1988)
10.5 P. Horsch, W. von der Linden, W.-D. Lucas: Sol. St. Commun. **62**, 359 (1987)
10.6 The self-energy is defined as the difference between the Hartree energies and the real ionization potentials and electron affinities. Thus the self-energy is zero at the Hartree zero; namely the energy of an electron at rest at infinity. However in experimental systems the reference level is normally the Fermi level. One sometimes meets texts in which the self-energy is taken to be zero at the Fermi level, but this is not correct. Only the imaginary part of the self-energy is zero at E_F in a metal
10.7 One other contribution to the self-energy is that from phonons. Little is known about this contribution, but for adsorbates phonon broadenings of the order of 600 meV have been observed (see e.g. J.C. Fuggle, D. Menzel: Surf. Sci. **79**, 1 (1979)). Effects of this size would be significant
10.8 B. Reihl, K.H. Frank, A. Otto: Z. Phys. B **62**, 473 (1986)
10.9 W. Speier, T.M. Hayes, J.W. Allen, J.B. Boyce, J.C. Fuggle: Phys. Rev. Lett. **55**, 1693 (1985)
10.10 E. Sobczak, J. Auleytner: Phys. Rev. B **37**, 6251 (1988)
10.11 A. Šimunek, J. Vackář, E. Sobczak: Phys. Rev. B **38**, 8515 (1988)
10.12 A. Šimunek, O. Šipr, J. Vackář: Phys. Rev. Lett. **63**, 2076 (1989)
10.13 E. Sobczak, J. Auleytner, S. Mobilio, A. Balerna, O. Smotlacha: Physica B **158**, 553 (1989)
10.14 V. Dose: Appl. Phys. **14**, 117 (1977)
10.15 J.B. Pendry: J. Phys. C **14**, 1381 (1981)
10.16 N.V. Smith: Rep. Prog. Phys. **51**, 1227 (1988)
10.17 P.D. Johnson, J.W. Davenport: Phys. Rev. B **31**, 7521 (1985)
10.18 E.A. Milne: Phil. Mag. **47**, 209 (1974)
10.19 J.W. Chamberlain: *Physics of Aurora and Airglow* International Geophysics Series **2** (Academic, New York 1961) p. 28

10.20 J.H. Scofield: J. Electron Spectr. and Rel. Phen. **8**, 129 (1976)
10.21 W. Speier, J.C. Fuggle, P. Durham, R. Zeller, R.J. Blake, P. Sterne: J. Phys. C **21**, 2621 (1988)
10.22 H. Winter, P.J. Durham, G.M. Stocks: J. Phys. F **14**, 1047 (1984)
10.23 A.L. Stewart: "The Quantal Calculation of Photoionization Cross Sections", in *Advances in Atomic and Molecular Physics*, Vol. **3**, ed. by D.R. Bates, I. Estremann (Academic, New York 1967)
10.24 I thank V. Dose for illuminating discussions on this point and for correcting some of my misconceptions
10.25 C.C. Chu, P.E. Best: Phys. Rev. B **19**, 3414 (1979)
10.26 K. Lawniczak-Jablonska: In *Inner Shell and X-ray Physics of Atoms and Solids*, ed. by D.J. Fabian, H. Kleinpoppen, L.M. Watson (Plenum, New York 1981) p. 521, and references therein
10.27 T. Fauster, F.J. Himpsel: Phys. Rev. B **30**, 1874 (1984)
10.28 W. Speier: J. Phys. C **21**, L1183 (1988)
10.29 J.J. Yeh, I. Lindau: At. Data and Nucl. Data Tables **32**, 1 (1985)
10.30 J.W. Cooper: Phys. Rev. **128**, 681 (1962)
10.31 The earliest reference to the use of the Cooper minimum for solids of which we know is W. Braun: In Proc. XIII Conf. on Physics of Semiconductors, ed. by F.G. Fumi (Tipografia, Rome 1976) p. 62
10.32 J.F. van Acker, P.W.J. Weijs, J.C. Fuggle, K. Horn, W. Wilke, H. Haak, H. Saalfield, H. Kuhlenbeck, W. Braun, G.P. Williams, D. Wesner, M. Strongin: Phys. Rev. B **38**, 10463 (1988)
10.33 R.J. Liefield, A.F. Burr, M.B. Chamberlain: Phys. Rev. A **9**, 316 (1974)
10.34 M.B. Chamberlain, A.F. Burr, R.J. Liefield: Phys. Rev. A **9**, 663 (1974)
10.35 F. Riehle: Ph.D. Thesis, Universität Karlsruhe (1977)
10.36 F. Riehle: Phys. Status Solid B **98**, 245 (1978)
10.37 F. Riehle: Jap. J. Appl. Phys. suppl. **17–2**, 314 (1978)
10.38 U. Fano: Phys. Rev. **124**, 1866 (1961)
10.39 G. Wendin: In *Giant Resonances in Atoms, Molecules and Solids*, ed. by J.P. Connerade et al. (Plenum, New York 1987) p. 195: K. Nuroh, G. Wendin: Phys. Rev. Lett. **39**, 48 (1977); Phys. Rev. B **24**, 5533 (1981)
10.40 There is a second way of viewing these resonance processes which is used by Wendin and others, particularly in lectures. In the random phase approximation (RPA), the resonances arise because the polarizability of the atoms and the response to the photon fields is enormously enhanced close to the thresholds for core excitations. Thus in RPA the whole atom resonates and many processes, including shake-up satellites in PS, Auger yields, etc. may be enhanced or depressed near thresholds
10.41 J.C. Fuggle, M. Campagna, Z. Zolnierek, R. Lässer, A. Plateau: Phys. Rev. Lett. **45**, 1597 (1980)
10.42 J.-M. Esteva, R.C. Karnatak, J.C. Fuggle, G.A. Sawatzky: Phys. Rev. Lett. **50**, 910 (1983)
10.43 Yongjun Hu, T.J. Wagener, Y. Gao, J.H. Weaver: Phys. Rev. B **38**, 12708 (1988)
10.44 T.J. Wagener, Yongjun Hu, Y. Gao, M.B. Jost, J.H. Weaver, N.D. Spencer, K.C. Goretta: Phys. Rev. B **39**, 2928 (1989)
10.45 B. Reihl: Private communication
10.46 J. Unguris, A. Seiler, R.J. Celotta, D.T. Pierce, P.D. Johnson, N.V. Smith: Phys. Rev. Lett. **49**, 1047 (1982)
10.47 H. Scheidt, M. Glöbl, V. Dose, J. Kirschner: Phys. Rev. Lett. **51**, 1688 (1983)
10.48 D.T. Pierce, A. Seiler, C.S. Feigerle, J.L. Pena, R.J. Celotta: J. Magn. Mat. **54–57**, 617 (1986)
10.49 H.R. Borsje, H.W.H.M Jongbloets, R.J.H. Kappert, J.C. Fuggle, S.F. Alvarado, R. Rochow, M. Campagna: Rev. Sci. Instrum. **61**, 765 (1990)
10.50 H.R. Borsje: Ph.D. Thesis, University of Nijmegen (1991)
10.51 D.T. Pierce, F. Meier: Phys. Rev. B **13**, 5484 (1976)
10.52 D.T. Pierce, R.J. Celotta, G.-C. Wang, W.N. Unertl, , A. Galejs, C.E. Kuyatt, S.R. Mielczarek: Rev. Sci. Instrum. **51**, 478 (1980)
10.53 M. Donath: Appl. Phys. A **49**, 351 (1989)
10.54 D.L. Webster: Phys. Rev. **9**, 220 (1917)
10.55 R. Turtle, R. Liefield: Phys. Rev. B **7**, 3411 (1973)
10.56 Note that one could in principle use the loss function obtained from direct measurent of electron energy loss in reflection. However, in that case the electrons pass through the surface "twice" and results in relative enhancement of the surface losses in the loss function thus obtained

10.57 Of course the inelastic losses extrinsic and intrinsic to the transition itself are not strictly separable in BIS, any more than they are in photoemission, because they interfere. This has been extensively discussed in the literature, e.g. by M. Šunjić, D. Šokčević: Solid State Commun. **15**, 165 (1974) and **18**, 373 (1976), C.O. Almbladh: Nuovo Cimento B **23**, 74 (1974) and J.C. Fuggle et al.: Phys. Rev. Lett. **44**, 1090 (1980). However the separation is conceptually useful and is not completely senseless at high energies, where one is aproaching the sudden limit, and the interference term is small
10.58 See e.g. P. Steiner, H. Höchst, S. Hüfner: in *Photoemission in Solids II*, ed. By L. Ley, M. Cardona, Topics Appl. Phys. (Springer, Berlin, Heidelberg 1979) p. 349ff
10.59 H.J.W.M. Hoekstra, W. Speier, R. Zeller, J.C. Fuggle: Phys. Rev. B **34**, 5177 (1986)
10.60 J.W.D. Connolly: Int. J. Quantum Chem. **35**, 807 (1970)
10.61 E. Ojala: Phys. Status Solidi B **119**, 269 (1983)
10.62 H.J. Levinson, F. Greuter, E.W. Plummer: Phys. Rev. B **27**, 727 (1983)
10.63 H. Raether: Springer Tracts in Modern Physics, Vol **38** (Springer, Berlin, Heidelberg 1965); *ibid.* Vol. 88 (1980) p. 1 and references therein
10.64 J.W. Allen, S.J. Oh, O. Gunnarsson, K. Schönhammer, M.B. Maple, M.S. Toriachvili, I. Lindau: Appl. Phys. **35**, 275 (1986)
10.65 V.L. Moruzzi, J.F. Janak, A.R. Williams: *Calculated Electronic Properties of Metals* (Pergamon, New York 1978)
10.66 W. Speier, J.C. Fuggle, R. Zeller, B. Ackermann, K. Szot, F.U. Hillebrecht, M. Campagna: Phys. Rev. B **30**, 6921 (1984)
10.67 J.C. Fuggle: In *Laboratory methods in Photoelectron Spectroscopy*, ed. by D. Briggs (Heyden, London 1978) p. 273
10.68 J.C. Fuggle: Unpublished results
10.69 A. Liebsch: Phys. Rev. Lett. **43**, 1431 (1979)
10.70 A. Liebsch: Phys. Rev. B **23**, 5203 (1981)
10.71 J.C. Fuggle, F.U. Hillebrecht, J.M. Esteva, R.C. Karnatak, O. Gunnarsson, K. Schönhammer: Phys. Rev. B **27**, 4637 (1983)
10.72 Actually U_{eff} may be larger in Mn as a result of multiplet interactions
10.73 I. Lindeyer, J.R. van Acker, J.C. Fuggle: To be published
10.74 O. Gunnarsson, K. Schönhammer: Phys. Rev. Lett. **50**, 604 (1983)
10.75 O. Gunnarsson, K. Schönhammer: Phys. Rev. Lett. B **28**, 4315 (1983)
10.76 O. Gunnarsson, K. Schönhammer: In *Handbook of the Physics and Chemistry of the Rare Earths*, ed. by K. Gschneidner et al. (Elsevier, Amsterdam 1987) p. 103
10.77 F.J. Pinski D.D. Johnson, G.M. Stocks: J. Appl. Phys. **57**, 3018 (1985)
10.78 E. Wuilloud, B. Delley, W.-D. Schneider, Y. Baer: Phys. Rev. Lett. **53**, 202 (1984)
10.79 Y.Baer, H.R. Ott, J.C. Fuggle, L.E. Long: Phys. Rev. B **24**, 5384 (1981)
10.80 F.U. Hillebrecht, J.C. Fuggle, G.A. Sawatzky, M. Campagna, O. Gunnarsson, K. Schönhammer: Phys.Rev. B **30**, 1777 (1984)
10.81 E. Wuilloud, H.R. Moser, W.-D. Schneider, Y. Baer: Phys. Rev. B **28**, 7354 (1983)
10.82 J.W. Allen, S.J. Oh, L.E. Cox, W.P. Ellis, M.S. Wire, Z. Fisk, J.L. Smith, B.B. Pate, I. Lindau, A.J. Arko: Phys. Rev. Lett. **54**, 2635 (1985)
10.83 C. Laubschat, G. Kaindl, W.-D. Schneider, B. Reihl, N. Martensson: Phys. Rev. B **33**, 6675 (1986)
10.84 Y. Baer, W.-D. Schneider: J. de Physique (Paris) **48-C9**, 967 (1987)
10.85 F.U. Hillebrecht, M. Campagna: In *Handbook of the Physics and Chemistry of the Rare Earths*, Vol. 10, ed. by K.A. Gschneidner et al. (Elsevier, Amsterdam 1984) 425 (1987)
10.86 Y. Baer: Physica B **102**, 106 (1980)
10.87 F.U. Hillebrecht, D.D. Sarma, N. Martensson: Phys. Rev. B **33**, 4376 (1986)
10.88 J.W. Allen, J.-S. Kang, Y. Lassailly, M.S. Torikachvili, W. Ellis, B. Pate, I. Lindau: Sol. St. Commun. **61**, 183 (1987)
10.89 C. Laubschat, W. Grentz, G. Kaindl: Phys. Rev. B **37**, 8082 (1988)
10.90 F.U. Hillebrecht, H.J. Trodahl, V. Sechovsky, B.T. Thole: Z. Phys. B **77**, 373 (1989)
10.91 F. Patthey, J.-M. Imer, W.-D. Schneider, H. Beck, Y. Baer, B. Delley: Phys. Rev. B (1990) in press
10.92 C. Laubschat, W. Grentz, G. Kaindl: Phys. Rev. B **36**, 8233 (1987)
10.93 In the standard many-body language of this field one would say that the impurity Anderson Hamiltonians allow one to extrapolate from the mixed valence to the Kondo limit
10.94 G.A. Sawatzky, J.W. Allen: Phys. Rev. Lett. **53**, 2339 (1984)

10.95 J. Ghijsen, L.H. Tjeng, J. van Elp, H. Eskes, J. Westerink, G.A. Sawatzky, M.T. Czyzyk: Phys. Rev. B **38**, 11322 (1988)
10.96 J.C. Fuggle, P.J.W. Weijs, R. Schoorl, G.A. Sawatzky, J. Fink, N. Nücker, P.J. Durham, W.M. Temmerman: Phys. Rev. B **37**, 123 (1988)
10.97 see F.Al Shamma, J.C. Fuggle: Physica C, **169**, 325 (1990) and references therein
10.98 W. Speier, D.D. Sarma, L. Kumar, R.A. de Groot, J.C. Fuggle: J. Phys.: Condens. Matter **1**, 9117 (1989)
10.99 W. Speier, E. v Leuken, J.C. Fuggle, D.D. Sarma, L. Kumar, B. Dauth, K.H. Buschow: Phys. Rev. B **39**, 6008 (1989)
10.100 see e.g. P.J.W. Weijs, M.T. Czyżyk, J.C. Fuggle, W. Speier, D.D. Sarma, K.H.J. Buschow: Z. Phys. B **78**, 423 (1990)
10.101 W.B. Jackson, S.M. Kelso, C.C. Tsai, J.W. Allen, S.-J. Oh: Phys. Rev. B **31**, 5187 (1985)
10.102 J.D. Johannopoulos, M.L. Cohen: Phys. Rev. B **7**, 504 (1977)
10.103 The momentum form of the matrix elements is used here. It is related to the electric vector form by the well-known commutation relationships, as summarized in [10.101]
10.104 M. Cini: Solid State Commun. **24**, 681 (1977)
10.105 G.A. Sawatzky: Phys. Rev. Lett. **39**, 504 (1977)
10.106 M. Cini: Phys. Rev. B **15**, 2788 (1978)

Appendix A. Electron Binding Energies

The following table reproduces data compiled by G. P. Williams of Brookhaven National Laboratory on electron binding energies for elements in their natural forms. Values have been taken from [A. 1] except as noted:

* Values taken from [A. 2]
† Values taken from [A. 3]
[a] One particle approximation not valid owing to the extremely short lifetime of the core hole
[b] Values derived indirectly from [A. 1]

The energies are given in electron volts, relative to the vacuum level for the rare gases and H_2, N_2, O_2, F_2 and Cl_2, relative to the Fermi level for metals, and relative to the top of the valence bands for semiconductors. Most of the values below ~1.5 keV are derived from photoemission measurements. For metals the threshold values found in XAS and XPS are normally identical. For other materials, and for transitions involving localized states, there may be up to a few volts difference between the core level binding energies derived from different techniques. For further details, see the discussions on core excitions in Chaps. 1, 3, 4 and 5.

Electron binding energies, in electron volts, for the elements in their natural forms

Element	K 1s	L_I 2s	L_{II} 2$p_{1/2}$	L_{III} 2$p_{3/2}$	M_I 3s	M_{II} 3$p_{1/2}$	M_{III} 3$p_{3/2}$	M_{IV} 3$d_{3/2}$	M_V 3$d_{5/2}$	N_I 4s	N_{II} 4$p_{1/2}$	N_{III} 4$p_{3/2}$
1 H	16*											
2 He	24.6*											
3 Li	54.7*											
4 Be	111.5*											
5 B	188*											
6 C	284.2*											
7 N	409.9*	37.3*										
8 O	543.1*	41.6*										
9 F	696.7*											
10 Ne	870.2*	48.5*	21.7*	21.6*								
11 Na	1070.8†	63.5†	30.4†	30.5†								
12 Mg	1303.0†	88.6*	49.6†	49.2†								
13 Al	1558.98*	117.8*	72.9*	72.5*								
14 Si	1839	149.7*b	99.8*	99.2*								
15 P	2149	189*	136*	135*								
16 S	2472	2309*b	163.6*	162.5*								
17 Cl	2833	270*	202*	200*								
18 Ar	3205.9*	326.3*	250.6*	248.4*	29.3*	15.9*	15.7*					
19 K	3608.4*	378.6*	297.3*	294.6*	34.8*	18.3*	18.3*					
20 Ca	4038.5*	438.4†	349.7†	346.2†	44.3†	25.4†	25.4†					
21 Sc	4492	498.0*	403.6*	398.7*	51.1*	28.3*	28.3*					
22 Ti	4966	560.9†	461.2†	453.8†	58.7†	32.6†	32.6†					
23 V	5465	626.7†	519.8†	512.1†	66.3†	37.2†	37.2†					
24 Cr	5989	695.7†	583.8†	574.1†	74.1†	42.2†	42.2†					
25 Mn	6539	769.1†	649.9†	638.7†	82.3†	47.2†	47.2†					
26 Fe	7112	844.6†	719.9†	706.8†	91.3†	52.7†	52.7†					
27 Co	7709	925.1†	793.3†	778.1†	101.0†	58.9†	58.9†					
28 Ni	8333	1008.6†	870.0†	852.7†	110.8†	68.0†	66.2†					
29 Cu	8979	1096.7†	952.3†	932.5†	122.5†	77.3†	75.1†					
30 Zn	9659	1196.2*	1044.9*	1021.8*	139.8*	91.4*	88.6*	10.2*	10.1*			
31 Ga	10367	1299.0*b	1143.2†	1116.4†	159.5†	103.5†	103.5†	18.7†	18.7†			
32 Ge	11103	1414.6*b	1248.1*b	1217.0*b	180.1*	124.9*	120.8*	29.0*	29.0*			
33 As	11867	1527.0*b	1359.1*b	1323.6*b	204.7*	146.2*	141.2*	41.7*	41.7*			
34 Se	12658	1652.0*b	1474.3*b	1433.9*b	229.6*	166.5*	160.7*	55.5*	54.6*			

35 Br	13474	1782*	1596*	1550*	257*	189*	182*	70*	69*			
36 Kr	14326	1921	1730.9*	1678.4*	292.8*	222.2*	214.4	95.0*	93.8*	27.5*	14.1*	14.1*
37 Rb	15200	2065	1864	1804	326.7*	248.7*	239.1*	113.0*	112*	30.5*	16.3*	15.3*
38 Sr	16105	2216	2007	1940	358.7†	280.3†	270.0†	136.0†	134.2†	38.9†	20.3†	20.3†
39 Y	17038	2373	2156	2080	392.0*b	310.6*	298.8*	157.7†	155.8†	43.8*	24.4*	23.1*
40 Zr	17998	2532	2307	2223	430.3†	343.5†	329.8†	181.1†	178.8†	50.6†	28.5†	27.7†
41 Nb	18986	2698	2465	2371	466.6†	376.1†	360.6†	205.0†	202.3†	56.4†	32.6†	30.8†
42 Mo	20000	2866	2625	2520	506.3†	410.6†	394.0†	231.1†	227.9†	63.2†	37.6†	35.5†
43 T_c	21044	3043	2793	2677	544*	447.6*	417.7*	257.6*	253.9*	69.5*	42.3*	39.9*
44 Ru	22117	3224	2967	2838	586.2†	483.3†	461.5†	284.2†	280.0†	75.0†	46.5†	43.2†
45 Rh	23220	3412	3146	3004	628.1†	521.3†	496.5†	311.9†	307.2†	81.4*b	50.5†	47.3†
46 Pd	24350	3604	3330	3173	671.6†	559.9†	532.3†	340.5†	335.2†	87.1*b	55.7†a	50.9†a
47 Ag	25514	3806	3524	3351	719.0†	603.8†	573.0†	374.0†	368.0†	97.0†	63.7†	58.3†
48 Cd	26711	4018	3727	3538	772.0†	652.6†	618.4†	411.9†	405.2†	109.8†	63.9†a	63.9†a
49 In	27940	4238	3938	3730	827.2†	703.2†	665.3†	451.4†	443.9†	122.7†	73.5†a	73.5†a
50 Sn	29200	4465	4156	3929	884.7†	756.5†	714.6†	493.2†	484.9†	137.1†	83.6†a	83.6†a
51 Sb	30491	4698	4380	4132	946†	812.7†	766.4†	537.5†	528.2†	153.2†	95.6†a	95.6†a
52 Te	31814	4939	4612	4341	1006†	870.8†	820.8†	583.4†	573.0†	169.4†	103.3†a	103.3†a
53 I	33169	5188	4852	4557	1072*	931*	875*	631*	620*	186*	123*	123*
54 Xe	34561	5453	5104	4782	1148.7*	1002.1*	940.6*	689.0*	676.4*	213.2*	146.7	145.5*
55 Cs	35985	5714	5359	5012	1211*b	1071*	1003*	740.5*	726.6*	232.3*	172.4*	161.3*
56 Ba	37441	5989	5624	5247	1293*b	1137*b	1063*b	795.7*	780.5*	253.5†	192	178.6†
57 La	38925	6266	5891	5483	1362*b	1209*b	1128*b	853*	836*	247.7*	205.8	196.0*
58 Ce	40443	6548	6164	5723	1436*b	1274*b	1187*b	902.4*	883.8*	291.0*	223.2	206.5*
59 Pr	41991	6835	6440	5964	1511	1337	1242	948.3*	928.8*	304.5	236.3	217.6
60 Nd	43569	7126	6722	6208	1575	1403	1297	1003.3*	980.4*	319.2*	243.3	224.6
61 Pm	45184	7428	7013	6459	—	1403	1357	1052	1027	—	242	242
62 Sm	46834	7737	7312	6716	1723	1541	1419.8	1110.9*	1083.4*	347.2*	265.6	247.4
63 Eu	48519	8052	7617	6977	1800	1614	1481	1158.6*	1127.5*	360	284	257
64 Gd	50239	8376	7930	7243	1881	1688	1544	1221.9*	1189.6*	378.6*	286	271
65 Tb	51996	8708	8252	7514	1968	1768	1611	1276.9*	1241.1*	396.0*	322.4*	284.1*
66 Dy	53789	9046	8581	7790	2047	1842	1676	1333	1292*	414.2*	333.5*	293.2*
67 Ho	55618	9394	8918	8071	2128	1923	1741	1392	1351	432.4*	343.5	308.2*
68 Er	57486	9751	9264	8358	2206	2006	1812	1453	1409	449.8*	366.2	320.2*
69 Tm	59390	10116	9617	8648	2307	2090	1885	1515	1468	470.9*	385.9*	332.6*
70 Yb	61332	10486	9978	8944	2398	2173	1950	1576	1528	480.5*	388.7*	339.7*

Electron binding energies (*Continued*)

Element	$N_{IV}4d_{3/2}$	$N_V4d_{5/2}$	$N_{VI}4f_{5/2}$	$N_{VII}4f_{7/2}$	O_I5s	$O_{II}5p_{1/2}$	$O_{III}5p_{3/2}$	$O_{IV}5d_{3/2}$	$O_V5d_{5/2}$	P_I6s	$P_{II}6p_{1/2}$	$P_{III}6p_{3/2}$
48 Cd	11.7†	10.7†										
49 In	17.7†	16.4†										
50 Sn	24.9†	23.9†										
51 Sb	33.3†	32.1†										
52 Te	41.9†	40.4†										
53 I	50*	50*										
54 Xe	69.5*	67.5*	—	—	23.3*	13.4*	12.1*					
55 Cs	79.8*	77.5*	—	—	22.7	14.2*	12.1*					
56 Ba	92.6†	89.9†	—	—	30.3†	17.0†	14.8†					
57 La	105.3*	102.5*	—	—	34.3*	19.3*	16.8*					
58 Cs	109*	—	0.1	0.1	37.8	19.8*	17.0*					
59 Pr	115.1*	115.1*	2.0	2.0	37.4	22.3	22.3					
60 Nd	120.5*	120.5*	1.5	1.5	37.5	21.1	21.1					
61 Pm	120	120	—	—	—	—	—					
62 Sm	129	129	5.2	5.2	37.4	21.3	21.3					
63 Eu	133	133	0	0	32	22	22					
64 Gd	—	127.7*	8.6*	8.6*	36	20	20					
65 Tb	150.5*	150.5*	7.7*	2.4*	45.6*	28.7*	22.6*					
66 Dy	153.6*	153.6*	8.0*	4.3*	49.9*	26.3	26.3					
67 Ho	160*	160*	8.6*	5.2*	49.3*	30.8*	24.1*					
68 Er	167.6*	167.6*	—	4.7*	50.6*	31.4*	24.7*					
69 Tm	175.5*	175.5*	—	4.6*	54.7*	31.8*	25.0*					
70 Yb	191.2*	182.4*	2.5*	1.3*	52.0*	30.3*	24.1*					

Element	K 1s	L_I 2s	L_{II} $2p_{1/2}$	L_{III} $2p_{3/2}$	M_I 3s	M_{II} $3p_{1/2}$	M_{III} $3p_{3/2}$	M_{IV} $3d_{3/2}$	M_V $3d_{5/2}$	N_I 4s	N_{II} $4p_{1/2}$	N_{III} $4p_{3/2}$
71 Lu	63314	10870	10349	9244	2491	2264	2024	1639	1589	506.8*	412.4*	359.2*
72 Hf	65351	11271	10739	9561	2601	2365	2107	1716	1662	538*	438.2†	380.7†
73 Ta	67416	11682	11136	9881	2708	2469	2194	1793	1735	563.4†	463.4†	400.9†
74 W	69525	12100	11544	10207	2820	2575	2281	1949	1809	594.1†	490.4†	423.6†
75 Re	71676	12527	11959	10535	2932	2682	2367	1949	1883	625.4†	518.7†	446.8†
76 Os	73871	12968	12385	10871	3049	2792	2457	2031	1960	658.2†	549.1†	470.7†
77 Ir	76111	13419	12824	11215	3174	2909	2551	2116	2040	691.1†	577.8†	495.8†
78 Pt	78395	13880	13273	11564	3296	3027	2645	2202	2122	725.4†	609.1†	519.4†
79 Au	80725	14353	13734	11919	3425	3148	2743	2291	2206	762.1†	642.7†	546.3†
80 Hg	83102	14839	14209	12284	3562	3279	2847	2385	2295	802.2†	680.2†	576.6†
81 Tl	85530	15347	14698	12658	3704	3416	2957	2485	2389	846.2†	720.5†	609.5†
82 Pb	88005	15861	15200	13055	3851	3554	—	3066	2586	891.8†	761.9†	643.5†
83 Bi	90526	16388	15711	13419	3999	3696	3177	2688	2580	939†	805.2†	678.8†
84 Po	93105	16939	16244	13814	4149	3854	3302	2798	2683	995*	851*	705*
85 At	95730	17493	16785	14214	4317	4008	3426	2909	2787	1042*	886*	740*
86 Rn	98404	18049	17337	14619	4482	4159	3538	3022	2892	1097*	929*	768*
87 Fr	101137	18639	17907	15031	4652	4327	3663	3136	3000	1153*	980*	810*
88 Ra	103922	19237	18484	15444	4822	4490	3792	3248	3105	1208*	1958*	879*
89 Ac	106755	19840	19083	15871	5002	4656	3909	3370	3219	1269*	1080*	890*
90 Th	109651	20472	19693	16300	5182	4830	4046	3491	3332	1330*	1168*	966.4†
91 Pa	112601	21105	20314	16733	5367	5001	4174	3611	3442	1387*	1224*	1007*
92 U	115606	21757	20948	17166	5548	5182	4303	3728	3552	1439*b	1271*b	1043†

Electron binding energies (*Continued*)

Element	$N_{IV}4d_{3/2}$	$N_{V}4d_{5/2}$	$N_{VI}4f_{5/2}$	$N_{VII}4f_{7/2}$	$O_{I}5s$	$O_{II}5p_{1/2}$	$O_{III}5p_{3/2}$	$O_{IV}5d_{3/2}$	$O_{V}5d_{5/2}$	$P_{I}6s$	$P_{II}6p_{1/2}$	$P_{III}6p_{3/2}$
71 Lu	206.1*	196.3*	8.9*	7.5*	57.3*	33.6*	26.7*					
72 Hf	220.0†	211.5†	15.9†	14.2†	64.2†	38*	29.9†					
73 Ta	237.9†	226.4†	23.5†	21.6†	69.7†	42.2*	32.7†					
74 W	255.9†	243.5†	33.6*	31.4†	75.6†	45.3*b	36.8†					
75 Re	273.9†	260.5†	42.9*	40.5†	83†	45.6†	34.6*b					
76 Os	293.1†	278.5†	52.4†	50.7†	83†	58*	44.5†					
77 Ir	311.9†	296.3†	63.8†	60.8†	95.2*b	63.0*b	48.0†					
78 Pt	331.6†	314.6†	74.5†	71.2†	101.7*b	65.3*b	51.7†					
79 Au	353.2†	335.1†	87.6†	83.9†	107.2*b	74.2†	57.2†					
80 Hg	378.2†	358.8†	104.0†	99.9†	127†	83.1†	64.5†	9.6†	7.8†			
81 Ti	405.7†	385.0†	122.2†	117.8†	136*b	94.6†	73.5†	14.7†	12.5†			
82 Pb	434.3†	412.2†	141.7†	136.9†	147*b	106.4†	83.3†	20.7†	18.1†			
83 Bi	464.0†	440.1†	162.3†	157.0†	159.3*b	119.0†	92.6†	26.9†	23.8†			
84 Po	500*	473*	184*	184*	177*	132*	104†	31*	31*			
85 At	533*	407*	210*	210*	195*	148*	115*	40*	40*			
86 Rn	567*	541*	238*	238*	214*	164*	127*	48*	48*	26		
87 Fr	603*	577*	268*	268*	234*	182*	140*	58*	58*	34	15	15
88 Ra	636*	603*	299*	299*	254*	200*	153*	68*	68*	44	19	19
89 Ac	675*	639*	319*	319*	272*	215*	167*	80*	80*	—	—	—
90 Th	712.1†	675.2†	342.4†	333.1†	290*a	229*a	182*a	92.5†	85.4†	41.4†	24.5†	16.6†
91 Pa	743*	708*	371*	360*	310*	232*	232*	94*	94*	—	—	—
92 U	778.3†	736.2†	388.2*	377.4†	321*ab	257*ab	192*ab	102.8†	94.2†	43.9†	26.8†	16.8†

References

[A. 1] J.A. Bearden, A.F. Burr: Rev. Mod. Phys. **39**, 125 (1967)
[A. 2] M. Cardona, L. Ley: eds. *Photoemission in Solids* (Springer Verlag, Berlin, Heidelberg 1978) The present table contains several corrections to the values appearing in that reference
[A. 3] J.C. Fuggle, N. Mårtensson: J. Electr. Spectrosc. Relat. Phenom. **21**, 275 (1980)

Appendix B. Core-Hole Lifetime Broadenings

The data presented in this appendix were compiled by F. Al Shamma, M. Abbate and J.C.Fuggle, Solid State Spectroscopy, University of Nijmegen, Toernooiveld, 6525 ED Nijmegen, The Netherlands

We include this compilation of core-hole lifetime broadenings in this book because they influence the experimental results in many spectroscopies. In particular they are important in XAS and EELS investigations where the features due to unoccupied state distributions and excited state multiplets will be smeared out by the core-hole linewidths. That is the reason why one likes to choose narrow core levels for near-edge XAS studies. There are also other factors leading to extra spectral broadening, including experimental resolution, phonon effects and the lifetime of the electron excited in the unoccupied state. Thus the values listed in the tables should be regarded as a lower limit to the broadening. But one should also consider non-Lorentzian broadening caused by other effects (for instance, non-resolved multiplet splitting, electron–hole pair creation, dispersional broadening, etc.).

Most of the values listed in the tables are taken from ab initio computations compiled from published work by *Krause* and *Oliver* [B. 1]. However it became clear about 10 years ago [B. 2] that the computed values were in severe error for some levels of elements in parts of the periodic table. These errors were due to difficulties in the computation of Coster-Kroning transition rates [B. 2]. Where appropriate we have replaced theoretical values by experimental data. In some others cases it is clear that the theoretical data is erroneous but no experimental data is available; in these cases we leave a blank in the table or use interpolated values which are given in italics.

Table B.1. K- and L-level lifetime broadenings in eV[3]

Element	Z	K	L_1	L_2	L_3
Ne	10	0.24	<0.1	0.000	0.000
Na	11	0.30	0.2[2]	0.000	0.000
Mg	12	0.36	0.4[2]	0.001	0.001
Al	13	0.42	0.7[2]	0.004	0.004
Si	14	0.48	1.03	0.015	0.014
P	15	0.53	1.26	0.03	0.03
S	16	0.59	1.49	0.05	0.05
Cl	17	0.64	1.2[2]	0.08	0.09
Ar	18	0.68	1.3[2]	0.13	0.13
K	19	0.74	1.40[2]	0.15	0.16
Ca	20	0.81	1.83[2]	0.17	0.17
Sc	21	0.86	2.21	0.19	0.19
Ti	22	0.94	2.34	0.24	0.1[4]
V	23	1.01	2.41	0.26	0.2[4]
Cr	24	1.08	2.54	0.29	0.2[4]
Mn	25	1.16	2.62	0.34	0.2[4]
Fe	26	1.25	2.76	0.37	0.2[4]
Co	27	1.33	2.79	0.43	0.3[4]
Ni	28	1.44	2.89	0.52	0.3[4]
Cu	29	1.55	3.06	0.62	0.4[4]
Zn	30	1.67	3.28	0.72	0.66[2]
Ga	31	1.82	2.8[2]	0.83	0.76
Ge	32	1.96	3.53	0.95	0.82
As	33	2.14	3.79	1.03	0.94
Se	34	2.33	3.94	1.13	1.00
Br	35	2.52	4.11	1.21	1.08
Kr	36	2.75	4.28	1.31	1.17
Rb	37	2.99	4.44	1.43	1.27
Sr	38	3.25	4.67	1.54	1.39
Y	39	3.52	4.71	1.65	1.50
Zr	40	3.84	4.78	1.78	1.57
Nb	41	4.14	4.94	1.87	1.66
Mo	42	4.52	4.25	1.97	1.78
Tc	43	4.91	4.36	2.08	1.91
Ru	44	5.33	4.58	2.23	2.00
Rh	45	5.77	4.73	2.35	2.13
Pd	46	6.24	4.93	2.43	2.25
Ag	47	6.75	4.88	2.57	2.40
Cd	48	7.28	4.87	2.62	2.50
In	49	7.91	5.00	2.72	2.65
Sn	50	8.49	2.97	2.84	2.75
Sb	51	9.16	3.13	3.00	2.87
Te	52	9.89	3.32	3.12	2.95
I	53	10.6	3.46	3.25	3.08
Xe	54	11.4	3.64	3.40	3.13
Cs	55	12.3	3.78	3.51	3.25
Ba	56	13.2	3.92	3.57	3.32
La	57	14.1	4.06	3.68	3.41
Ce	58	15.1	4.21	3.80	3.48
Pr	59	16.2	4.34	3.89	3.60
Nd	60	17.3	4.52	3.97	3.65
Pm	61	18.5	4.67	4.06	3.75
Sm	62	19.7	4.80	4.15	3.86
Eu	63	21.0	4.91	4.23	3.91
Gd	64	22.3	5.05	4.32	4.01

Table B.1. (*Continued*)

Element	Z	K	L_1	L_2	L_3
Tb	65	23.8	5.19	4.43	4.12
Dy	66	25.2	5.25	4.55	4.17
Ho	67	26.8	5.33	4.66	4.26
Er	68	28.4	5.43	4.73	4.35
Tm	69	30.1	5.47	4.79	4.48
Yb	70	31.9	5.53	4.82	4.60
Lu	71	33.7	5.54	4.92	4.68
Hf	72	35.7	5.63	5.02	4.80
Ta	73	37.7	5.58	5.15	4.88
W	74	39.9	5.61	5.33	4.98
Re	75	42.1	6.18	5.48	5.04
Os	76	44.4	7.25	5.59	5.16
Ir	77	46.8	8.30	5.69	5.25
Pt	78	49.3	9.39	5.86	5.31
Au	79	52.0	10.5	6.00	5.41
Hg	80	54.6	11.3	6.17	5.50
Ti	81	57.4	12.0	6.32	5.65
Pb	82	60.4	12.2	6.48	5.81
Bi	83	63.4	12.4	6.67	5.98
Po	84	66.6	12.6	6.83	6.13
At	85	69.8	12.8	7.01	6.29
Rn	86	73.3	13.1	7.20	6.41
Fr	87	76.8	13.3	7.47	6.65
Ra	88	80.4	13.4	7.68	6.82
Ac	89	84.1	13.6	7.95	6.98
Th	90	88.0	13.7	8.18	7.13
Pa	91	91.9	14.3	8.75	7.33
U	92	96.1	14.0	9.32	7.43

Table B.2. M-level lifetime broadenings in eV[5]

Element	Z	M_1	$M_3(M_2)$	M_4	M_5
Ca	20	1.2			
Sc	21	1.3			
Ti	22	2.2	0.1		
V	23	2.2	0.2		
Cr	24	2.6	*0.3*		
Mn	25	2.1	0.4		
Fe	26	2.2	0.6		
Co	27	2.5	1.3		
Ni	28	2.3	1.3		
Cu	29	1.9	1.6		
Zn	30	1.7	2.0		
Ga	31	2.2	*2.0*		
Ge	32	2.0	2.1	0.06	0.04
As	33	3.4		*0.07*	*0.06*

Table B.2. (*Continued*)

Table B.2. (*Continued*)

Element	Z	M_1	$M_3(M_2)$	M_4	M_5
Se	34			*0.07*	*0.07*
Br	35		1.3	0.08	0.09
Kr	36	5.2	1.0	0.10	0.10
Rb	37			*0.09*	*0.09*
Sr	38			*0.09*	*0.08*
Y	39		2.4	0.09	0.07
Zr	40	6.0	2.0	0.09	0.07
Nb	41	5.8	2.1[6]	0.11	0.10
Mo	42		2.2[6]	0.16	0.14
Ru	44	7.2	2.4[6]	0.24	0.24
Rh	45		2.5[6]	0.32	0.30
Pd	46	9.8	2.7[6]	0.40	0.38
Ag	47	9.0	2.8[6]	0.47	0.44
Cd	48		3.1[6]	0.50	0.46
In	49	8.8	3.3[6]	0.54	0.50
Sn	50	9.5	3.4[6]	0.58	0.54
Sb	51		3.8[6]	0.60	0.57
Te	52		4.1[6]	0.6[6]	0.74
Xe	54	10.2	3.2	0.5	0.5

Table B.3. N-level lifetime broadenings in eV[5]

Element	Z	N_1	N_2	N_3	N_4	N_5	$N_{6,7}$
Nb	41	2.8[6]	1.5[6]	1.0[6]			
Mo	42	3.8	2.2[6]	1.6[6]			
Ru	44	3.2[6]	3.3[6]	2.5[6]			
Rh	45	3.4[6]	4.2[6]	3.4[6]			
Pd	46	3.8	4.7[6]	4.7[6]			
Ag	47	3.0	8.5	7.5			
Cd	48	4.0	12	10			
In	49	2.8	15	14			
Sn	50	1.3	17	17	0.14	0.14	
Sb	51	3.3[6]		15	*0.12*	*0.12*	
Te	52	3.0[6]		8.4	*0.10*	*0.10*	
Xe	54	2.7	2.6	2.6	0.08	0.08	
Cs	55		*2.7*	*2.6*	*0.09*	*0.09*	
Ba	56		*2.7*	*2.7*	*0.10*	*0.10*	
La	57		2.8	2.7	0.13	0.13	
Yb	70		6.5	4.9			
Ta	73		7.4	5.0			0.05
W	74		4.6	4.7	4.0[8]	3.9[8]	0.07
Re	75	9.2	6.8[7]	4.8	4.2[8]	3.8[8]	*0.10*
Os	76	8.0	8.1[7]	5.3[7]	4.2[8]	3.8[8]	*0.13*
Ir	77			4.2	4.0[8]	3.5[8]	0.15
Pt	78	7.9	6.1	4.0	4.0[8]	3.6[8]	*0.19*
Au	79	7.6	5.2	4.5	4.4[8]	3.9[8]	0.23
Hg	80				4.0[8]	3.6[8]	0.3[8]
Tl	81	11.2	6.5	5.6	4.2[8]	3.8[8]	*0.20*
Pb	82	8.8	7.6	5.1	3.7[8]	3.6[8]	*0.16*
Bi	83		6.5	5.6		3.8[8]	0.13
Th	90			7.6	5.3	5.2	0.15
U	92			6.5	4.3	4.2	0.28

Table B.4. Some O-level lifetime broadenings in eV[5]

Element	O_1	O_2	O_3	O_4	O_5
Xe	0.05	<0.05	<0.05		
Pt		1.8 ± 0.2			
Au		>2.4	2.5 ± 0.4		
Tl		3.2 ± 0.5	3.4 ± 0.5	0.2 ± 0.2	0.2 ± 0.2
Pb		3.2 ± 0.5	3.4 ± 0.5		
Bi		2.0 ± 0.5	3.6 ± 0.6	0.15 ± 0.15	0.2 ± 0.15
Th		2.2 ± 1	2.8 ± 0.8	0.6 ± 0.2	0.6 ± 0.2
U		4.5 ± 1	3.0 ± 0.5	0.9 ± 0.3	1.1 ± 0.3

Table B.5. Some P-level lifetime broadenings in eV[5]

Element	P_1	P_2	P_3
Th	1.9 ± 0.5	0.8 ± 0.4	0.9 ± 0.2
U	2.3 ± 0.5	1.5 ± 0.5	1.9 ± 0.3

References

[B.1] M.O. Krause, J.H. Oliver: J. Phys. Chem. Ref. Data **8**, 329 (1979)
[B.2] J.C. Fuggle, S.F. Alvarado: Phys. Rev. A **22**, 1615 (1980) and references therein
[B.3] All values from [B.1] except where specified
[B.4] R. Nyholm, N. Mårtensson, A. Lebugle, U. Axelsson: J. Phys. F **11**, 1727 (1981)
[B.5] All values from [B.2] except where specified
[B.6] N. Mårtensson, R. Nyholm: Phys. Rev. B **24**, 7121 (1981)
[B.7] A. Berndtsson, R. Nyholm, N Mårtensson, R. Nilsson, J. Hedman: Phys. Stat. Sol. (b) **93**, K103 (1979)
[B.8] R. Nyholm, N. Mårtensson: Phys. Rev. B **36**, 20 (1987)

Subject Index

Topics in Applied Physics Founded by Helmut K. V. Lotsch

1 **Dye Lasers** 3rd Ed. Editor: F. P. Schäfer

2 **Laser Spectroscopy** of Atoms and Molecules. Editor: H. Walther

3 **Numerical and Asymptotic Techniques in Electromagnetics** Editor: R. Mittra

4 **Interactions on Metal Surfaces** Editor: R. Gomer

5 **Mössbauer Spectroscopy** Editor: U. Gonser

6 **Picture Processing and Digital Filtering** 2nd Ed. Editor: T. S. Huang

7 **Integrated Optics** 2nd Ed. Editor: T. Tamir

8 **Light Scattering in Solids I** 2nd Ed. Editor: M. Cardona

9 **Laser Speckle** and Related Phenomena 2nd Ed. Editor: J. C. Dainty

10 **Transient Electromagnetic Fields** Editor: L. B. Felsen

11 **Digital Picture Analysis** Editor: A. Rosenfeld

12 **Turbulence** 2nd Ed. Editor: P. Bradshaw

13 **High-Resolution Laser Spectroscopy** Editor: K. Shimoda

14 **Laser Monitoring of the Atmosphere** Editor: E. D. Hinkley

15 **Radiationless Processes** in Molecules and Condensed Phases. Editor: F. K. Fong

16 **Nonlinear Infrared Generation** Editor: Y.-R. Shen

17 **Electroluminescence** Editor: J. I. Pankove

18 **Ultrashort Light Pulses** Picosecond Techniques and Applications 2nd Ed. Editor: S. L. Shapiro

19 **Optical and Infrared Detectors** 2nd Ed. Editor: R. J. Keyes

20 **Holographic Recording Materials** Editor: H. M. Smith

21 **Solid Electrolytes** Editor: S. Geller

22 **X-Ray Optics.** Applications to Solids Editor: H.-J. Queisser

23 **Optical Data Processing.** Applications Editor: D. Casasent

24 **Acoustic Surface Waves** Editor: A. A. Oliner

25 **Laser Beam Propagation in the Atmosphere** Editor: J. W. Strohbehn

26 **Photoemission in Solids I.** General Principles Editors: M. Cardona and L. Ley

27 **Photoemission in Solids II.** Case Studies Editors: L. Ley and M. Cardona

28 **Hydrogen in Metals I.** Basic Properties Editors: G. Alefeld and J. Völkl

29 **Hydrogen in Metals II** Application-Oriented Properties Editors: G. Alefeld and J. Völkl

30 **Excimer Lasers** 2nd Ed. Editor: Ch. K. Rhodes

31 **Solar Energy Conversion.** Solid State Physics Aspects. Editor: B. O. Seraphin

32 **Image Reconstruction from Projections** Implementation and Applications Editor: G. T. Herman

33 **Electrets** 2nd Ed. Editor: G. M. Sessler

34 **Nonlinear Methods of Spectral Analysis** 2nd Ed. Editor: S. Haykin

35 **Uranium Enrichment** Editor: S. Villani

36 **Amorphous Semiconductors** 2nd Ed. Editor: M. H. Brodsky

37 **Thermally Stimulated Relaxation in Solids** Editor: P. Bräunlich

38 **Charge-Coupled Devices** Editor: D. F. Barbe

39 **Semiconductor Devices** for Optical Communication. 2nd Ed. Editor: H. Kressel

40 **Display Devices** Editor: J. I. Pankove

41 **The Computer in Optical Research** Methods and Applications. Editor: B. R. Frieden

42 **Two-Dimensional Digital Signal Processing I** Linear Filters. Editor: T. S. Huang

43 **Two-Dimensional Digital Signal Processing II** Transforms and Median Filters. Editor: T. S. Huang

44 **Turbulent Reacting Flows** Editors: P. A. Libby and F. A. Williams

45 **Hydrodynamic Instabilities and the Transition to Turbulence** 2nd Ed. Editors: H. L. Swinney and J. P. Gollub

46 **Glassy Metals I** Editors: H.-J. Güntherodt and H. Beck

47 **Sputtering by Particle Bombardment I** Editor: R. Behrisch

48 **Optical Information Processing** Fundamentals. Editor: S. H. Lee

49 **Laser Spectroscopy of Solids** 2nd Ed. Editors: W. M. Yen and P. M. Selzer

50 **Light Scattering in Solids II.** Basic Concepts and Instrumentation Editors: M. Cardona and G. Güntherodt

51 **Light Scattering in Solids III.** Recent Results Editors: M. Cardona and G. Güntherodt

52 **Sputtering by Particle Bombardment II** Sputtering of Alloys and Compounds, Electron and Neutron Sputtering, Surface Topography Editor: R. Behrisch

53 **Glassy Metals II.** Atomic Structure and Dynamics, Electronic Structure, Magnetic Properties Editors: H. Beck and H.-J. Güntherodt

54 **Light Scattering in Solids IV.** Electronic Scattering, Spin Effects, SERS, and Morphic Effects Editors: M. Cardona and G. Güntherodt

55 **The Physics of Hydrogenated Amorphous Silicon I** Structure, Preparation, and Devices Editors: J. D. Joannopoulos and G. Lucovsky

56 **The Physics of Hydrogenated Amorphous Silicon II** Electronic and Vibrational Properties Editors: J. D. Joannopoulos and G. Lucovsky